Advances in Geometric Modeling

Advances in Geometric Modeling

Editor:

Gerard Ian Prudhomme

ArclerPress

www.arclerpress.com

Advances in Geometric Modeling

Editor: Gerard Ian Prudhomme

708 3rd Avenue, 6th Floor
New York
NY 10017
United States of America
www.arclerpress.com

ISBN: 978-1-68094-469-3

Library of Congress Control Number: 2016951352

About the Editor

Gerard Ian Prudhomme

Gerard I. Prudhomme has a graduate degree (M.S.) for Computer Science from University College London (UCL). He has also worked as a software programmer and tech writer for different Fortune 500 companies, and studied at UCL, Harvard, and Oxford.

Contents

List of Contributors

Filip Biljecki
3D Geoinformation, Delft University of Technology, Delft, The Netherlands

Ken Arroyo Ohori
3D Geoinformation, Delft University of Technology, Delft, The Netherlands

Hugo Ledoux
3D Geoinformation, Delft University of Technology, Delft, The Netherlands

Ravi Peters
3D Geoinformation, Delft University of Technology, Delft, The Netherlands

Jantien Stoter
3D Geoinformation, Delft University of Technology, Delft, The Netherlands

Chuong V. Nguyen
Computational Informatics, Commonwealth Scientific and Industrial Research Organization (CSIRO), Canberra, ACT, Australia,

David R. Lovell
Computational Informatics, Commonwealth Scientific and Industrial Research Organization (CSIRO), Canberra, ACT, Australia,

Matt Adcock
Computational Informatics, Commonwealth Scientific and Industrial Research Organization (CSIRO), Canberra, ACT, Australia,

John La Salle
Ecosystem Sciences, Commonwealth Scientific and Industrial Research Organization (CSIRO), Canberra, ACT, Australia,
Atlas of Living Australia, Canberra, ACT, Australia

Daniel J. Kelley
Waisman Laboratory for Brain Imaging and Behavior, Waisman Center, Madison, Wisconsin, United States of America,
Medical Scientist Training Program, University of Wisconsin School of Medicine and Public Health, University of Wisconsin, Madison, Wisconsin, United States of America,
Neuroscience Training Program, University of Wisconsin School of Medicine and Public Health, University of Wisconsin, Madison, Wisconsin, United States of America,
Howard Hughes Medical Institute (HHMI) Teaching Fellows Program, Wisconsin Program for Scientific Teaching, Department of Bacteriology, University of Wisconsin, Madison, Wisconsin, United States of America,

Mohammed Farhoud
Paul P. Carbone Comprehensive Cancer Center, University of Wisconsin, Madison, Wisconsin, United States of America,
Biology New Media Center, University of Wisconsin Biotechnology Center, Madison, Wisconsin, United States of America,

M. Elizabeth Meyerand
Department of Medical Physics, University of Wisconsin, Madison, Wisconsin, United States of America,

David L. Nelson
Department of Biochemistry, College of Agricultural and Life Science, University of Wisconsin, Madison, Wisconsin, United States of America,

Lincoln F. Ramirez
Department of Neurological Surgery, University of Wisconsin, Madison, Wisconsin, United States of America

Robert J. Dempsey
Department of Neurological Surgery, University of Wisconsin, Madison, Wisconsin, United States of America

Alan J. Wolf
Biology New Media Center, University of Wisconsin Biotechnology Center, Madison, Wisconsin, United States of America,

Andrew L. Alexander
Waisman Laboratory for Brain Imaging and Behavior, Waisman Center, Madison, Wisconsin, United States of America, Department of Medical Physics, University of Wisconsin, Madison, Wisconsin, United States of America,

Richard J. Davidson
Waisman Laboratory for Brain Imaging and Behavior, Waisman Center, Madison, Wisconsin, United States of America,

Casey M. Holliday
Integrative Anatomy, Department of Pathology and Anatomical Sciences, University of Missouri, Columbia, Missouri, United States of America

Henry P. Tsai
Integrative Anatomy, Department of Pathology and Anatomical Sciences, University of Missouri, Columbia, Missouri, United States of America

Rebecca J. Skiljan
Integrative Anatomy, Department of Pathology and Anatomical Sciences, University of Missouri, Columbia, Missouri, United States of America

Ian D. George
Integrative Anatomy, Department of Pathology and Anatomical Sciences, University of Missouri, Columbia, Missouri, United States of America

Sami Pathan
Integrative Anatomy, Department of Pathology and Anatomical Sciences, University of Missouri, Columbia, Missouri, United States of America

Thomas van de Kamp
ANKA/Institute for Photon Science and Synchrotron Radiation, Karlsruhe Institute of Technology (KIT), Eggenstein-Leopoldshafen, Germany
State Museum of Natural History (SMNK), Karlsruhe, Germany

Tomy dos Santos Rolo
ANKA/Institute for Photon Science and Synchrotron Radiation, Karlsruhe Institute of Technology (KIT), Eggenstein-Leopoldshafen, Germany,

Patrik Vagovic
ANKA/Institute for Photon Science and Synchrotron Radiation, Karlsruhe Institute of Technology (KIT), Eggenstein-Leopoldshafen, Germany,

Tilo Baumbach
ANKA/Institute for Photon Science and Synchrotron Radiation, Karlsruhe Institute of Technology (KIT), Eggenstein-Leopoldshafen, Germany,

Alexander Riedel
ANKA/Institute for Photon Science and Synchrotron Radiation, Karlsruhe Institute of Technology (KIT), Eggenstein-Leopoldshafen, Germany,
State Museum of Natural History (SMNK), Karlsruhe, Germany

Kara L. Crossley
Institute of Neuroscience, Newcastle University, United Kingdom

Piers L. Cornelissen
Institute of Neuroscience, Newcastle University, United Kingdom
Department of Psychology, University of Northumbria, United Kingdom

Martin J. Tove´e
Institute of Neuroscience, Newcastle University, United Kingdom

Carolin Helbig
Department of Environmental Informatics, Helmholtz Centre for Environmental Research (UFZ), Leipzig, Germany, Faculty of Environmental Sciences, Technical University Dresden, Dresden, Germany,

Lars Bilke
Department of Environmental Informatics, Helmholtz Centre for Environmental Research (UFZ), Leipzig, Germany,

Hans-Stefan Bauer
Institute of Physics and Meteorology, University of Hohenheim, Stuttgart, Germany,

Michael Böttinger
German Climate Computing Center (DKRZ), Hamburg, Germany

Olaf Kolditz
Department of Environmental Informatics, Helmholtz Centre for Environmental Research (UFZ), Leipzig, Germany,
Faculty of Environmental Sciences, Technical University Dresden, Dresden, Germany,

Hwa Jen Yap
Department of Mechanical Engineering, Faculty of Engineering, University of Malaya, Kuala Lumpur, Malaysia,

Zahari Taha
Faculty of Mechanical Engineering, Universiti Malaysia Pahang, Pekan, Pahang, Malaysia,

Siti Zawiah Md Dawal
Department of Mechanical Engineering, Faculty of Engineering, University of Malaya, Kuala Lumpur, Malaysia,

Siow-Wee Chang
Bioinformatics Program, Institute of Biological Sciences, Faculty of Science, University of Malaya, Kuala Lumpur, Malaysia

Peter Claes
Medical Image Computing, ESAT/PSI, Department of Electrical Engineering, KU Leuven, Medical Imaging Research Center, KU Leuven & UZ Leuven, iMinds-KU Leuven Future Health Department, Leuven, Belgium,

Denise K. Liberton
Department of Anthropology, Penn State University, University Park, Pennsylvania, United States of America,

Katleen Daniels
Medical Image Computing, ESAT/PSI, Department of Electrical Engineering, KU Leuven, Medical Imaging Research Center, KU Leuven & UZ Leuven, iMinds-KU Leuven Future Health Department, Leuven, Belgium,

Kerri Matthes Rosana
Department of Anthropology, Penn State University, University Park, Pennsylvania, United States of America,

Ellen E. Quillen
Department of Anthropology, Penn State University, University Park, Pennsylvania, United States of America,

Laurel N. Pearson
Department of Anthropology, Penn State University, University Park, Pennsylvania, United States of America,

Brian McEvoy
Smurfit Institute of Genetics, Dublin, Ireland,

Marc Bauchet
Department of Anthropology, Penn State University, University Park, Pennsylvania, United States of America,

Arslan A. Zaidi
Department of Anthropology, Penn State University, University Park, Pennsylvania, United States of America,

Wei Yao
Department of Anthropology, Penn State University, University Park, Pennsylvania, United States of America,

Hua Tang
Department of Genetics, Stanford University, Palo Alto, California, United States of America,

Gregory S. Barsh
Department of Genetics, Stanford University, Palo Alto, California, United States of America,
HudsonAlpha Institute for Biotechnology, Huntsville, Alabama, United States of America,

Devin M. Absher
HudsonAlpha Institute for Biotechnology, Huntsville, Alabama, United States of America,

David A. Puts
Department of Anthropology, Penn State University, University Park, Pennsylvania, United States of America,

Jorge Rocha
CIBIO: Centro de Investigac¸a˜o em Biodiversidade e Recursos Gene´ticos, Universidade do Porto, Porto, Portugal,
Departamento de Biologia, Faculdade de Cie^ncias, Universidade do Porto, Porto, Portugal,

Sandra Beleza
Department of Genetics, Stanford University, Palo Alto, California, United States of America,
IPATIMUP: Instituto de Patologia e Imunologia Molecular da Universidade do Porto, Porto, Portugal,

Rinaldo W. Pereira
Programa de Po´s-Graduac¸a˜o em Cie^ncias Geno^micas e Biotecnologia, Universidade Cato´lica de Brası´lia, Brasilia, Brasil,

Gareth Baynam
School of Paediatrics and Child Health, University of Western Australia, Perth, Australia,
Institute for Immunology and Infectious Diseases, Murdoch University, Perth, Australia,
Genetic Services of Western Australia, King Edward Memorial Hospital, Perth, Australia,

Paul Suetens
Medical Image Computing, ESAT/PSI, Department of Electrical Engineering, KU Leuven, Medical Imaging Research Center, KU Leuven & UZ Leuven, iMinds-KU Leuven Future Health Department, Leuven, Belgium,

Dirk Vandermeulen
Medical Image Computing, ESAT/PSI, Department of Electrical Engineering, KU Leuven, Medical Imaging Research Center, KU Leuven & UZ Leuven, iMinds-KU Leuven Future Health Department, Leuven, Belgium,

Jennifer K. Wagner
Center for the Integration of Genetic Healthcare Technologies, University of Pennsylvania, Philadelphia, Pennsylvania, United States of America,

James S. Boster
Department of Anthropology, University of Connecticut, Storrs, Connecticut, United States of America

Mark D. Shriver
Department of Anthropology, Penn State University, University Park, Pennsylvania, United States of America,

Janice L. Pappas
Museum of Paleontology, University of Michigan, Ann Arbor, Michigan, United States of America

Daniel J. Miller
Museum of Paleontology, University of Michigan, Ann Arbor, Michigan, United States of America

Fre´de´ ric Boudon
Virtual Plants Inria team, UMR AGAP, CIRAD, INRIA, INRA, Montpellier, France,

Je´ ro^ me Chopard
Virtual Plants Inria team, UMR AGAP, CIRAD, INRIA, INRA, Montpellier, France,

Olivier Ali
Virtual Plants Inria team, UMR AGAP, CIRAD, INRIA, INRA, Montpellier, France
Laboratoire de Reproduction et De´veloppement des Plantes, Universite´ de Lyon 1, ENSLyon, INRA, CNRS, Lyon, France

Benjamin Gilles
Laboratoire d'Informatique, de Robotique et de Microe´lectronique de Montpellier, Universite´ Montpellier 2, CNRS, Montpellier, France

Olivier Hamant
Laboratoire de Reproduction et De´veloppement des Plantes, Universite´ de Lyon 1, ENSLyon, INRA, CNRS, Lyon, France,

Arezki Boudaoud
Laboratoire de Reproduction et De´veloppement des Plantes, Universite´ de Lyon 1, ENSLyon, INRA, CNRS, Lyon, France,

Jan Traas
Laboratoire de Reproduction et De´veloppement des Plantes, Universite´ de Lyon 1, ENSLyon, INRA, CNRS, Lyon, France,

Christophe Godin
Virtual Plants Inria team, UMR AGAP, CIRAD, INRIA, INRA, Montpellier, France,

Demetrios Gatziolis
USDA Forest Service, Pacific Northwest Research Station, Portland, Oregon, United States of America,

Jean F. Lienard
Department of Mathematics, Washington State University Vancouver, Vancouver, Washington, United States of America,

Andre Vogs
Department of Mathematics, Washington State University Vancouver, Vancouver, Washington, United States of America,
Intel Inc., Hillsboro, Oregon, United States of America

Nikolay S. Strigul
Department of Mathematics, Washington State University Vancouver, Vancouver, Washington, United States of America,

Peter Husen
Department of Biochemistry and Molecular Biology, University of Southern Denmark, Odense, Denmark,

Kirill Tarasov
Zora Biosciences Oy, Espoo, Finland,

Maciej Katafiasz
Department of Biochemistry and Molecular Biology, University of Southern Denmark, Odense, Denmark,

Elena Sokol
Department of Biochemistry and Molecular Biology, University of Southern Denmark, Odense, Denmark,

Johannes Vogt
Institute for Microscopic Anatomy and Neurobiology, Johannes Gutenberg University Mainz, Mainz, Germany

Jan Baumgart
Institute for Microscopic Anatomy and Neurobiology, Johannes Gutenberg University Mainz, Mainz, Germany

Robert Nitsch
Institute for Microscopic Anatomy and Neurobiology, Johannes Gutenberg University Mainz, Mainz, Germany

Kim Ekroos
Zora Biosciences Oy, Espoo, Finland,

Christer S. Ejsing
Department of Biochemistry and Molecular Biology, University of Southern Denmark, Odense, Denmark,

Jue Li
Institute of Cardiovascular Sciences, University of Manchester, Core Technology Facility, Manchester, United Kingdom

Shin Inada
Institute of Cardiovascular Sciences, University of Manchester, Core Technology Facility, Manchester, United Kingdom

Jurgen E. Schneider
Institute of Cardiovascular Sciences, University of Manchester, Core Technology Facility, Manchester, United Kingdom

Henggui Zhang
Institute of Cardiovascular Sciences, University of Manchester, Core Technology Facility, Manchester, United Kingdom

Halina Dobrzynski
Institute of Cardiovascular Sciences, University of Manchester, Core Technology Facility, Manchester, United Kingdom

Mark R. Boyett
Institute of Cardiovascular Sciences, University of Manchester, Core Technology Facility, Manchester, United Kingdom

Maria V. Sanchez-Vives
Institucio´ Catalana de Recerca i Estudis Avanc¸ats, Barcelona, Spain,
IDIBAPS, Barcelona, Spain,

Bernhard Spanlang
Universitat Polite`cnica de Catalunya, Barcelona, Spain,

Antonio Frisoli
PERCRO, CEIICP, Scuola Superiore Sant'Anna, Pisa, Italy,

Massimo Bergamasco
PERCRO, CEIICP, Scuola Superiore Sant'Anna, Pisa, Italy,

Mel Slater
Institucio´ Catalana de Recerca i Estudis Avanc¸ats, Barcelona, Spain,
University of Barcelona, Barcelona, Spain,
University College London, London, United Kingdom

Luke Bashford
Department of Bioengineering, Imperial College London, London, United Kingdom
Bernstein Centre Freiburg, University of Freiburg, Freiburg, Germany University of Bologna, ITALY

Carsten Mehring
Bernstein Centre Freiburg, University of Freiburg, Freiburg, Germany University of Bologna, ITALY

Clare K. Purvis
PGSP-Stanford Consortium, Department of Psychiatry & Behavioral Sciences, Stanford University School of Medicine, Stanford, California, United States of America,

Megan Jones
Department of Psychiatry & Behavioral Sciences, Stanford University School of Medicine, Stanford, California, United States of America,

Jakki O. Bailey
Department of Communication, Stanford University, Stanford, California, United States of America

Jeremy Bailenson
Department of Communication, Stanford University, Stanford, California, United States of America

C. Barr Taylor
Department of Psychiatry & Behavioral Sciences, Stanford University School of Medicine, Stanford, California, United States of America,

Inge A. Hoevenaren
Department of Plastic, Reconstructive and Hand Surgery, Radboud University Nijmegen Medical Centre, Nijmegen, the Netherlands,

J. Meulstee
Department of Oral and Maxillofacial Surgery, Radboud University Nijmegen Medical Centre, Nijmegen, the Netherlands,

E. Krikken
Department of Oral and Maxillofacial Surgery, Radboud University Nijmegen Medical Centre, Nijmegen, the Netherlands,
Technical medicine, University of Twente, Enschede, the Netherlands

S. J. Bergé
Department of Oral and Maxillofacial Surgery, Radboud University Nijmegen Medical Centre, Nijmegen, the Netherlands,

D. J. O. Ulrich
Department of Plastic, Reconstructive and Hand Surgery, Radboud University Nijmegen Medical Centre, Nijmegen, the Netherlands,

Thomas J. J. Maal
Department of Oral and Maxillofacial Surgery, Radboud University Nijmegen Medical Centre, Nijmegen, the Netherlands,
Barbier M
Discovery Sciences, Janssen Pharmaceutical companies of Johnson & Johnson, Beerse, Belgium.

Jaensch S
Discovery Sciences, Janssen Pharmaceutical companies of Johnson & Johnson, Beerse, Belgium.

Cornelissen F
Pharma R&D IT, Janssen Pharmaceutical companies of Johnson & Johnson, Beerse, Belgium.

Vidic S
Discovery Sciences, Janssen Pharmaceutical companies of Johnson & Johnson, Beerse, Belgium.
Department of Urology, Erasmus MC Rotterdam, Rotterdam, The Netherlands.
Faculty of Mathematics, Natural Sciences and Information Technologies, University of Primorska, Koper, Slovenia.

Gjerde K
Discovery Sciences, Janssen Pharmaceutical companies of Johnson & Johnson, Beerse, Belgium.
Department of Urology, Erasmus MC Rotterdam, Rotterdam, The Netherlands.

de Hoogt R
Discovery Sciences, Janssen Pharmaceutical companies of Johnson & Johnson, Beerse, Belgium.

Graeser R
Translational Medicine and Clinical Pharmacology, Boehringer Ingelheim, Ingelheim am Rhein, Germany.

Gustin E
Discovery Sciences, Janssen Pharmaceutical companies of Johnson & Johnson, Beerse, Belgium.

Chong YT
Discovery Sciences, Janssen Pharmaceutical companies of Johnson & Johnson, Beerse, Belgium.

James B. Hengenius
Department of Biological Sciences, Purdue University, West Lafayette, Indiana, United States of America,

Michael Gribskov
Department of Biological Sciences, Purdue University, West Lafayette, Indiana, United States of America,

Ann E. Rundell
Department of Biomedical Engineering, Purdue University,West Lafayette, Indiana, United States of America,

Charless C. Fowlkes
Department of Computer Science, University of California Irvine, Irvine, California, United States of America,

David M. Umulis
Department of Agricultural and Biological Engineering, Purdue University, West Lafayette, Indiana, United States of America

Kinga Nyíri
Department of Applied Biotechnology and Food Sciences, Budapest University of Technology and Economics, Budapest, 1111, Hungary,
Institute of Enzymology, Research Centre for Natural Sciences, Hungarian Academy of Sciences, Budapest, 1117, Hungary,

Bianka Kőhegyi
Department of Applied Biotechnology and Food Sciences, Budapest University of Technology and Economics, Budapest, 1111, Hungary,
Institute of Enzymology, Research Centre for Natural Sciences, Hungarian Academy of Sciences, Budapest, 1117, Hungary,

András Micsonai
Department of Biochemistry and MTA-ELTE NAP B Neuroimmunology Research Group, Institute of Biology, Eötvös Loránd University, Budapest, 1117, Hungary

József Kardos
Department of Biochemistry and MTA-ELTE NAP B Neuroimmunology Research Group, Institute of Biology, Eötvös Loránd University, Budapest, 1117, Hungary

Beata G. Vertessy
Department of Applied Biotechnology and Food Sciences, Budapest University of Technology and Economics, Budapest, 1111, Hungary,
Institute of Enzymology, Research Centre for Natural Sciences, Hungarian Academy of Sciences, Budapest, 1117, Hungary,

Preface

First of all, what exactly is geometric modeling? CAD (computer-aided design) software programs are utilized by designers, technical engineers, technicians, and others to produce accurate technical illustrations. There are a variety of functions of these software programs, probably the most widely utilized function is geometric modeling. The software programs permit the arithmetical description of the item to be shown as a picture on the monitor of the PC.

The PC compatible arithmetical explanation of the geometry of the item is known as geometric modeling. CAD software programs may be utilized by engineers to produce geometric modeling in both two-dimensional (2D), or even three-dimensional (3D), versions.The first chapter gives an outline of 3D semantic city models. Chapter 2 scrutinizes a reasonably-priced platform of off-the-shelf equipment utilized to pull together natural-colour 3D varieties of insects. Chapter 3 analyzes a manner to manufacture cerebral cortex, as well as skull reproductions, and additionally substantiate that the stereolithograph version conserves the volume of the VR method. Chapter 4 suggests a model of an engaging 3D version of the jawbone muscle mass of the American Alligator. Chapter 5 puts emphasis on exterior surface mesh designs in accordance with segmented datasets. Chapter 6 evaluates an involved 3D software application permitting users to generate a digital photorealistic physique. Chapter 7 peruses development of a versatile workflow that integrates diverse highly developed visualization application elements to be able to conceal the intricacy of three dimensional information visualization. Chapter 8 views technology formulated to allocate the configuration layout for an industrialized robotic work cell. Chapter 9 evaluates an innovative approach to determine facial structure in human population samples with combined Western African as well as European ancestry. Chapter 10 gives an outline of how surface area geometry of a life form corresponds to the perimeter of its actual three-dimensional structure. Chapter 11 showcase an example of morphogenesis in the shape of a three-dimensional digital structure. Chapter 12 views three-dimensional graphical stylizations of specific trees. Chapter 13 shows how improvements in electronic biotelemetry systems are allowing the analysis of authoritative information on the actions of free-ranging wild animals. Chapter 14 features a three-dimensional form of the right atrium. Chapter 15 scrutinizes a digital three-dimensional reflection of an arm where suitable synchronous visuotactile stimulus is supplied. Chapter 16 is designed to develop brain-computer interfaces and virtual reality. Chapter 17 scrutinizes the utility of virtual reality and body image disturbance. Chapter 18 deals with three-dimensional stereophotogrammetry. Chapter 19 showcases how three-dimensional image stacks can give details about the framework of volumetric cultures. Chapter 20 analyzes authentic 3D geometry accommodated to entire 3D gap gene expression information. Chapter 21 studies a 3D structural model for virulence.

Editor
Gerard Ian Prudhomme
Hawaii, US

CHAPTER 1

POPULATION ESTIMATION USING A 3D CITY MODEL: A MULTI-SCALE COUNTRY-WIDE STUDY IN THE NETHERLANDS

Filip Biljecki, Ken Arroyo Ohori, Hugo Ledoux, Ravi Peters, and Jantien Stoter

3D Geoinformation, Delft University of Technology, Delft, The Netherlands

ABSTRACT

The remote estimation of a region's population has for decades been a key application of geographic information science in demography. Most studies have used 2D data (maps, satellite imagery) to estimate population avoiding field surveys and questionnaires. As the availability of semantic 3D city models is constantly increasing, we investigate to what extent they can be used for the same purpose. Based on the assumption that housing space is a proxy for the number of its residents, we use two methods to estimate the population with 3D city models in two directions: (1) disaggregation (areal interpolation) to estimate the population of small administrative entities (e.g. neighbourhoods) from that of larger ones (e.g. municipalities); and (2) a statistical modelling approach to estimate the population of large entities from a sample composed of their smaller ones (e.g. one acquired by a government register). Starting from a complete Dutch census dataset at the neighbourhood level and a 3D model of all 9.9 million buildings in the Netherlands, we compare the population estimates obtained by both methods with the actual population as reported in the census, and use it to evaluate the quality that can be achieved by estimations at different administrative levels. We also analyse how the volume-based estimation enabled by 3D city models fares in comparison to 2D methods using building footprints and floor areas, as well as how it is affected by different levels of semantic detail in a 3D city model. We conclude that 3D city models are useful for estimations of large areas (e.g. for a country), and that the 3D approach has clear advantages over the 2D approach.

INTRODUCTION

Geographic information science (GIS) and demography have long been closely related, and GIS techniques are ubiquitous in mapping, analysing, and filling gaps in demographic data. In particular, geostatistical techniques are often used to estimate a region's population in the absence of reliable or complete census data [1, 2].

2D GIS datasets (e.g. satellite imagery and maps) have been used extensively in the past 50 years for this purpose, as several of them have been found to be reasonable proxies for population [2–21]. For instance, Bakillah et al. [22] and Doll et al. [23] estimate the population based on the concentration of surrounding points of interest (e.g. restaurants); Anderson et al. [24] and Sutton [25] use night-time imagery following the hypothesis that city lights indicate the magnitude of the urban

Citation: Biljecki F, Arroyo Ohori K, Ledoux H, Peters R, Stoter J. Population Estimation Using a 3D City Model: A Multi-Scale Country-Wide Study in the Netherlands. Bachschmid MM, ed. PLoS ONE. 2016;11(6):e0156808. doi:10.1371/journal.pone.0156808.

extent, which in turn indicates the population. Pozzi and Small [26] infer the population density from a vegetation cover map, based on the idea that less vegetation means more people; Xie [27] finds the relation between the density of road network and population; Steiger et al. [28] analyse georeferenced Twitter data to locate clusters indicating home- and work-related social activities that can serve as a proxy to estimate the residential and workplace population census data; and Lwin et al. [29] do a similar work using geolocated mobile phone usage data.

Among all these methods, many successful approaches rely on 2D datasets (maps) containing *building footprints* (e.g. derived from cadastral records or satellite imagery). The simplest approaches rely on the total number of buildings in a region or the total area of building footprints in it [30–32]. These methods perform reasonably well in homogeneous areas, but they exhibit significant errors in areas where buildings have a great variation in the number of storeys.

With the advancement of remote sensing technologies, such as lidar and aerial photogrammetry [33–37], it is now possible to automatically and remotely measure the height of a building, which can be used to obtain a volumetric representation of a building (3D city model) that is useful for population estimates. In fact, several researchers have indicated that the volume of buildings and the floorspace provide a strong cue for its population [5, 15, 22, 30, 38–46]. For example, Lu et al. [39] use multiple regression models to perform a study in Denver, Colorado, based on both footprint areas and building volumes. Lwin and Murayama [44] and Alahmadi et al. [46, 47] estimate the number of floors from an elevation dataset, and multiply it with the footprint area to get the approximate internal area of the apartments. Their results indicate that the volume-based approach gives more accurate results than the area of the footprints due to heterogeneous building morphologies.

However, despite the frequent indication that volume-based methods can improve on the estimates of area-based methods, there has been no large-scale study that conclusively proves that this is true. Existing studies have several gaps: they usually focus on single metropolitan areas, which can be relatively homogeneous; they seldom compare the accuracy of different approaches within the same region; they derive a building's volume based on a raster dataset, which limits its accuracy; they do not consider how this approach scales between larger and smaller areas; and they do not consider how the level of detail of the used volumetric representation affects the accuracy of the result.

The goal of this paper is to bridge these gaps. We investigate *to what extent 3D city models can be used to estimate the population of a region* by performing a multi-scale country-wide study in the Netherlands. As the Dutch government provides both highly accurate census and building data, we consider that the Netherlands serves as an excellent case study, both for the experiments and the validation of the methods.

We therefore evaluate the use of 3D city models in population estimation in two directions: (1) disaggregation (areal interpolation) to estimate the population of small administrative entities (e.g. neighbourhoods) from that of larger ones (e.g. municipalities); and (2) a statistical modelling approach to estimate the population of large entities from a sample composed of their smaller ones (e.g. one acquired by a government register). We compare the population estimates obtained by both methods with the actual population as reported in the census, and use it to evaluate the quality that can be achieved by estimations at different administrative levels. We also analyse how the volume-based estimation enabled by 3D city models fares in comparison to 2D methods using building footprints and floor areas, as well as how it is affected by different levels of semantic detail (information on building use) in a 3D city model. We conclude that 3D city models are useful for large scale estimations (e.g. for a country), and that the 3D approach has clear advantages over the 2D approach.

MATERIALS AND METHODS

Census Data

The Netherlands is decomposed hierarchically into 12 provinces (not covered further in this paper), 393 municipalities, 2816 districts, and 12237 neighbourhoods. The population of each of the entities is known thanks to the open dataset of Statistics Netherlands—CBS (*Centraal Bureau voor de Statistiek*). As shown inFig 1, the dataset consists of sets of polygons representing statistical units—the population within each polygon is stored as an attribute. We use this dataset to validate our results, and its subset to train one of the methods. The properties of statistical units across the country vary (see Fig 2), covering widely heterogeneous household sizes, population densities, and dwelling sizes, among others.

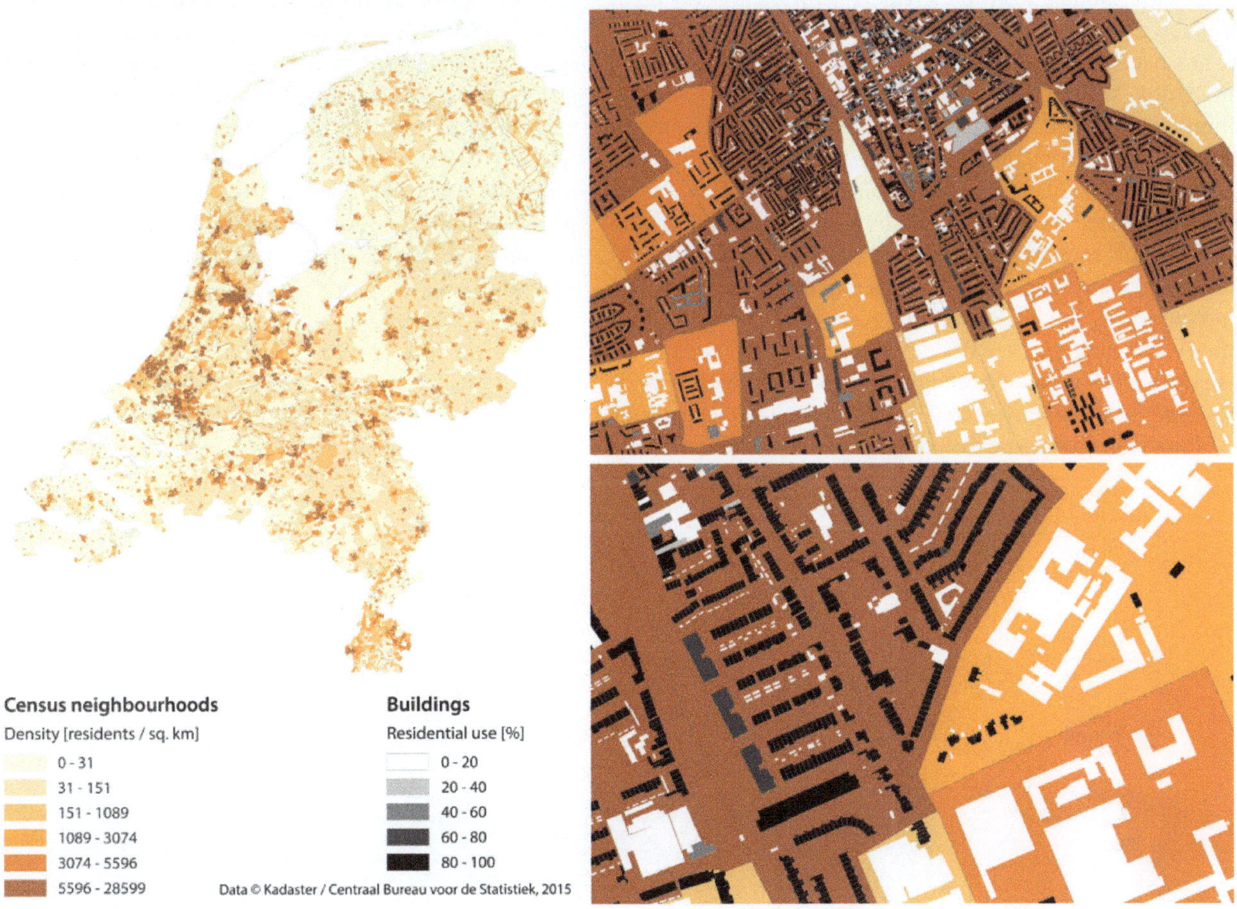

Figure1: Datasets used in this research: census neighbourhoods with building footprints.

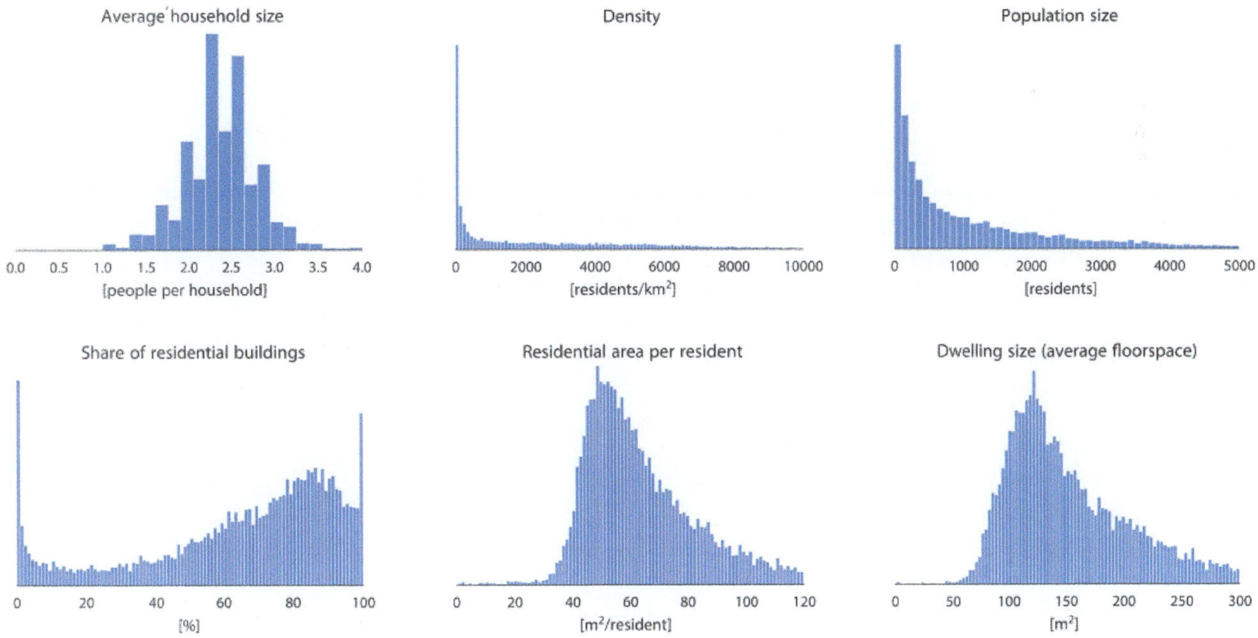

Figure 2: Census neighbourhoods statistics.

3D city model of the Netherlands

3D city models are digital representations of the urban environment, focusing on buildings [48–50]. They are used for many different purposes [50], e.g. the prediction of noise pollution [51]. Their key advantage over 2D maps is that they provide volumetric data, which is beneficial for applications that take advantage of the height or volume of buildings, such as energy demand estimations [52, 53] and visibility analyses [54, 55]. Population estimation is clearly such a case, as high-rise residential buildings are very likely to contain more inhabitants per unit area than low-rise buildings.

3D city models can be created with many different techniques, e.g. from airborne laser scanning, and considerable work has been devoted to their automatic generation [56–58]. In this study, we generate a country-wide 3D city model by combining two open datasets from the Netherlands government: (i) building data from the national register of addresses and buildings (BAG—*Basisregistraties Adressen en Gebouwen*, which is collected and maintained by each municipality, and disseminated as country-wide dataset through the national portal of *Kadaster*, the national mapping agency of the Netherlands; and processed by the NLExtract project)—containing the base geometry, building use, and floorspace information (see Fig 1); and (ii) elevation data—the Height Model of the Netherlands (AHN—*Actueel Hoogtebestand Nederland*), which contains 639 billion elevation points covering the whole country (see S3 Fig for an illustration). The 3D model creation is done using a process called extrusion, where the building footprint is lifted to a certain height to obtain a simple volumetric model [59, 60], yielding so-called block models of buildings (LOD1 according to the CityGML standard [61, 62]). For this purpose we have used the software 3dfier, developed by our group and released under an open-source licence (https://github.com/tudelft3d/3dfier). The software analyses all elevation points whose projection is within the footprint of a building, and determines the elevation at the building base and a single value for the height for the building. The height of the building is set to the median of all elevation points, which is considered optimal for building volume estimations [63]. A visual representation of these building block models is given in Fig 3.

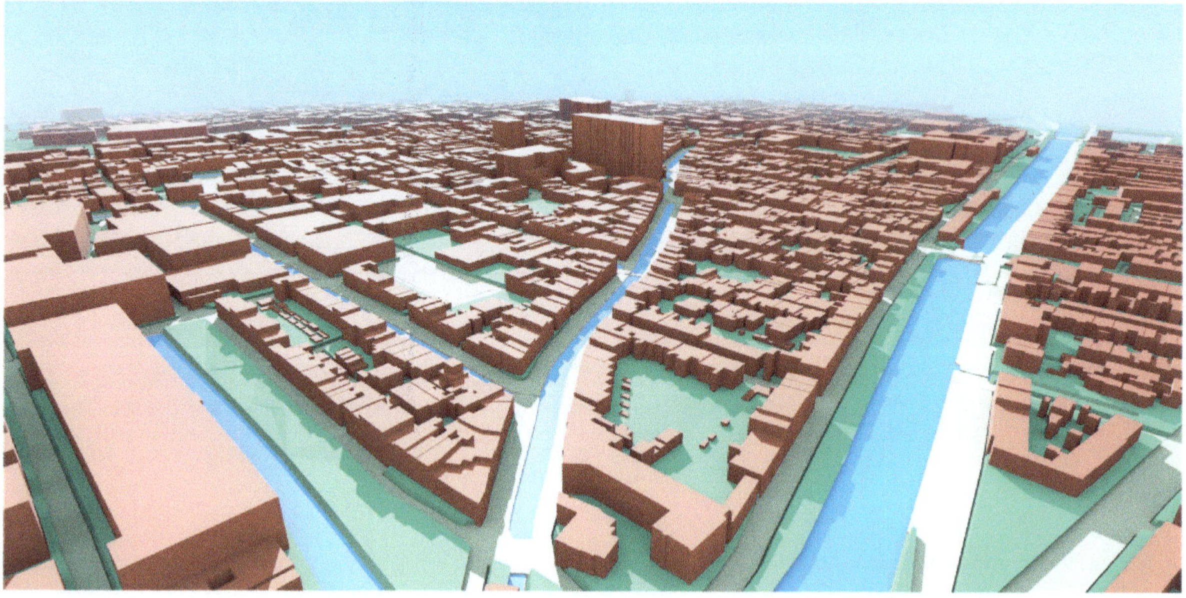

Figure 3: Example of the 3D city model.

3D city models come in different levels of detail (LODs) and with heterogeneous quality [62, 64, 65], both in terms of geometry and in terms of semantic information (e.g. a building's use) [66]. Thus, in order to test how different LODs of a 3D city model affect the population estimations, we construct 9 different LODs using various combinations of different levels of detail in a building's geometric and semantic information (Fig 4). In this way, we can directly compare the quality of the estimations given by the area-based (footprint and floor area) and volume-based approaches.

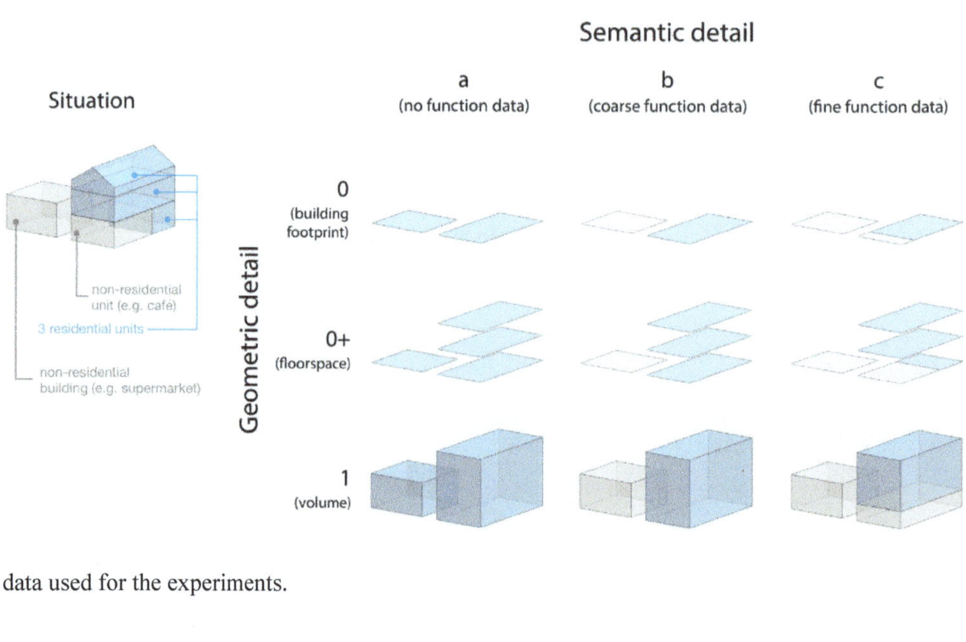

Figure 4: Multi-LOD data used for the experiments.

We consider three geometric LODs: (LOD0) 2D building footprints (the traditional area-based approach without height measurements); (LOD0+) building floorspace (area-based approach in which the vertical extent of the building is available); and (LOD1) volumetric 3D block models (from which the volume of a building can be calculated). For LOD0+, we rely on accurate indoor measurements from the Dutch cadastre, which is a dataset that is rarely available elsewhere. However, it should also be noted there is recent work focused on its automatic reconstruction [67, 68].

The general hypothesis used in this paper, and in related work, is simple: the larger the building, the more people reside in it; and the larger the living capacity of a district, the more populous it is. However, we argue that other building properties should be taken into account as well. The occupancy of a building also depends on its type, e.g. a cathedral, indoor arena, or a factory can be very large but at the same time they house zero inhabitants. Therefore, only residential buildings must be taken into account. This is further complicated by mixed-use buildings, which are composed of non-residential and of residential units, e.g. a three-storey building, where the ground floor is occupied by non-residential space (e.g. a restaurant and a shop), and the remaining two floors by residential units (fairly common in the Netherlands). However, such information is not always available, hence we pay special attention to the semantic aspect of data. Therefore, for the semantic part, we distinguish three levels of detail: (a) no data about the function of the building, and hence all buildings are treated equally; (b) a building is either residential or non-residential [42, 69]; and (c) fractional building use, where the share of the residential use within a building is known.

The possible combinations of the three geometric LODs and the three semantic LODs result in the 9 LODs used in this study, e.g. LOD1[b] denotes a block model with the singular information on the building use.

Existing methods for population estimation

The estimation of population with GIS data and techniques has been extensively reviewed by numerous authors [22, 41, 70–72]. Generally two groups of methods are recognised [70], both of which are used in this paper (Fig 5):

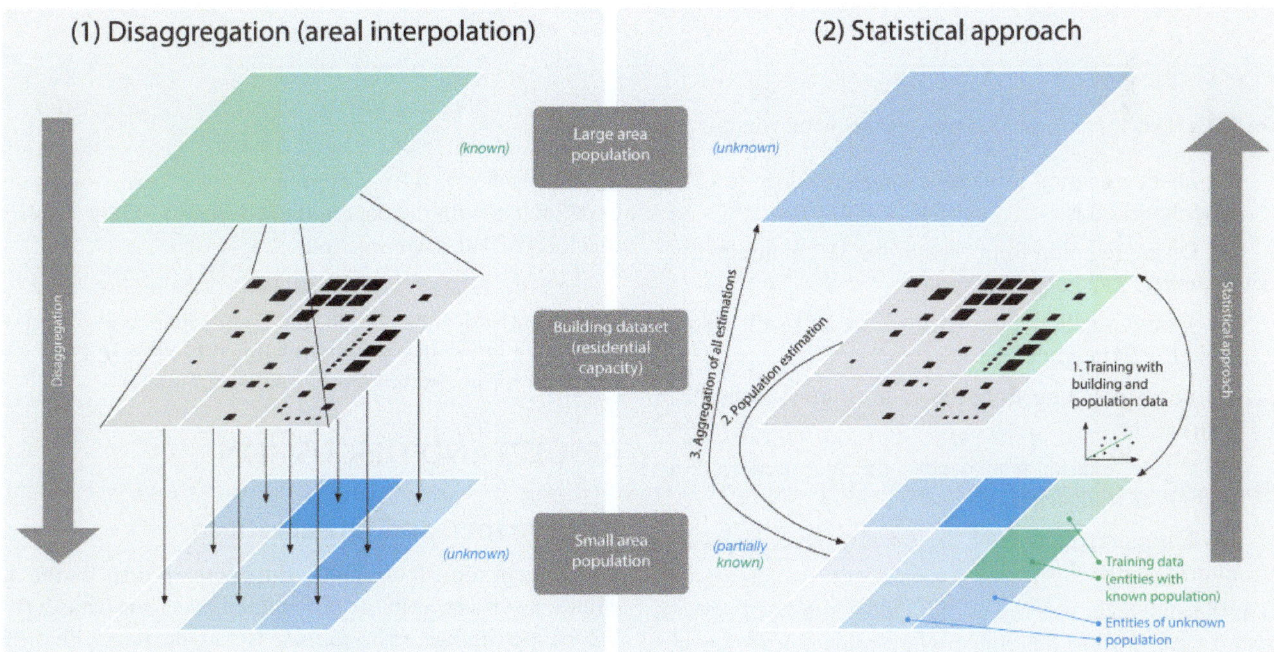

Figure 5: The two population estimation methods used.

- Disaggregation (areal interpolation): this is a top-down approach where the population of a larger administrative unit or zone (e.g. region, municipality, census district) is distributed across smaller units (e.g. neighbourhood), usually by weighting it according to different factors which hint at the population [73–78]. This approach is typically used when the population of a large entity is known (e.g. a city), but the one of its composing entities is not known (e.g. its neighbourhoods). The disaggregation can be done by simply distributing the population among administrative sub-zones, but it also can be aided by dasymetric mapping to shape smaller surfaces in such a way that variation within each surface is minimised [79]. This is especially useful when the smaller units are political subdivision of the larger (parent) unit often found in choropleth maps (e.g. interpolation from a province to the containing municipalities), because such regions may contain variations in the population density. Dasymetric mapping therefore results in (sub-)units that are more homogeneous [80–83].
- Statistical modelling approach: first the relationships between population and socio-economic and morphological variables associated with the population density are inferred, e.g. land use [9, 84], proximity to transportation network [71], and distance from the central business district [85]. The deduced relationships are then applied to estimate the population count

of unknown areas. In this approach multiple linear regression is most commonly used. The advantage of this bottom-up approach is that a sampling census has to be carried out for only a small area. It is useful in the scenario when only the population of a subset (e.g. a city) of a large area (e.g. a province) is known.

Our Proposed Method Using 3D City Models

For our population estimation study, we test three indicators to determine the disaggregation weights and the statistical relationships: (i) area of the 2D building footprints (in m^2), (ii) area of the building floorspace (in m^2), and (iii) building volume (in m^3). Each of these is tested at three levels of semantic detail, resulting in the 9 aforementioned LODs of the input datasets.

In order to diminish residential and socio-economic variations across a large area, but also to test the performance of different estimation scenarios, we use multiple scales of estimations, as shown in Fig 6. In the disaggregation approach 6 scales are analysed:

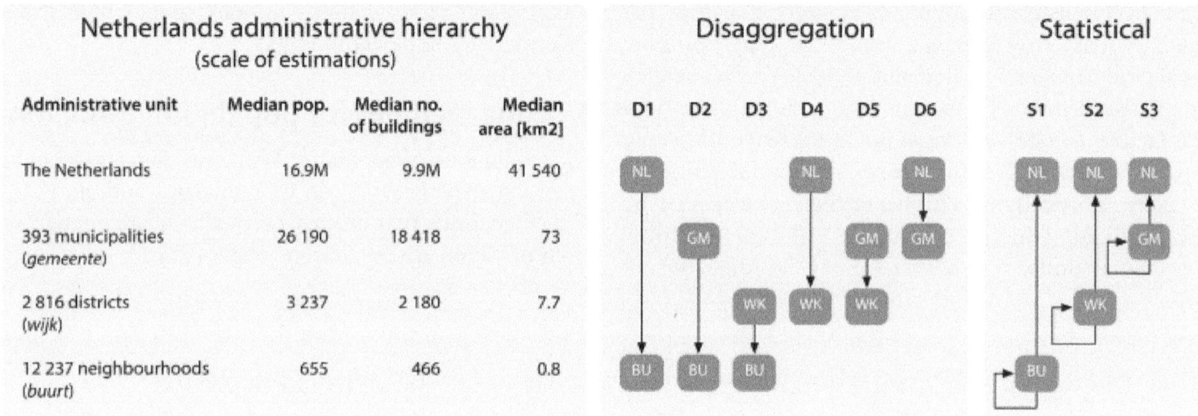

Figure 6: The Dutch statistical hierarchy, and our hybrid multi-scale approach.

- **D1** Disaggregation from the country level to its 12237 neighbourhoods.
- **D2** Disaggregation from each of the 393 municipalities to their 12237 neighbourhoods.
- **D3** Disaggregation from each of the 2816 districts to their 12237 neighbourhoods.
- **D4** Disaggregation from the country level to its 2816 districts.
- **D5** Disaggregation from each of the 393 municipalities to their 2816 districts.
- **D6** Disaggregation from the country level to its 393 municipalities.

On the statistical side we use a random subset of 10% of each statistical level to determine with ordinary least squares the relationships between building space and population, and apply them for three different experiments:

- **S1** Estimation of the population of the test neighbourhoods (i.e. the remaining 90%).
- **S2** Estimation of the population of the test districts (i.e. the remaining 90%).
- **S3** Estimation of the population of the test municipalities (i.e. the remaining 90%).

Furthermore, in the statistical approaches (S1, S2, and S3) we also estimate the population of the Netherlands. This means that we test the suitability of carrying out the census for 10% of the country (training dataset), and estimating the population of the rest of a country (test dataset).

In each of the 9 approaches we carry out separate experiments with the data in the 9 different LODs. This results in a total of 108 experiments.

As in related work [44, 86], we ignore very small buildings (footprint smaller than 20 m^2) such as sheds, garages, etc. which are unlikely to be inhabited (visible in Fig 1 as tiny white footprints in the overly residential areas).

RESULTS AND DISCUSSION

Performance and Observations

We perform the experiments, and compare them to the actual values, as observed in the governmental census dataset (CBS). We use percentage error because we are dealing with different scales of data (e.g. an error of 1000 residents is not of the same magnitude on the neighbourhood or city level). Furthermore, because of large errors in some statistical units (explained later), instead of the usual mean absolute error and root-mean-square error we use the median absolute error. As in related work [71, 87], we observe that estimations in areas with small populations is prone to a high relative error (see S1 Fig), hence medians are a good option here. The results of all experiments are given in Table 1. Because of many different models and types of data, we focus on the most important results only, however, the elaborated observations are similar with the rest of the models. It should be noticed that both the disaggregation and statistical approach exhibit congruent behaviour in most cases.

Table 1: Median absolute percentage errors in the population estimates resulting from our experiments

	a	b	c	a	b	c	a	b	c
					(1) Disaggregation				
		D1 (n = 12237)			D2 (n = 12237)			D3 (n = 12237)	
0	61.9	41.9	42.4	53.9	25.5	25.7	42.7	17.7	17.7
0+	39.8	20.8	20.8	37.2	16.2	16.4	29.1	12.0	12.0
1	56.4	25.5	25.8	53.0	20.8	20.7	42.4	15.6	15.3
		D4 (n = 3237)			D5 (n = 3237)			D6 (n = 393)	
0	56.5	37.7	38.2	34.3	15.5	15.5	32.0	25.3	25.5
0+	25.8	16.9	16.5	21.3	9.3	9.2	13.2	11.5	11.4
1	43.5	20.0	20.5	32.0	11.8	11.9	22.1	13.2	13.2
					(2) Statistical approach (local units)				
		S1 (n = 12237)			S2 (n = 3237)			S3 (n = 393)	
0	85.4	42.0	42.2	56.9	53.1	52.8	74.0	38.7	38.8
0+	35.4	18.3	18.5	41.5	28.9	28.5	20.6	12.6	12.2
1	66.8	24.3	24.8	49.8	26.1	28.6	28.9	9.5	9.3
					(2) Statistical approach (country level)				
		S1 (n = 1)			S2 (n = 1)			S3 (n = 1)	
0	0.6	1.4	1.4	2.7	5.6	5.7	21.5	1.3	1.7
0+	9.3	2.0	2.2	2.7	0.6	0.5	7.9	1.9	1.9
1	4.1	1.2	1.3	3.1	1.9	1.9	11.7	2.0	1.8

The order of errors in each 3x3 matrix is expressed in the same order as the LODs in Fig 4.

The results exhibit a large degree of variation between the accuracy depending on the approach, level of detail of the data, and the scale of the estimations. The smallest error of the volume-based disaggregation approach is in D5/LOD1^b (the disaggregation from municipalities to districts) and it equals 11.8%. The smallest error in the statistical approach was observed in S3 (estimation of the population of cities), resulting in an error of 9.3%. We observe and conclude the following:

- 3D city models and the volume-based approach provide a substantial advantage over traditional 2D maps and the area-based approach because they capture the vertical extent of the building. However, the estimations carried out with 3D models are still less accurate than when using floorspace information. We think that volume does not add value over floorspace because two flats of the same floorspace but of different volumes (e.g. ceiling height of 2.5 m vs 4 m) generally do not host a different number of residents, unlike what the method would predict. It should be noted, however, that floorspace information is difficult to acquire automatically and it is generally not available.
- In most cases, semantic information on the use of buildings provides a substantial improvement in the estimations over data without such information. This helps to exclude non-residential units, which can significantly skew the estimations. Such behaviour is visible as outliers in the scatter plots in Fig 7(other observations will be discussed in the continuation). Population estimation without information on the building function is practically unusable in most cases, especially in industrial neighbourhoods (in our experiments we have seen overestimations of more than 5000%). In fact, the results show that in this use case, semantic information is typically more important than the geometric detail (e.g. cf. error of 41.9% in D1/LOD0^b—semantically enriched 2D footprints vs error of 56.4% in D1/LOD1^a—plain 3D buildings).

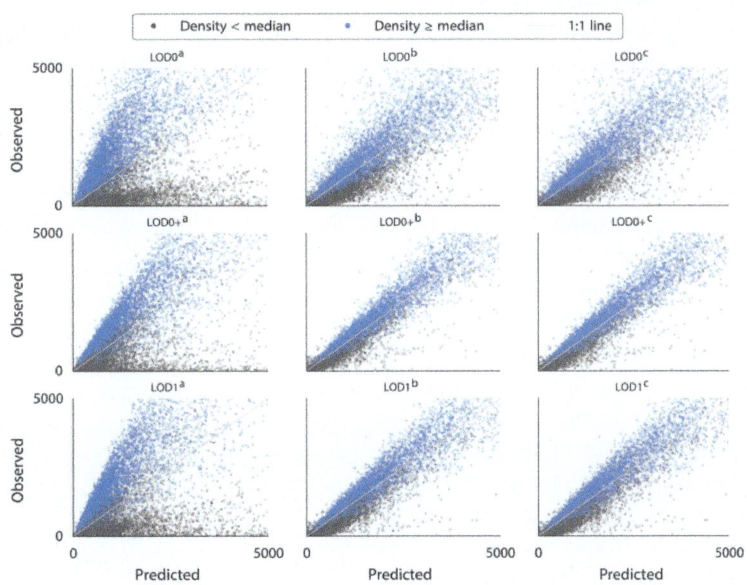

Figure 7: Observed (actual data from the government census) vs predicted scatter plots of the 9 input datasets in the D1 method.

- While semantic data is crucial, it appears that there is inconsistent added value of the detailed (fractional) semantic information versus only binary information. It seems that the difference between binary and fractional semantic information becomes negligible at the neighbourhood level. In fact, in some estimations (e.g. D4/LOD1) the estimations with fractional semantic information (D4/LOD1^c) are slightly less accurate than when using binary semantic information (D4/LOD1^b).

In the floorspace data (LOD0+) there is generally a small improvement of using fractional semantic information rather than binary. A possible reason is that the volume-based estimations are more sensitive to errors in the input dataset.

For the purposes described in this paper, it does not seem worthwhile to collect detailed building usage, as the binary information suffices. Because such information may be automatically derived from the building morphology, aerial imagery, land use maps, etc. [88–94], this insight is beneficial for estimations that need to be carried out on a large extent where cadastral data is not available.

- Different scales of estimations show different performance and different suitability for the different methods. In the disaggregation, the method works best in hierarchically close units: compare D3 (districts to neighbourhoods) with D1 (country to neighbourhoods). This is because such relations exhibit less difference in housing variations. Furthermore, it seems that disaggregating data to units higher in the hierarchy is more accurate than to units of a finer scale, because larger units such as districts and municipalities capture larger residential differences than small neighbourhoods, i.e. the variation among smaller units is greater than that among larger units. For example, two municipalities may have equal population but within municipalities the population differences among districts may be relatively large (e.g. rural vs urban zones). On the other hand differences among neighbourhoods in a district may be small.
- The statistical approach is of comparable accuracy to the disaggregation because it is also based on coefficients uniform for the whole country, which hide massive disparities among different neighbourhoods and provinces, and it is therefore equally affected by the differences in living standards and residential choices. However, for the largest extent (country), the statistical approach is impressively accurate: in the S1/LOD1^b experiment (statistical approach applied on neighbourhoods with the semantic volume-based LOD1 block model) the population of the Netherlands based on a subset of 10% neighbourhoods has been estimated to 17 100 292, just a 1.2% overestimation from the true figure. The floorspace-based (LOD0+^{b}) data fares even better with a deviation of 0.5% in S2. This finding gives confidence in the use of 3D city models for estimating the population of large areas such as countries, especially in developing countries since the data required for such estimations can be derived automatically and remotely from airborne sensors [95]. However, it should be noted that the model S1/LOD0^a (building footprints without information on the building use) performed best with an error of 0.6%. It is hard to explain the reason why in this particular model lesser data gave better results, because all errors (induced by different LODs, uncertainty in the input data, different residential choices, etc.) are aggregated in a single number that cannot be decomposed.
- We have noticed that the models tend to overestimate the population in rural areas, and underestimate it in urban areas (see the coloured points in Fig 7). This finding is similar to the observations in related work [38]. The differences are caused by the varying utilisation of living space, which differ between less and more densely populated entities. We use this finding in the succeeding sections for additional insights and we take advantage of it to improve the statistical approach (models S1, S2, and S3).

Sources of Error

After analysing the errors we observe different causes of errors. The residential differences (e.g. residential space per resident) is the principal cause of the residuals (the errors very strongly correlate with the average space per resident; $r > 0.99$). There is a variable level of occupancy and variable utilisation of space within each building, i.e. living space per inhabitant considerably varies based on social, economical, and other factors. Some households live in large houses, while others in small studios and dormitories, rendering significant differences in the residential density [43], and presenting a problem for population estimation with remotely sensed data [46]. Furthermore, these differences are also caused by non-residential space within residential units, such as storage rooms, utility rooms, common rooms, gyms, garages, etc., which increase the building size and considered dwelling space, but due to the shortcomings of the data cannot be accounted as non-residential space. It is usually not possible to assume that these characteristics are equally distributed in each entity, as they are not constant among different neighbourhoods and also on larger extent such as among municipalities [68, 96]. This fact is also visible in Fig 2. Therefore it is important to consider different environments when calibrating the method, and accept imperfections as one model cannot fit all situations within a large area such as a province or country.

We had expected that these differences would cancel out within the statistical entities (since one typically contains hundreds of houses, see Fig 1), however, the difference between units, including larger ones such as cities, is still gross. One would assume that a city contains a fair diversity of different configurations, but it turns out that each city has a unique setting which cannot be applied to another one.

Furthermore, another variation of the dwelling density is caused by vacant residential buildings (e.g. empty houses for sale, vacation homes). In our method we can only assume that the vacancy rate is homogeneous in our area of study, consistently with other researchers (e.g. when estimating the energy demand [97]), however, that assumption might deviate from the reality.

When using the data without information on building use (i.e. LODx^a) many large errors were found in industrial neighbourhoods with huge building volumes, highlighting the importance of using semantics. When using the semantically enriched buildings, the results improved substantially.

However, errors in the input data on building use have also caused errors in the estimation of the population. For instance, we have noted that in an industrial neighbourhood a large factory was mislabelled as a residential building, so the population has been pointlessly disaggregated in an empty building, inducing a substantial error. The input datasets that we used were very accurate [98, 99], but occasional small errors induced gross errors in the estimations. Furthermore, it is worthwhile to mention that there were peculiar cases which also caused discrepancies, such as a small neighbourhood with a prison as its sole building. Its inmates are counted as residents in the CBS dataset, but the prison building is not classified as residential in the cadastral dataset, hence the estimate of the neighbourhood exhibited a large error—the population was predicted to be 0, while in reality it is 75.

The 3D geometric aspect (calculated volume) may induce errors to the estimations as well. It has been suggested [64] that geometric errors in 3D city models (e.g. inconsistencies caused by vegetation in the elevation dataset) may substantially influence spatial analyses, especially the computation of the volume [100].

The related work in analysing error propagation in population estimation is limited to 2D [101–103]. For future work it would be interesting to investigate the influence of errors in the input data when using volume-based approaches.

Analysis of Errors and Enhancement of the Statistical Approach

We have analysed the errors with demographic and other indicators for each statistical unit in order to understand them better and to potentially improve our methods.

We have analysed the income of each neighbourhood and did not find a correlation with errors. We had presumed that low-income neighbourhoods might have less space per resident, as income may be related to living standard and may drive residential purchasing choices. However, that is not the case, because there are cheap but large country-side properties, and expensive small flats in cities such as Amsterdam, invalidating our assumption.

In the previous section we have noted the particular behaviour of errors with respect to the different population densities of estimated areas. There is a clear difference between more and less urbanised areas caused by the different utilisation of dwelling space (see Fig 8). It is clear that the data on the population density could be used to improve the estimations, but as such it is not available prior to the estimation of population (otherwise we would not need to conduct the estimations).

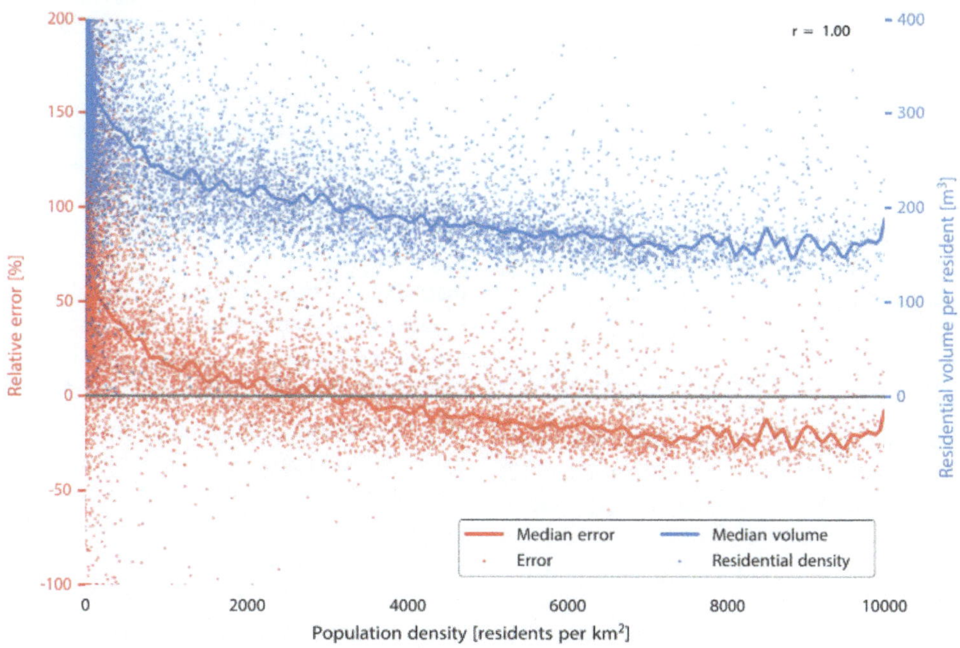

Figure 8: The relations between the errors, population density, and living space per statistical neighbourhood.

However, we have realised that there is another indicator that it is associated with the population density, and which is available prior to the estimations: the average building height in a neighbourhood is associated with the population density (see S2 Fig), and consequently to the living space. Therefore, for each neighbourhood we have calculated the average building height (easily available since we have 3D city models), and we have incorporated it in our multiple linear regression model (which now contains two variables: the total building space in the statistical unit, and the average height of buildings in the unit). We have not applied this enhancement to the LOD0 approach in which vertical measurements are not available.

The statistical experiments show that there is an improvement to the models: a reduction of errors by a few percent on average has been observed in the models S1, S2, and S3. Note that the results presented in the previous section are of those with the enhanced models, and that the disaggregation method was not enhanced because of its inherently different approach in which there is no training data.

While we believe that the presented prediction models *might* be further augmented to improve the estimations with additional variables and 2D GIS data such as land use, in this paper we have used only 3D models to determine how accurate the predictions can be if relied solely on them. Adding such additional variables is avoided because of a contradictory situation: if such data is available, it is likely that accurate census data is also available, rendering such estimations unnecessary.

CONCLUSIONS AND OUTLOOK

In this study we have used a 3D city model to estimate the population of 12.2 thousand neighbourhoods, 2816 districts, and 393 municipalities in the Netherlands, and of the Netherlands itself. Our results indicate that in certain circumstances 3D city models can give a good approximation of the population, and that, in most cases, 3D city models add value over traditionally used 2D datasets, but also that they are not accurate enough to replace accurate census techniques employed by governments. Furthermore, there were certain instances when 2D data (even without the information on building use, e.g. S1/LOD0[a]) performed better than 3D data, which is beneficial because such data is simpler to acquire. The main reason why this method is useful is because it does not require expensive and time consuming field surveys and other means of collecting population counts as the data can be acquired automatically and remotely, and it can be carried out more frequently, in contrast to official censuses (usually conducted every decade).

One of the strengths of our work over previous studies is that we carried out a country-wide analysis, in which differences between neighbourhoods are more emphasised. Our study is multi-LOD (both area-based and volume-based approaches have been evaluated, along with multiple grades of semantic information), multi-scale (for assessing the suitability of mapping statistical units of different sizes), and multi-method (both the weighted disaggregation and statistical approaches have been employed).

Remote estimation of population with GIS could be applied in areas where census information is not available or it is not reliable, and serves two purposes: (1) as a potential solution to estimate the population count of large areas where a census is not available, or as an intercensal estimate; and (2) for refining the population on a finer scale (e.g. disaggregation of an accurate census of a city among its neighbourhoods).

Our approach is easily applicable in other countries. Governments have started to publicly release building footprints and other GIS data [89], and where data is available many 3D city models have been generated [104–107]. Alternatively, 3D city models may be generated from volunteered geoinformation [108], ensuring the applicability of our method elsewhere. While in this study for the building use we used datasets from the cadastre, it is worth noting that such data can also be derived manually from aerial images, and automatically from the building morphology and other characteristics, or from volunteered geoinformation [88–94]. Such an approach provides an enhancement over previous research, since in related work coarse datasets have traditionally been used, e.g. Kressler and Steinnocher [5] and Silván-Cárdenas et al. [41] distinguish residential buildings from non-residential ones with a zoning map.

Concerning the first application, estimating the population count of large areas where a census is not available, in the 21st century there are still many places around the world where the census has not been carried out in decades, and such remote sensing methods can help to bridge the gap [109, 110]. For instance, Myanmar did not have a reliable census until two years ago, and in the meantime the authorities were dealing with information which turned out to be significantly erroneous [111, 112], something unthinkable in developed countries nowadays. Obviously, low income countries cannot boast about 3D city models, however, with the development of remote sensing technologies, and surge of volunteered geoinformation and their quality [113, 114], the generation of 3D city models is becoming increasingly simpler and cheaper [107,108, 115]. Therefore we expect that in the near future country-wide 3D city models will not be a luxury exclusive to developed countries.

With respect to the second purpose, the derived data on the number of residents on a finer scale is beneficial for a multitude of applications [116], such as disaster management (e.g. in flooded areas) [117, 118], analysing accessibility [119], public health [1, 120], crime mapping [121], environmental risk [80, 122], infrastructure planning and transportation sustainability [123], epidemiology [124], territorial classification [125], assessing exposure to noise [51, 126, 127], optimising network coverage (e.g. television) to cover more people [128, 129], for finding areas for landing of stratospheric balloons [129], marketing strategies [44], estimating the quantity of waste [130], estimating energy consumption [131], and in urban simulations [132].

We have also discovered that this method can also be used to detect potential errors in authoritative census and building data (e.g. we have detected erroneous semantic information for some commercial buildings by analysing the large errors in population estimates). Furthermore, we envisage that such method could be used for detecting false residencies (e.g. a large number of people registered in a particular neighbourhood for tax-related reasons, triggering an alert by the population that exceeds the housing capacity in that area).

The results indicate that the estimations are hampered by socio-economic disparities between neighbourhoods, and that population estimation is more reliable when focused on statistical units with a closer proximity. However, this limitation does not seem to affect the estimation of the national population, in which case our method has particularly excelled.

For future work it would be worthwhile to advance the sampling method of the training data in the statistical approach to investigate whether that leads to more accurate estimates. For instance, stratified sampling [133] could be employed instead of the simple random sampling which is used now. Such sampling method could stratify entities based on different characteristics obtainable from 3D city models, such as predominant building types in a neighbourhood, and apply different statistical models to each stratum.

SUPPORTING INFORMATION

S1 Figure Less populated districts exhibit large relative errors, promoting the use of medians.

In relative terms, the estimation is more accurate when carried out in more populous areas. These are the results from the experiments S1/LOD1[c]. The two histograms show the data divided in two bins (the left one of the statistical units with the population smaller than the median value of all units (710 residents), and the one on the right the units with the population higher than the median). Not to be confused with Fig 8 which shows the relation of errors to the population density (however, notice that in this case as well the methods tend to underestimate the population in more populated areas).

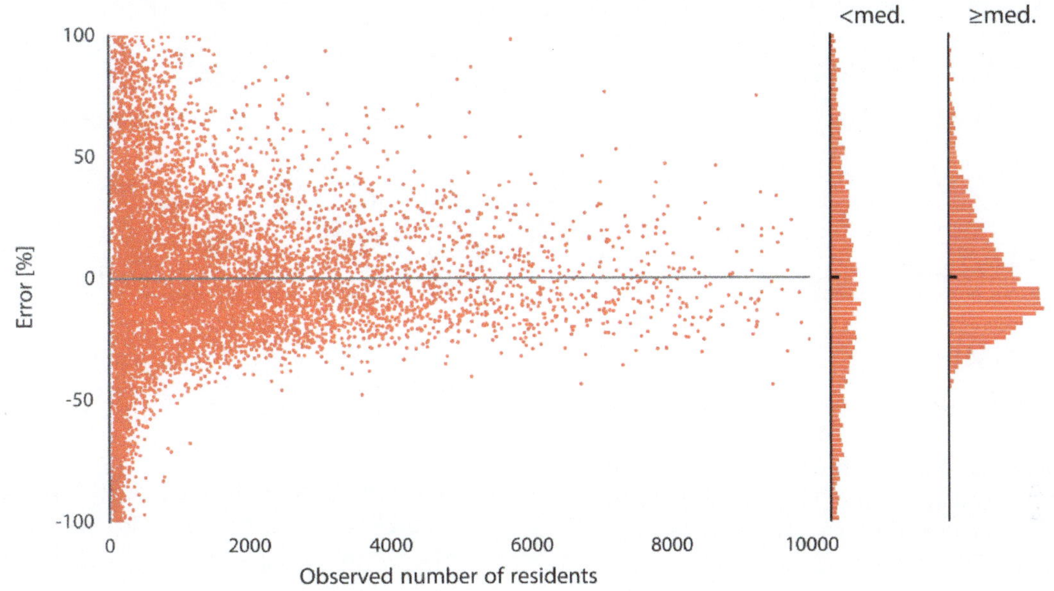

S2 Figure Association of the population density and vertical extent of the neighbourhood.

While the population density is not available for adjusting our models, we have taken advantage of the vertical extent which hints at the population density, and in turn helps in adjusting the prediction between urban and rural areas.

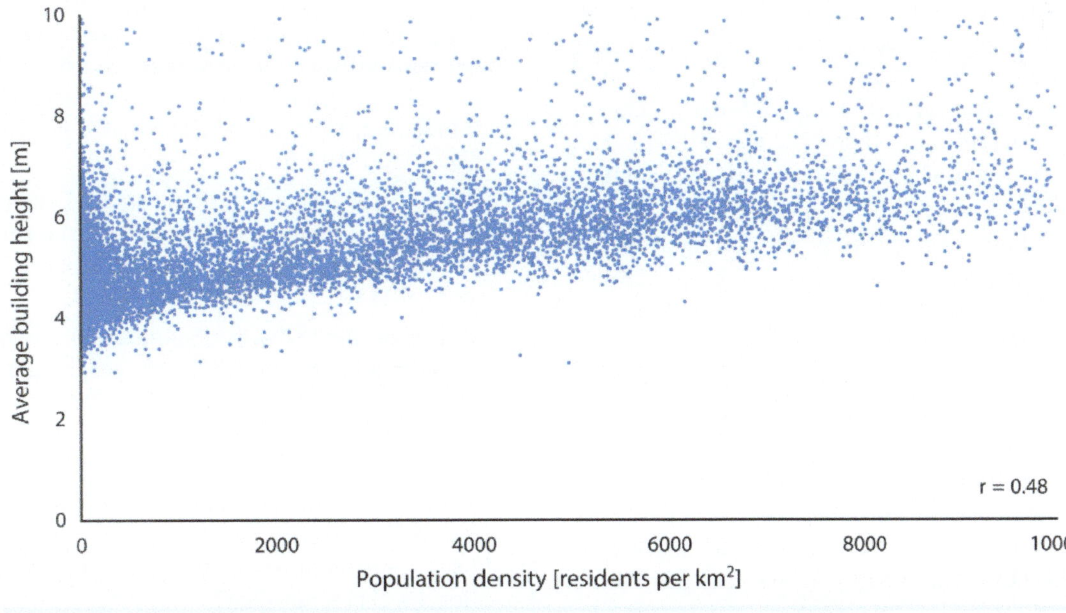

S3 Figure Elevation dataset (AHN) used to generate the 3D city model.

The point cloud was obtained with airborne laser scanning, and the colours represent the elevation. The spatial extent and angle of view correspond to the one shown in Fig 3. The accuracy of the points is within a few centimetres [98]. The whole dataset contains 639B points [134]. Data (c) Actueel Hoogtebestand Nederland.

ACKNOWLEDGMENTS

We gratefully acknowledge the availability of open data of the Government of the Netherlands, and the work of the NLExtract project. We appreciate the constructive comments of the anonymous reviewers, and the information and clarifications provided by Just van den Broecke, Thomas Spoorenberg, Dominique Laurent, and Pieter Bresters.

REFERENCES

1. Anderson W, Guikema S, Zaitchik B, Pan W. Methods for Estimating Population Density in Data-Limited Areas: Evaluating Regression and Tree-Based Models in Peru. PLOS ONE. 2014. July;9(7):e100037 doi:10.1371/journal.pone.0100037
2. Hillson R, Alejandre JD, Jacobsen KH, Ansumana R, Bockarie AS, Bangura U, et al. Methods for Determining the Uncertainty of Population Estimates Derived from Satellite Imagery and Limited Survey Data: A Case Study of Bo City, Sierra Leone. PLOS ONE. 2014. November;9(11):e112241 doi:10.1371/journal.pone.0112241
3. Welch R. Monitoring urban population and energy utilization patterns from satellite Data. Remote sensing of Environment. 1980. February;9(1):1–9. doi: 10.1016/0034-4257(80)90043-7
4. Lu D, Weng Q, Li G. Residential population estimation using a remote sensing derived impervious surface approach. International Journal of Remote Sensing. 2006. August;27(16):3553–3570. doi:10.1080/01431160600617202
5. Kressler F, Steinnocher K. Object-oriented analysis of image and LiDAR data and its potential for a dasymetric mapping application In: On segment based image fusion. Springer Berlin; Heidelberg; 2008. p. 611–624.
6. Lo CP. Automated population and dwelling unit estimation from high-resolution satellite images: a GIS approach. International Journal of Remote Sensing. 1995. January;16(1):17–34. doi:10.1080/01431169508954369
7. Lo CP. Population Estimation Using Geographically Weighted Regression. GIScience & Remote Sensing. 2013. May;45(2):131–148. doi: 10.2747/1548-1603.45.2.131
8. Tobler WR. Satellite confirmation of settlement size coefficients. Area. 1969;1(3):30–34.
9. Kraus SP, Senger LW, Ryerson JM. Estimating population from photographically determined residential land use types. Remote sensing of Environment. 1974. January;3(1):35–42. doi: 10.1016/0034-4257(74)90036-4
10. Lo CP, Welch R. Chinese Urban Population Estimates. Annals of the Association of American Geographers. 1977. June;67(2):246–253. doi: 10.1111/j.1467-8306.1977.tb01137.x
11. Al-garni AM. Mathematical predictive models for population estimation in urban areas using space products and GIS technology. Mathematical and Computer Modelling. 1995. July;22(1):95–107. doi:10.1016/0895-7177(95)00104-A
12. Wu C, Murray AT. Population Estimation Using Landsat Enhanced Thematic Mapper Imagery. Geographical Analysis. 2007. January;39(1):26–43. doi: 10.1111/j.1538-4632.2006.00694.x
13. Zhan FB, Tapia Silva FO, Santillana M. Estimating small-area population growth using geographic-knowledge-guided cellular automata. International Journal of Remote Sensing. 2010. November;31(21):5689–5707. doi: 10.1080/01431161.2010.496802
14. Langford M. An Evaluation of Small Area Population Estimation Techniques Using Open Access Ancillary Data. Geographical Analysis. 2013. July;45(3):324–344. doi: 10.1111/gean.12012
15. Xie Z. A Framework for Interpolating the Population Surface at the Residential-Housing-Unit Level. GIScience & Remote Sensing. 2013. May;43(3):233–251. doi: 10.2747/1548-1603.43.3.233

16. Deng C, Wu C, Wang L. Improving the housing-unit method for small-area population estimation using remote-sensing and GIS information. International Journal of Remote Sensing. 2010. November;31(21):5673–5688. doi: 10.1080/01431161.2010.496806

17. Wu C, Murray AT. A cokriging method for estimating population density in urban areas. Computers, Environment and Urban Systems. 2005. September;29(5):558–579. doi:10.1016/j.compenvurbsys.2005.01.006

18. Bagan H, Yamagata Y. Landsat analysis of urban growth: How Tokyo became the world's largest megacity during the last 40 years. Remote sensing of Environment. 2012. December;127:210–222. doi:10.1016/j.rse.2012.09.011

19. Yuan Y, Smith RM, Limp WF. Remodeling census population with spatial information from LandSat TM imagery. Computers, Environment and Urban Systems. 1997. May;21(3–4):245–258. doi:10.1016/S0198-9715(97)01003-X

20. Stevens FR, Gaughan AE, Linard C, Tatem AJ. Disaggregating Census Data for Population Mapping Using Random Forests with Remotely-Sensed and Ancillary Data. PLOS ONE. 2015. February;10(2):e0107042 doi: 10.1371/journal.pone.0107042

21. Gaughan AE, Stevens FR, Linard C, Jia P, Tatem AJ. High Resolution Population Distribution Maps for Southeast Asia in 2010 and 2015. PLOS ONE. 2013. February;8(2):e55882 doi:10.1371/journal.pone.0055882

22. Bakillah M, Liang S, Mobasheri A, Jokar Arsanjani J, Zipf A. Fine-resolution population mapping using OpenStreetMap points-of-interest. International Journal of Geographical Information Science. 2014. September;28(9):1940–1963. doi: 10.1080/13658816.2014.909045

23. Doll CNH, Muller JP, Morley JG. Mapping regional economic activity from night-time light satellite imagery. Ecological Economics. 2006. April;57(1):75–92. doi: 10.1016/j.ecolecon.2005.03.007

24. Anderson SJ, Tuttle BT, Powell RL, Sutton PC. Characterizing relationships between population density and nighttime imagery for Denver, Colorado: issues of scale and representation. International Journal of Remote Sensing. 2010. November;31(21):5733–5746. doi: 10.1080/01431161.2010.496798

25. Sutton P. Modeling population density with night-time satellite imagery and GIS. Computers, Environment and Urban Systems. 1997. May;21(3–4):227–244. doi: 10.1016/S0198-9715(97)01005-3

26. Pozzi F, Small C. Analysis of urban land cover and population density in the United States.Photogrammetric Engineering and Remote Sensing. 2005. June;71(6):719–726. doi:10.14358/PERS.71.6.719

27. Xie Y. The overlaid network algorithms for areal interpolation problem. Computers, Environment and Urban Systems. 1995. July;19(4):287–306. doi: 10.1016/0198-9715(95)00028-3

28. Steiger E, Westerholt R, Resch B, Zipf A. Twitter as an indicator for whereabouts of people? Correlating Twitter with UK census data. Computers, Environment and Urban Systems. 2015. November;54:255–265. doi: 10.1016/j.compenvurbsys.2015.09.007

29. Lwin KK, Sugiura K, Zettsu K. Space–time multiple regression model for grid-based population estimation in urban areas. International Journal of Geographical Information Science. 2016;30(8):1579–1593. doi: 10.1080/13658816.2016.1143099

30. Wu Ss, Wang L, Qiu X. Incorporating GIS Building Data and Census Housing Statistics for Sub-Block-Level Population Estimation. The Professional Geographer. 2008. January;60(1):121–135. doi:10.1080/00330120701724251

31. Harvey JT. Estimating census district populations from satellite imagery: Some approaches and limitations. International Journal of Remote Sensing. 2010. November;23(10):2071–2095. doi:10.1080/01431160110075901

32. Lwin KK, Murayama Y. Estimation of Building Population from LIDAR Derived Digital Volume Model In: Spatial Analysis and Modeling in Geographical Transformation Process. Dordrecht: Springer Netherlands; 2011. p. 87–98.

33. Suveg I, Vosselman G. Reconstruction of 3D building models from aerial images and maps. ISPRS Journal of Photogrammetry and Remote Sensing. 2004. January;58(3–4):202–224. doi:10.1016/j.isprsjprs.2003.09.006

34. Musialski P, Wonka P, Aliaga DG, Wimmer M, van Gool L, Purgathofer W. A Survey of Urban Reconstruction. Computer Graphics Forum. 2013. May;32(6):146–177. doi: 10.1111/cgf.12077

35. Truong-Hong L, Laefer DF. Quantitative evaluation strategies for urban 3D model generation from remote sensing data. Computers and Graphics. 2015. June;49:82–91. doi: 10.1016/j.cag.2015.03.001

36. Serna A, Marcotegui B. Detection, segmentation and classification of 3D urban objects using mathematical morphology and supervised learning. ISPRS Journal of Photogrammetry and Remote Sensing. 2014. July;93:243–255. doi: 10.1016/j.isprsjprs.2014.03.015

37. Rottensteiner F, Sohn G, Gerke M, Wegner JD, Breitkopf U, Jung J. Results of the ISPRS benchmark on urban object detection and 3D building reconstruction. ISPRS Journal of Photogrammetry and Remote Sensing. 2014. July;93:256–271.

38. Dong P, Ramesh S, Nepali A. Evaluation of small-area population estimation using LiDAR, Landsat TM and parcel data. International Journal of Remote Sensing. 2010. November;31(21):5571–5586. doi:10.1080/01431161.2010.496804

39. Lu Z, Im J, Quackenbush L. A Volumetric Approach to Population Estimation Using Lidar Remote Sensing. Photogrammetric Engineering and Remote Sensing. 2011. November;77(11):1145–1156. doi:10.14358/PERS.77.11.1145

40. Lu Z, Im J, Quackenbush L, Halligan K. Population estimation based on multi-sensor data fusion. International Journal of Remote Sensing. 2010.

November;31(21):5587–5604. doi:10.1080/01431161.2010.496801

41. Silván-Cárdenas JL, Wang L, Rogerson P, Wu C, Feng T, Kamphaus BD. Assessing fine-spatial-resolution remote sensing for small-area population estimation. International Journal of Remote Sensing. 2010. November;31(21):5605–5634. doi: 10.1080/01431161.2010.496800
42. Ural S, Hussain E, Shan J. Building population mapping with aerial imagery and GIS data. International Journal of Applied Earth Observation and Geoinformation. 2011. December;13(6):841–852. doi:10.1016/j.jag.2011.06.004
43. Sridharan H, Qiu F. A Spatially Disaggregated Areal Interpolation Model Using Light Detection and Ranging-Derived Building Volumes. Geographical Analysis. 2013. July;45(3):238–258. doi:10.1111/gean.12010
44. Lwin K, Murayama Y. A GIS Approach to Estimation of Building Population for Micro-spatial Analysis. Transactions in GIS. 2009. August;13(4):401–414. doi: 10.1111/j.1467-9671.2009.01171.x
45. Qiu F, Sridharan H, Chun Y. Spatial Autoregressive Model for Population Estimation at the Census Block Level Using LIDAR-derived Building Volume Information. Cartography and Geographic Information Science. 2010. January;37(3):239–257. doi: 10.1559/152304010792194949
46. Alahmadi M, Atkinson P, Martin D. Estimating the spatial distribution of the population of Riyadh, Saudi Arabia using remotely sensed built land cover and height data. Computers, Environment and Urban Systems. 2013. September;41:167–176. doi: 10.1016/j.compenvurbsys.2013.06.002
47. Alahmadi M, Atkinson PM, Martin D. A Comparison of Small-Area Population Estimation Techniques Using Built-Area and Height Data, Riyadh, Saudi Arabia. IEEE Journal of Selected Topics in Applied Earth Observations and Remote Sensing. 2016;9(5):1959–1969. doi: 10.1109/JSTARS.2014.2374175
48. Kolbe TH. Representing and exchanging 3D city models with CityGML In: Zlatanova S, Lee J, editors.3D Geo-Information Sciences. Springer Berlin; Heidelberg; 2009. p. 15–31.
49. Billen R, Cutting-Decelle AF, Marina O, de Almeida JP, M C, Falquet G, et al. 3D City Models and urban information: Current issues and perspectives In: 3D City Models and urban information: Current issues and perspectives—European COST Action TU0801. Les Ulis, France: EDP Sciences; 2014. p. I–118.
50. Biljecki F, Stoter J, Ledoux H, Zlatanova S, Çöltekin A. Applications of 3D City Models: State of the Art Review. ISPRS International Journal of Geo-Information. 2015. December;4(4):2842–2889. doi:10.3390/ijgi4042842
51. Stoter J, de Kluijver H, Kurakula V. 3D noise mapping in urban areas. International Journal of Geographical Information Science. 2008;22(8):907–924. doi: 10.1080/13658810701739039
52. Kaden R, Kolbe TH. Simulation-Based Total Energy Demand Estimation of Buildings using Semantic 3D City Models. International Journal of 3-D Information Modeling. 2014;3(2):35–53. doi:10.4018/ij3dim.2014040103
53. Bahu JM, Koch A, Kremers E, Murshed SM. Towards a 3D Spatial Urban Energy Modelling Approach. International Journal of 3-D Information Modeling. 2015. July;3(3):1–16. doi: 10.4018/ij3dim.2014070101
54. Fisher-Gewirtzman D, Shashkov A, Doytsher Y. Voxel based volumetric visibility analysis of urban environments. Survey Review. 2013. November;45(333):451–461. doi: 10.1179/1752270613Y.0000000059
55. Bartie P, Reitsma F, Kingham S, Mills S. Advancing visibility modelling algorithms for urban environments. Computers, Environment and Urban Systems. 2010. November;34(6):518–531. doi:10.1016/j.compenvurbsys.2010.06.002
56. Xiong B, Jancosek M, Oude Elberink S, Vosselman G. Flexible building primitives for 3D building modeling. ISPRS Journal of Photogrammetry and Remote Sensing. 2015. March;101:275–290. doi:10.1016/j.isprsjprs.2015.01.002
57. Haala N, Kada M. An update on automatic 3D building reconstruction. ISPRS Journal of Photogrammetry and Remote Sensing. 2010. November;65(6):570–580. doi: 10.1016/j.isprsjprs.2010.09.006
58. Sirmacek B, Taubenbock H, Reinartz P, Ehlers M. Performance Evaluation for 3-D City Model Generation of Six Different DSMs From Air- and Spaceborne Sensors. IEEE Journal of Selected Topics in Applied Earth Observations and Remote Sensing. 2012;5(1):59–70. doi: 10.1109/JSTARS.2011.2178399
59. Arroyo Ohori K, Ledoux H, Stoter J. A dimension-independent extrusion algorithm using generalised maps. International Journal of Geographical Information Science. 2015. March;29(7):1166–1186. doi:10.1080/13658816.2015.1010535
60. Ledoux H, Meijers M. Topologically consistent 3D city models obtained by extrusion. International Journal of Geographical Information Science. 2011. April;25(4):557–574. doi:10.1080/13658811003623277
61. Gröger G, Plümer L. CityGML—Interoperable semantic 3D city models. ISPRS Journal of Photogrammetry and Remote Sensing. 2012. July;71:12–33. doi: 10.1016/j.isprsjprs.2012.04.004
62. Biljecki F, Ledoux H, Stoter J, Zhao J. Formalisation of the level of detail in 3D city modelling. Computers, Environment and Urban Systems. 2014. November;48:1–15. doi:10.1016/j.compenvurbsys.2014.05.004
63. Biljecki F, Ledoux H, Stoter J, Vosselman G. The variants of an LOD of a 3D building model and their influence on spatial analyses. ISPRS Journal of Photogrammetry and Remote Sensing. 2016;116:42–54. doi:10.1016/j.isprsjprs.2016.03.003
64. Biljecki F, Heuvelink GBM, Ledoux H, Stoter J. Propagation of positional error in 3D GIS:

estimation of the solar irradiation of building roofs. International Journal of Geographical Information Science. 2015. December;29(12):2269–2294. doi: 10.1080/13658816.2015.1073292

65. Arroyo Ohori K, Ledoux H, Biljecki F, Stoter J. Modeling a 3D City Model and Its Levels of Detail as a True 4D Model. ISPRS International Journal of Geo-Information. 2015. September;4(3):1055–1075. doi:10.3390/ijgi4031055

66. Stadler A, Kolbe TH. Spatio-semantic coherence in the integration of 3D city models. Int Arch Photogramm Remote Sens Spatial Inf Sci. 2007. June;XXXVI-2/C43:8.

67. Boeters R, Arroyo Ohori K, Biljecki F, Zlatanova S. Automatically enhancing CityGML LOD2 models with a corresponding indoor geometry. International Journal of Geographical Information Science. 2015. December;29(12):2248–2268. doi: 10.1080/13658816.2015.1072201

68. Shiravi S, Zhong M, Beykaei SA, Hunt JD, Abraham JE. An assessment of the utility of LiDAR data in extracting base-year floorspace and a comparison with the census-based approach. Environment and Planning B: Planning and Design. 2015;42(4):708–729.

69. Xie Y, Weng A, Weng Q. Population Estimation of Urban Residential Communities Using Remotely Sensed Morphologic Data. IEEE Geoscience and Remote Sensing Letters. 2015. May;12(5):1111–1115. doi: 10.1109/LGRS.2014.2385597

70. Wu Ss, Qiu X, Wang L. Population Estimation Methods in GIS and Remote Sensing: A Review.GIScience & Remote Sensing. 2005. March;42(1):80–96. doi: 10.2747/1548-1603.42.1.80

71. Brinegar SJ, Popick SJ. A Comparative Analysis of Small Area Population Estimation Methods. Cartography and Geographic Information Science. 2013. March;37(4):273–284. doi:10.1559/152304010793454327

72. Mennis J. Dasymetric Mapping for Estimating Population in Small Areas. Geography Compass. 2009. March;3(2):727–745. doi: 10.1111/j.1749-8198.2009.00220.x

73. Goodchild MF, Lam NSN. Areal Interpolation—a Variant of the Traditional Spatial Problem. Geo-Processing. 1980;1(3):297–312.

74. Flowerdew R, Green M, Kehris E. Using areal interpolation methods in geographic information systems. Papers in Regional Science. 1991. July;70(3):303–315.

75. Liu XH, Kyriakidis P C, Goodchild MF. Population-density estimation using regression and area-to-point residual kriging. International Journal of Geographical Information Science. 2008. March;22(4):431–447. doi: 10.1080/13658810701492225

76. Langford M. Obtaining population estimates in non-census reporting zones: An evaluation of the 3-class dasymetric method. Computers, Environment and Urban Systems. 2006. March;30(2):161–180. doi:10.1016/j.compenvurbsys.2004.07.001

77. Zandbergen PA, Ignizio DA. Comparison of Dasymetric Mapping Techniques for Small-Area Population Estimates. Cartography and Geographic Information Science. 2013. March;37(3):199–214. doi:10.1559/152304010792194985

78. Mennis J. Generating Surface Models of Population Using Dasymetric Mapping?. The Professional Geographer. 2003;55(1):31–42.

79. Wright JK. A Method of Mapping Densities of Population: With Cape Cod as an Example. Geographical Review. 1936. January;26(1):103 doi: 10.2307/209467

80. Maantay JA, Maroko AR, Herrmann C. Mapping Population Distribution in the Urban Environment: The Cadastral-based Expert Dasymetric System (CEDS). Cartography and Geographic Information Science. 2007. January;34(2):77–102. doi: 10.1559/152304007781002190

81. Eicher CL, Brewer CA. Dasymetric mapping and areal interpolation: Implementation and evaluation.Cartography and Geographic Information Science. 2001;28(2):125–138. doi:10.1559/152304001782173727

82. Mennis J, Hultgren T. Intelligent Dasymetric Mapping and Its Application to Areal Interpolation. Cartography and Geographic Information Science. 2006. January;33(3):179–194. doi:10.1559/152304006779077309

83. Holt JB, Lo CP, Hodler TW. Dasymetric Estimation of Population Density and Areal Interpolation of Census Data. Cartography and Geographic Information Science. 2004. January;31(2):103–121. doi:10.1559/1523040041649407

84. Alahmadi M, Atkinson P, Martin D. Fine spatial resolution residential land-use data for small-area population mapping: a case study in Riyadh, Saudi Arabia. International Journal of Remote Sensing. 2015. September;36(17):4315–4331. doi: 10.1080/01431161.2015.1079666

85. Liu X, Clarke K. Estimation of Residential Population Using High Resolution Satellite Imagery. In: Proceedings of the 3rd Symposium in Remote Sensing of Urban Areas. Istanbul, Turkey; 2002. p. 153–160.

86. Greger K. Spatio-Temporal Building Population Estimation for Highly Urbanized Areas Using GIS. Transactions in GIS. 2015. February;19(1):129–150. doi: 10.1111/tgis.12086

87. Zoraghein H, Leyk S, Ruther M, Buttenfield BP. Exploiting temporal information in parcel data to refine small area population estimates. Computers, Environment and Urban Systems. 2016. July;58:19–28. doi:10.1016/j.compenvurbsys.2016.03.004

88. Henn A, Römer C, Gröger G, Plümer L. Automatic classification of building types in 3D city models. Geoinformatica. 2012. April;16(2):281–306. doi: 10.1007/s10707-011-0131-x

89. Hecht R, Meinel G, Buchroithner M. Automatic identification of building types based on topographic databases—a comparison of different data sources. International Journal

of Cartography. 2015. August;1(1):18–31. doi: 10.1080/23729333.2015.1055644

90. Kunze C, Hecht R. Semantic enrichment of building data with volunteered geographic information to improve mappings of dwelling units and population. Computers, Environment and Urban Systems. 2015. September;53:4–18. doi: 10.1016/j.compenvurbsys.2015.04.002
91. Neidhart H, Sester M. Identifying building types and building clusters using 3-D laser scanning and GIS-data. Int Arch Photogramm Remote Sens Spatial Inf Sci. 2004;XXXV/B4:715–720.
92. Belgiu M, Tomljenovic I, Lampoltshammer T, Blaschke T, Höfle B. Ontology-Based Classification of Building Types Detected from Airborne Laser Scanning Data. Remote Sensing. 2014. February;6(2):1347–1366. doi: 10.3390/rs6021347
93. Hermosilla T, Ruiz LA, Recio JA, Cambra-López M. Assessing contextual descriptive features for plot-based classification of urban areas. Landscape and Urban Planning. 2012. May;106(1):124–137. doi:10.1016/j.landurbplan.2012.02.008
94. Hermosilla T, Ruiz LA, Recio JA, Balsa-Barreiro J. Land-use mapping of Valencia city area from aerial images and LiDAR data. In: GEOProcessing 2012: The Fourth International Conference on Advanced Geographic Information Systems, Applications, and Services. Valencia, Spain; 2012. p. 232–237.
95. Xiong B, Oude Elberink S, Vosselman G. A graph edit dictionary for correcting errors in roof topology graphs reconstructed from point clouds. ISPRS Journal of Photogrammetry and Remote Sensing. 2014. July;93:227–242. doi: 10.1016/j.isprsjprs.2014.01.007
96. Swanson DA, Hough GC Jr. An Evaluation of Persons per Household (PPH) Estimates Generated by the American Community Survey: A Demographic Perspective. Population Research and Policy Review. 2012;31(2):235–266. doi: 10.1007/s11113-012-9227-8
97. Nouvel R, Mastrucci A, Leopold U, Baume O, Coors V, Eicker U. Combining GIS-based statistical and engineering urban heat consumption models: Towards a new framework for multi-scale policy support.Energy and Buildings. 2015;107:204–212. doi: 10.1016/j.enbuild.2015.08.021
98. van der Sande C, Soudarissanane S, Khoshelham K. Assessment of Relative Accuracy of AHN-2 Laser Scanning Data Using Planar Features. Sensors. 2010. September;10(9):8198–8214. doi:10.3390/s100908198
99. VROM. Geometrie en de BAG (Verdiepingsdocument). Dutch Ministry of Housing, Spatial Planning and the Environment; 2008. Available from: http://www.kadaster.nl/web/artikel/download/BAG-verdiepingsdocument-geometrie-1.htm.
100. Biljecki F, Ledoux H, Stoter J. Error propagation in the computation of volumes in 3D city models with the Monte Carlo method. ISPRS Ann Photogramm Remote Sens Spatial Inf Sci. 2014. October;II-2:31–39. doi: 10.5194/isprsannals-II-2-31-2014
101. Fisher PF, Langford M. Modeling Sensitivity to Accuracy in Classified Imagery: A Study of Areal Interpolation by Dasymetric Mapping. The Professional Geographer. 1996. August;48(3):299–309. doi:10.1111/j.0033-0124.1996.00299.x
102. Fisher PF, Langford M. Modelling the Errors in Areal Interpolation between Zonal Systems by Monte Carlo Simulation. Environment and Planning A. 1995. February;27(2):211–224. doi: 10.1068/a270211
103. Sadahiro Y. Accuracy of areal interpolation: A comparison of alternative methods. Journal of Geographical Systems. 1999. December;1(4):323–346. doi: 10.1007/s101090050017
104. Kolbe TH, Burger B, Cantzler B. CityGML goes to Broadway In: Photogrammetric Week'15. Stuttgart, Germany; 2015. p. 343–356.
105. Aringer K, Roschlaub R. Bavarian 3D Building Model and Update Concept Based on LiDAR, Image Matching and Cadastre Information In: Innovations in 3D Geo-Information Sciences. Springer International Publishing; 2014. p. 143–157.
106. Stoter J, Roensdorf C, Home R, Capstick D, Streilein A, Kellenberger T, et al. 3D Modelling with National Coverage: Bridging the Gap Between Research and Practice. In: Advances in 3D Geo-Information Sciences. Cham: Springer International Publishing; 2015. p. 207–225.
107. Zhu L, Lehtomäki M, Hyyppä J, Puttonen E, Krooks A, Hyyppä H. Automated 3D Scene Reconstruction from Open Geospatial Data Sources: Airborne Laser Scanning and a 2D Topographic Database. Remote Sensing. 2015. May;7(6):6710–6740. doi: 10.3390/rs70606710
108. Goetz M. Towards generating highly detailed 3D CityGML models from OpenStreetMap. International Journal of Geographical Information Science. 2013. May;27(5):845–865. doi:10.1080/13658816.2012.721552
109. Tatem AJ, Noor AM, von Hagen C, Di Gregorio A, Hay SI. High Resolution Population Maps for Low Income Nations: Combining Land Cover and Census in East Africa. PLOS ONE. 2007. December;2(12):e1298 doi: 10.1371/journal.pone.0001298
110. Linard C, Gilbert M, Snow RW, Noor AM, Tatem AJ. Population Distribution, Settlement Patterns and Accessibility across Africa in 2010. PLOS ONE. 2012. February;7(2):e31743 doi:10.1371/journal.pone.0031743
111. Spoorenberg T. Provisional results of the 2014 census of Myanmar: The surprise that wasn't. Asian Population Studies. 2014. October;11(1):4–6. doi: 10.1080/17441730.2014.972084
112. Spoorenberg T. Myanmar's first census in more than 30 years: A radical revision of the official population count. Population & Societies. 2015. November;527:1–4.
113. Hecht R, Kunze C, Hahmann S. Measuring Completeness of Building Footprints in OpenStreetMap over Space and Time. ISPRS International Journal of Geo-Information. 2013. December;2(4):1066–1091. doi: 10.3390/ijgi2041066

114. van Winden K, Biljecki F, van der Spek S. Automatic Update of Road Attributes by Mining GPS Tracks. Transactions in GIS. 2016;p. n/a–n/a. doi: 10.1111/tgis.12186

115. Stoter J, Ledoux H, Zlatanova S, Biljecki F. Towards sustainable and clean 3D Geoinformation In: Kolbe TH, Bill R, Donaubauer A, editors. Geoinformationssysteme 2016: Beiträge zur 3. Münchner GI-Runde. Munich, Germany; 2016. p. 100–113.

116. Chen K. An approach to linking remotely sensed data and areal census data. International Journal of Remote Sensing. 2002. January;23(1):37–48. doi: 10.1080/01431160010014297

117. Schneiderbauer S, Ehrlich D. Population Density Estimations for Disaster Management: Case Study Rural Zimbabwe In: Geo-information for Disaster Management. Springer Berlin; Heidelberg; 2005. p. 901–921.

118. Akbar M, Aliabadi S, Patel R, Watts M. A fully automated and integrated multi-scale forecasting scheme for emergency preparedness. Environmental Modelling & Software. 2013. January;39:24–38.

119. Langford M, Higgs G. Measuring Potential Access to Primary Healthcare Services: The Influence of Alternative Spatial Representations of Population. The Professional Geographer. 2006. August;58(3):294–306. doi: 10.1111/j.1467-9272.2006.00569.x

120. Hay SI, Noor AM, Nelson A, Tatem AJ. The accuracy of human population maps for public health application. Tropical Medicine and International Health. 2005. October;10(10):1073–1086. doi:10.1111/j.1365-3156.2005.01487.x

121. Poulsen E, Kennedy LW. Using Dasymetric Mapping for Spatially Aggregated Crime Data. Journal of Quantitative Criminology. 2004. September;20(3):243–262. doi: 10.1023/B:JOQC.0000037733.74321.14

122. Lin J, Cromley RG. Evaluating geo-located Twitter data as a control layer for areal interpolation of population. Applied Geography. 2015. March;58:41–47. doi: 10.1016/j.apgeog.2015.01.006

123. Meinel G, Hecht R, Herold H. Analyzing building stock using topographic maps and GIS. Building Research & Information. 2009. November;37(5–6):468–482. doi: 10.1080/09613210903159833

124. Vine MF, Degnan D, Hanchette C. Geographic information systems: their use in environmental epidemiologic research. Environmental Health Perspectives. 1997. June;105(6):598–605. doi:10.1289/ehp.97105598

125. Wandl A, Nadin V, Zonneveld W, Rooij R. Beyond urban–rural classifications: Characterising and mapping territories-in-between across Europe. Landscape and Urban Planning. 2014. October;130:50–63. doi: 10.1016/j.landurbplan.2014.06.010

126. de Kluijver H, Stoter J. Noise mapping and GIS: optimising quality and efficiency of noise effect studies. Computers, Environment and Urban Systems. 2003;27(1):85–102. doi: 10.1016/S0198-9715(01)00038-2

127. Ögren M, Barregard L. Road Traffic Noise Exposure in Gothenburg 1975–2010. PLOS ONE. 2016. May;11(5):e0155328 doi: 10.1371/journal.pone.0155328

128. Tutschku K. Demand-based radio network planning of cellular mobile communication systems. In: IEEE INFOCOM'98 Conference on Computer Communications Seventeenth Annual Joint Conference of the IEEE Computer and Communications Societies. San Francisco, CA, United States: IEEE; 1998. p. 1054–1061.

129. INSPIRE Thematic Working Group Buildings. D2.8.III.2 INSPIRE Data Specification on Buildings—Technical Guidelines; 2013.

130. Kohler N, Hassler U. The building stock as a research object. Building Research & Information. 2002. July;30(4):226–236. doi: 10.1080/09613210110102238

131. Kavgic M, Mavrogianni A, Mumovic D, Summerfield A, Stevanovic Z, Djurovic-Petrovic M. A review of bottom-up building stock models for energy consumption in the residential sector. Building and Environment. 2010. July;45(7):1683–1697. doi: 10.1016/j.buildenv.2010.01.021

132. Hargreaves AJ. Representing the dwelling stock as 3D generic tiles estimated from average residential density. Computers, Environment and Urban Systems. 2015. November;54:280–300. doi:10.1016/j.compenvurbsys.2015.08.001

133. Levy PS, Lemeshow S. Stratification and Stratified Random Sampling In: Sampling of Populations: Methods and Applications. 4th ed Hoboken, NJ, USA: John Wiley & Sons, Inc; 2008. p. 121–142.

134. van Oosterom P, Martinez-Rubi O, Ivanova M, Horhammer M, Geringer D, Ravada S, et al. Massive point cloud data management: Design, implementation and execution of a point cloud benchmark.Computers and Graphics. 2015. June;49(C):92–125. doi: 10.1016/j.cag.2015.01.007

CHAPTER 2

CAPTURING NATURAL-COLOUR 3D MODELS OF INSECTS FOR SPECIES DISCOVERY AND DIAGNOSTICS

Chuong V. Nguyen[1], David R. Lovell,[1] Matt Adcock,[1] and John La Salle[2,3]

[1]Computational Informatics, Commonwealth Scientific and Industrial Research Organization (CSIRO), Canberra, ACT, Australia,

[2]Ecosystem Sciences, Commonwealth Scientific and Industrial Research Organization (CSIRO), Canberra, ACT, Australia,

[3]Atlas of Living Australia, Canberra, ACT, Australia

ABSTRACT

Collections of biological specimens are fundamental to scientific understanding and characterization of natural diversity—past, present and future. This paper presents a system for liberating useful information from physical collections by bringing specimens into the digital domain so they can be more readily shared, analyzed, annotated and compared. It focuses on insects and is strongly motivated by the desire to accelerate and augment current practices in insect taxonomy which predominantly use text, 2D diagrams and images to describe and characterize species. While these traditional kinds of descriptions are informative and useful, they cannot cover insect specimens "from all angles" and precious specimens are still exchanged between researchers and collections for this reason. Furthermore, insects can be complex in structure and pose many challenges to computer vision systems. We present a new prototype for a practical, cost-effective system of offthe-shelf components to acquire natural-colour 3D models of insects from around 3 mm to 30 mm in length. ("Naturalcolour" is used to contrast with "false-colour", i.e., colour generated from, or applied to, gray-scale data post-acquisition.) Colour images are captured from different angles and focal depths using a digital single lens reflex (DSLR) camera rig and two-axis turntable. These 2D images are processed into 3D reconstructions using software based on a visual hull algorithm. The resulting models are compact (around 10 megabytes), afford excellent optical resolution, and can be readily embedded into documents and web pages, as well as viewed on mobile devices. The system is portable, safe, relatively affordable, and complements the sort of volumetric data that can be acquired by computed tomography. This system provides a new way to augment the description and documentation of insect species holotypes, reducing the need to handle or ship specimens. It opens up new opportunities to collect data for research, education, art, entertainment, biodiversity assessment and biosecurity control.

INTRODUCTION

Technology has a critical role to play in accelerating the understanding of biological diversity and, for decades, scientists have strived to create accurate 3D duplicates of plants and animal specimens [1]. This paper describes a novel method of using technology to liberate information about physical specimens by bringing them into the digital domain as natural-colour 3D models—consistent with ideas and directions articulated by several other authors [2]–[10]. In particular, the proof of concept system we present fits well with the suggestion of Wheeler *et al.* [11] to "engineer and deploy a network of automated instruments capable of rapidly creating 3D images

Citation: Nguyen CV, Lovell DR, Adcock M, La Salle J. Capturing Natural-Colour 3D Models of Insects for Species Discovery and Diagnostics. López-Vaamonde C, ed. PLoS ONE. 2014;9(4):e94346. doi:10.1371/journal.pone.0094346.

of type specimens" as part of a larger strategy of dealing with the massive backlog of insect types that are not yet digitized in any form. High resolution 3D scans, as well as being useful as versatile replicas, also have the potential to act as a common frame of reference for other data relating to the original insect such as annotations, auxiliary image collections, and measurements. These additional aspects are vital for the ways taxonomists convey the various morphological characters that distinguish a new species from those previously discovered.

Our work is focused on the digitization of insect species, building on research and development at the Australian National Insect Collection (ANIC) which currently holds over 12 million specimens, and is growing by around 100,000 specimens every year. Our mission is to enable high-quality 3D models of insects to be acquired quickly and cheaply, for ANIC to use as a component of its digitization strategy. Like many Natural History collections around the globe, the ANIC maintains many (thousands) Holotypes - each the single specimen of a species that is used to define the characteristic features of that species. Holotypes exist as a physical object carefully protected from damage through handling. Digital colour 3D models of sufficient detail will enable collections managers to liberate these precious specimens for the research work they are intended to fulfill.

Micro Computed Tomography (Micro CT) is currently a key method [12], [13], able to create micron-accurate volumetric models of millimeter-scale objects and their internal structure. However, like recent 3D reconstructions from scanning electron microscope (SEM) micrographs [14], [15], Micro CT is unable to capture important information about the surface of the object: its natural colour. Exposure and reconstruction times can be long (tens of hours) and, as an X-ray imaging method, Micro CT generally demands special safety equipment. Current systems cost in the hundred-thousand dollar range and, while more compact desktop models are available, these are still not especially portable.

The inability of X-ray based methods for insect digitization to capture colour led us to consider image-based 3D reconstruction techniques as reviewed in [16], [17]. These methods have been successfully applied to the reconstruction of 3D cityscapes and other (generally fairly simple) objects [18]–[20]. Some small biological specimens have been digitized [21]–[23] but the methods used do not specifically cater for the complex structures and challenging surface optical properties of insects. Human-in-the-loop approaches have been proposed for insect modeling [24] as have methods (limited to simple insect geometries) for inferring 3D insect shape from a single 2D image [25]. Experiments [26], [27] with laser scanning systems like [28] have suggested that this approach has difficulties with the fine structures and the small scale of many insects, as well as reflective, transparent or iridescent surfaces.

One way to avoid these difficulties is to steer clear of 3D reconstruction altogether and simply present 2D images obtained from different viewing angles [29]. While this method of 3D visualization is popular for museum collections it does not provide the quantitative information (e.g., 3D morphology) needed to analyze and compare insect specimens. Furthermore large amounts of data are involved: many high-resolution images are needed to give a convincing illusion of looking at an actual 3D object. This makes smooth, realistic interaction difficult and precludes straightforward email exchange or embedding of the object data.

In summary, there is a lack of existing systems that could capture the 3D structure and surface optical properties of small, intricate insect specimens at sufficient resolution for ANIC and other collections to digitize, share, analyze and compare their holdings. The rest of the paper describes our prototype system and its operation, and how it has achieved these design objectives.

MATERIALS AND METHODS

Here we provide overviews of the digitization process and equipment. A video [30] as depicted in Figure 1shows the main components of the system and the digitization process in action.

Figure 1: 3D visualisation of a granary weevil on web as part of a video showing an overview of the 3D scanning process.

Process Overview

In high-level terms, our system and work-flow involve three main steps (Figure 2):

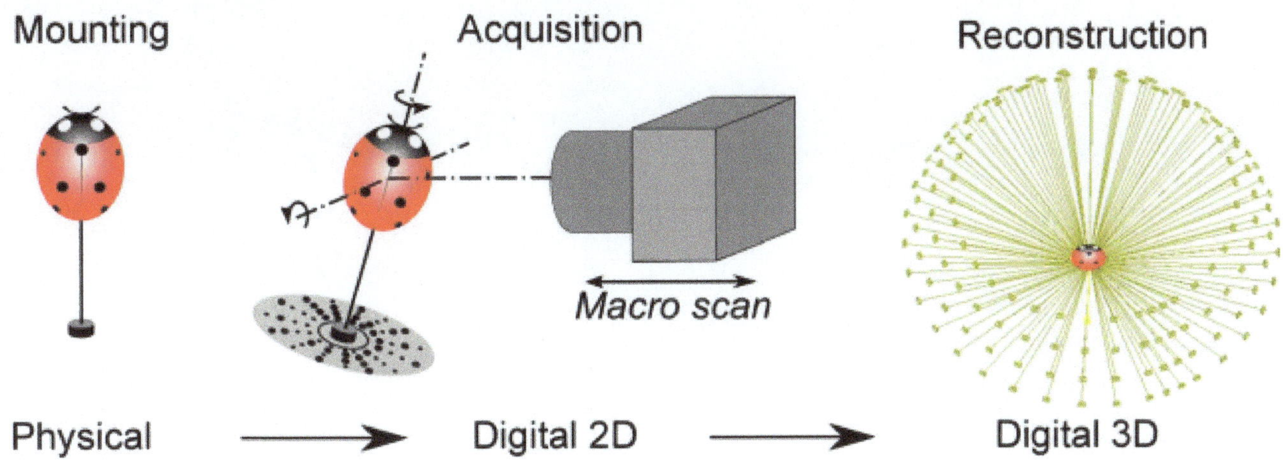

Figure 2: The three main steps to create a natural-colour 3D model of specimen.

- ***Mounting***: the physical specimen is pinned onto a pre-printed mat used later by the reconstruction software to estimate camera pose (viewing angle and position).
- ***Acquisition***: 2D images of the specimen are automatically acquired from different orientations (and focal depths for small insects). This step marks the transition from the physical to the digital domain.
- ***Reconstruction***: In which a 3D model is inferred from multiple 2D images. For small insects, this involves multi-focus image stacking before the general steps of extracting camera pose, shape and colour.

The system has two modes of acquisition, depending on the specimen size. Insects larger than 10 mm are captured in *normal-mode* in which the depth of focus of the normal DSLR camera lens is enough to keep the whole specimen in focus at any viewing angle. Insects smaller than 10 mm are captured in *macro-mode*using a high-magnification lens. Because of the shallow depth of focus of this lens, multiple images are captured at different distances from the specimen and processed into a single in-focus image.

Equipment Overview

Figures 3 and 4 show normal- and macro-mode setups. The main hardware components of the system are:

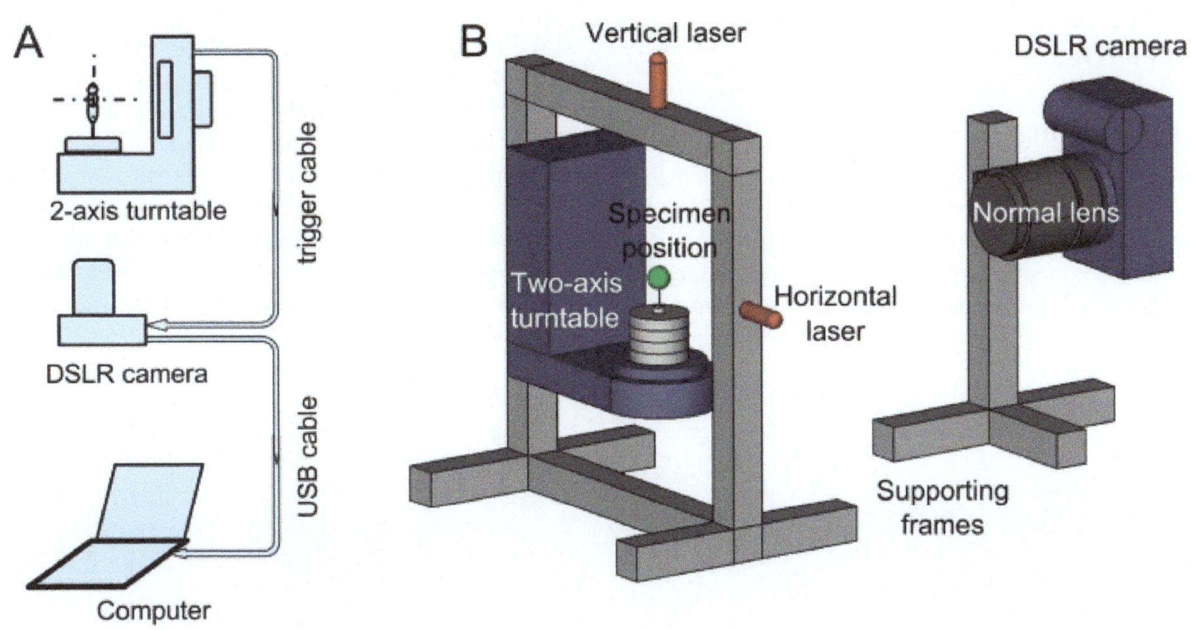

Figure 3: Connections (A) and hardware (B) for *normal-mode* **image acquisition.**

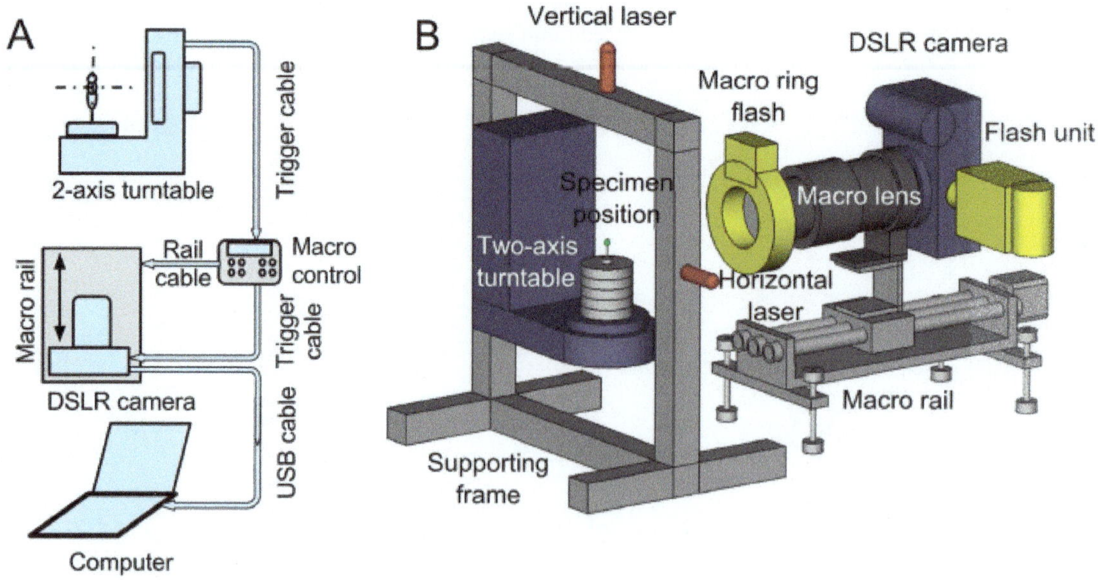

Figure 4: Connections (A) and hardware (B) for *macro-mode* **image acquisition.**

- A two-axis turntable to present views of the specimen from different angles of rotation
- A macro-rail to vary the distance between the camera and specimen in macro-mode
- A camera with macro lens and flash.
- Two laser pointers for specimen alignment
- A computer for 2D image processing and 3D reconstruction.

It is noted that in macro-mode our system uses a macro-rail to capture multi-focus images exactly at predefined depths, as opposed to refocusing the camera lens. A camera flash is needed to eliminate motion blur due to camera shutter›s vibration when capturing at high magnification.

To minimize cost and development time we sought to use off-the-shelf components wherever practicable. These are described in detail in Supplementary Information S1.

Process in Detail

Step 1: Mounting

Collections usually store and display insects larger than ~10 mm by pinning them so that the insect's long axis is horizontal and the pin vertical. Insects smaller than ~10mm are usually either pinned or glued in cards. This paper however focuses on pinned insects and issues arising from this mounting method. Pinning insects horizontally allows many insects to be stored in wide, flat display drawers but creates a few problems for our system:

- The pin becomes part of the 3D model and must be edited or segmented out in post-reconstruction
- Editing can often not fully remove evidence of the pin
- Images of the underside of the specimen can be difficult or impossible to capture, leading to an incomplete 3D model.

Re-pinning the insect so its long axis is vertical helps with image acquisition but risks damaging the specimen, including parts, such as genitalia, that are important for the identification of some species. For some specimens, these affected parts can be isolated through dissection and scanned separately.

After the specimen is pinned, the pin is glued to a small magnet (Figure 5C) that will hold the pin in position on the turntable. Next, a specially patterned mat (Figure 5B), required by the reconstruction software (3DSOM [31]), is attached to provide information about camera pose and position relative to the specimen. Generally the suitable size of the pattern is about one to two times the length of the insect to be scanned. Scanning smaller insects requires smaller patterns to be printed. Currently, modern laser printers with 1200 dpi printing resolution can produce patterned mats as small as 5 mm in diameter. Printing smaller patterns that are sharp enough to be recognised by the reconstruction software is currently a technical challenge.

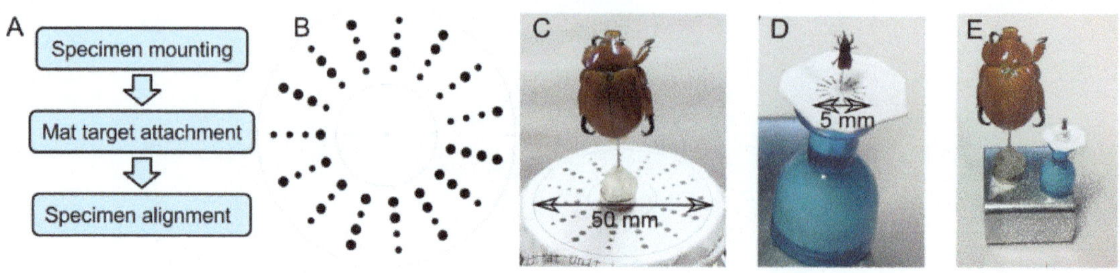

Figure 5: Preparing insect specimen for scanning.

Finally, the whole assembly is placed on the two-axis turntable and positioned (with the assistance of horizontal and vertical laser pointers) so the specimen is centered on the intersection of the axes of tilt and rotation. The lasers are aligned to the rotation axes of the turntable. A specimen is manually aligned to each of the laser beams such that each beam hits the centre of the insect's body.

Step 2: Acquisition

In essence, the acquisition process is about automatically obtaining 2D images of the specimen in different poses. As far as the relationship between the camera and specimen goes, this system has three degrees of freedom: pan, tilt and (in macro-mode) distance along the specimen-camera axis. With the specimen mounted at the intersection of the pan and tilt axes of the turntable, this amounts to rotating the turntable through a range of pan and tilt angles, capturing an image at each step (Figure 6A). In macro-mode there is an additional "inner loop" of translating the camera to acquire partially focused images at different distances from the specimen for later processing into a single image with all parts of the specimen fully in focus (Figure 6B).

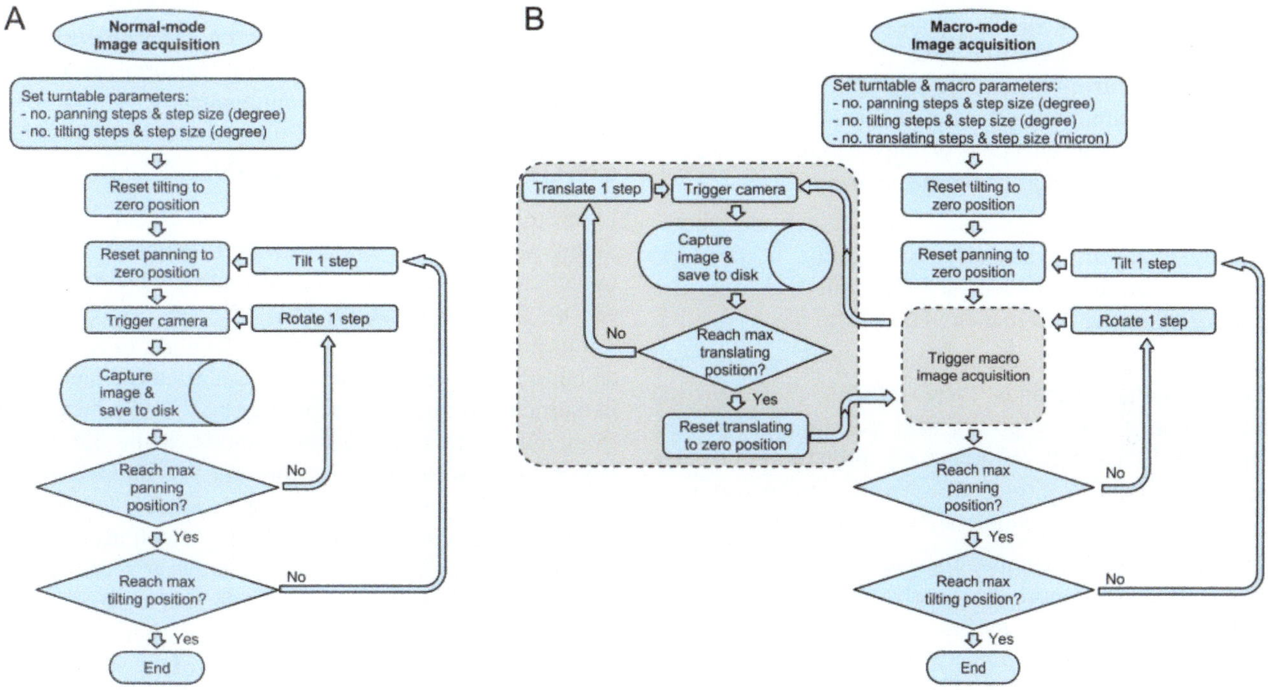

Figure 6: Automated image acquisition process.

There are many ways to automate the acquisition process. The desire to use off-the-shelf components led us to use the GigaPan Panorama Robot EPIC 100 [32] for mounting the *specimen*. The GigaPan is designed for mounting and controlling a *camera*—and this led to the GigaPan robot also acting as the acquisition controller. In other words, it is the turntable that triggers the macro-rail. The macro-rail moves and triggers the camera which triggers its flash and takes an image. Supplementary Information S1 contains more detail about this set-up.

In normal-mode, using rotation and axis tilt, the set-up captures 144 individual images. In macro-mode, the additional up to 31 images required at each step mean that the system can capture up to 4,464 separate images per specimen. Capturing more images is also possible.

Step 3: Reconstruction

The third and final step of the digitization process is where the 2D digital information acquired from a physical specimen is manipulated to produce a 3D digital model (Figure 7).

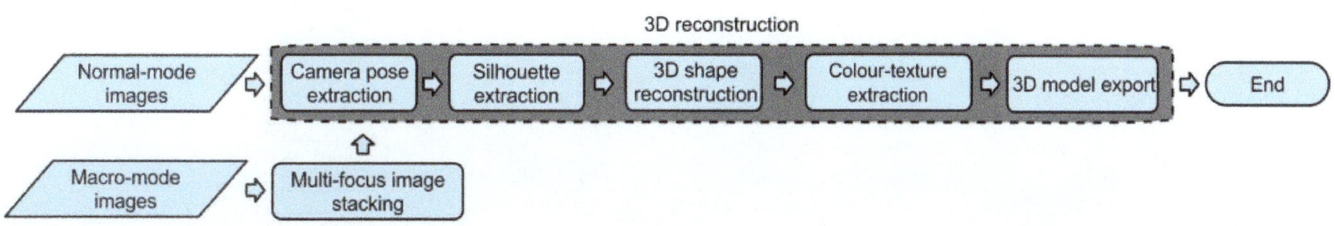

Figure 7: Image processing pipeline for normal-mode and macro-mode images.

In macro-mode, the stack of partially focused images acquired at different specimen-camera distances must be combined into a single in-focus image for a given viewing angle. We used Helicon Focus [33] for this because of its ability to exploit multiple CPU cores. Single core open-source alternatives are available [34],[35].

Armed with a set of in-focus 2D images of an object from different viewing angles, there are two main 3D reconstruction techniques that could be applied:

Visual hull: (also known as *volume carving*) algorithms [36], [37] project the silhouette of the object into a virtual volume at each viewing angle, carving away the volume outside the silhouette to leave a 3D visual hull which approximates the shape of the actual object. This approach does not recover concave surfaces, but photo-consistency can be used to correct this to an extent[38]. The extent of improvement by photo-consistency is limited for some insects due to strong speculiar reflections on the outer-surface and fine body structures such as legs, antennae, spikes and hairs.

Multi-view stereo: algorithms generally rely on photo-consistency measures to identify the location of common features seen in different views [39], [40] and can also incorporate silhouette information [41].

Both strategies are computationally intensive and the computational demands increase with reconstruction resolution. Image clustering [18], [42] and improved feature descriptors [20] have been previously proposed to enable reconstructions to better exploit the very high image resolution produced by professional photography cameras.

Our initial investigations indicated that the visual-hull-based method could more accurately reconstruct some of the thin structures found in insects (e.g., legs, antennae, wings) and insect surfaces with strong specular reflections. 3DSOM [31] was used to provide off-the-shelf visual-hull-based reconstruction as it produced the best quality output of the different approaches [42]–[44].

Figure 7 sets out the detail of the reconstruction process, including the extraction of the camera pose in each input image. 3DSOM initially estimates this information from the target pattern captured in the image and further refines these estimates during 3D reconstruction. Specimen silhouettes are extracted from input images. Once the 3D geometry of the specimen›s surface is reconstructed, texture colour is extracted from the images and added to the model. The resulting 3D model can be then exported to different formats—including HTML (with WebGL, Flash or Java), X3D, 3DS (AutoDesk), and STL (STereoLithography)—for subsequent viewing, analysis or embedding into documents. X3D is a convenient format as it is supported by popular 3D visualisation software, and a X3D file can included as an embedded object or as XML inline in an HTML5 file for 3D web visualisation. InstantReality›s [45] tool "aopt" can perform this conversion X3D to 3D-supported HTML automatically.

RESULTS AND DISCUSSION

Figure 8 shows high-resolution natural-colour 3D models of insects ranging from 3 mm to 30 mm in length. These 3D insect models are also available for interactive viewing at [46] and can be downloaded at[47]–[54]. The smallest of these—the 3 mm granary weevil—proved challenging to resolve due to an out-of-focus problem when its images were captured at 2 × magnification. The 3D model of granary weevil was obtained from images captured in macro-mode, while 3D models of larger insects were obtained from images captured in normal-mode. The 3D visualisation of insect models is based on the open-source X3DOM framework [55] which uses WebGL for plug-in-less display within a web browser (such as Firefox and Chrome). The file size of models, including 3D mesh and texture, depends on the desired visualisation quality and the complexity of the geometry and colour of the actual specimen. For the 3D models shown at[46], the file size ranges from 5 to 24 megabytes, with number of vertices from 80,000 to 130,000 and texture resolution from 4 to 16 megapixels.

Figure 8: Various 3D insect models.

Figure 9 illustrates the effectiveness of macro-mode image acquisition as compared to normal-mode image acquisition when applied to very small insects such as the granary-weevil. A Canon EF-65 mm macro lens was employed in both cases. In normal-mode, a stencil with a ∅2 mm hole had to be attached immediately in front of the camera (Figure 9A) to reduce the effective aperture and increase the depth of focus. In both cases a flash was used to mitigate the effects of wobble due to the camera shutter movement. With a flash, the exposure time of an image is effectively the very short duration of the flash when it triggers, and therefore it minimizes any motion blur. Flash energy in macro mode was $\frac{1}{128}$ of full power and in normal-mode (for the ∅2 mm aperture) it was $\frac{1}{16}$ of full power. The results shown in Figure 9 clearly illustrate the improvements of macro-mode. The macro-mode model was reconstructed with multi-focus stacking of 31 images from each view, each captured with an F/8 lens aperture at increments of 0.25 mm along the specimen-camera axis.

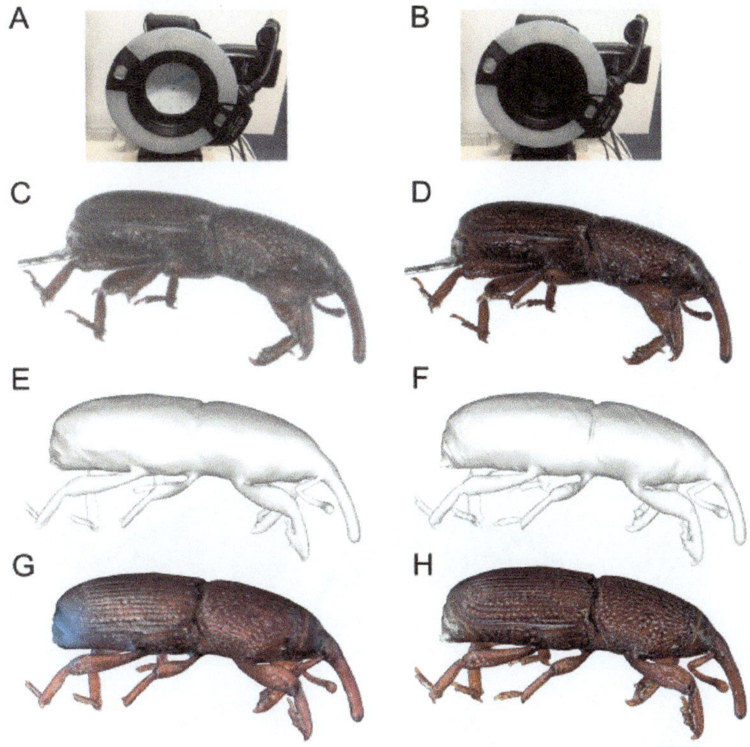

Figure 9: Comparison of natural-colour 3D reconstructions using (A) a small aperture and (B) a F/8 aperture with multi-focus image stacking.

Figure 10 provides a qualitative comparison of a natural-colour 3D model obtained using our system and a Micro CT model of a different specimen of the same species. While the 5.7μm resolution Micro CT clearly captures more details of the surface geometry than our optical approach (including the missing antenna socket in inset A), there are features that it cannot resolve at these resolutions because they are to do with variation in the colour of the specimen (e.g., the compound eye in inset B). One option could be to develop ways to combine the strengths of both approaches: fertile ground for further research.

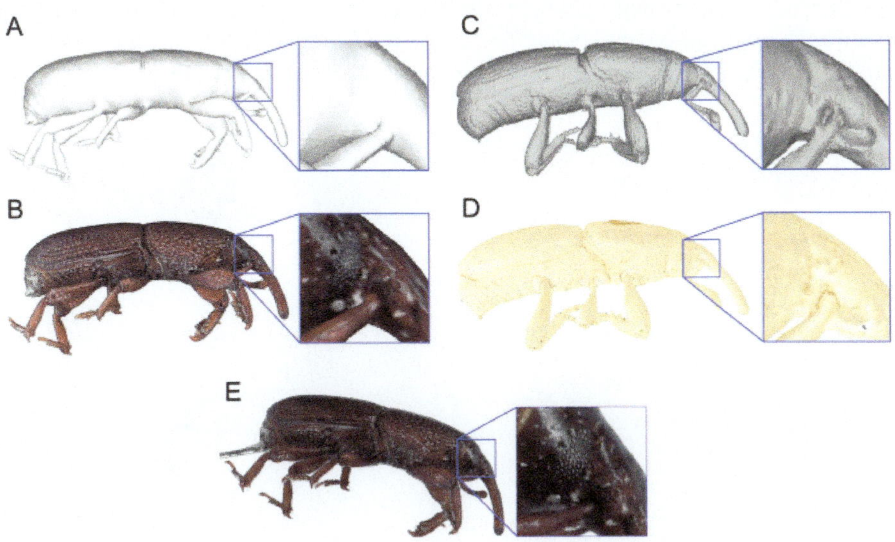

Figure 10: Comparison of a natural-colour 3D model, a Micro CT reconstruction and 2D image at a similar angle.

By convention, insect specimens are often mounted horizontally. However this mounting orientation may not be ideal for 3D reconstruction. To investigate the effect of mounting orientation on reconstruction quality, we acquired images of a specimen mounted horizontally, then vertically (Figure 11). For the structure of that particular specimen, vertical mounting gave markedly better reconstruction of both geometry and colour, avoiding occlusions and capturing texture in more detail. Increasing the number and variety of poses by acquiring images at different tilt angles improved the reconstructions of both vertically and horizontally mounted insects (Figure 12). Even in this case, vertical mounting afforded more detail in geometry and colour. We therefore note that the best mounting orientation is specimen dependent: visual hull reconstruction of geometry improves the more surface normals are captured in silhouette, while colour and texture improve the more surface normals are captured parallel to the camera viewing axis.

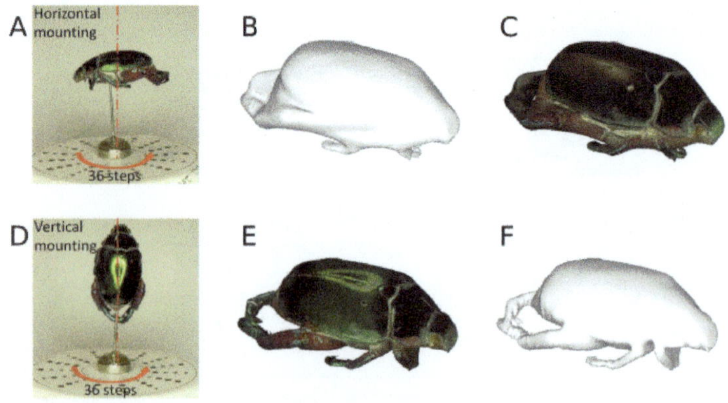

Figure 11: The impact of mounting orientation on reconstruction quality.

Figure 12: Impacts of mounting orientation and tilt on reconstruction quality.

Further surface geometry issues arise as the structures of specimens become more complex. Wings, for example, can be especially challenging as shown in Figure 13(A–C) where self-occlusion causes poor reconstruction of the wings. Fortunately, additional informative views can be obtained to alleviate this problem (Figure 13D–F). Ideally, some of these additional views will be captured tangentially to the wing surface to ensure the reconstructed wings have the correct thickness.

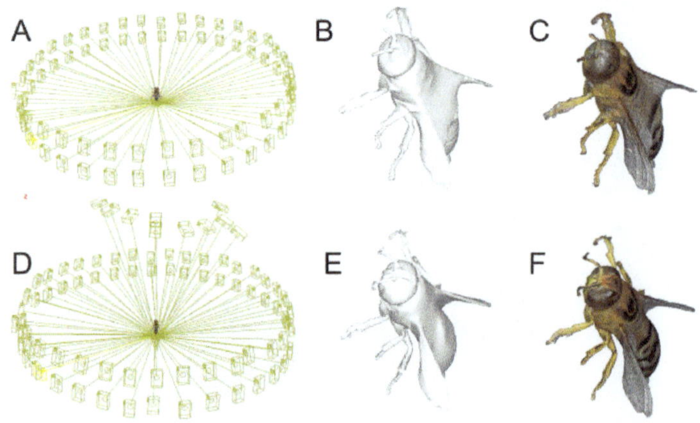

Figure 13: Additional camera poses can improve wing reconstruction.

We explored ways to achieve an informative mounting orientation even when the specimen cannot be re-pinned (e.g., when the specimen is too precious to handle, or the pin too firmly embedded to remove without certain damage). Previously, we mentioned that vertical orientation provides better quality than the horizontal orientation. However, repinning the specimen to have a vertical orientation causes damage, while keeping the horizontal orientation produces a lower-quality 3D model. To avoid this trade-off, the normally-pinned insect can be attached to a second pin (in this case using yellow Blu-Tack) so that the specimen is rotated on its long axis (Figure 14A). Then, the pins and the Blu-Tack need to be removed digitally to produce a clean final 3D model of the specimen. There are two methods to do this. The first method involves editing the Blu-Tack and mounting pins out of the set of 2D images (Figure 14B) during background removal prior to reconstruction. However, this method does not work well with image views where the pins and Blu-Tack occlude parts of the insect and the resulting reconstruction shows contaminated texture colour (Figure 14C). The second method is to keep the pins and Blu-Tack with the specimen during 3D reconstruction (Figure 14D and E) *then* remove them from the 3D model using a mesh editor. Overall, this second strategy produces the better result (Figure 14F).

Figure 14: Two methods to deal with an insect whose pin cannot be removed.

In this paper, we have shown that high resolution, natural-colour 3D digitization system for insects and other small specimens can be implemented using readily available components with hardware and software cost under AUD8000. As well as being cost effective, the system produces digital 3D models that are fairly efficient in terms of the ratio of information to data. The file size of the 3D granary weevil model shown in Figure 9H is around 10 megabytes. It was reconstructed from 18 megapixel 2D JPEG images (2–4 megabytes/image) taken at 144 different angles and 31 different distances creating 10–17 gigabytes of 2D image data in all for a single specimen. By stacking each set of 31 multi-focus images into a single in-focus one, the image data is reduced approximately 20 times. By transforming this 2D data into a 3D model, the system further achieves a 30:1 compression of data. This level of compression enables useful information about the specimen to be exchanged via email, presented in web pages and embedded in 3D PDF documents.

This work raises a number of research challenges and opportunities for further improvement, including:

- Eliminating the need for the printed mat: 3DSOM requires this mat to estimate the camera pose of individual images. We have reached the lower size limit of what we can straightforwardly print and attach to specimens. Furthermore, the range of poses is limited to those in which the mat is viewable. There are reconstruction methods that do not need this kind of pattern to estimate camera pose (e.g.,[56], [57]), relying instead on feature matching and bundle adjustment. However, the accuracy of these estimates depend strongly on the geometry of the specimen and other objects captured in the images.
- Detailed features, such as hairs and surface roughness, demand higher 2D image and 3D model resolution and a concomitant increase in the memory and computation needed to store and visualize the model. Our strategy is to leverage the high resolution 2D image corresponding to a particular pose of interest, reminiscent of the approach used in [29].
- Concave surfaces: current photo-consistency based methods to resolve concavities can be challenged by the specular reflective properties of many insects.
- Transparent wings and membranes pose challenges for acquisition, reconstruction, and for representation and rendering of the resulting 3D model.
- View- and lighting-dependent appearance such as iridescence or sub-surface light scattering is also difficult to capture, represent and render.

- 3D annotation standards, strategies and software are not yet as developed as 2D approaches. The ability to augment 3D models with additional information is important for taxonomy and other scientific ends, as well as engaging a broader range of end users.

Despite these future challenges, we believe that the proof-of-concept prototype presented in this paper demonstrates that natural-colour 3D model digitization is feasible and affordable enough for insect collections to implement and apply right now.

An initial investigation of the usefulness of 3D insect models, as described in Supplementary Information S1, showed that the quality of 3D insect models were good enough to provide sufficient information for species identification, and allow for easier specimen examination than the actual specimen being viewed under a microscope.

The specific usage scenarios for wider communities such as quarantine officer or educator. A quarantine officer can use 3D models of invasive insects while on duty to improve the speed and the accuracy of identification process. The challenges and possible solutions by using 3D models in quarantine control have been discussed in [58]. For educators, 3D models of insects can be used as rich education materials, allowing students to interact with insects without the need to access to fragile specimens.

SUPPORTING INFORMATION

Supplementary Information S1

Traditional insect mounting

Figure 1 shows insects mounted horizontally with a pin going through the body from the back. This mounting technique gives a strong hold on the insect body and facilitate specimen handling but provides limited access for 3D scanning.

Image processing and 3D reconstruction software

Software used for image processing and 3D reconstruction are:

- Helicon Focus [1] for multi-focus image stacking to extend depth of _eld. There are alternative open-source software such as CombineZP [2] and Huggins & Enfuse [3]. However we found Helicon easier to use and able to exploit multi-core processing.
- 3DSOM™ software [4] for 3D reconstruction from multiple view images based on visual hull technique. The 3D reconstruction pipeline used by 3DSOMTM is described in [5]. There are other 3D reconstruction software including commercial software Agisoft Photo Scan [6]; and open-source softwares Bundler [7] and Patch-based Multi-view Stereo [7, 8], and visual hull mesh software [9].

These two off the-shelf software packages, together, cost around AUD1700 based on 2013 prices.

Image Acquisition Equipment

We use the following off-the-shelf components:

- A GigaPan™ Panorama Robot EPIC 100 [10]. Normally this device is used to control the tilt and pan of a camera to capture panorama images. In this project, we turn it side-ways and use it as a two-axis turntable to control the tilt and pan of the specimen to be imaged. The GigaPan™ Panorama Robot connects to a camera via a cable and triggers a camera to capture one or more images per rotation angle.
- A Canon™ 600D camera or better. It provides low-noise, high resolution images (18MP). It is highly customizable with several ports for external trigger input, ash trigger output, and a USB port for remote tethering. One problem with many DSLR cameras is that they rely on a mechanical shutter which can cause mechanical vibration and image blur. The image blur due to mechanical shutter vibration can be alleviated by using a camera ash triggered between shutter movement.
- A Canon™ EF 100mm macro lens for normal-mode and Canon™ MP-65mm for macro mode. The EF-100mm lens provides magnification of 1x or smaller, while the MP-E 65mm lens provides magnification from 1 x to 5x. These macro lenses also have much less optical distortion than some other lenses. A low-cost solution to make normal lens achieve higher magnification is to add a macro extension tube between the camera and the lens. However this often increases optical distortions [11]. The depth of field of MP-65mm lens as function of magnification and aperture can be obtained from its user manual.
- A Viltrox™ JY-670 macro ring ash [12]. This ash provides illumination and reduces exposure time. Short exposure time is critical for macro photography to reduce image blurring due to vibration from the camera's mechanical shutter. Additionally, a Tronix Speed Fire ash power supply [13] is used for fast ash charging.
- A StackShot™ macro-rail [14]. This device enables us to capture high-magnification partially- focused images at predetermined depth intervals with high position accuracy. This macro-rail can run automatically to capture images along a single direction. The process of movement and image acquisition using macro-rail starts with a press of \Up" button on the rail's control panel. To synchronize StackShot™ macro-rail and GigaPan™ robot, we built a circuit interface, shown in Figure 2, to convert the robot's trigger signal to a press-button effect. The step size of the rail is set to be approximately 70% of the depth of field of the macro lens to allow for adjacent overlapping required for stacking multi-focus images.
- Two 1mW laser pointers [15] for specimen alignment. The laser beams are used to locate the specimen at the center of rotation of the two-axis turntable so that the specimen stays at the same position while being rotated and imaged.

- Aluminium frames [16] to hold the turntable, laser pointers, camera, and macro-rails in position.
- Button magnets [17], plastic spacers [18], epoxy glue and pins to mount insect specimens.

The system does not need any special software for acquiring images. The EOS Utility software accompanying the camera is needed to remote-tether the camera and transfer images to a computer during image acquisition. Based on 2013 prices, we estimate the total hardware to cost around AUD3200 to AUD5600. Examples of actual acquisition setups are shown in Figure 3.

Estimation of Processing Time

The reconstruction process described in this paper is a proof of concept and not yet optimized for speed and for large-scale digitization. A time estimation for each stage of the 3D reconstruction procedure for the insects shown in paper is provided in Table 1. The longhorn beetle, Christmas beetle and Amycterine ground weevil take relatively similar amount of time to obtain a 3D model. The sand wasp has increased reconstruction time due to the extra time required to manually correct the errors of background removal around the wings, and to refine camera pose during reconstruction (hairs significantly increase the time for pose refinement). The granary weevil, due to its small size, takes much more time to mount using a microscope, acquire multi-focus images and perform 3D reconstruction. It is often the case that the reconstruction stage is repeated several times to find and _x errors in background removal, or to iterate the refinement process until the result is acceptable. Therefore, the times provided in Table 1 should only be considered as indicative.

Informal Feedback on Using A 3D Insect Model for Species Identification

One possible use case for 3D insect models is species identification. In drafting this paper, we needed to fully identify the Christmas beetle. We asked CSIRO entomologist Mr Tom Weir to try to do so from the photograph, the 3D model, and finally the actual specimen. In addition to identifying the specimen as a male Anoplognathus viriditarsus from the 3D model alone, Mr Weir provided feedback which we summarize as:

- The single top-view photograph does not capture the key features needed to identify the species, just enough information to determine its genus. The features are located at different parts of the insects such as abdomen, head, mouth, claws, rear end, etc. Multiple images captured at particular angles are required to show all these features.
- The 3D model provides useful extra information and allows views from any angle. There are still missing features such as the mouth area and hairy surface on the head/nose due to resolution of geometries and texture colours. However the available features are roughly enough to identify its species. Examination and species identification would be facilitated if the texture information was provided more clearly, or high-resolution images of the features were attached to the model.
- The actual specimen obviously provides all the information needed but needs examination under a microscope for some features such as the mouth area and hair surface on the head. However, out-of-focus effect and other physical restrictions makes the use of microscope to view the actual specimen more cumbersome than to view the 3D model of specimen on a mobile device (such as iPad).
- Identification involves matching the given specimen to its corresponding specimen in the Australian National Insect Collection and comparing it to closely related species.
- For identification purposes, the natural-colour 3D model is much more helpful than the Micro CT 3D model. The better geometry accuracy of the Micro CT model does not assist in this instance.
- Higher resolution of the natural-colour 3D model is desirable to provide details of key identification features. This is particularly important for features, such as hairs, that never get included in the 3D models. One solution would be to attach high resolution photographs to locations of the key features.
- To make the 3D model useful for species identification, it is important to know in advance all key features and the identification procedure and level. Ideally, this information would be arranged in a way that directly supports fast, correct identification. 3D visualization and annotation could make the identification procedure more obvious that the instructions currently provided in text and 2D illustrations.

ACKNOWLEDGMENTS

We thank Ms Eleanor Gates-Stuart (CSIRO Computational Informatics), Dr Rolf Oberprieler, Ms Jennings Debbie, Dr Beth Mantle, Ms Nicole Fisher, Ms Cate Lemann, Ms Anne Hastings, Mr Eric Hines and Mr Tom Weir (CSIRO Ecosystem Sciences) for their kind assistance during the course of the project. Special thanks goes to Dr Sherry Mayo (CSIRO Materials Science & Engineering) for providing Micro CT model of the granary weevil for comparison. We also thanks Dr Vince Smith (Natural History Museum - London), Prof Pavel Stoev (National Museum of Natural History - Sofia) and another anonymous reviewer for their valuable feedback to our manuscript. We gratefully acknowledge CSIRO's Office of the Chief Executive Postdoctoral Scheme, Transformational Biology initiative, and Business and Service Analytics Program for funding this work.

REFERENCES

1. McNally RS, Buschini N (1993) The Harvard Glass Flowers: Materials and Techniques. Journal of the American Institute for Conservation 32: 231–240
2. Godfray HCJ (2002) Challenges for taxonomy. Nature 417: 17–9 [PubMed]
3. La Salle J, Wheeler Q, Jackway P, Winterton S, Hobern D, et al. (2009) Accelerating taxonomic discovery through automated character extraction. Zootaxa 2217: 43–55

4. Beaman R, Cellinese N (2012) Mass digitization of scientific collections: New opportunities to transform the use of biological specimens and underwrite biodiversity science. ZooKeys 209: 7–17 [PMC free article][PubMed]
5. Blagoderov V, Kitching I, Livermore L, Simonsen T, Smith V (2012) No specimen left behind: industrial scale digitization of natural history collections. ZooKeys 209: 133–146 [PMC free article] [PubMed]
6. Mantle BL, La Salle J, Fisher N (2012) Whole-drawer imaging for digital management and curation of a large entomological collection. ZooKeys 163: 147–63 [PMC free article] [PubMed]
7. Smith V, Blagoderov V (2012) Bringing collections out of the dark. ZooKeys 209: 1–6 [PMC free article] [PubMed]
8. Ang Y, Puniamoorthy J, Pont AC, Bartak M, Blanckenhorn WU, et al. (2013) A plea for digital reference collections and other science-based digitization initiatives in taxonomy: Sepsidnet as exemplar. Systematic Entomology 38: 637–644
9. Balke M, Schmidt S, Hausmann A, Toussaint E, Bergsten J, et al. (2013) Biodiversity into your hands - A call for a virtual global natural history 'metacollection'. Frontiers in zoology 10: 55. [PMC free article][PubMed]
10. Johnson L, Mantle BL, Gardner JL, Backwell PRY (2013) Morphometric measurements of drag-onywings: the accuracy of pinned, scanned and detachedmeasurement methods. ZooKeys 276: 77–84[PMC free article] [PubMed]
11. Wheeler Q, Bourgoin T, Coddington J, Gostony T, Hamilton A, et al. (2012) Nomenclatural benchmarking: the roles of digital typification and telemicroscopy. ZooKeys 202: 193–202[PMC free article] [PubMed]
12. Metscher BD (2009) MicroCT for comparative morphology: simple staining methods allow high-contrast 3D imaging of diverse non-mineralized animal tissues. BMC physiology 9: 11. [PMC free article] [PubMed]
13. Faulwetter S, Vasileiadou A, Kouratoras M, Dailianis T, Arvanitidis C (2013) Micro-computed tomography: Introducing new dimensions to taxonomy. ZooKeys 263: 1–45 [PMC free article] [PubMed]
14. Akkari N, Koon-Bong Cheung D, Enghoff H, Stoev P (2013) Revolving SEM images visualising 3D taxonomic characters: application to six species of the millipede genus Ommatoiulus Latzel, 1884, with description of seven new species and an interactive key to the Tunisian members of the genus (Diplopoda, Julida, Jul. ZooKeys 328: 5–45 [PMC free article] [PubMed]
15. Koon-Bong Cheung D, Brunke A, Akkari N, Souza CM, Pape T (2013) Rotational Scanning Elec-tron Micrographs (rSEM): A novel and accessible tool to visualize and communicate complex morphology. ZooKeys 328: 47–57 [PMC free article] [PubMed]
16. Hartley R, Zisserman A (2004) Multiple View Geometry in Computer Vision. Cambridge University Press, 2 edition, 672 pp. doi:10.1017/CBO9780511811685. Available:http://www.robots.ox.ac.uk/~vgg/hzbook/
17. Szeliski R (2010) Computer Vision: Algorithms and Applications, volume 5. Springer, 979 pp. doi:10.1007/978-1-84882-935-0. Available: http://research.microsoft.com/en-us/um/people/szeliski/book/drafts/szelski_20080330am_draft.pdf
18. Furukawa Y, Ponce J (2010) Accurate, dense, and robust multiview stereopsis. IEEE Transactions on Pattern Analysis and Machine Intelligence 32: 1362–76 [PubMed]
19. Hernández C, Vogiatzis G (2010) Shape from photographs: a multi-view stereo pipeline. In: Cipolla R, Battiato S, Farinella GM, editors, Computer Vision: Detection, Recongnition and Reconstruction, Springer Berlin Heidelberg. pp. 281–312. doi:10.1007/978-3-642-12848-6 11
20. Tola E, Strecha C, Fua P (2011) Efficient large-scale multi-view stereo for ultra high-resolution image sets. Machine Vision and Applications 23: 903–920
21. Atsushi K, Sueyasu H, Funayama Y, Maekawa T (2011) System for reconstruction of three-dimensional micro objects from multiple photographic images. Computer-Aided Design 43: 1045–1055
22. Gallo A, Muzzupappa M, Bruno F (2013) 3D reconstruction of small sized objects from a sequence of multi-focused images. Journal of Cultural Heritage in press.
23. Chalmers A, Mudge M, Paulo Santos L, Ritz M, Langguth F, et al. (2012) High resolution acqui-sition of detailed surfaces with lens-shifted structured light. Computers & Graphics 36: 16–27
24. Murakawa J, Yoon I, Hong T, Lank E (2006) Parts, image, and sketch based 3D modeling method. In: EUROGRAPHICS Workshop on Sketch-Based Interfaces and Modeling (2006). pp. 67–74. doi: 10.2312/SBM/SBM06/067-074. Available: http://dl.acm.org/citation.cfm?id=2386286
25. Zhang X, Gao Y, Caelli T (2010) Primitive-based 3D structure inference from a single 2D image for insect modeling: Towards an electronic field guide for insect identification. Control Automation Robotics & Vision (ICARCV).
26. Polo ME, Felicísimo AM (2012) Analysis of uncertainty and repeatability of a low-cost 3d laser scanner. Sensors 12: 9046–9054 [PMC free article] [PubMed]
27. Kuzminsky SC, Gardiner MS (2012) Three-dimensional laser scanning: potential uses for museum conservation and scientific research. Journal of Archaeological Science 39: 2744–2751
28. NextEngine (2013). NextEngine 3D Scanner. Available: http://www.nextengine.com
29. Ortery (2013). Photosimile 5000 - Automated 3D Product Photography Stu-dio. Available:http://ortery.com/Photography-Equipment/3D-Photography-Equipment/Photosimile-5000-Automated-3D-Product-Photography
30. Nguyen C, Adcock M, Lovell D, Mantle B, La Salle J (2014). Insect collections going 3D. doi:10.4225/08/531FE0104C11A. Available: http://dx.doi.org/10.4225/08/531FE0104C11A View-able

athttp://www.youtube.com/watch?v=THvfu6shJjg&list=UUak3NZxjNnWip327vYL8RLA orhttp://publish.viostream.com/play/xbz189zsuhn8

31. Creative Dimension Software Ltd (2013). 3DSOM. Available: http://www.3dsom.com
32. GigaPan Systems (2013). The GigaPan Store - Products, software and accessories to create gi-gapixel panoramas. Available: http://gigapan.com/cms/shop/store
33. HeliconSoft (2013). Helicon Focus. Available: http://www.heliconsoft.com/heliconsoft-products/helicon-focus/
34. Hadley A (2010). CombineZP. Available: http://www.hadleyweb.pwp.blueyonder.co.uk/CZP/Newshtm.
35. D›Angelo P (2013). Hugin & Enfuse. Available: http://hugin.sourceforge.net/tutorials/index shtml.
36. Laurentini A (1994) The visual hull concept for silhouette-based image understanding. IEEE Transactions on Pattern Analysis and Machine Intelligence 16: 150–162
37. Franco JS, Boyer E (2003) Exact polyhedral visual hulls. In: British Machine Vision Conference (BMVC›03). volume 1 , pp. 329–338. doi:10.5244/C.17.32. Available: http://hal.inria.fr/inria-00349075
38. Kutulakos K, Seitz S (2000) A theory of shape by space carving. International Journal of Computer Vision 38: 199–218
39. Goesele M, Curless B, Seitz S (2006) Multi-View Stereo Revisited. In: 2006 IEEE Computer Society Conference on Computer Vision and Pattern Recognition - Volume 2 (CVPR›06). IEEE, volume 2, pp. 2402–2409. doi:10.1109/CVPR.2006.199. Available: http://ieeexplore.ieee.org/lpdocs/epic03/wrapper.htm?arnumber=1641048
40. Seitz SM, Curless B, Diebel J, Scharstein D, Szeliski R (2006) A comparison and evaluation of multi-view stereo reconstruction algorithms. In: Proceedings of IEEE Conference on Computer Vision and Pattern Recognition (CVPR). volume 1 , pp. 519–528. doi:10.1109/CVPR.2006.19. Available:http://ieeexplore.ieee.org/xpls/abs_all.jsp?arnumber=1640800
41. Sinha S, Pollefeys M (2005) Multi-view reconstruction using photo-consistency and exact silhouette constraints: a maximum-ow formulation. In: Tenth IEEE International Conference on Computer Vision (ICCV›05) Volume 1. IEEE, volume 1, pp. 349–356 Vol. 1. doi:10.1109/ICCV.2005.159. Available:http://ieeexplore.ieee.org/lpdocs/epic03/wrapper.htm?arnumber=1541277
42. Furukawa Y (2010). Patch-based Multi-view Stereo Software. Available: http://www.di.ens.fr/pmvs/
43. Snavely N (2010). Bundler: Structure from Motion (SfM) for Unordered Image Collections. Available:http://www.cs.cornell.edu/~snavely/bundler/
44. Forbes K (2009). Visual Hulls from Uncalibrated Snapshots. Available: http://www.dip.ee.uct.ac za/~kforbes/DoubleMirror/DoubleMirror.html.
45. InstantReality (2014). Instantreality. Available: http://www.instantreality.org/
46. Nguyen CV, Lovell DR, Adcock M, La Salle J (2013). Demo of natural-color 3D insect models. Available: http://www2.ala.org.au/chuong/
47. Nguyen C, Adcock M, Lovell D, Oberprieler R, Mantle B, et al. .. (2014). 3d reconstruction from multi-view images of a black palm weevil. doi:10.4225/08/531E56D86DDD5. Available:http://dx.doi.org/10.4225/08/531E56D86DDD5
48. Nguyen C, Adcock M, Lovell D, Jennings D, Oberprieler R, et al. .. (2014). 3d reconstruction from multi-view images of a granary weevil. doi:10.4225/08/531E570A8D614. Available:http://dx.doi.org/10.4225/08/531E570A8D614
49. Nguyen C, Adcock M, Lovell D, Weir T, Mantle B, et al. .. (2014). 3d reconstruction from multi-view images of a jewel beetle. doi:10.4225/08/531E57227A11B. Available:http://dx.doi.org/10.4225/08/531E57227A11B
50. Nguyen C, Adcock M, Lovell D, Lemann C, Oberprieler R, et al. .. (2014). 3d reconstruction from multi-view images of a amycterine ground weevil. doi:10.4225/08/531E56A9C6A60. Available:http://dx.doi.org/10.4225/08/531E56A9C6A60
51. Nguyen C, Adcock M, Lovell D, Mantle B, La Salle J (2014). 3d reconstruction from multi-view images of a black field cricket. doi:10.4225/08/531E56C1CD687. Available: http://dx.doi.org/104225/08/531E56C1CD687.
52. NguyenC, AdcockM, LovellD, MantleB, LaSalleJ(2014). 3d reconstruction from multi-view images of a sand wasp. doi:10.4225/08/531FCA8D36F44. Available:http://dx.doi.org/10.4225/08/531FCA8D36F44
53. Nguyen C, Adcock M, Lovell D, Lemann C, Escalona Garcia H, et al. .. (2014). 3d reconstruction from multi-view images of a longhorn beetle. doi:10.4225/08/531E573D7F06C. Available:http://dx.doi.org/10.4225/08/531E573D7F06C
54. Nguyen C, Adcock M, Lovell D, Gates-Stuart E, Weir T, et al. .. (2014). 3d reconstruction from multi-view images of a christmas beetle. doi:10.4225/08/531E56EFBD298. Available:http://dx.doi.org/10.4225/08/531E56EFBD298
55. X3DOM (2014). x3dom Instant 3D HTML way! Available: http://www.x3dom.org/?page_id=2
56. Agisoft (2013). Agisoft PhotoScan. Available: http://www.agisoft.ru/
57. Autodesk (2013). 123D Catch. Available: http://www.123dapp.com/catch
58. Nguyen CV, Lovell D, Oberprieler R, Jennings D, Adcock M, et al. .. (2013) Vir-tual 3d models of insects for accelerated quarantine controlvirtual 3d models of in-sects for accelerated quarantine control. In: 2013 IEEE International Conference on Computer Vision Workshops (ICCVW 2013). pp. 161–167. doi:10.1109/ICCVW.2013.27. Available: http://www.cv-foundation.org/openaccess/content_iccv_workshops_2013/W06/papers/Nguyen_Virtual_3D_Models_2013_ICCV_paper.pdf

CHAPTER 3

CREATING PHYSICAL 3D STEREOLITHOGRAPH MODELS OF BRAIN AND SKULL

Daniel J. Kelley[1,2,3,4], Mohammed Farhoud[5,6], M. Elizabeth Meyerand[7] , David L. Nelson[8] , Lincoln F. Ramirez[9] , Robert J. Dempsey[9] , Alan J. Wolf[6] , Andrew L. Alexander[1,7], Richard J. Davidson[1]

[1]Waisman Laboratory for Brain Imaging and Behavior, Waisman Center, Madison, Wisconsin, United States of America,

[2]Medical Scientist Training Program, University of Wisconsin School of Medicine and Public Health, University of Wisconsin, Madison, Wisconsin, United States of America,

[3]Neuroscience Training Program, University of Wisconsin School of Medicine and Public Health, University of Wisconsin, Madison, Wisconsin, United States of America,

[4]Howard Hughes Medical Institute (HHMI) Teaching Fellows Program, Wisconsin Program for Scientific Teaching, Department of Bacteriology, University of Wisconsin, Madison, Wisconsin, United States of America,

[5]Paul P. Carbone Comprehensive Cancer Center, University of Wisconsin, Madison, Wisconsin, United States of America,

[6]Biology New Media Center, University of Wisconsin Biotechnology Center, Madison, Wisconsin, United States of America,

[7]Department of Medical Physics, University of Wisconsin, Madison, Wisconsin, United States of America,

[8]Department of Biochemistry, College of Agricultural and Life Science, University of Wisconsin, Madison, Wisconsin, United States of America,

[9]Department of Neurological Surgery, University of Wisconsin, Madison, Wisconsin, United States of America

ABSTRACT

The human brain and skull are three dimensional (3D) anatomical structures with complex surfaces. However, medical images are often two dimensional (2D) and provide incomplete visualization of structural morphology. To overcome this loss in dimension, we developed and validated a freely available, semi-automated pathway to build 3D virtual reality (VR) and handheld, stereolithograph models. To evaluate whether surface visualization in 3D was more informative than in 2D, undergraduate students (n = 50) used the Gillespie scale to rate 3D VR and physical models of both a living patient-volunteer's brain and the skull of Phineas Gage, a historically famous railroad worker whose misfortune with a projectile tamping iron provided the first evidence of a structure-function relationship in brain. Using our processing pathway, we successfully fabricated human brain and skull replicas and validated that the stereolithograph model preserved the scale of the VR model. Based on the Gillespie ratings, students indicated that the biological utility and quality of visual information at the surface of VR and stereolithograph models were greater than the 2D images from which they were

Citation: Kelley DJ, Farhoud M, Meyerand ME, et al. Creating Physical 3D Stereolithograph Models of Brain and Skull. Isalan M, ed. PLoS ONE. 2007;2(10):e1119. doi:10.1371/journal.pone.0001119.

derived. The method we developed is useful to create VR and stereolithograph 3D models from medical images and can be used to model hard or soft tissue in living or preserved specimens. Compared to 2D images, VR and stereolithograph models provide an extra dimension that enhances both the quality of visual information and utility of surface visualization in neuroscience and medicine.

INTRODUCTION

Leonardo da Vinci was the first to model brain structure by injecting molten wax into the ventricle of an oxen brain [1]. Since then, neuroanatomical models have shown utility in neuroscience and medicine in areas such as education, diagnosis, and surgical planning [2], [3]. Currently, classical modelling techniques are being supplanted by modern methods that emphasize three-dimensional anatomical relationships using imaging techniques [4], [5]. In this report, we describe and evaluate a modern method to replicate an individual›s anatomy as a physical, hand-held model by using reverse engineering and rapid prototyping stereolithography [6].

Reverse engineering is a process in which a 3D physical object is scanned using an MRI or CT, for example, and the images are used to render a three-dimensional virtual model. Rapid prototyping is a new technology that uses this virtual model to print or fabricate a physical model. Stereolithography is a rapid prototyping technique with several variants; and, in our application, layers of plaster can form a solid, physical model after a binding agent is applied to each printed layer.

Although other hand-made and computerized modelling approaches have been used to model neuroanatomy[7], [8], rapidly prototyped models preserve anatomical scale and anatomical relationships, are three-dimensional, are understood by visual and tactile learners, do not require software training, and can be produced to the anatomical specifications of the individual patient. Compared to computerized virtual models which have the capability to virtually dissect an object in an infinite number of ways, physical models can be intuitively held and rotated, can be interactively manipulated regardless of complexity, and are accessible without the need for computers or advanced training.

In this report, we developed a streamlined procedure to replicate human brain and skull morphology using freely available software to produce CT/MRI-assisted reverse engineered VR and stereolithograph models; validated the scaling precision of our modelling pathway for quality control; and evaluated 2D images, virtual models (VM), and stereolithograph models (SM) of brain and skull for the quality of visual information and utility of these models.

ANALYSIS

Algorithm

We developed an open-source processing pathway to visualize VR and stereolithograph models (Figure 1). With University of Wisconsin-Madison IRB approval, reverse engineering of brain began with 3D SPGR MRI acquisition of living human brain images on a 3.0 Tesla GE Signa (General Electric Medical Systems; Waukesha, WI) scanner (114 axial slices 1.3 mm thick; 256×256 resolution; pixel size 0.9375×0.9375). DICOM images were loaded into the NIH Analysis of Functional Neuroimaging (AFNI) software package[9] for automated inhomogeneity correction and brain isolation through skull stripping. Although a mesh is commonly used to model the brain's surface (as in Figure 2b), this method failed to print because the mesh lacks information about the thickness of the surface and is too thin to print. For this reason, we preserved the volume dimension of this dataset by generating an isosurface which preserves surface morphology and has enough thickness to be printed. After exporting these files in an ANALYZE format, these images were reconstructed in three dimensions and saved in the visualization tool kit (.vtk) format using 3dSlicer [10]. The reconstructed images were converted to the virtual reality markup language (VRML) for three dimensional printing using isosurface visualization without scalar coloring in MayaVi [11]. We viewed and navigated the VRML brain model with VRMLView (Systems in Motion; Norway). The virtual whole brain isosurface (Figure 2a) and cortical mesh clearly delineated sulci from gyri (Figure 2b).

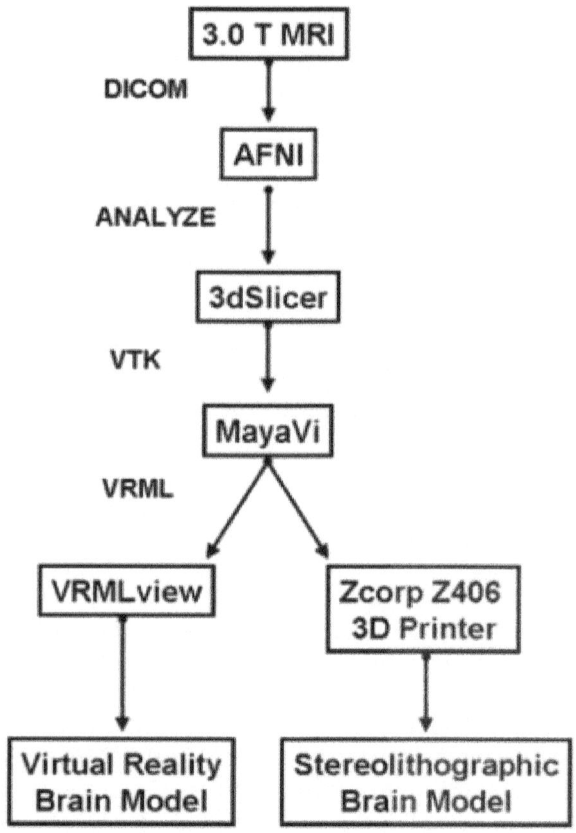

Figure 1: MRI-based Reverse Engineering Processing Pathway.

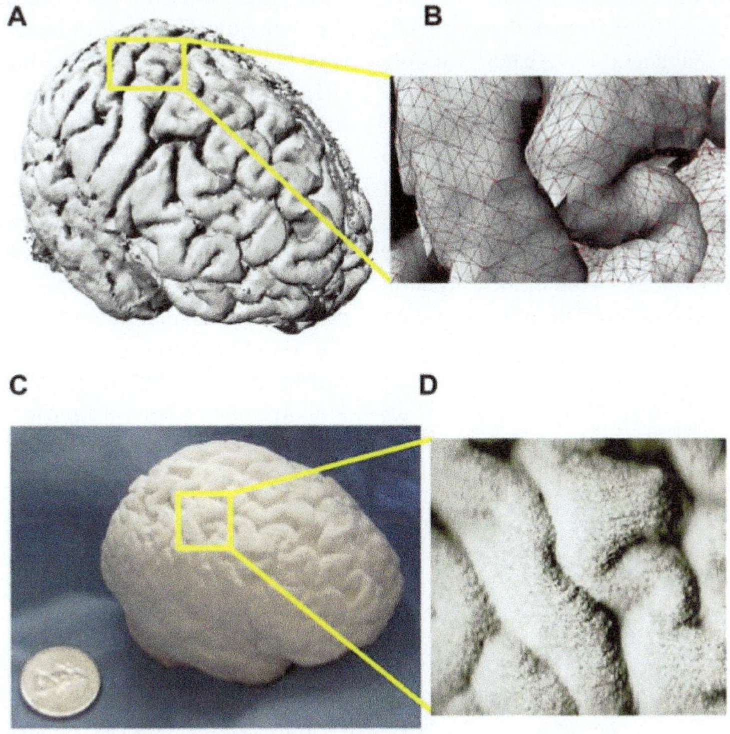

Figure 2: VR and Rapidly Prototyped Stereolithographic Human Brain Model.

Validation

We fabricated an intact, half scale, whole brain 3D stereolithograph model (SM) which preserved cortical morphology using the VRML model to drive the 3D Printing (3DP) [6] process on a Zcorp Z406 3D printer (Z Corporation; MA, USA) at the UW-Madison New Media Center. The whole brain 3D stereolithograph model (Figure 2c) preserved the cortical morphology (Figure 2d; compare to Figure 2b) of the virtual reality markup language (VRML) model.

Maximum length (mm) measurements along the primary right-left (RL), anterior-posterior (AP), and superior-inferior (SI) axes were respectively taken for the VM and SM models with AFNI using DICOM images and with ImageJ [12] using images collected with an HP (Hewlett Packard, USA) PSC 1210 scanner. The primary axes length measures for the scaled stereolithograph model (RL,AP,SI: 67.2, 90.1, 63.2 mm) were as expected based on the virtual brain model (RL,AP,SI: 134.1, 179.1, 133.9 mm). The SM brain was reduced 3.75 mm (5.9%) in the SI dimension during model construction because the model was printed in layers from the inferior to superior direction and the weight of the model caused compression along this dimension. Temperature, humidity, and seasonal calibration also contributed to error in the SI dimension. The printer operator can correct this printing artifact by adjusting the length of the model in the dimension of model growth.

For validation, we tested whether the average virtual to stereolithograph ratios for the primary axes lengths $\left[mean\left(\frac{RL_{VM}}{RL_{SM}}+\frac{AP_{VM}}{AP_{SM}}+\frac{SI_{VM}}{SI_{SM}}\right)\right]$ preserved the 2:1 VM:SM axis scaling factor and the stereolithograph model average axisproportions $\left[mean\left(\frac{\frac{AP}{RL}_{VM}}{\frac{AP}{RL}_{SM}}+\frac{\frac{AP}{SI}_{VM}}{\frac{AP}{SI}_{SM}}+\frac{\frac{RL}{SI}_{VM}}{\frac{RL}{SI}_{SM}}\right)\right]$ preserved the 1:1 VM:SM ratio. These measures indicated that our half-scale stereolithograph brain model preserved the dimensions of the virtual model (Figure 3).

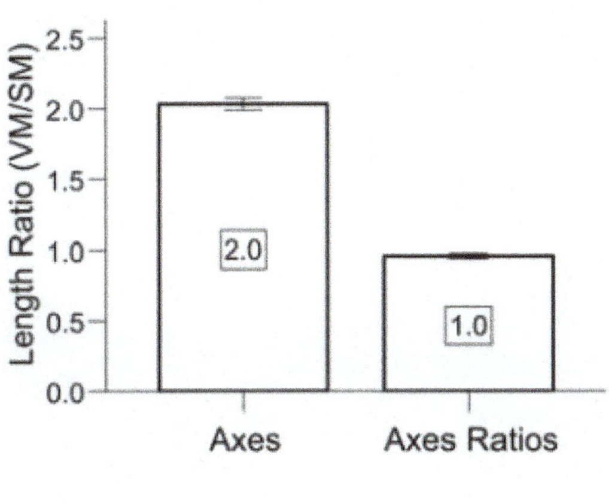

Figure 3: Validation of Processing Pathway.

We then applied our method to produce a VRML (Figure 4a) and physical 3D model (Figure 4b) of Phineas Gage's skull. As the raw CT scans of Phineas Gage's skull were not available because scan requests from prior authors failed, we produced the 3D model from a video clip of multi-plane, computed tomography (CT) images taken of Phineas Gage's skull in 2004 [13]. The accurate physical reproduction of Phineas› skull required a virtual model with the correct dimensions; however, voxel size was not reported with the video clip [13] and a prior publication did not report measurements taken of the authentic Phineas skull [14]. Accurate replication of the virtual model was possible using a voxel size (RL,AP,SI: 0.48,0.8,0.8 mm) in which the VR skull dimensions matched the anatomical dimension measures of the authentic Phineas Gage skull (Table 1) provided by Dominic Hall, Curator of the Warren Anatomical Museum at Harvard Medical School which houses the authentic skull of Phineas Gage. The mpeg video clip was brought into AFNI, cropped to isolate sagittal images (RL,AP,SI: 458,352,342 voxels), exported in ANALYZE format, entered into our processing pathway (Figure 1), and volume rendered in Video S1. Caliper measurements (12″ digital caliper; Neiko Tools, USA) indicated that the half-scaled dimensions of the stereolithograph skull model were as expected (RL,AP,SI: 66.0,90.6,56.6 mm) and the printer operator successfully compensated for compression in the RL print out dimension. The stereolithograph skull model preserved the VM:SM average axes scaling factor of 2:1 [Mean+/−SEM=1.99+/−0.00; t(2)=−1.73;p=0.23, 2-tailed] and the 1:1 average axes proportion [Mean+/−SEM=1.0+/−0.00; t(2)=−0.32;p=0.78, 2-tailed].

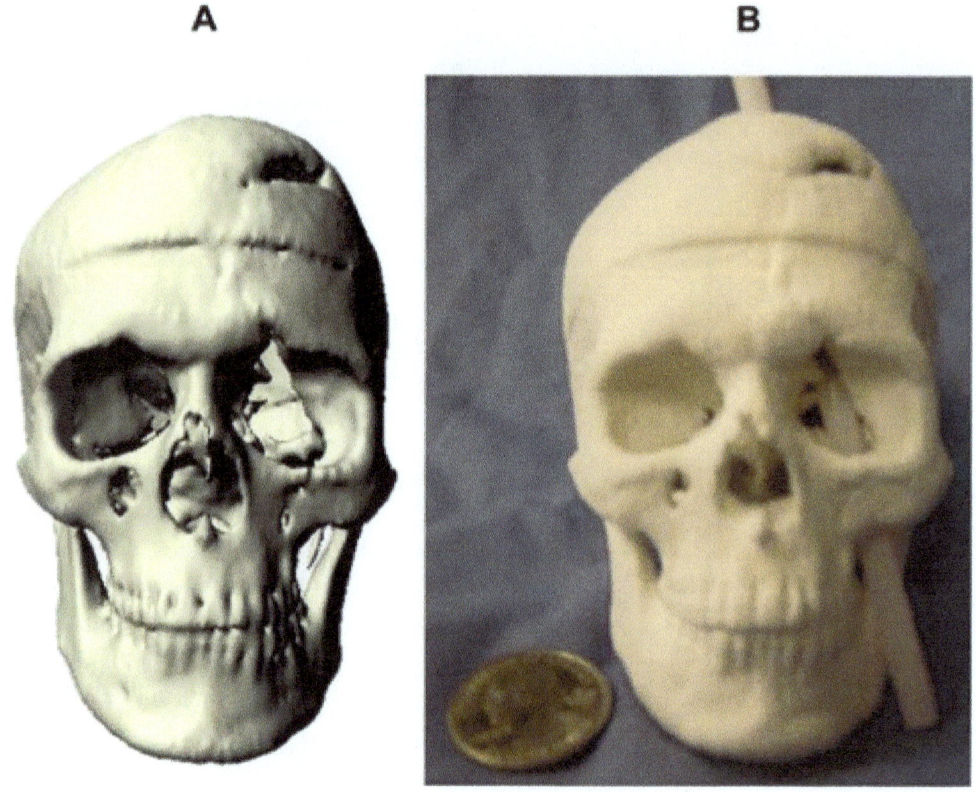

Figure 4: Stereolithograph of Phineas Gage's Skull.

Table 1: Dimension Estimates of Authentic Phineas Gage Skull

Primary Axis Dimension	Anatomical Landmarks	Length Estimates (mm)
Right:Left (RL)	Right Extrema to Left Extrema	131
Anterior:Posterior (AP)	Glabella to Posterior Extrema	180
Superior:Inferior (SI)	Nasion to Gnathion	113

Evaluation

Students enrolled in the University of Wisconsin-Madison course entitled "Ways of Knowing Biology," voluntarily and anonymously evaluated the quality of visual information contained in the 2D, VRML, and stereolithograph models of brain and Phineas Gage's skull. We used the Gillespie rating scale (Table S1) to assess students' perception of biological image models relative to a baseline model [3]. In our assessment, 2D images were used as the baseline for assessing VRML and stereolithograph models for the quality of visual information and biological utility. Gillespie ratings were coded with 1=Inferior, 2=Similar/Equivalent, 3=Superior (similar information more rapidly assimilated), 4=Superior (additional information provided). Since all models were derived from the same data, differences in perceived visual information were a function of model type.

The Gillespie rating survey indicated that the 3D models were superior in surface visualization quality (Table 2) compared to 2D images (Gillespie rating=2) for the brain VRML [t(49)=37.5; p=0.00], brain stereolithograph [t(49)=12.7; p=0.00], Phineas VRML [t(49)=20.8;p=0.00], and the Phineas stereolithograph models [t(49)=14.1; p=0.00]. The Wilcoxon signed ranks test indicated that the quality of visualization using VRML is superior to stereolithograph models for brain [Z=−4.88; p=0.00; n=50] and skull [Z=−2.29; p=0.022; n=50] and that biological image utility was greater for VRML than 2D models [Z=−6.13; p=0.00; n=50], was greater for stereolithograph than 2D models [Z=−4.97; p=0.00; n=50], and was greater for VRML than stereolithograph models [Z=−3.20; p=0.001; n=50].

Table 2: Gillespie Ratings of Visual Information and Biological Utility.

Question	Topic	Mean	SD	1	2	3	4
VI	**Quality of Visual Information, VI**						
1	Brain VRML	3.86	0.35	0	0	7	43
2	Brain Stereolithograph	3.10	0.61	1	4	34	11
3	Phineas VRML	3.58	0.54	0	1	19	30
4	Phineas Stereolithograph	3.28	0.64	0	5	26	19
BU	**Biological Utility, BU**						
1	2D	2.20	0.70	6	30	12	2
2	VRML	3.74	0.56	0	3	7	40
3	Stereolithograph	3.18	0.80	2	6	23	19

Questionnaire results reporting mean, standard deviation, and distribution of responses in each Likert-scaled category.

DISCUSSION

The reverse engineering and rapid prototyping pathway we developed has medical applications in biomodel guided stereotactic surgery [15], cranioplasty [16], aneurysm research and repair [17], and craniofacial reconstruction [18]. In neuroscience, this pathway has applications in producing phantoms of living or preserved specimens and in both basic science and imaging education. Even though the 3D models were derived from the same data as 2D images, 3D models display anatomical relationships in an extra dimension to enhance the quality of visual information and model utility compared to 2D images.

Efficient replication of the intact whole brain with high resolution was previously unattainable using rapid prototyping techniques due to limitations that have largely been overcome through advances in neuroimaging software, computer hardware, and higher field magnets. Disadvantages that remain for rapid prototyping include monetary cost (approximately 50 cents/cc, excluding the cost of MR imaging and data post-processing) and time to print (approximately 2.5 vertical cm/hour). Our half-scaled 219 cc physical brain model cost approximately $110 and took 2.5 hours to print. The 1756 cc full brain would have cost about $880 and taken 5.5 hours to print. The cost would have been significantly more using other rapid prototyping technologies.

We developed and validated a freely available method to model brain and skull anatomy in three dimensions using images collected with multiple modalities. Virtual reality and stereolithograph models are advantageous for observing surface morphology compared to two-dimensional planar images and can be manipulated using a data glove or bare hand, respectively. A disadvantage of the stereolithography approach compared to the VR approach is the lack of subcortical visualization and the inability to visualize the interior skull regions. Future adaptations of this technique should take advantage of 3D tissue printers which are currently being developed and the current color printing capability of 3D printers to delineate brain and skull regions.

SUPPORTING INFORMATION

Table S1: Gillespie Rating Survey

Rate the quality of visual information contained in the following models.

Model	Inferior	Similar/ Equivalent	Superior (similar information more rapidly assimilated)	Superior (additional information provided)
Brain Cortical Surface				
2D	*Baseline*			
VRML				
Stereolithograph				
Phineas Gage Skull Injury				
2D	*Baseline*			
VRML				
Stereolithograph				

Evaluate the usefulness of different model types for understanding biology:

	Not at all	A little	Somewhat	A lot	A great deal
2D					
VRML					
Stereolithograph					

REFERENCES

1. Pevsner J. Leonardo da Vinci's contributions to neuroscience. Trends Neurosci. 2002;25:217–220.
2. Meakin JR, Shepherd DE, Hukins DW. Short communication: fused deposition models from CT scans. Br J Radiol. 2004;77:504–507.
3. Muller A, Krishnan KG, Uhl E, Mast G. The application of rapid prototyping techniques in cranial reconstruction and preoperative planning in neurosurgery. J Craniofac Surg. 2003;14:899–914.
4. Lukic IK, Gluncic V, Ivkic G, Hubenstorf M, Marusic A. Virtual dissection: a lesson from the 18th century. Lancet. 2003;362:2110–2113.
5. ChenJC,AmarAP,LevyML,ApuzzoML.Thedevelopment of anatomic art and sciences: the ceroplastica anatomic models of La Specola. Neurosurgery. 1999;45:883–891; discussion 891–882.
6. Noorani R. Hoboken, N.J.: Wiley; 2006. Rapid prototyping : principles and applications. p. xxi, 377.
7. Eftekhar B, Ghodsi M, Ketabchi E, Ghazvini AR. Play dough as an educational tool for visualization of complicated cerebral aneurysm anatomy. BMC Med Educ. 2005;5:15. [PMC free article]
8. Panchaphongsaphak B, Burgkart R, Riener R. BrainTrain: brain simulator for medical VR application. Stud Health Technol Inform. 2005;111:378–384.
9. Cox RW. AFNI: software for analysis and visualization of functional magnetic resonance neuroimages.Comput Biomed Res. 1996;29:162–173.
10. Gering DT, Nabavi A, Kikinis R, Hata N, O'Donnell LJ, et al. An integrated visualization system for surgical planning and guidance using image fusion and an open MR. J Magn Reson Imaging. 2001;13:967–975.
11. MayaVi RamachandranP. Aeronautical Society of India; A free tool for CFD data visualization; 2001.
12. Abramoff MD, Magelhaes PJ, Ram SJ. Image Processing with ImageJ. Biophotonics International.2004;11:36–42.
13. Ratiu P, Talos IF. Images in clinical medicine. The tale of Phineas Gage, digitally remastered. N Engl J Med. 2004;351:e21.
14. Damasio H, Grabowski T, Frank R, Galaburda AM, Damasio AR. The return of Phineas Gage:

clues about the brain from the skull of a famous patient. Science. 1994;264:1102–1105.

15. D'Urso PS, Hall BI, Atkinson RL, Weidmann MJ, Redmond MJ. Biomodel-guided stereotaxy. Neurosurgery. 1999;44:1084–1093; discussion 1093–1084.

16. Wurm G, Tomancok B, Holl K, Trenkler J. Prospective study on cranioplasty with individual carbon fiber reinforced polymer (CFRP) implants produced by means of stereolithography. Surg Neurol.2004;62:510–521.

17. Wurm G, Tomancok B, Pogady P, Holl K, Trenkler J. Cerebrovascular stereolithographic biomodeling for aneurysm surgery. Technical note. J Neurosurg. 2004;100:139–145.

18. McGurk M, Amis AA, Potamianos P, Goodger NM. Rapid prototyping techniques for anatomical modelling in medicine. Ann R Coll Surg Engl. 1997;79:169–174.

CHAPTER 4

A 3D INTERACTIVE MODEL AND ATLAS OF THE JAW MUSCULATURE OF ALLIGATOR MISSISSIPPIENSIS

Casey M. Holliday, Henry P. Tsai, Rebecca J. Skiljan, Ian D. George, Sami Pathan

Integrative Anatomy, Department of Pathology and Anatomical Sciences, University of Missouri, Columbia, Missouri, United States of America

ABSTRACT

Modern imaging and dissemination methods enable morphologists to share complex, three-dimensional (3D) data in ways not previously possible. Here we present a 3D interactive model of the jaw musculature of the American Alligator (Alligator mississippiensis). Alligator and crocodylian jaw musculature is notoriously challenging to inspect and interpret because of the derived nature of the feeding apparatus. Using Iodine-contrast enhanced microCT imaging, a segmented model of jaw muscles, trigeminal nerve, brain and skull are presented as a cross-sectional atlas and 3D, interactive pdf of the rendered model. Modern 3D dissemination methods like this 3D Alligator hold great potential for morphologists to share anatomical information to scientists, educators, and the public in an easily downloadable format.

INTRODUCTION

Modern technology has afforded morphologists and evolutionary biologists the ability to visualize and share the anatomy of species using a variety of imaging modalities such as serial histological reconstruction, computed tomography (CT) scanning, magnetic resonance imaging (MRI), confocal microscopy, and laser scanning among others [1], [2]. Recently, contrast techniques, such as Iodine Potassium Iodide (I_2KI) and Phosphotungstic acid (PTA), coupled with CT imaging have enabled the visualization of soft tissues in ways not previously possible [3]–[5]. These techniques are particularly beneficial to studying musculoskeletal anatomy because the contrast techniques clarify soft tissue anatomy and organization more vividly than MRI. Segmentation analysis of these serial images then facilitates the creation of three-dimensional (3D) models that can be easily rendered into high-resolution, interactive 3D pdfs or even printed models using rapid prototyping methods [6]. Finally, the recent surge in biomechanical analyses, such as finite element modeling and multi-body dynamics analysis, has made disseminating accurate, three-dimensionally accurate complex anatomy a necessity as collaborative projects investigating the functional morphology of animal systems mature [6]–[8].

Accurate three-dimensional representations of jaw muscles, like those of *Alligator* are beneficial to many researchers including those interested in crocodylian natural history and feeding ecology, evolutionary morphology and vertebrate paleontology, and biomechanics. Interpretations from an early version of this particular model have already been applied to one investigation into the soft tissue anatomy of *Alligator* [9],[10]. But also, these small-sized, web-hosted, easily downloadable, interactive files are easily accessible by

Citation: Holliday CM, Tsai HP, Skiljan RJ, George ID, Pathan S. A 3D Interactive Model and Atlas of the Jaw Musculature of Alligator mississippiensis. Farke AA, ed. PLoS ONE. 2013;8(6):e62806. doi:10.1371/journal.pone.0062806.

the general public, undergraduate students, and K-12 students interested in modern anatomical studies as well as charismatic extant archosaurs. Examples of shared, web-based vertebrate imaging resources can be found at Digimorph (http://www.Digimorph.org/); Aves3D (www.Aves3D.org); Paleoview (http://paleoview3d.marshall.edu/); and the 3D Alligator (web.missouri.edu/~hollidayca/3DAnatomy/Alligator3D; http://www.oucom.ohiou.edu/dbms-witmer/3D_gator.htm). This contribution follows in the footsteps of these cutting edge, shared anatomical resources.

Here we employed I_2KI contrast staining and microCT imaging to develop the most anatomically accurate segmented three-dimensional model and interactive 3D pdf of the jaw muscles, nervous tissues, and skull of the American alligator (*Alligator mississippiensis*) (Figs. 1–3). Alligator, and crocodylian jaw musculature has been notoriously challenging to study for several reasons. First, the characteristically flat skull makes the adductor muscles, or jaw-closing muscles logistically difficult to access deep to the temporal bars, large quadrate, and robust mandible. Second, individual crocodylian muscles are generally poorly-defined compared to those of lizards, birds and mammals making their individual identification challenging [11], [12]. Third, this hypertrophied amalgam of jaw musculature has resulted in conflicting interpretations of muscle homology and evolution [13]–[16] (see Holliday and Witmer [2007] [11] for a review). In addition to the jaw musculature, we include the brain and trigeminal nerve divisions as they pass among the musculature. The relative positions of the jaw muscles and the divisions of the trigeminal nerve are classic topological characters used to interpret anatomy and homology of vertebrate jaw muscles [13]. We intend this contribution to serve as a visual atlas rather than a text-based description as the jaw muscle anatomy of *Alligator* and crocodylians have been described in numerous other papers [11]–[16].

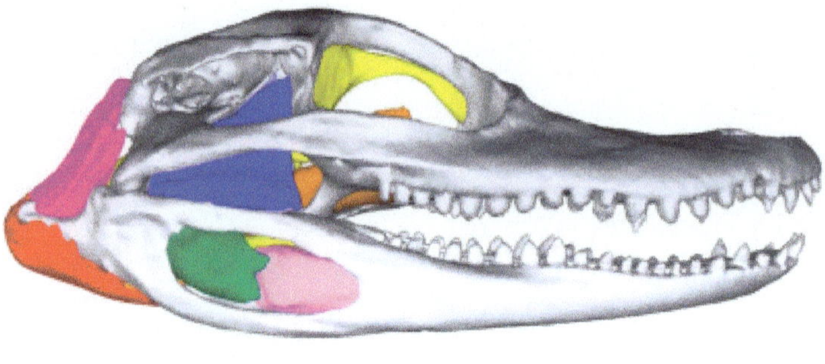

Figure 1: 3D interactive model of the jaw musculature of *Alligator mississippiensis* **as modeled from I2KI staining and CT-scanning.**

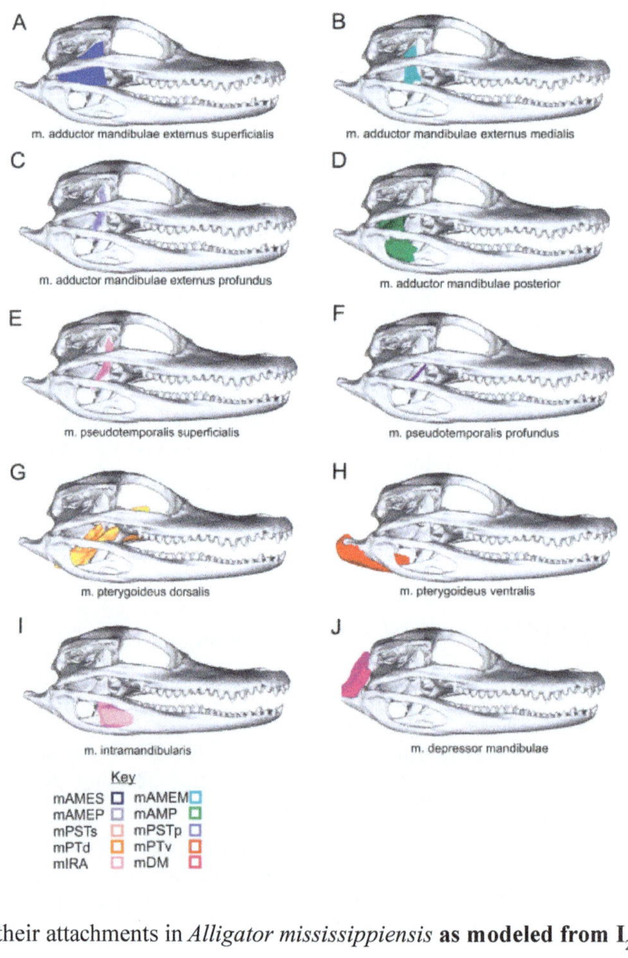

Figure 2: Individual jaw muscles and their attachments in *Alligator mississippiensis* **as modeled from** I_2KI staining and CT-scanning.

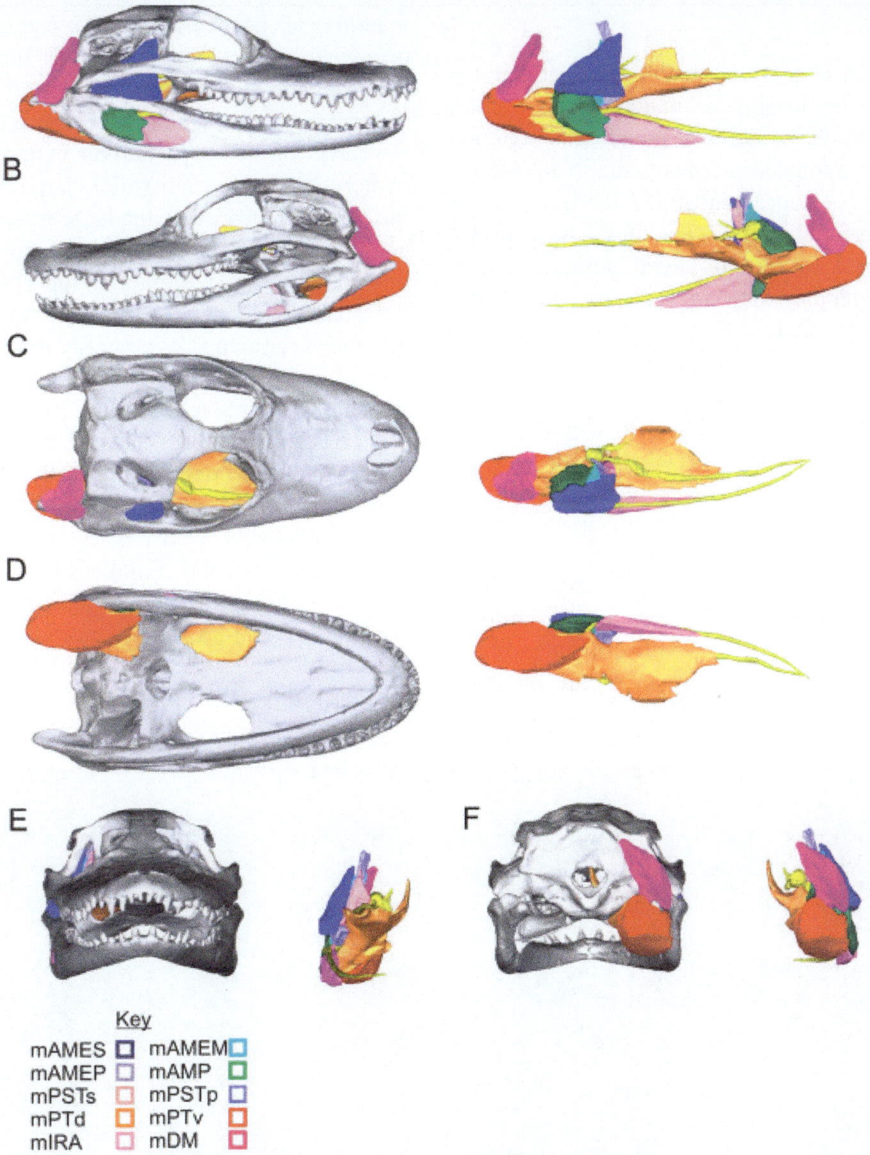

Figure 3: The 3D model of the jaw muscles of *Alligator mississippiensis* showing the relationship between the skull, muscles and branches of the trigeminal nerve.

RESULTS AND DISCUSSION

Measurements of the skull and muscle volumes, as well as commonly used abbreviations are found in Table 1. Abbreviations for muscles are also provided in the text following their first use.

Table 1: List of the materials in the 3D Alligator mode relevant l, the volumes of the segmented portions, and their abbreviations

Material	Abbreviation	Volume (mm³)	Skull length (mm)	Skull Width (mm)
Cranium		3915.78	56.71	29.83
Mandibles		1360.33		
m. adductor mandibulae externus superficialis	mAMES	134.65		
m. adductor mandibulae externus medialis	mAMEM	39.03		
m. adductor mandibulae externus profundus	mAMEP	24.07		
m. pseudotemporalis superficialis	mPSTs	33.47		
m. pseudotemporalis profundus	mPSTp	2.79		
Cartilago transiliens		12.98		
m. intramandibularis	mIRA	95.32		
m. adductor mandibulae posterior	mAMP	219.01		
m. pterygoideus ventralis	mPTv	397.41		
m. pterygoideus dorsalis	mPTd	606.78		
m. depressor mandibulae	mDM	110.27		
Brain		949.83		
Trigeminal ganglion		15.05		
Ophthalmic nerve*		2.63		
Maxillary nerve		26.55		
Mandibular nerve		39.37		

*ophthalmic nerve was only partially segmented as it became impossible to see the nerve between the eye and orbit.

Ventral View

In the transverse sections (Fig. 4), beginning at the dorsalmost section, one can see the origins of many of the vertical adductor muscles including m. adductor mandibulae externus profundus(mAMEP), m. adductor externus medialis(mAMEM), m. adductor externus superficialis(mAMES), and m. pseudotemporalis superficialis(mPSTs) as they pass ventrally from the dorsotemporal fossa and lateral surfaces of the quadrate and laterosphenoid. Moving ventrally, the section passes through the trigeminal ganglion and the ophthalmic and maxillary nerves as the pass rostrally towards the face. The adductor muscles begin to expand in cross-section as they continue towards their mandibular attachments. The attachments of m. pterygoideus dorsalis(mPTd) can be seen on the cartilaginous interorbital septum. Caudally, m. depressor mandibulae passes from the occipital surface of the skull towards the retroarticular process. In the next section, the expansive origin of m. pterygoideus dorsalis is seen passing from the cavichonchal recess and is bounded dorsally by the maxillary nerve. Caudally, the mandibular nerve passes towards the Meckelian fossa of the mandible and is surrounded by musculature including m. adductor mandibulae posterior(mAMP) caudally and the pterygoid origin of m. pterygoideus dorsalis and the origin on the small m. pseudotemporalis profundus(mPTSp) medially. Further ventrally, the mandibular nerve begins to curve rostrally towards the corner of the mouth and the expansive pterygoid origins of m pterygoideus ventralis(mPTv)are visible. In this section, near the corner of the mouth, the large sesamoid cartilago transiliens can be seen linking mm. adductor mandibulae externus profundus, pseudotemporalis superficialis, and pterygoideus dorsalis dorsally with m. intramandibularis(mIRA) ventrally. The next section passes ventrally through the dorsal part of the mandible. The insertions of m. adductor mandibulae posterior into the mandibular fossa and m. pterygoideus dorsalis into the medial surface of the articular are visible. The mandibular nerve is seen passing laterally to m. intramandibular within the rostral portion of the Meckelian fossa. Caudally, m. pterygoideus ventralis is visible passing around the retroarticular process. Finally, the ventralmost section shows the mandibular nerve passing rostrally towards the chin. The bellies of mm. intramandibularis, adductor mandibular posterior and pterygoideus dorsalis and ventralis are visible as they attach to the mandible.

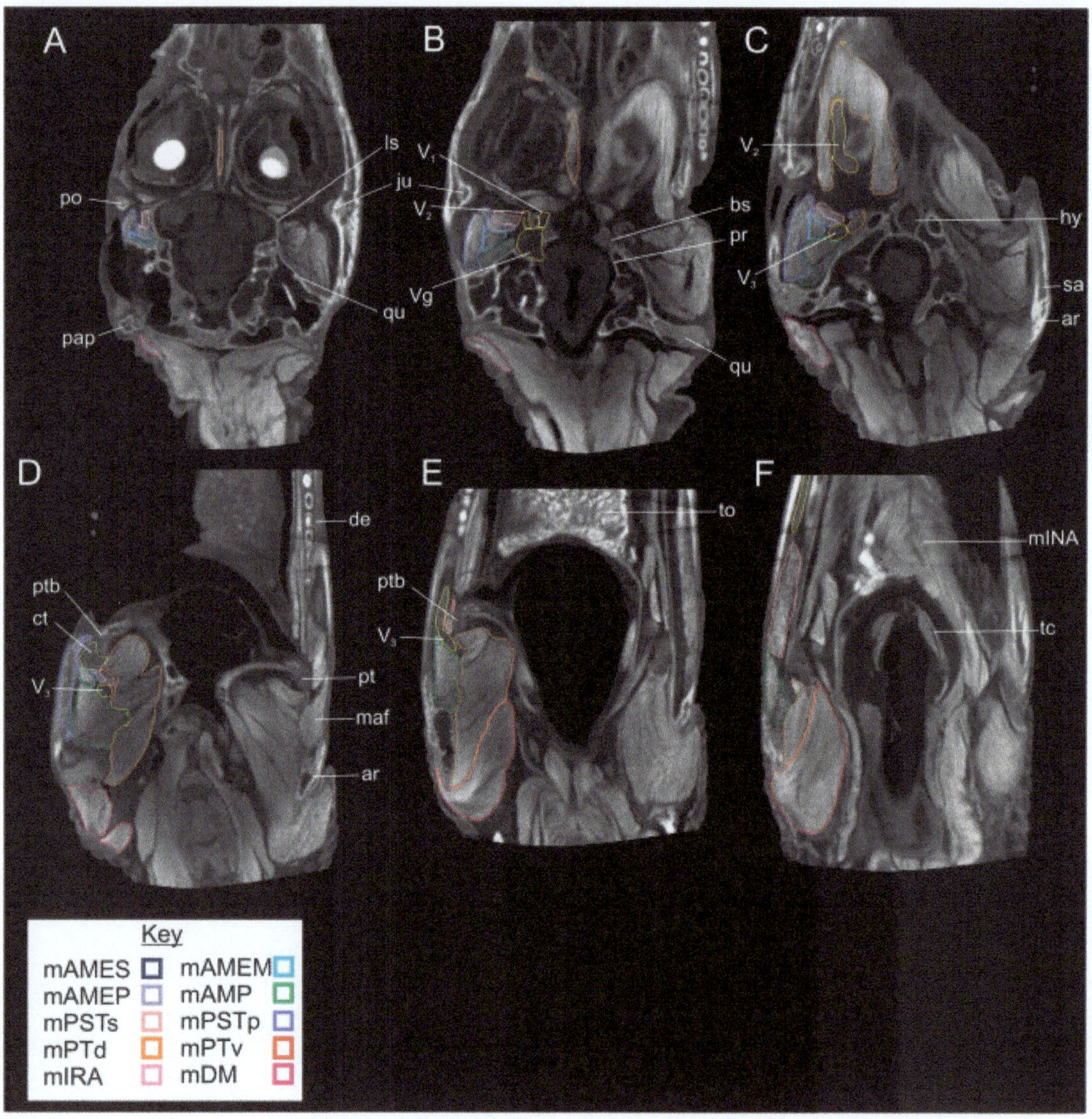

Figure 4: Ventral views of transverse sections through the segmented CT data showing labeled individual right jaw muscles on the left, and unlabeled left structures on the right.

Rostral View

In the axial sections (Fig. 5), beginning at the rostralmost section, m. pterygoideus dorsalis and the maxillary nerve are visible passing ventral to the orbit. Ventrally, the mandibular nerve is visible passing dorsal to m. intramandibularis within the Meckelian fossa. In the next section caudally, many of the adductor muscles are visible as they pass from the lateral wall of the braincase towards the mandible. The cartilago transiliens is visible linking bellies of m. adductor mandiblue externus and m. pseudotemporalis superficialis with m. intramandibularis as it wraps laterally around the pterygoid buttress. Medially, the ophthalmic and maxillary nerves are separated by the laterosphenoid lateral bridge. In the next section, the laminar pattern of the adductor muscles is easily appreciated with m. adductor mandibulae externus superficialis located laterally followed medially by the other vertical adductors and finally m. pseudotemporalis profundus and m. pterygoideus dorsalis. The next section occurs in a plane through the trigeminal ganglion and the mandibular nerve. Ventral to the nerve are m. pterygoideus dorsalis and the beginnings of m. pterygoideus ventralis as it comes off of the caudal edge of the pterygoid buttress. The next section caudally shows m. adductor mandibulae externus superficialis attaching to the dorsal surface of the surangular, m. adductor mandibulae posterior passing from the surface of the quadrate to the mandibular fossa and then mm. pterygoideus dorsalis and ventralis passing caudally towards the caudal end of the mandible. Finally, the caudalmost section passes through the neck and retroarticular process. The majority of m. depressor mandibulae is visible as it passes from the caudal surface of the paroccipital process to the dorsal surface of the retroarticular process. Ventrally the fibers of m. pterygoideus ventralis are visible passing underneath the mandible.

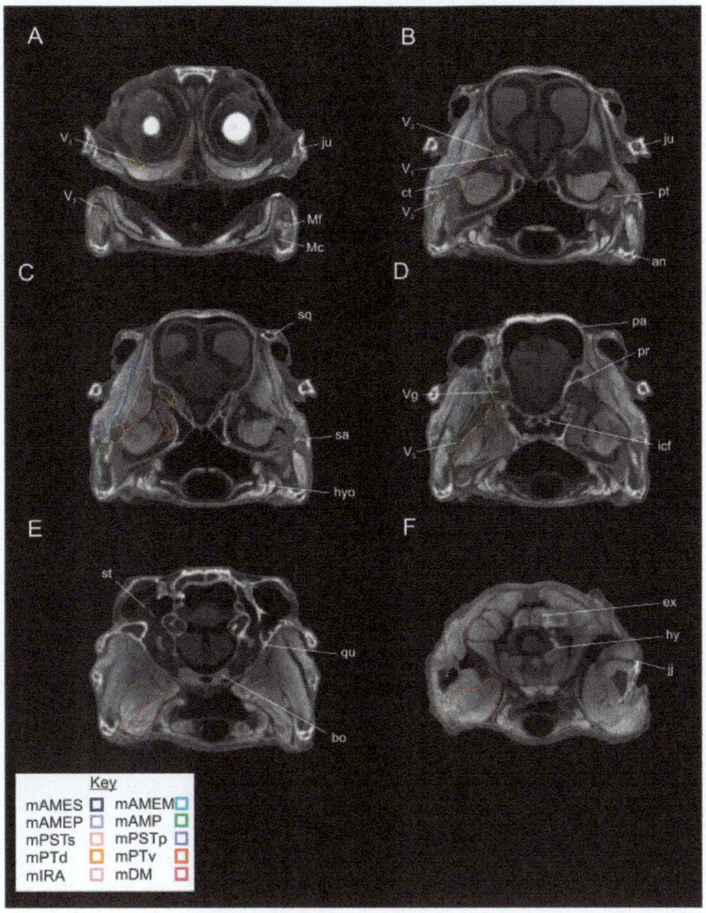

Figure 5: Rostral views of axial sections through the segmented CT data showing labeled individual right jaw muscles on the left, and unlabeled left structures on the right.

Lateral View

In the parasagittal sections (Fig. 6), beginning at the medialmost section, the trigeminal ganglion, maxillary nerve and majority of the origin of m. pterygoideus dorsalis are visible. In the next section lateral, the mandibular nerve passes ventrolaterally towards the mandible and is surrounded by numerous adductor muscles. The origin of m. adductor mandibulae externus profundus is visible in the dorsotemporal fossa as is the origin of m. pseudotemporalis superficialis on the lateral surface of the laterosphenoid. The large belly of m. pterygoideus ventralis is visible originating from the caudal edge of the pterygoid buttress. The cartilago transiliens is visible on the dorsolateral edge of the pterygoid buttress. The mandibular nerve and m. intramandibularis are visible within the Meckelian fossa. Finally, the lateral most parasagittal section shows m. adductor mandibulae posterior filling the mandibular fossa medial to the mandibular fenestra. Caudally, m. depressor mandibulae and m. pterygoideus ventralis respectively attach to the dorsal and ventral portions of the retroarticular process. Dorsally, the obliquely vertical fibers of m. adductor mandibulae externus superficialis and medialis are visible passing towards the surangular.

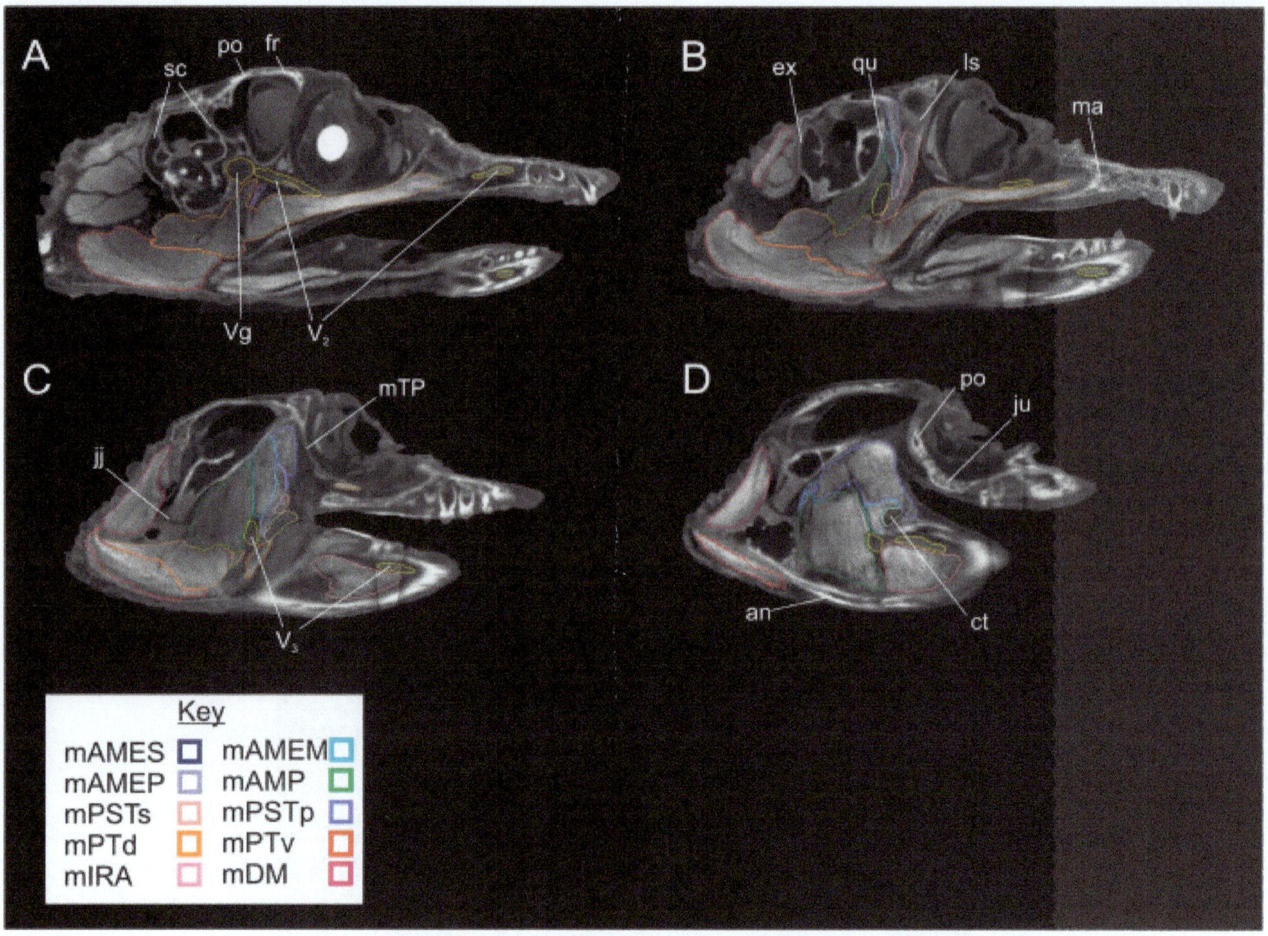

Figure 6: Lateral views of parasagittal sections through the segmented CT data showing labeled individual right jaw muscles.

CONCLUSIONS

Alligator jaw musculature, like that of other vertebrates, is three-dimensionally complex and difficult to illustrate in two-dimensional media. Here we provided the first volumetric model of *Alligator*, and crocodylian musculature based on raw imaging data. This downloadable, 3D model of *Alligator* jaw muscles, skull and nervous tissues shows the potential the next generation of anatomical tools have for morphologists, biologists, and the general public.

MATERIALS AND METHODS

One 12-month-old fresh-frozen cadaveric *Alligator mississippiensis* was obtained from Rockefeller State Refuge, Grand Chenier, Louisiana, US and accessioned into the University of Missouri Vertebrate Collections as MU AL031. The head was removed, fixed in 10% neutral buffered formalin and stored in 70% ethanol. The head was then immersed in 10% Iodine Potassium Iodide (i.e., I_2KI; Lugol's Iodine) for 5 weeks with 2 intervening CT scans to check on diffusion success. The final scan was conducted on a Siemens Inveon MicroCT scanner using 80 kV, 500 mA, and slice thickness of 83 microns at the University of Missouri Biomolecular Imaging Center. Scan data were imported as DICOM files into Amira v 5.2 (Visage Imaging) for segmentation. Anatomical structures were segmented manually using both magic wand and paintbrush tools by the authors. A defect in the scanner resulted in an obvious stitching artifact along the z-axis of the scan visible passing through the junction of the face and braincase. This did not significantly impact the interpretation of the data. Individual segmented structures were saved as STL files, smoothed and cleaned in Geomagic 12, and then imported into Adobe Acrobat 3D v.8 Toolkit, further reduced, and organized into a 3D model. This file was then imported into Adobe 8 3D and organized into its final format. Screen captures of specific slices of the segmented model were taken from Amira to show the outlines of muscles and soft tissues.

Ethics Statement

A cadaveric alligator was provided by Ruth Elsey, Rockefeller Wildlife Refuge, Grand Chenier, Louisiana under their permits.

ACKNOWLEDGMENTS

We thank Ashley Szcodroski and Tim Hoffman of the University of Missouri Biomolecular Imaging Center for access to the microCT scanner, Ruth Elsey and Rockefeller Wildlife Refuge for cadaveric specimens, Integrative Anatomy, and the University of Missouri Department of Pathology and Anatomical Sciences.

REFERENCES

1. Handschuh S, Thomas S, Metscher B (2010) Showing their true colors: a practical approach to volume rendering from serial sections. BMC Dev Biol 101: 41.

2. Johnston P (2011) Cross-Sectional Imaging in Comparative Vertebrate Morphology - The Intracranial Joint of the Coelacanth *Latimeria chalumnae,* Computed Tomography - Special Applications, Luca Saba (Ed.), InTech. pp. 259–274.
3. Metscher BD (2009a) Micro CT for comparative morphology: Simply staining methods allow high-contrast 3D imaging of diverse non-mineralized animal tissues. BMC Physiol 9(1): 11.
4. Metscher BD (2009b) MicroCT for developmental biology: A versatile tool for high-contrast imaging at histological resolutions. Dev Dynam 238(3): 632–640
5. Jeffery NS, Stephenson RS, Gallagher JA, Jarvis JC, Cox PG (2011) Micro-computed tomography with iodine staining resolves the arrangement of muscle fibres. J Biomech 44(1): 189–192
6. Ruthensteiner B, Heß M (2008) Embedding 3D models of biological specimens in PDF publications.Microsc Res Technique 71: 778–786
7. Dumont ER, Davis JL, Grosse IR, Burrows AM (2011) Finite element analysis of performance in the skulls of marmosets and tamarins. J Anat. 218(1): 151–62
8. Curtis N, Jones MEH, Evans SE, Shi J, O›Higgins P, et al. (2010) Predicting muscle activation patterns from motion and anatomy: modelling the skull of *Sphenodon*(Diapsida: Rhynchocephalia) J Roy Soc Interface. 7: 153–160
9. Tsai HP, Holliday CM (2011) Ontogeny of the Alligator cartilago transiliens and its significance for sauropsid jaw muscle evolution. PLoS ONE 6(9): e24935.
10. George ID, Holliday CM (2013) Trigeminal nerve morphology in *Alligator mississippiensis* and its significance for crocodyliform facial sensation and evolution. Anat Rec 296(4): 670–680
11. Holliday CM, Witmer LM (2007) Archosaur adductor chamber evolution: integration of musculoskeletal and topological criteria in jaw muscle homology. J Morphol 268: 457–484
12. Bona P, Desojo JB (2011) Osteology and cranial musculature of *Caiman latirostris* (Crocodylia: Alligatoridae). J Morphol 272(7): 780–795
13. Lakjer T (1926) Die Trigeminus-Versorgte Kaumuskulature der Sauropsiden. Auf kosten der Carlsbergstiftung, Kopenhagen. 154 p.
14. Schumacher GH (1973) The head muscles and hyolaryngeal skeleton of turtles and crocodilians. In: Gans C, Parsons TS, editors. Biology of the Reptilia, Vol. 4: Morphology D. New York: Academic Press. pp 101–200.
15. Busbey AB III (1989) Form and function of the feeding apparatus of *Alligator mississippiensis*. J Morphol 202: 99–127
16. Iordansky NN (2000) Jaw muscle of the crocodiles: structure, synonymy, and some implications on homology and functions. Russ J Herpetol7: 41–50

CHAPTER 5

THREE-DIMENSIONAL RECONSTRUCTIONS COME TO LIFE – INTERACTIVE 3D PDF ANIMATIONS IN FUNCTIONAL MORPHOLOGY

Thomas van de Kamp[1,2], Tomy dos Santos Rolo[1], Patrik Vagovic[1], Tilo Baumbach[1], Alexander Riedel[2,1]

[1]ANKA/Institute for Photon Science and Synchrotron Radiation, Karlsruhe Institute of Technology (KIT), Eggenstein-Leopoldshafen, Germany,

[2]State Museum of Natural History (SMNK), Karlsruhe, Germany

ABSTRACT

Digital surface mesh models based on segmented datasets have become an integral part of studies on animal anatomy and functional morphology; usually, they are published as static images, movies or as interactive PDF files. We demonstrate the use of animated 3D models embedded in PDF documents, which combine the advantages of both movie and interactivity, based on the example of preserved Trigonopterus weevils. The method is particularly suitable to simulate joints with largely deterministic movements due to precise form closure. We illustrate the function of an individual screw-and-nut type hip joint and proceed to the complex movements of the entire insect attaining a defence position. This posture is achieved by a specific cascade of movements: Head and legs interlock mutually and with specific features of thorax and the first abdominal ventrite, presumably to increase the mechanical stability of the beetle and to maintain the defence position with minimal muscle activity. The deterministic interaction of accurately fitting body parts follows a defined sequence, which resembles a piece of engineering.

INTRODUCTION

Functional morphology of animals usually relies on observations of living specimens and/or the interpretation of morphological characters found in dead ones [1]. In recent years, the arrival of three-dimensional (3D) imaging techniques significantly extended the pool of available methods for morphological studies [2]–[5]. Digital models based on segmented datasets allow the analysis of both external and internal structures [6], and by providing a "digital copy" they facilitate a non-destructive examination of minute, brittle, and irreplaceable samples. Some animations of 3D data have been published recently as 2D movies [7]–[11].

Animated 3D PDF (portable document format) files, however, provide a much broader range of interactivity as opposed to movies, as the perspective can be chosen and varied and/or complex models can be masked to show only selected parts of interest, e.g. distinct muscle groups or parts of the skeleton [12], [13]. Most software applications used for image stack segmentation do not offer sufficient functionality to move polygon meshes with respect to each other. Herein, we describe an approach to analyse and illustrate complex motion systems by animating 3D mesh models of static specimens with the help of 3D animation software.

We illustrate the workflow (Figure 1) based on µCT (synchrotron X-ray microtomography) data of*Trigonopterus* weevils [14]: First, the hind leg's screw-

Citation: Van de Kamp T, dos Santos Rolo T, Vagovič P, Baumbach T, Riedel A. Three-Dimensional Reconstructions Come to Life – Interactive 3D PDF Animations in Functional Morphology. Evans AR, ed. PLoS ONE. 2014;9(7):e102355. doi:10.1371/journal.pone.0102355.

and-nut type joint [15] is animated (Figure S1); we proceed with the animation of the entire weevil, i.e., a motion system comprising 44 components (Figure S2), to clarify the functional morphology of its defensive behaviour. The latter involves death-feigning, also known as thanatosis [16]. When preparing preserved specimens we found it hard to move their rostrum and legs from thanatosis into a walking position. Movements appeared mechanically blocked and it was impossible to identify the blocking mechanism by manual examination.

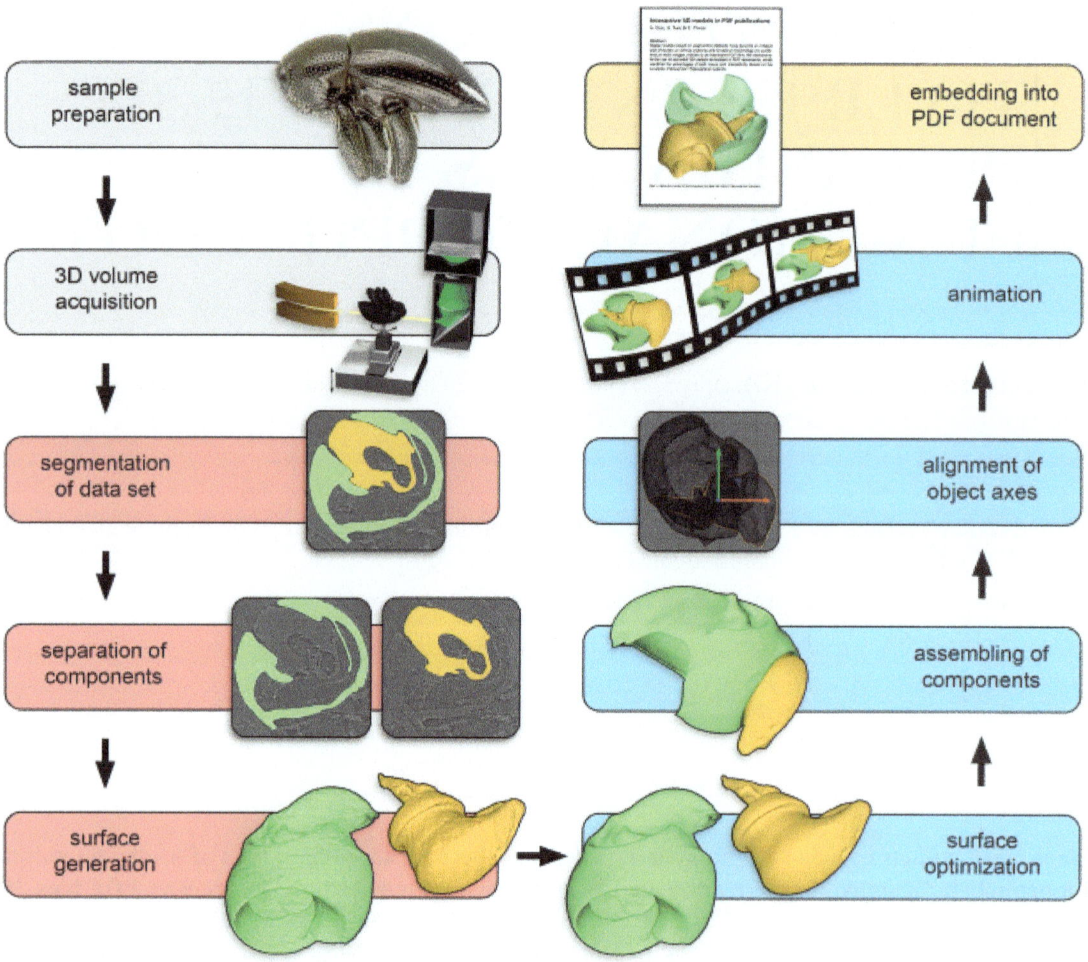

Figure 1: Flow diagram of the steps creating an interactive animated 3D model, based on the example of a screw-and-nut type hip joint of the weevil Trigonopterus oblongus.

Trigonopterus Fauvel is a genus of wingless weevils dwelling in primary forests of Southeast Asia and Melanesia. Its hundreds of species are spread over its range, many of them still undescribed. New Guinea appears to be a centre of its diversity with more than 300 species recorded [17], [18]. Specimens are found sitting on foliage or in the litter of forest floors, but little is known of their biology. A compact thanatosis position may be a character that gained evolutionary significance with *Trigonopterus*' inability to fly. Thus, a full understanding of the passive defence mechanisms may lead to a better understanding of the genus' extraordinary diversity.

MATERIALS AND METHODS

Samples

We scanned two complete specimens of *Trigonopterus vandekampi* Riedel [19] of similar body size and one specimen of *Trigonopterus oblongus* (Pascoe). One specimen of *T. vandekampi* was in walking, the others in thanatosis position. All specimens had been fixed in 100% ethanol and were critical point dried.

Synchrotron-based X-ray Microtomography

Tomographic scans were performed at the microtomographic station at the TOPO-TOMO beamline of the ANKA synchrotron radiation facility located at Karlsruhe Institute of Technology, Germany. The tomographic 180° scans were taken using a filtered white beam with the spectrum peak at about 20 keV. An indirect detector system based on scintillating screen, diffraction limited optical microscope and CCD detector was used for the acquisition of the frames. For converting X-rays into visible light an LSO terbium doped scintillator was employed.

In the case of *T. oblongus*, a magnification of 18× resulted in an effective pixel size of 0.5 μm. An exposure time of 240 ms per frame was used to record 2,500 projections. Both specimens of *T. vandekampi* were scanned with a magnification of 10× and an effective pixel size of 0.9 μm.

An exposure time of 2 s per frame was used to record 1,500 projections. For all scans, a PCO 4000 14 bit CCD camera system with a resolution of 4,008×2,672 pixels served for recording the frames. Before reconstruction, the frames were processed with the phase retrieval ImageJ plugin ANKAphase [20]. Volume reconstruction was done with the PyHST software developed at the European Synchrotron Radiation Facility in Grenoble, France. Microtomographic image data are deposited in MorphDBase (accession numbers A_Riedel_20140623-M-10.1, A_Riedel_20140623-M-19.1, A_Riedel_20140623-M-15.1, A_Riedel_20140623-M-13.1, A_Riedel_20140623-M-16.1, A_Riedel_20140623-M-17.1)

Segmentation

Body sclerites were segmented and converted into individual surface components (polygon meshes), as done in other recent studies [21]–[23] following the procedure described in [12]. Soft tissue and connecting cuticle were not segmented unless hard to delimit from sclerites, i.e. at the attachment points of tendons. The 3D volumes were imported into Amira (version 5.4.2; FEI Visualization Sciences Group) or Avizo (version 6.2.1; FEI Visualization Sciences Group). The Image slices were segmented manually to create polygon meshes (surface models). Initially, every tenth slice was segmented with subsequent interpolation on interjacent slices. For delicate structures, smaller steps were taken to minimize interpolation errors. The interpolated labels were checked; errors and artefacts were corrected manually. After segmenting the objects each morphological structure was isolated. The *smooth labels* dialog was used for smoothing the labels (size 5; mode: 3D volume) and polygon meshes of the structures' surfaces were created with the *SurfaceGen*module at default settings.

Optimization of Polygon Meshes

Polygon meshes from segmented image volumes typically contain millions of polygons and numerous segmentation artefacts showing the traces of individual layers. A smooth surface facilitates reduction of the polygon count without losing too many structural details. Thus, an iterative series of surface smoothing and polygon reduction is most effective in removing segmentation artefacts and simultaneously reducing polygon count to 0.1% (Figure S3) thus greatly helping data handling in the downstream process.

For this study, the polygon count of the original meshes was reduced to 10% in Amira/Avizo. The files were subsequently saved in the Wavefront format (OBJ) to allow import into CINEMA 4D (versions 12 & 14; Maxon Computer GmbH) for subsequent smoothing and polygon reduction. The parameters were set with respect to the polygon count and the general shape of the objects.

Axis Alignment, Motion Analysis and Animation

Surface meshes may be animated using any suitable 3D program from a wide choice of software. For embedding an animated model into a 3D PDF document, the data have to be saved as Universal 3D Files (U3D) using e.g. Deep Exploration. Here, we used CINEMA 4D (Version 14) in the case of *T. oblongus*and Deep Exploration (Version 6; Right Hemisphere®; Note S2) to animate the joints of *T. vandekampi*.

Before animation, all meshes were assembled in CINEMA 4D with each component separately editable. Based on the position of the segmented sclerites in the original image stack, the individual components are automatically placed at their correct positions in the software's coordinate system. For the complex model of*T. vandekampi*, symmetric appendices (i.e. antennae and legs) were duplicated and mirrored. Object hierarchies were created and meshes of the different body parts were coloured.

Most joints of the heavily sclerotized weevil show a precise form closure of its components, so possible movements could be simulated by interactively moving one component towards its counterpart until the joint reaches the fully bent, respectively depressed position, yet avoiding any overlap of the adjacent surfaces. The joint's motion could be approximated by iterative trial and error. First, an appropriate position for the animation axes had to be found for each component of the joint. The axes were aligned by using the software's object axis tool (Figure 2). The position of an object axis was altered from three 2D perspectives (bottom, right and front view) to determine the optimal position in three-dimensional space. Positioning of the axis is highly sensitive and a tilting of only 0.1° from the ideal position may visibly increase artificial overlap of surfaces.

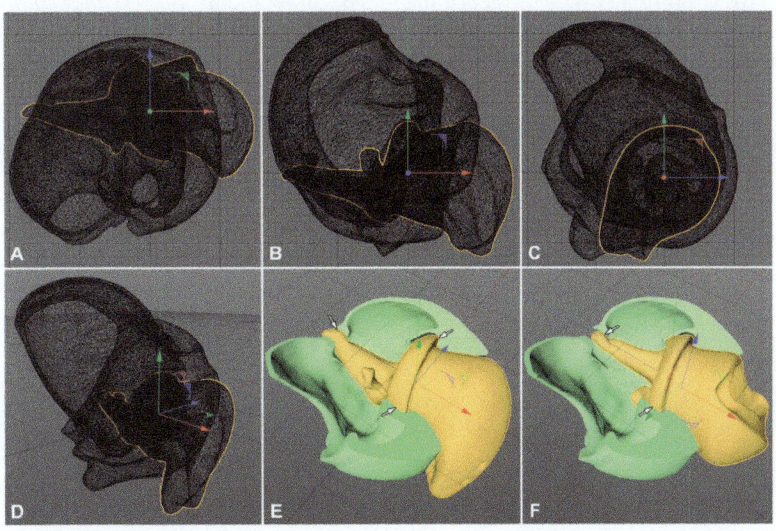

Figure 2: Axis alignment and animation of the screw joint of *Trigonopterus oblongus* in CINEMA 4D.

Then, one component was moved relative to the other finding its terminal positions, i.e. its fully extended and its fully depressed position, and for each a keyframe was created, thus defining the beginning and the end of the motion. Intermediate frames were interpolated automatically using linear interpolation setting. In joints with simple movements, e.g. a rotation around one stable axis, these two terminal keyframes were enough to simulate the joint's motion satisfactorily. However, in most cases the position of an animated component required realignment during the movement, and between two and six additional keyframes at intermediate positions had to ensure a precise simulation. During this process of approximating an optimal simulation, invalid arrangements could be detected by overlapping surfaces with display settings to isoparms in different 2D perspectives (e.g. bottom, right, front (Figure 2 A–C). In addition, the joints were temporarily cut to reveal any unrealistic friction of surfaces (Figure 2 E,F). Hard, guiding surfaces and soft structures, e.g. membranes or flexible tendons, which are pushed aside during movement in the living animal, had to be distinguished by the investigator.

The specimen with extended legs was segmented in part to verify the terminal position of the metacoxa. Its cavity is anteriorly open, so its movement is not strictly confined by the thorax (as is the case in pro- and mesocoxa), and thus required empiric measurement of its position with legs extended. From both positions, groups of polygon models composed of the metacoxa, metatrochanter, metafemur and parts of thorax and abdomen were loaded into the same scene and scaled to the same size. The walking position group was moved until thorax and abdomen overlapped with the ones from thanatosis. Thus assigning the final positions of the hind leg, we simulated its movement from walking position to thanatosis. Based on our field observations, the whole process of attaining thanatosis position in *Trigonopterus* takes about one second, i.e. it is faster than the eye can follow in detail. Thus, we decreased the motion speed of our animation. The precise timing of each joint's motion is considered a working hypothesis, since no video recording of the process is available. The adduction of all joints starts simultaneously as is the case in many other weevils falling into thanatosis.

Between 120 and 180 frames for the animation of each joint allowed smooth interpolation and an overall animation time of several seconds at 30 fps (Figure S2). The model of the screw joint of *T. oblongus*, which was animated in CINEMA 4D, was saved as a COLLADA 1.4 file (DAE). and imported into Deep Exploration.

For both models Deep Exploration was used to colour the mesh components and to create the final hierarchies for the meshes. Animation speed was set to 30 fps. Each model including materials and animations was subsequently saved as a Universal 3D File (U3D), containing both mesh geometry and animation sequences.It can be opened and displayed with suitable software, e.g. Deep Exploration, but for a wide dissemination the PDF format is preferable.

Embedding into PDF Files

New documents were created with Adobe Acrobat (version 9 Pro Extended; Note S2) and the U3D meshes were implemented with the *3D tool*. Using default *Activation Settings* and assigning a *Poster Image* from default view, the 3D visualization parameters were set as follows: white background, CAD optimized lights, solid rendering style and default 3D conversion settings. For the reconstruction of the coxa-trochanteral joints of *T. oblongus*, the animation style was set to *Bounce*, whereas it was set to *Loop* for the animated reconstruction of *T. vandekampi*. After starting the 3D view by clicking on the poster image, several views were created using the *Manage Views* option from the 3D toolbar. Annotations were added to the documents, which were subsequently saved as Portable Document Format files (PDF). Animated models are deposited at Dryad (http://doi.org/10.5061/dryad.56kf4).

RESULTS

Animation of a Screw Joint

For the isolated metacoxal screw joint, each coxa and trochanter were segmented separately (Figure S3). The terminal keyframes were set at 0 and 120, and four additional keyframes were needed to ensure realistic simulation for an arbitrary animation time of four seconds. The animation shows a rotation of 130 degrees with a translatory movement of 65 μm. Besides its larger size, the metacoxal joint of *T. oblongus* appears similar or identical to that of *T. vandekampi*.

Animation of a Complex System - Thanatosis of a Trigonopterus Weevil

A digital model of *T. vandekampi* suitable to answer our questions pertaining to the functional morphology of thanatosis was created by segmenting the major body sclerites (Table S1) of the specimen in thanatosis and by animating 50 individual articulations (Table S2). The noteautomatic placement of the individual components (i.e. the corresponding joint partners) in a consistent coordinate system as assigned by the software Amira resulted in an accurate animation of the assembled virtual beetle (Figure S2).

Trigonopterus Weevil's Cascade of Movements to Attain Thanatosis

The movements of *T. vandekampi* from walking position to thanatosis and reverse follow a defined sequence (Figure 3A). Some movements of the head, thorax and the appendices may partly happen simultaneously, but there are some benchmarks (Note S1) that must be passed by one component before another component can proceed for mechanical reasons. If this sequence of motions is violated, the weevil is unable to attain a perfect thanatosis position. The functional morphology is designed in a way to maintain the thanatosis position by the interaction of multiple body parts which mechanically block an unwanted opening of appendages. The following sequence of movements and mechanisms is hypothesized based on our animated model and on extensive field observations of the defence behaviour of cryptorhynchine weevils:

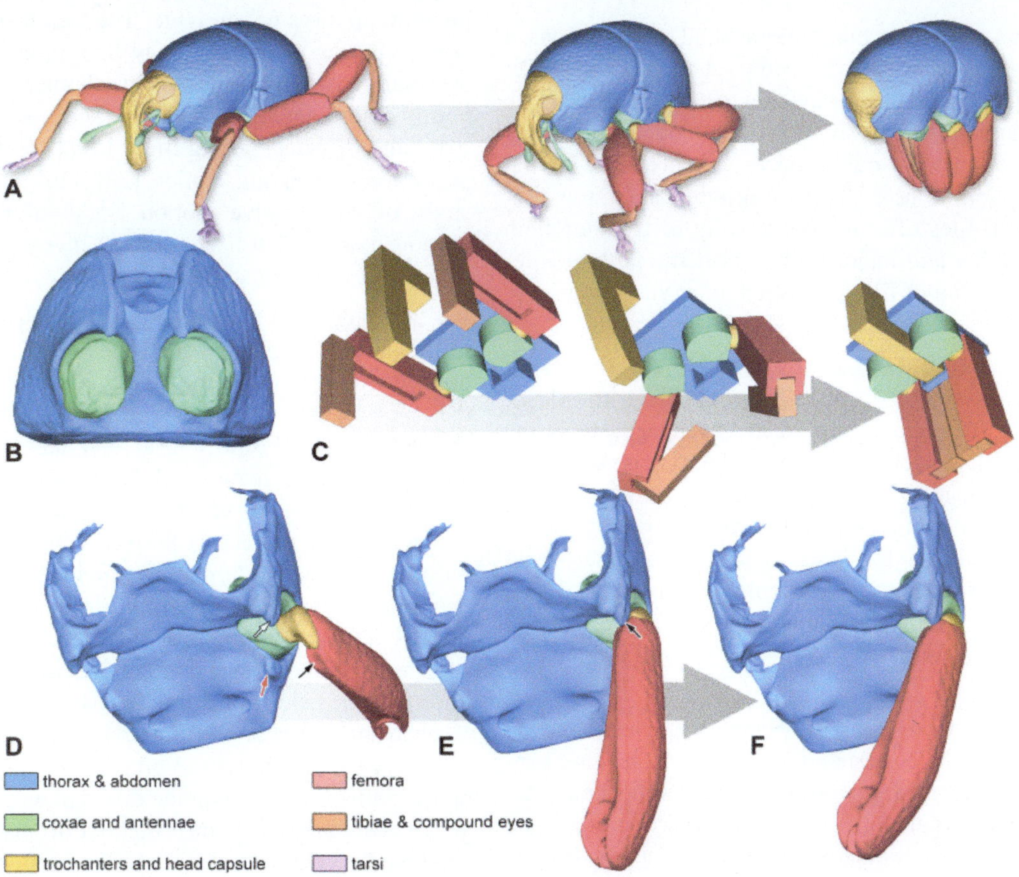

Figure 3: **Blocking mechanisms of legs in** *Trigonopterus vandekampi*.

- The tarsi are lifted and nestled backwards along the posterior face of the tibial apices which causes the weevil to lose its hold and fall to the side. The bent tibiae fit into the ventrally sulcate femora, their ventral edge overlapped by the anteroventral ridge of the femora. Thus, each leg forms a compact unit with potential stress relieved from the tibia-femoral joint.
- The compact femur-tibia-units are pressed together with the left and right legs touching medially. Almost no interspaces are visible between the legs in lateral aspect. Now the tibiae cannot be unfolded since their movement is blocked: the protibiae are blocking each other mesially; the mesotibiae are blocked by the overlapping profemora, and the metatibiae are blocked by the overlapping mesofemora. To allow unfolding of tibiae and tarsi, the pro- and mesocoxae have to be rotated outwards by approx. 10°.
- The prothoracic acetabula are mesially bordering a comparatively narrow thoracic canal; the procoxae are somewhat D-shaped in cross-section, with their flattened inner faces forming portions of the thoracic canal's lateral wall (Figure 3B). As the rostrum fits tightly into the thoracic canal and the movement of the procoxa is confined to rotation, the latter is mechanically inhibited by the retracted rostrum. Head and prothorax now form one functional unit.
- The head/prothorax-complex is retracted. Now, the ventral rim of the mesothoracic receptacle overlaps the retracted rostrum's tip ventrally. Thus, the head capsule must be moved forward ca. 57 µm along the beetle's body axis to allow an outward rotation of the procoxae. During thanatosis the antennae are almost fully concealed in the thoracic canal with the rostrum completely covering the opening of the thoracic canal.

Because outward rotation of the trochanteral screw joints is combined with a translatory movement (approx. 0.19 µm/°), the rotation of the protrochanters – and profemora, respectively – is inhibited by the midlegs while the prothorax is pressed against the mesothorax. The posterior surface of the profemur is concave at middle but swollen at the base. This swelling fits tightly into the concave anterior face of the mesocoxa thus blocking the rotation of the latter. The dorsal edge of the mesofemur basally forms an angulation which is posteriorly blocked by the intercoxal ridge of the metathorax tightly opposing it. Since the rotation axis of the mesotrochanter (translation: 0.19 µm/°) is almost perpendicular to the body axis, any elevating rotation is effectively blocked. Such a rotation, which is necessary to bring the leg into walking position, is only possible if the mesocoxa is turned to the side. The metacoxa differs markedly from pro- and mesocoxa as it tilts around two pivotal points. When the metatrochanter is rotating inwards and approaching its resting position (Figures 3D–F), the metafemur is pressed against the intercoxal ridge of the metathorax (Figure 3E). The leverage created causes the metacoxa to swing backwards (Figure 3F) and the metacoxa-trochanteral joint comes to a dead stop. In this position, coxa, trochanter and femur form a functional unit. The metafemur is maximally approximated to the body by the translation of the coxa-trochanteral screw joint, which is largest in the hind leg (0.24 µm/°).

DISCUSSION

In recent years, complex morphological 3D models based on segmented datasets have been published as PDF files [12], [22], [24] which allow the user to handle and examine relevant structures interactively. Other 3D models containing motion information were published as animated movies but without the option of user-interactivity other than stop-and-go. In fact, PDF files offer the opportunity to combine both motion information and interactivity. Furthermore, their file-size is only a fraction of files published in movie formats e.g. in MOV or MP4.

The recording of 3D data by e.g. CT for studies of functional morphology is ideally coupled with direct movierecording of motion, both taken simultaneously in the best case [7]–[10]. However, such an ideal setting is not always possible: the organisms of interest may be long extinct, too rare, or too shy to observe in a laboratory setting. High-resolution µCT recording suitable for *in vivo* imaging of small-sized specimens still pose radiation doses killing most insects within a few seconds [25]. Obviously, there remains a wide field of conditions where simultaneous recording of both motion and 3D data is impossible.

In many arthropod joints, movements are restricted by the morphology of the corresponding rigid parts, leaving very little play due to precise form closure of the components [26]. Simulations can be performed by interactively moving one component towards its counterpart until the joint reaches one endpoint. Some joints may involve uncertainty where exactly this endpoint is located, but in the described case where the limbs always reach a clearly defined and stable terminal position this was not an issue. The lack of information on the precise timing of motions may be a more serious drawback, especially when it concerns the simulation of complex and highly coordinated movements, such as the movement of two pairs of wings during flight [27],[28] or six pairs of legs performing a running motion [29], [30]. However, while the study of coordinative motion is out of reach without real motion data, it is still possible to investigate the qualitative movement of an isolated limb.

Although these limitations may appear quite restrictive, in the case of *Trigonopterus* weevils the attempt used in the present study proved to be highly effective for understanding the mechanisms of the weevils' defensive morphology (Note S1). The beetle's head and legs interlock mutually and with specific features of thorax and the first abdominal ventrite, presumably to increase its mechanical stability in thanatosis. The protective posture is maintained by minimal muscle activity, and largely by the mechanical interaction of exoskeletal parts. The deterministic interaction of accurately fitting body parts follows a defined sequence, which resembles a piece of engineering and in fact a closer analysis could be of interest to the field of biomimetics. Most aspects of the complex mechanisms could be illustrated in a single PDF 3D model of relatively small data size. While being completely interactive, predefined views illustrate the different mechanisms described above. This underlines the potential of animated 3D models: preserved or extinct species can be brought to life again, at least in the digital world.

SUPPORTING INFORMATION

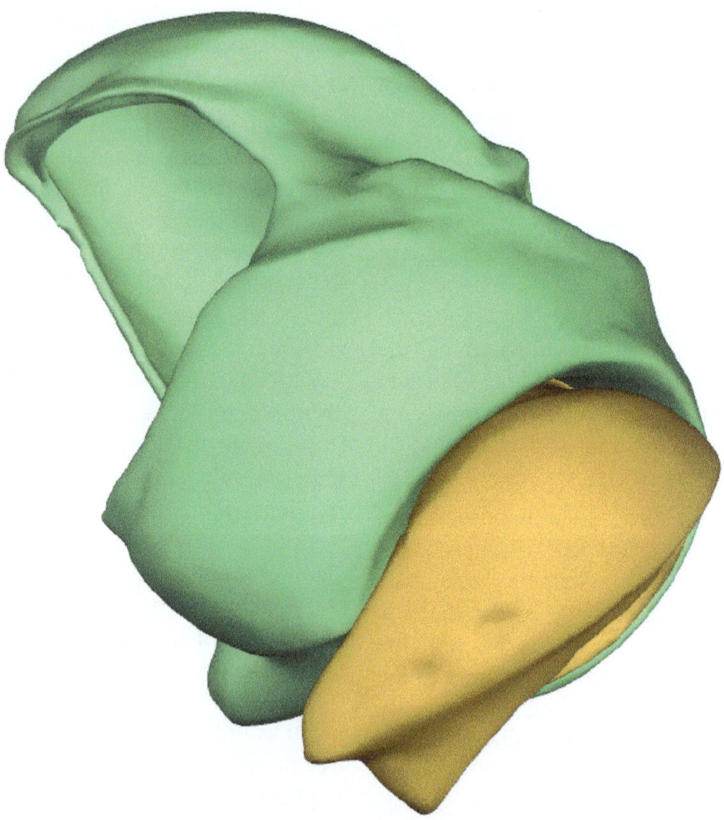

Figure S1: Interactive animated 3D reconstruction of the metacoxal joint of *Trigonopterus oblongus*.

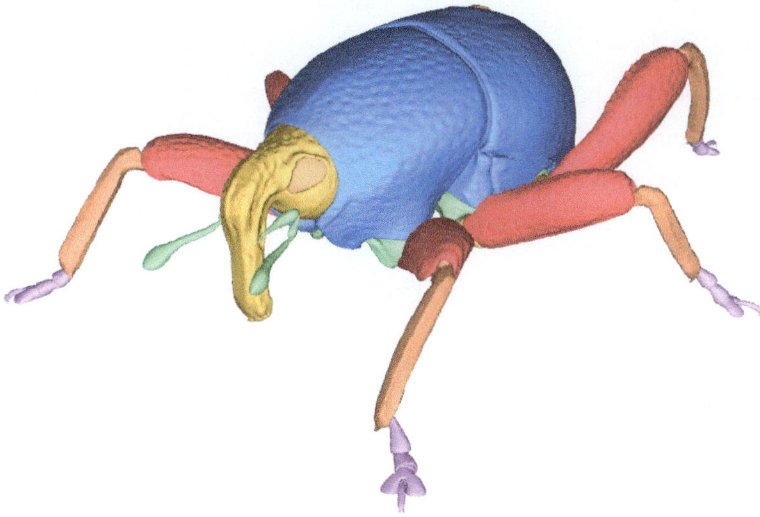

Figure S2: Interactive animated 3D reconstruction of Trigonopterus vandekampi simulating the movements from walking position to thanatosis posture. Default views illustrating the blocking mechanisms are provided.

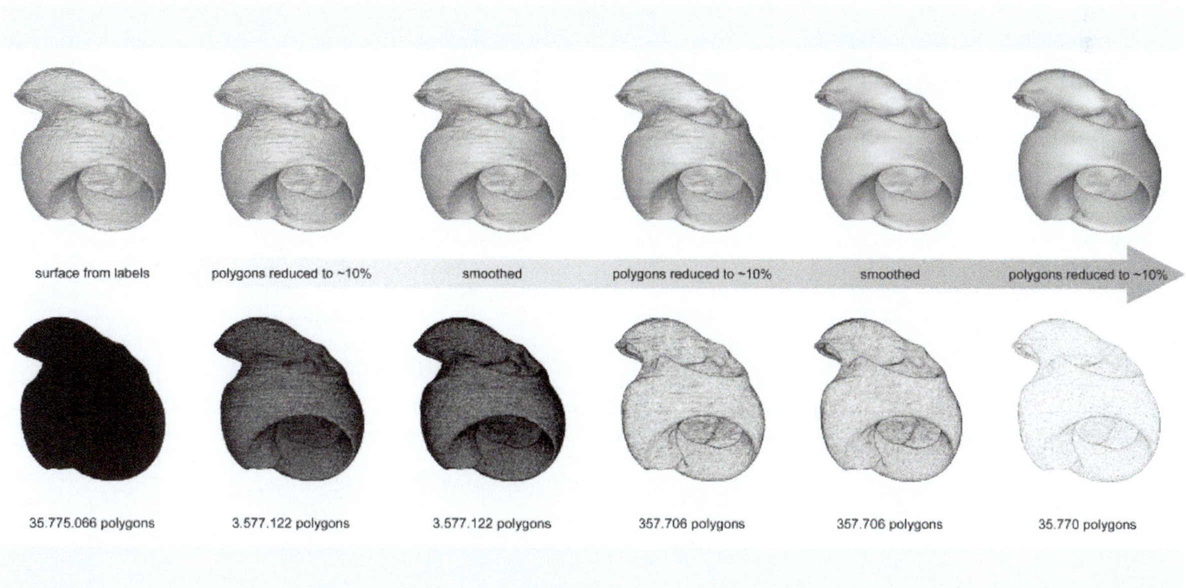

Figure S3: Optimization of polygon meshes, exemplified with the metacoxa of Trigonopterus oblongus, showing surface (top) and corresponding mesh (bottom). By a consecutive series of polygon reduction and smoothing, the polygon count – and thus the file size – was reduced to ca. 1/ 1,000 of its original value without compromising the surface structure while simultaneously reducing labelling artefacts.

Table S1: List of separate polygon meshes created from labeled exoskeleton parts of Trigonopterus vandekampi

animated articulations	joint type	QTY
prothorax-head capsule	form & force guided	1
head capsule-scape	ball-and-socket	2
scape-pedicel	hinge	2
pedicel-flagellum	ball-and-socket	2

prothorax-mesothorax	form & force guided	1
prothorax-procoxa	rotative	2
procoxa-protrochanter	screw-and-nut	2
profemur-protibia	hinge	2
protibia-protarsus 1	ball-and-socket	2
protarsomere 1-protarsomere 2	freely movable	2
Protarsomere 2-protarsomere 3	freely movable	2
protarsomere 4-propretarsus	freely movable	2
mesothorax-mesocoxa	rotative	2
mesocoxa-mesotrochanter	screw-and-nut	2
mesofemur-mesotibia	hinge joint	2
mesotibia-mesotarsus 1	ball-and-socket	2
mesottarsomere 1-mesotarsomere 2	freely movable	2
mesotarsomere 2-mesotarsomere 3	freely movable	2
mesotarsomere 4-mesopretarsus	freely movable	2
metathorax-metacoxa	dicondylic	2
metacoxa-metatrochanter	screw-and-nut	2
metafemur-metatibia	hinge	2
metatibia-metatarsomere 1	ball-and-socket	2
metatarsomere 1-metatarsomere 2	freely movable	2
metatarsomere 2-metatarsomere 3	freely movable	2
metatarsomere 4-metapretarsus	freely movable	2

Table S2: List of the 50 individual articulations animated to create the moving interactive model of *Trigonopterus vandekampi*. Note that femora and trochanters do not share movable articulations in the species. Joints between tarsomeres 3 and the minute tarsomeres 4 were neglected.

structure	component
head	head capsule
	compound eyes
	scape
	pedicel
	flagella
prothorax	prothoracic sclerites
	cryptopleuron
	procoxa
	protrochanter
	profemur
	protibia
	protarsomere 1
	protarsomere 2
	protarsomere 3
	protarsomere 4
	propraetarsus

	propraetarsal claws
mesothorax	mesothoracic sclerites
	elytra (as single object)
	mesocoxa
	mesotrochanter
	mesofemur
	mesotibia
	mesotarsomere 1
	mesotarsomere 2
	mesotarsomere 3
	mesotarsomere 4
	mesopraetarsus
	mesopraetarsal claws
metathorax	metathoracic sclerites
	metacoxa
	metatrochanter
	metafemur
	metatibia
	metatarsomere 1
	metatarsomere 2
	metatarsomere 3
	metatarsomere 4
	metapraetarsus
	metapraetarsal claws
abdomen	abdominal sclerites 1 & 2
	abdominal sclerite 3
	abdominal sclerite 4
	abdominal sclerite 5

Note S1: Benchmarks of thanatosis cascade

- The head must be tilted less than 6° dorsad from its ventral terminal position, so the rostrum gets pushed into the thoracic receptacle during the retraction of the prothorax. the shape of the receptacle would ensure that the rostrum is passively placed into the 0° terminal position.
- The antennae must be folded to the ventral face of the rostrum while the head is ca. 20° dorsad from its ventral terminal position, to avoid them being trapped between the rostrum and the edge of the prothoracic canal.
- The rotations of the head and the procoxae should be completed at the same time. the rostrum only fits into the thoracic canal when the procoxae are rotated less then ca. 8° from their terminal position.
- The femora must be at least 34° from the thanatosis position to allow the fully extended tibiae being folded.
- The profemur must be in its final position before the retraction of the prothorax, otherwise its base collides with the mesocoxa. the final 10° rotation of the procoxa could be achieved passively by pressure from the mesothorax.
- The midlegs must be in their final position before the retraction of the prothorax. the final 8° rotation of the mesocoxa could be achieved passively by pressure from the profemur.
- the folding of the metatibia should be completed before or simultaneous with the folding of the midlegs to avoid collision with the mesofemora.

Note S2: Software changes

Right Hemisphere® Deep Exploration

In 2011, Right Hemisphere® was acquired by SAP Deutschland AG & Co. KG. The software package Deep Exploration that was employed in our study, is now called SAP Visual Enterprise Author.

Adobe® Acrobat®

Embedding interactive 3D models into PDF documents was first introduced by Adobe in 2005 with the release of Acrobat 3D 7.0. In Version 9.0 (2008), 3D support was now part of Acrobat Pro Extended (Windows only), which we used in our study. With the arrival of Acrobat X (2010), Adobe suspended the creation of 3D PDF documents and outsourced 3D support to the company Tetra 4D, which now provides the 3D PDF converter plugins for Acrobat necessary to create 3D PDF documents.

ACKNOWLEDGMENTS

We thank S. Scharf for helping with the segmentation of data sets, D. Pelliccia for assistance during the tomographic scans, and R. Hofmann and R. Heine for helpful discussions. B. Ruthensteiner and A. R. Evans reviewed the manuscript and their comments lead to many improvements. The ANKA Synchrotron Radiation Facility is acknowledged for providing beamtime.

REFERENCES

1. Homberger DG (1988) Models and tests in functional morphology: the significance of description and integration. Amer Zool 28: 217–229doi:10.1093/icb/28.1.217
2. Zill S, Frazier SF, Neff D, Quimby L, Carney M, et al. (2000) Three-dimensional graphic reconstruction of the insect exoskeleton through confocal imaging of endogenous fluorescence. Microsc Res Techniq 48: 367–384doi:;10.1002/(SICI)1097-0029(20000315)48:6<367::AID-JEMT7>3.0.CO;2-Y
3. Corfield JR, Wild JM, Cowan BR, Parsons S, Kubke MF (2008) MRI of postmortem specimens of endangered species for comparative brain anatomy. Nat Protoc 3: 597–605doi:10.1038/nprot.2008.17
4. Westneat MW, Socha JJ, Lee W-K (2008) Advances in biological structure, function, and physiology using X-ray imaging. Ann Rev Physiol 70: 119–142doi:10.1146/annurev.physiol.70.113006.100434
5. Handschuh S, Baeumler N, Schwaha T, Ruthensteiner B (2013) A correlative approach for combining microCT light and transmission electron microscopy in a single 3D scenario. Front Zool 10: 44.doi:10.1186/1742-9994-10-44
6. Betz O, Wegst U, Weide D, Heethoff M, Helfen L, et al. (2007) Imaging applications of synchrotron X-ray phase-contrast microtomography in biological morphology and biomaterials science. I. General aspects of the technique and its advantages in the analysis of millimetre-sized arthropod structure. J Microsc 227: 51–71doi:10.1111/j.1365-2818.2007.01785.x
7. Sahara W, Sugamoto K, Murai M, Tanaka H, Yoshikawa H (2006) 3D kinematic analysis of the acromioclavicular joint during arm abduction using vertically open MRI. J Orthop Res 24: 1823–1831doi:10.1002/jor.20208
8. Brainerd EL, Baier DB, Gatesy SM, Hedrick TL, Metzger KA, et al. (2010) X-ray reconstruction of moving morphology (XROMM): precision, accuracy and applications in comparative biomechanics research.J Exp Zool Part A 313: 262–279doi:10.1002/jez.589
9. Gatesy SM, Baier DB, Jenkins FA, Dial KP (2013) Scientific rotoscoping: a morphology-based method of 32-2D motion analysis and visualization. J Exp Zool Part A 313: 244–261doi:10.1002/jez.588
10. Baier DB, Gatesy SM, Dial KP (2013) Three-dimensional, high-resolution skeletal kinematics of the avian wing and shoulder during ascending flapping flight and uphill flap-running. PLOS ONE 8: e63982. doi:10.1371/journal.pone.0063982
11. Lauridsen H, Hansen K, Wang T, Agger P, Andersen L (2011) Inside Out: Modern imaging techniques to reveal animal anatomy. PLOS ONE 6: e17879.doi:10.1371/journal.pone.0017879
12. Ruthensteiner B, Heβ M (2008) Embedding 3D models of biological specimens in PDF publications.Microsc Res Techniq 71: 778–786doi:10.1002/jemt.20618)
13. Murienne J, Ziegler A, Ruthensteiner B (2008) A 3D revolution in communicating science. Nature 453: 450. doi:10.1038/453450d
14. Riedel A, Sagata K, Surbakti S, Tänzler R, Balke M (2013) One hundred and one new species of*Trigonopterus* weevils from New Guinea. ZooKeys 280: 1–150doi:10.3897/zookeys.280.3906
15. van de Kamp T, Vagovič P, Baumbach T, Riedel A (2011) A biological screw in a beetle›s leg. Science333: 52.doi:10.1126/science.1204245
16. Bleich OE (1928) Thanatose und Hypnose bei Coleopteren. Experimentelle Untersuchungen. Z Morphol Oekol Tiere 10: 1–61doi:10.1007/BF00419278
17. Riedel A, Daawia D, Balke M (2010) Deep cox1 divergence and hyperdiversity of Trigonopterus weevils in a New Guinea mountain range (Coleoptera, Curculionidae). Zool. Scripta 39: 63–74doi:10.1111/j.1463-6409.2009.00404.x
18. Tänzler R, Sagata K, Surbakti S, Balke M, Riedel A (2012) DNA barcoding for community ecology - how to tackle a hyperdiverse, mostly undescribed Melanesian fauna. PLoS ONE 7: e28832.doi:10.1371/journal.pone.0028832
19. Riedel A (2010) One of a thousand - a new species of *Trigonopterus* (Coleoptera, Curculionidae, Cryptorhynchinae) from New Guinea. Zootaxa 2403: 59–68doi:not available
20. Weitkamp T, Haas D, Wegrzynek D, Rack A (2011) ANKAphase: software for single-distance phase retrieval from inline X-ray phase-contrast radiographs. J Synchrotron Radiat 18: 617–629doi:10.1107/S0909049511002895
21. Witmer LM, Ridgely RC (2008) The paranasal air sinuses of predatory and armored dinosaurs (Archosauria: Theropoda and Ankylosauria) and their contribution to cephalic structure. Anat Rec 291: 1362–1388doi:10.1002/ar.20794
22. Ziegler A, Ogurreck M, Steinke T, Beckmann F, Prohaska S, et al. (2010) Opportunities and challenges for digital morphology. Biol Direct 5: 45.doi:10.1186/1745-6150-5-45

23. Weide D, Thayer MK, Betz O (2012) Comparative morphology of the tentorium and hypopharyngeal–premental sclerites in sporophagous and non-sporophagous adult Aleocharinae (Coleoptera: Staphylinidae).Acta Zool 95: 84–110doi:10.1111/azo.12011
24. Baeumler N, Haszprunar G, Ruthensteiner B (2008) 3D interactive microanatomy of Omalogyra atomus (Philippi, 1841) (Gastropoda, Heterobranchia, Omalogyridae). Zoosymposia 1: 101–118doi:10.11646/zoosymposia.1.1.9
25. dos Santos Rolo T, Ershov A, van de Kamp T, Baumbach T (2014) In vivo X-ray cine-tomography for tracking morphological dynamics. Proc Natl Acad Sci USA 111: 3921–3926doi:10.1073/pnas.1308650111
26. Bögelsack G, Karner M, Schilling C (2000) On technomorphic modelling and classification of biological joints. Theory Biosc 119: 104–121doi:10.1007/s12064-000-0007-3
27. Willmott AP, Ellington CP (1997) The mechanics of flight in the hawkmoth Manduca sexta. I. Kinematics of hovering and forward flight. J Exp Biol 200: 2705–2722doi:not available
28. Willmott AP, Ellington CP (1997) The mechanics of flight in the hawkmoth Manduca sexta. II. Aerodynamic consequences of kinematic and morphological variation. J Exp Biol 200: 2723–2745doi:not available
29. Full RJ, Tu MS (1991) Mechanics of a rapid running insect - two-, four- and six-legged locomotion. J Exp Biol 156: 215–231doi:not available
30. Spence AJ, Revzen S, Seipel J, Mullens C, Full R (2010) Insects running on elastic surfaces. J Exp Biol213: 1907–1920doi:not available

CHAPTER 6

WHAT IS AN ATTRACTIVE BODY? USING AN INTERACTIVE 3D PROGRAM TO CREATE THE IDEAL BODY FOR YOU AND YOUR PARTNER

Kara L. Crossley[1] , Piers L. Cornelissen[1,2], Martin J. Tove´e[1]

[1]Institute of Neuroscience, Newcastle University, United Kingdom

[2]Department of Psychology, University of Northumbria, United Kingdom

ABSTRACT

What is the ideal body size and shape that we want for ourselves and our partners? What are the important physical features in this ideal? And do both genders agree on what is an attractive body? To answer these questions we used a 3D interactive software system which allows our participants to produce a photorealistic, virtual male or female body. Forty female and forty male heterosexual Caucasian observers (females mean age 19.10 years, s.d. 1.01; 40 males mean age 19.84, s.d. 1.66) set their own ideal size and shape, and the size and shape of their ideal partner using the DAZ studio image manipulation programme. In this programme the shape and size of a 3D body can be altered along 94 independent dimensions, allowing each participant to create the exact size and shape of the body they want. The volume (and thus the weight assuming a standard density) and the circumference of the bust, waist and hips of these 3D models can then be measured. The ideal female body set by women (BMI = 18.9, WHR = 0.70, WCR = 0.67) was very similar to the ideal partner set by men, particularly in their BMI (BMI = 18.8, WHR = 0.73, WCR = 0.69). This was a lower BMI than the actual BMI of 39 of the 40 women. The ideal male body set by the men (BMI = 25.9, WHR = 0.87, WCR = 0.74) was very similar to the ideal partner set by the women (BMI = 24.5, WHR = 0.86, WCR = 0.77). This was a lower BMI than the actual BMI of roughly half of the men and a higher BMI than the other half. The results suggest a consistent preference for an ideal male and female body size and shape across both genders. The results also suggest that both BMI and torso shape are important components for the creation of the ideal body.

INTRODUCTION

What makes a human body attractive to the opposite sex? In evolutionary psychology terms it is a judgment of a potential partner's health and reproductive potential [1], [2]. In this context it is important that we are able to detect and accurately assess the physical cues that indicate that one individual is more attractive (i.e., fitter and with a better reproductive potential) than another, and then use these cues to choose the partner who is most likely to enhance our chances of successful reproduction [1]–[3]. As a result there should be a strong selective pressure to detect and accurately evaluate reliable cues to health and fertility in potential partners. However, there remains considerable debate over which cues are used to judge human physical attractiveness, their relative importance and whether these cues differ between men and women.

Citation: Crossley KL, Cornelissen PL, Tovée MJ. What Is an Attractive Body? Using an Interactive 3D Program to Create the Ideal Body for You and Your Partner. Howe PD, ed. PLoS ONE. 2012;7(11):e50601. doi:10.1371/journal.pone.0050601.

Previous studies that have attempted to define the importance of these physical cues have had a significant limitation. These studies have used line-drawings, photographs and, more rarely, video clips and 3D laser scans as test stimuli [4]–[18]. Typically, observers are asked to rate a set of images that vary on a number of anthropometric dimensions. However, these studies all suffer from the same intrinsic methodological limitation that they require their participants to rate bodies from the limited set of alternatives presented to them. Unfortunately, the ideal combination of features may not be included in the set of images with which they are presented. Thus, their apparent preference may actually be for a suboptimal body size and shape. To try and overcome this problem some researchers have presented participants with silhouettes or photographs in interactive computer programmes which allows the simple alteration of certain body features [19]–[21]. However, these techniques are obviously limited in the range of shape changes that can be made and the realism of the bodies produced. Additionally, the 2D representation of the bodies limits what can be seen of the change in the physical dimensions produced by the programme. It can be difficult to extrapolate from a 2D representation of a body to its 3D shape [22].

To overcome these important methodological limitations, we have used an interactive 3D software programme to determine male and female participants' perceptions of their ideal body and their ideal partners' body size and shape. The participants could alter the virtual 3D image of the body in more than 90 independent dimensions allowing very subtle changes in body shape. The body could be rotated through 360° to allow our participants to examine the body from different viewpoints. The scaled volume of these 3D models can then be measured and, assuming they have a standard body density, their body weight can then be estimated. Additionally, the scaled circumference of the chest, waist and hips of each body can be measured to allow the waist-to-hip ratio (WHR) and the waist-to-chest ratio (WCR) to be calculated. By taking anthropometric measures from all our participants, we can determine whether the participants' own physical dimensions influence their choice of their own ideal body. This morphing technique allows us to answer two key questions:

What is the ideal body size and shape? For women, several studies have suggested that the ideal body is based on a curvaceous body, with a curvy lower torso (indexed by the WHR) but also a curvaceous upper body (WCR) [11], [13], [23]. Set against this, is an alternative hypothesis which postulates that the primary predictor of female attractiveness is overall body fat (usually measured as the Body Mass Index or BMI) [4],[12], [16], [17]. Changes in BMI have a strong impact on both health [24], [25] and reproductive potential[26]–[28], and a low WHR and WCR (i.e., a curvaceous body) is believed to correspond to the optimal fat distribution for high fertility [23], [29]. So there are clear reasons why both these features might impact on attractiveness judgements.

A similar difference of opinion exists for what is the main determinant of male attractiveness. Some studies assert that upper body shape (a broad upper body and a narrow waist, the classic V-shape) is the primary predictor of attractiveness, whereas others point to BMI as the key feature [5], [8], [10], [30]. It has been suggested that this v-shaped torso represents a muscular, strong body type that would be an advantage in our ancestral environment and therefore be sexually selected [10], [31]. BMI is an important predictor of male health and mortality [32], [33], and a narrow waist circumference is also important in long-term health and so should also be associated with a low WHR [34], [35].

By asking both men and women to set their ideal bodies we can determine which features they change and how their ideal body differs from their actual bodies. We can see whether they change shape or size or both.

Do men and women share body ideals? A number of studies have suggested a difference between the genders for the ideal body size and shape of a particular gender (for example, men may prefer a more curvaceous, heavier female body than women think they do) [36]–[38] and eye-tracking studies have suggested significantly different patterns of eye-movements between the genders when assessing female attractiveness [39]. However, mate selection theory predicts that an individual will have a very precise and accurate idea of what the opposite sex find attractive [3]. This allows them to judge their own relative value, with respect to their peer group, and match this value with the value of a prospective mate. So mate selection theory predicts that there will not be any difference between men and women in their ideals for both genders. There is some evidence to support this hypothesis in rating studies which have suggested the same ideals are held by both genders [8], [40], [41]. Our technique will allow us to accurately determine whether there are gender differences in body preferences, even if they are comparatively subtle and would not be detected in the choice between bodies within an image set. However large the image set, it cannot provide a continuous smooth change along all feature dimensions and so can only provide a comparatively coarse grained assessment of attractiveness ideals.

MATERIALS AND METHODS

Ethics Statement

The study was reviewed and approved by the School of Psychology Ethics Committee of Newcastle University.

Participants

A total of 80 heterosexual Caucasian undergraduate students aged 18–21 (40 females mean age 19.10 years, s.d. 1.01; 40 males mean age 19.84, s.d. 1.66) were recruited from Newcastle and Northumbria Universities. Participation was voluntary. However some students gained course credit. All participants gave informed consent and the aims and procedure of the study were explained beforehand. None of the students had previous experience with using the software.

Protocol

The participants used a 3D modelling software package (Daz Studio 3.1 from Daz3d.com) which allows the adjustment of photo-realistic male and female 3D models on a flat panel screen in order to modify different aspects of the body's features (see fig 1). The female 3D model used was Victoria 4.2 and the male model was Michael 4.0. The program allows the body to be rotated to allow a 360° view of the model. Along one side of the model is a set of 94 graphic sliders with which different aspects of individual body parts can be altered (using the 'Body morphs' and 'Body morphs++' add-on packages from Daz3D). When the slider is adjusted, the model simultaneously changes, providing immediate visual feedback. Sliders could be adjusted as many times as necessary and no time limit was set, so the participants could take as much time as they wished to satisfy themselves that the model was as accurate a representation as possible. The model was positioned so that the head was not visible and did not play a role in the judgements.

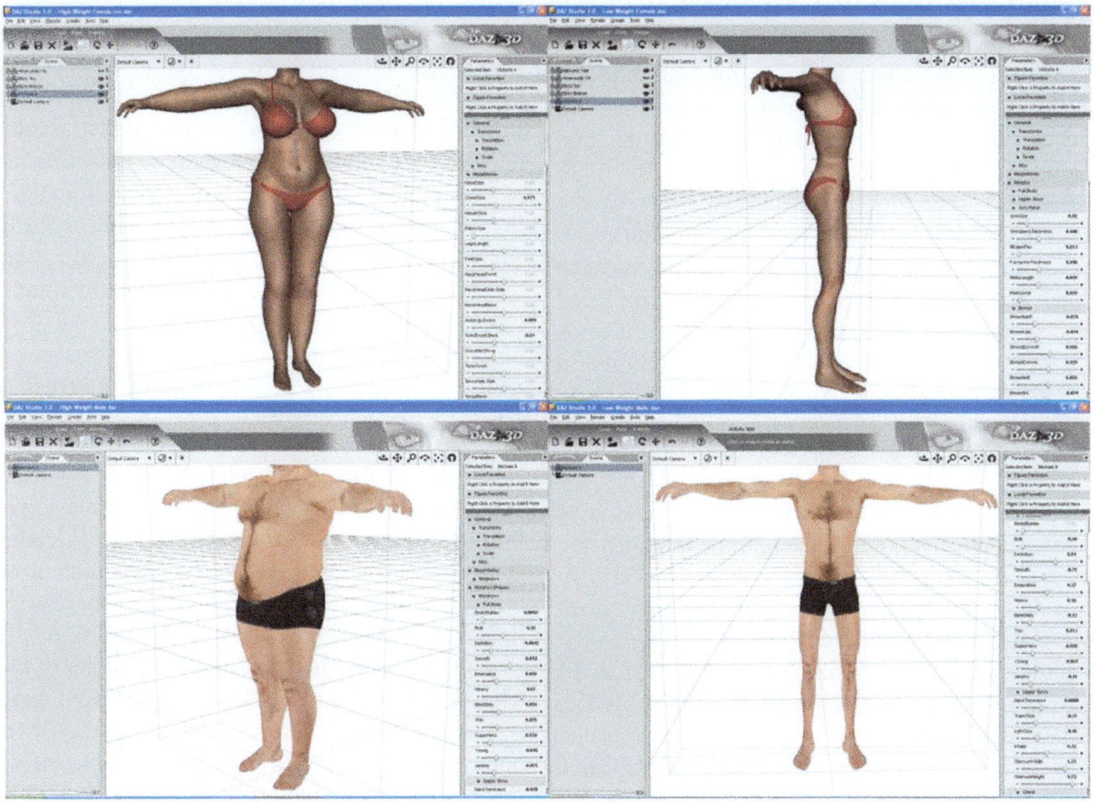

Figure 1: An example of the Daz3D interface, with examples of male and female bodies created in the software package.

Each participant created a total of four 3D bodies; two that represented their ideal body and two that represented their ideal partner's body. In each of the two conditions, the participants began with a 'heavy' body and then a 'thin' body, or vice versa. The order was counterbalanced between participants. The two estimates were averaged to render a final model. The use of fat/thin bodies as a starting point was to reduce potential anchor effects which might have occurred if participants had just begun by adjusting a normal weight body. The female "thin" body had a BMI of 14.9 and the "large" body had a BMI of 26.6. The male "thin" body had a BMI of 16.5 and the "large" body had a BMI of 37.7.

All the participants were tested on the same PC in the Body Image Lab at the Institute of Neuroscience. Participants were asked to adjust the sliders until they were satisfied that the model looked like their ideal body and then they were asked to produce their ideal partner's body. No time limit was placed upon them. Although there are 94 sliders, many of them are used for comparatively subtle adjustments to features such as the length of the ring finger on the left hand, and were not used. We ourselves had not altered these minor features in the "heavy" and "thin" bodies and we had left them at the default setting. Instead, most participants used a core set of sliders (mean 36.2 sliders, s.d. 7.8) which changed features, such as stomach depth and hip width.

After completion, a set of anthropometric measures were taken from the participants by the lead author (K.L.C.). Height was measured using the Marsden/Invicta Free Standing Height Measure and weight was measured using the Weight Watchers 8944U Heavy Duty Body Fat Analyser Scale. Using a standard tape measure, the waist and hip circumferences were measured, along with bust and under-bust circumferences if female, and chest circumference if male, following the protocols outlined in the Health Survey for England[42].

The 3D Body Analysis

The final 3D models were exported from Daz Studio, once clothing had been removed, and reopened in 3ds Max (autodesk.com), where they were set either to the height of the participant (for their own 'ideal') or to the height of the average British man (1.78 m) or woman (1.64 m) (for 'ideal partner'). First, the volumes of the 3D models were calculated by the software, scaling the body volume relative to the body height entered by the experimenter. Once the volumes were known, the weights of the models were estimated by multiplying their volumes by the density of either the average young adult female body (1.04 g/cm^3) or the average young adult male body (1.06 g/cm^3) [43], [44]. Finally, the BMI of each model was calculated as its weight (kg) divided by its height (m) squared.

Next, 3ds Max was used to slice through each model at predetermined points along its length to measure the circumference of the bodies at the chest, waist and hips in male models, and the bust, under-bust, waist and hips in female models. The software scaled the circumferences (measured in cm) to the dimensions that the bodies would have if they were real. However, the circumference measures generated by 3ds Max for the hips in male bodies and bust and hips in female bodies tend to be larger than the same measurements taken from real bodies. This is because 3ds max calculates the path length around - each slice which includes, for example, the cleft in the bust or buttocks. In comparison, a tape measure looped around the bust or hips will straddle these gaps, and so will produce a shorter distance. To compensate for these effects, we screen grabbed the cross-sectional slices of the bust or hips in 3ds Max and imported them into ImageJ (http://rsbweb.nih.gov/ij/). There we used the lasso drawing tools to replicate the path that a tape measure would take when placed around the bust or buttocks, and used the measurement tools to calculate the path length which better reflected a real world measurement with a tape measure.

Test-Retest Reliability

A potential weakness of this methodology is the question of whether the participants can reliably manipulate the software controls to produce the body size and shape that they want. To answer this question we ran a test-retest experiment in which participants were asked to repeat the modelling task. We asked 15 Caucasian female participants (average age 22.57, s.d. 2.76) and 15 Caucasian male participants (average age 23.21, s.d. 2.84) to set their ideal body size and shape using the same methodology as described above. They then repeated the same tasks the following day. The Pearson's correlation between the BMI values of the models that participants set on the two days was highly statistically significant ($r=0.99$, $p<.001$) and a paired-samples t-test of the bodies' BMIs showed no significant differences between the settings for the bodies on the two days (paired T-test: $t(25)=1.69$, $p=.103$).

RESULTS

In this results section we first show that that there are significant differences in size and shape between the actual bodies of the participants and their ideals. We then show that these ideal bodies differ from the expected shape of real bodies of the same BMI, implying an explicit choice for specific sizes *and* shapes in their ideal bodies. Finally, we show that the ideal size and shape for both a male and a female body is shared by both our male and female participants (i.e. there is no gender based difference on what constitutes an attractive male or female body).

Comparisons of Participants' Actual BMI versus Ideal BMI

A summary of the anthropometric data from the participants' actual and ideal bodies are shown in table 1 and examples of the ideal bodies are shown in figure 2. A comparison of the BMI values for the male participants' actual body and their male ideal body showed a significant increase in the BMI of the ideal body (paired T-test: $t(39)=-2.26$, $p=.029$; effect size $r=0.34$; power to detect at two-sided alpha of 0.05=0.59)[45]. This difference is also significant for the female participants who showed a significant reduction in the BMI of their ideal body (paired T-test: $t(39)=7.49$, $p<.0001$; effect size $r=0.77$; power to detect at two-sided alpha of 0.05>.99).

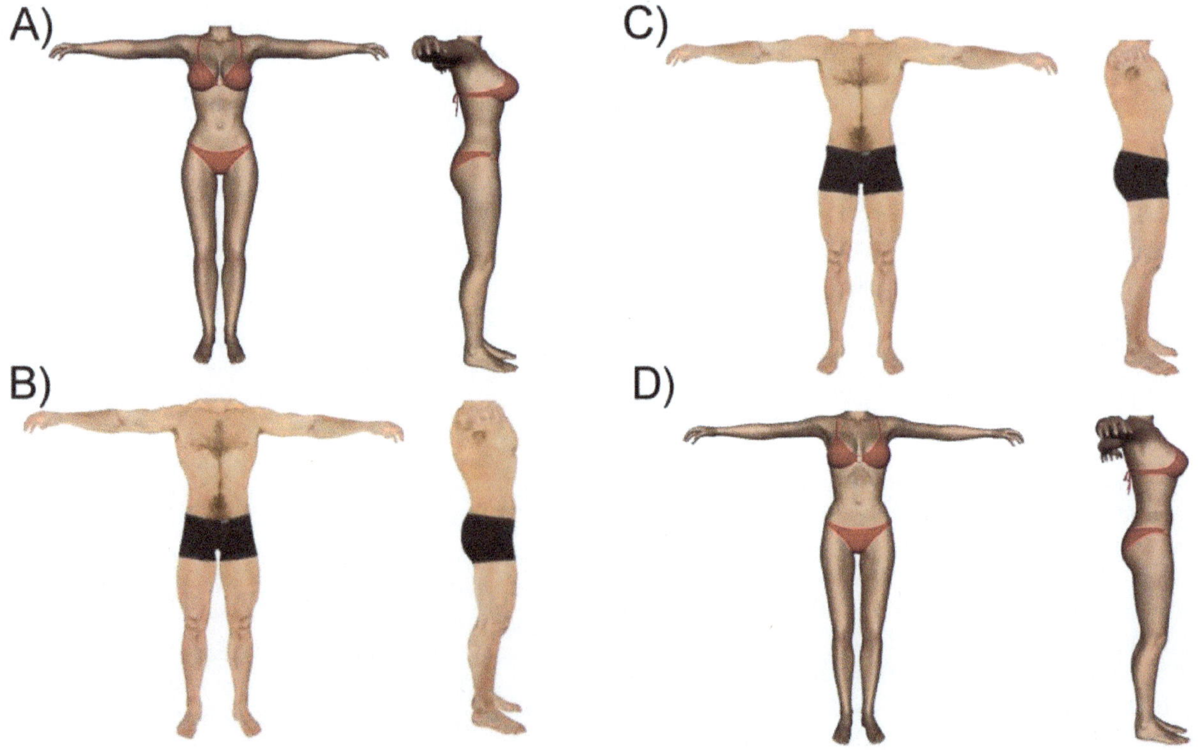

Figure 2: Figure 2 shows examples of the bodies set by the female participants (A & B) and the male participants (C & D).

Table 1: **Summarises the anthropometric measures taken from the male and female bodies in this study**

	BMI	Bust/Chest	Under Bust	Waist	Hips	WCR	WHR
Female Actual Body							
Average	21.7	87.4	75.93	72.91	99.4	0.86	0.73
SD	2.07	5.17	5.6	5.48	5.36	0.2	0.19
Female's Ideal Body							
Average	18.85	93.97	68.33	61.12	87.89	0.67	0.70
SD	1.75	8.24	4.06	3.38	6.52	0.09	0.04
Male's Ideal Female Body							
Average	18.82	90.02	69.2	61.95	84.82	0.69	0.73
SD	1.56	4.73	5.79	5.79	4.92	0.05	0.04
Male Actual Body							
Average	24.54	97.74	-	86.12	98.76	0.88	0.87
SD	3.38	9.21	-	9.47	7.93	0.04	0.06
Male's Ideal Body							
Average	25.86	111.26	-	82.00	91.17	0.74	0.87
SD	3.95	9.44	-	9.17	9.59	0.05	0.04
Female's Ideal Male Body							
Average	24.46	104.16	-	80.57	90.81	0.77	0.86
SD	2.9	7.43	-	7.22	7.15	0.05	0.03

General Patterns of Shape Change Estimated by WHR and WCR Comparing Male with Female Actual versus Ideal Bodies

Fig. 3A shows a plot of the actual and ideal WHRs of male and female observers. It shows that WHRs are generally larger for male bodies than for female bodies. Moreover, males appear to prefer a more tubular shape in their lower torso, indexed by a higher WHR, as their ideal. In comparison, females appear to desire a curvier lower torso shape, as indexed by lower WHR values for their ideal. Fig. 3B shows an equivalent plot for WCR. Again, male observers have higher WCR's overall compared to females. However, both genders appear to desire larger circumference chests than waists by about the same proportion in their ideal figures.

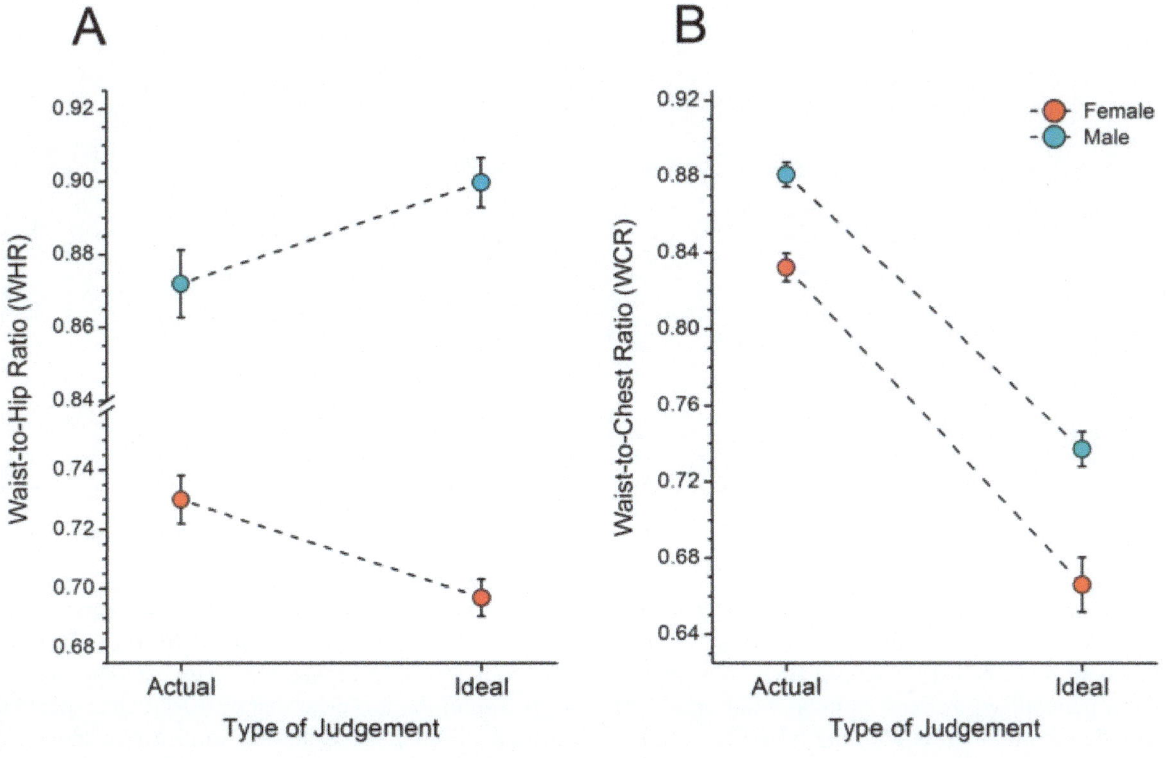

Figure 3: Fig. 3A shows a plot of the average actual and ideal WHRs of male and female observers.

To quantify these effects we computed a between-subjects (i.e. gender: male versus female) ANOVA and a within-subjects (i.e. condition: actual versus ideal) mixed ANOVA separately for WHR and WCR.

For WHR, we found a statistically significant main effect of gender ($F_{1,117}$=506.48, p<.0001), but not of condition ($F_{1,117}$=0.26, p=.613). The interaction between gender and condition was statistically significant ($F_{1,117}$=17.34, p<.0001). In order to identify which differences contributed to the interaction term, we computed post-hoc differences of least square means using the Tukey-Kramer correction to compensate for multiple statistical comparisons. We found that the female participants' ideal WHRs were significantly less than their actual WHRs (t_{117}=3.30, p=.0069; r=0.29) and that the males' ideal WHRs were significantly higher than their actual WHRs (t_{117}=−2.59, p=.05; r=0.23).

For WCR, we found a statistically significant main effect of gender ($F_{1,\ 156}$=37.15, p<.0001) and condition ($F_{1,\ 156}$=249.13, p<.0001). However, the interaction between gender and condition was not statistically significant ($F_{1,156}$=1.31, p=.254). We computed post-hoc differences of least square means using the Tukey-Kramer correction to compensate for multiple statistical comparisons and found that females' and males' ideal WCRs were significantly lower than their actual WCRs (t_{156}=11.97, p<.0001; r=0.69 and t_{156}=10.35, p<.0001; r=0.64 respectively).

The Non-Linear Co-Variation of Body Mass and Shape

Our analysis suggests that the ideal body size and shape of both the male's and female's ideals differs from the corresponding actual bodies. However, a possible confound is that in real life, body shape and body size tend to co-vary in a non-linear way (i.e. a body with a particular BMI will have a particular shape), with different parts of the body changing size at different rates with changing BMI. We have already illustrated this relationship in women's bodies in several previous studies [17], [46], [47] and we can illustrate this co-variation here in male and female bodies by plotting the torso width of a set of 122 young Caucasian men (average age 27.4, s.d. 11.9) and 60 young Caucasian women (average age 26.1 years, s.d. 6.7) who agreed to be photographed to provide stimuli for a number of studies of physical attractiveness (see [8], [46], [48]). The widths of 31 slices taken through the torso of 2D frontal images of the participants were obtained, along with their respective BMIs (see Fig. 4). The location for each slice was standardized across participants by equally dividing the distance between fixed anatomical landmarks (the acromio-clavicular joint and the perineum) into 30 equal partitions. This is illustrated in Fig 4, which shows a plot of the width of the right side of the torso, starting from the midline, for the average male and female body at five different BMI levels.

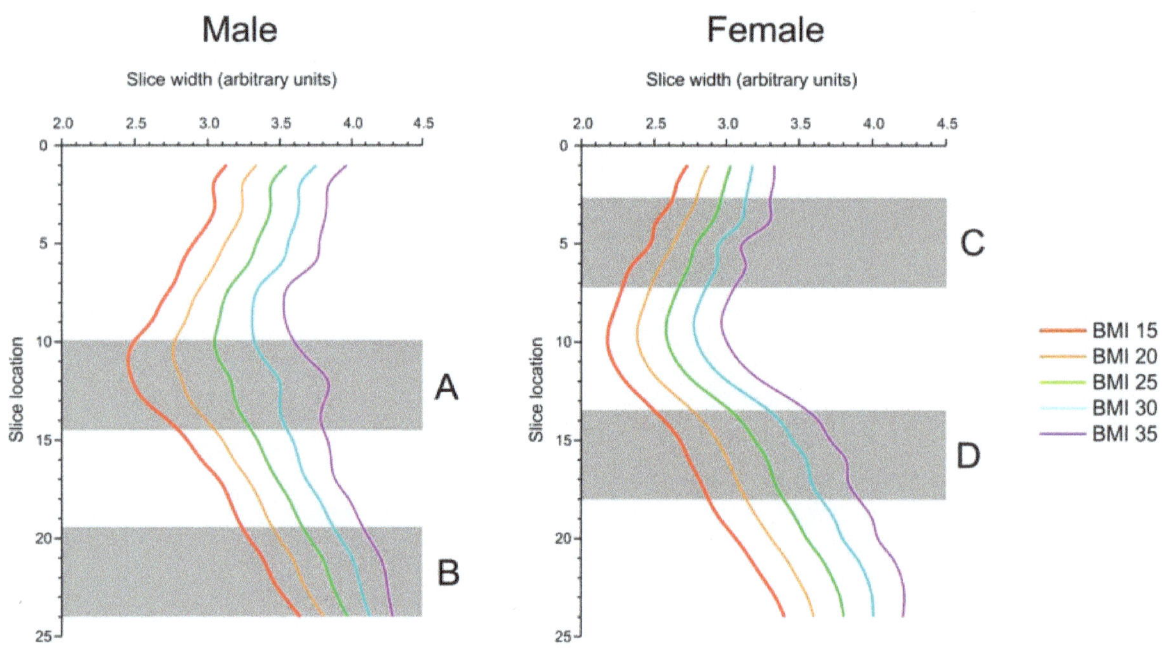

Figure 4: Plots of the width of the right side of the torso, starting from the midline, for the average male and female bodies at five different BMI levels.

A simple regression can then be used to estimate the relationship between each slice width and BMI. The key feature to appreciate about fig. 4 is that increasing BMI is associated not only with a generalized increase in torso width, reflected by the systematic separation of one profile from the next, but also with a non-linear component to the change in body shape. This non-linear component is illustrated by considering, for example, the male torso outline in sub-regions A (near the waist) and B (the lower hip) in Fig. 4. In region A, as BMI increases from 15 to 35, the contour of the waist changes from convex to concave. Over the same BMI rage, the slope of the line from lower to higher hip slices becomes less and less steep. Therefore, it is clear that by selecting an ideal body with a different BMI, participants are implicitly selecting a complex change in the shape of the ideal body. There are similar non-linear shape changes in the female torso that can be seen in sub-regions C (the upper chest) and D (upper hip) of Fig. 4.

In the current study we seek to answer the question of how different are people's own ideal body shapes compared to the shape they currently have, as well as the ideal body they would seek in a partner. The complex shape changes illustrated in Fig. 4 that occur as a result of changing BMI demonstrate that this question needs to be carefully refined. It could be that when people pick an ideal body shape, what they are really doing is picking a body which for them represents a body with an ideal BMI. Not only is this choice necessarily associated with a change in the width of the body, but also there are additional shape changes caused by the fact that fat is not deposited equally around the body. Therefore, in addition to any width changes represented in the ideal body, there are also non-linear shape changes associated with a change in BMI as illustrated in Fig. 4. An alternative possibility, when people are asked to pick an ideal body shape, is that they may choose a shape which goes beyond any changes attributable to a change in BMI alone, including the linear and non-linear components. Therefore, in the analysis that follows, we address this confounding problem directly.

Since the BMI of both genders' ideals is different from their actual BMI, we can calculate what proportion of the change in torso shape of their ideal body is attributable just to the change in BMI alone. In other words we can predict the component of shape change in the ideal which is predicted by the BMI of the ideal body shape selected. We can then compute the difference between the bust/chest, under-bust, waist and hip circumferences of the ideal image and the equivalent circumferences computed on the basis of the BMI of the ideal and then ask whether, on average, these are significantly different from zero. If this population of differences is not significantly different from zero, this suggests that the shape of the body that participants choose as their ideal is no different from merely choosing a higher or lower BMI. However, if the population of differences in circumferences is significantly different from zero, this means that the shape of the bodies that participants choose as their ideal is different from what they would achieve by merely selecting a higher or lower BMI.

The regression analyses to estimate the BMI shape change effect are based on circumference measures taken from 120 male and 120 female volunteers. The females were measured at bust, under-bust, waist and hips and the males at chest, waist and hips. The average age of the female volunteers was 20.3 years s.d. 3.5 and the average age of the male volunteers was 20.7 years s.d. 2.1. For each gender, we computed separately the regression between BMI and chest/bust, under bust waist and hip respectively, and then used these regression equations to estimate the expected circumferences in the ideal bodies chosen, based purely on their BMI.

Are the Circumferences of Ideal Male Bodies Different from those Expected from their BMIs?

Table 2 below shows the descriptive statistics for the difference between the circumferences of the 3D model settings for the ideal male body shapes set by both male and female participants and those predicted from the BMI of the ideal models. Table 2 shows that both male and female participants set ideal body shapes which have chest circumferences substantially larger than the chest circumference attributable to the lighter BMI ideal set in section above. Moreover, the commensurate waist and hip circumferences are both substantially smaller than the values predicted on the basis of the ideal BMI that was selected in each case.

Table 2: Summary of the comparison between the ideal male body set by the participants and the body predicted by the BMI

Group	Body Slice	Average Difference in Circumference (cm)	t-test value	p value	r value	Power
Male's ideal male body	Chest	11.04 (0.86)	12.76	<.0001	0.90	>.99
	Waist	−12.92 (0.69)	−18.67	<.0001	0.95	>.99
	Hips	−9.64 (0.59)	−16.46	<.0001	0.93	>.99
Female's ideal male body	Chest	6.93 (0.92)	7.52	<.0001	0.77	>.99
	Waist	−10.99 (0.52)	−21.08	<.0001	0.96	>.99
	Hips	−7.60 (0.46)	−16.55	<.0001	0.94	>.99

To further explore this result we carried out t-tests for each set of circumferences (i.e. chest, waist, hips) for the populations of differences (see table 2), where the null hypothesis was a mean of zero. All are statistically significant at $p<.05$, even after applying a Bonferroni correction for multiple comparisons.

Are there Differences in the Circumferences of the Ideal Male Bodies set by the Male and Female Participants?

The results from the T-tests of location above show that, the average shape of the ideal female bodies set by male and female participants differs significantly from the shape that would be predicted based solely on the BMI of the ideals. Next, we test whether the shapes of these ideals differ when comparing the settings made by male versus female participants. To address this question, we used a 2-factor, repeated-measures GLMM, where factor 1 was the gender of the participant (male, female) and factor 2 was the circumference (chest, waist and hip). There was no main effect of gender ($F_{1,234}=0.01$, $P=.938$). The main effect of circumference was significant ($F_{2,234}=523.42$, $p<.0001$) as was the interaction between gender and circumference ($F_{2,234}=12.78$, $p<.0001$). To determine which individual ideal shape measures differed between male and female participants, we calculated post-hoc differences of least square means using the Tukey-Kramer correction to compensate for multiple statistical comparisons. The difference between male and female settings of chest circumference was statistically significant ($p<.0001$), whereas the differences for waist and hip were not.

Is the Ideal Male Body Different in Size and Shape for the Male and Female Participants?

An independent t-test shows that the ideal male BMI set by the female participants is not significantly different from that set by the male participants (t(78)=1.81, p=0.074; effect size r=0.20; power to detect at two-sided alpha of 0.05=0.44). The WHR of the two bodies were also not significantly different: t(78)=1.43, p=.229; effect size r=0.16; power to detect at two-sided alpha of 0.05=0.20), but WCR was significantly different (t(78)= −3.09, p=.003; effect size r=0.33; power to detect at two-sided alpha of 0.05=0.67).

Are the Circumferences of Ideal Female Bodies different from those Expected from their BMIs?

Table 3 below shows the descriptive statistics for the difference between the circumferences of the 3D model settings for the ideal female body shapes set by both male and female participants and those predicted from the BMI of the ideal models. Table 3 shows that both male and female participants set ideal body shapes which have bust circumferences substantially larger than the bust circumference attributable to the lighter BMI ideal set above. Moreover, the commensurate under-bust, waist and hip circumferences are substantially smaller than the values predicted on the basis of the ideal BMI that was selected in each case.

Table 3: Summary of the comparison between the ideal female body set by the participants and the body predicted by the BMI

Group	Body Slice	Average Difference in Circumference (cm)	t-test value	p value	r value	Power
Female's ideal Female body	Bust	10.78 (2.97)	3.62	.0008	0.50	>.94
	Under Bust	−3.73 (0.60)	−6.19	<0.0001	0.70	>.99
	Waist	−6.43 (0.47)	−13.68	<.0001	0.91	>.99
	Hips	−0.63 (0.77)	−0.81	.42	0.13	.12
Male's ideal Female body	Bust	6.88 (0.81)	8.46	<.0001	0.80	>.99
	Under Bust	−2.81 (0.81)	−3.48	.001	0.50	.92
	Waist	−5.53 (0.48)	−11.56	<.0001	0.88	>.99
	Hips	−3.64 (0.57)	−6.40	<.0001	0.72	>.92

T-tests of location for the populations of differences, where the null hypothesis was a mean of zero, are all statistically significant at p<.05, even after applying a Bonferroni correction for multiple comparisons, with the exception of female settings for the hip circumference.

Are there Differences in the Circumferences of the Ideal Female bodies set by the Male and Female Participants?

The results from the T-tests of location above show that the average shape of the ideal female bodies set by male and female participants differs significantly from the shape that would be predicted based solely on the BMI of the ideals. Next, we test whether these ideal body shapes differ when comparing the settings made by male versus female participants. As before, to address this question, we used a 2-factor repeated measures GLMM, where factor 1 was the gender of the participant (male, female) and factor 2 was the circumference (bust, under-bust, waist and hip). There was no main effect of gender ($F_{1,78}$=1.67, P=.201). The main effect of circumference was significant ($F_{3,234}$=63.68, p<.0001), but there was no significant interaction between gender and circumference ($F_{3,234}$=2.43, p=.066).

Is the Ideal Female Body Different in Size and Shape for the Male and Female Participants?

An independent t-test shows that the ideal female BMI set by the female participants is not significantly different from the ideal female BMI set by the male participants (t(78)=0.09, p=0.93; effect size r=0.01; power to detect at two-sided alpha of 0.05=0.05). The WBR of the two bodies were also not significantly different (t(78)= −3.64, p<.001; effect size r=0.38; power to detect at two-sided alpha of 0.05=0.91), but WHR was significantly different (t(78)= −3.64, p<.001; effect size r=0.38; power to detect at two-sided alpha of 0.05=0.91).

DISCUSSION

What is the Ideal Female Body Size and Shape?

Both male and female participants created an ideal body that was significantly different in body size relative to their own. The female participants significantly reduced the body size and the male participants increased it. Although some studies have suggested BMI is the primary predictor of female attractiveness and that shape is of marginal importance (e.g. [4], [12], [17], [46], this study suggests that body shape is a significant factor, at least with respect to the perception and creation of ideals. Shape and body mass co-vary (e.g. [17],[47], [49], [50], but by controlling for the expected changes which occur with changing BMI, we show that both male and female participants nevertheless produce ideals with a specific shape which is independent of the ideal's BMI.

The female participants' ideal female body has a BMI which is significantly lower than their actual BMI. Consistent with this lowered BMI, there is a general narrowing of the torso, with the hips, waist and chest (excluding the bust) reducing in circumference (i.e. the volume of the body is reduced). The actual BMI values of the female participants all fall within the normal BMI range (18.5-24.9), with the majority around the middle part of this scale [51], [52]. While their ideal female body is also just within the normal range, it is only just above the underweight category. However, this is consistent with previous studies in which photographs of women's bodies have been rated for attractiveness which have suggested an ideal BMI of as low as 18–20 for Western male and female observers [16], [40]. Only 1 of the 40 female participants wanted an ideal BMI above their actual BMI. This low ideal BMI is similar to the BMI reported for female models appearing in the media [53], [54], a result consistent with the hypothesis that low BMI women in the media influence body size preferences [55]–[58] and contribute to the high proportion of women who show dieting and weight loss behaviours even though they have a normal BMI [59]–[61]. The participants in our experiment are university students and are therefore a relatively young group who may be more sensitive to media influence on body ideals than older people (e.g. [62], [63]). However in previous attractiveness studies which have used participants with wide age ranges we have not found differences in their ideal size and shape [64] suggesting the findings in the current study are representative of the general population.

In contrast to the narrowing of the rest of the female body, the "ideal" bust increases in size (as indexed by bust circumference). Previous studies have linked relative bust size to circulating estrogen levels, with the suggestion that a large bust and a narrow waist should indicate high levels of estrogen and therefore be regarded as attractive [25]. A number of studies have suggested that female bodies with a larger bust are considered to be more attractive [20], [65] and breast augmentation is the most common cosmetic surgical procedure in the UK and US [66]. The large bust and low BMI set by both the male and female participants also reflects the size and shape of glamour models in men's magazines which are often taken as a proxy for a cultural ideal of female beauty [53], [54].

The increase in bust size and narrowing of the torso between the female participants' actual body and their ideal changes the upper body shape (as indexed by WCR and illustrated in figure 3). The female participants also narrow their hips as well as their waist, but because there is a relatively greater narrowing of the waist, the lower torso also increases in curvature (as indexed by WHR). There is less change in the WHR than in WCR, but this may be because the WHR of the participants' actual body is already quite close to a value of 0.7 which has been suggested to be optimal for health and fertility and thus also for attractiveness [11], [13].

What is the Ideal Male Body Size and Shape?

Unlike the female thin ideal body, the ideal male body is comparatively heavy, falling at the boundary of the normal to overweight categories of the BMI scale. However, these are not bodies that look over-weight, but instead are big and muscular. In fact, our calculation of their BMI is probably an under estimation, because we are assuming that the bodies have the average density for young men (i.e. the average balance of fat to muscle). As muscle is approximately 20% denser than fat, this would under-estimate the mass of a more muscular body such as the male ideals set in this experiment. This result is consistent with previous studies, which have suggested that muscularity (and the associated perception of dominance and strength) is the primary determinant of male attractiveness [8], [10], [30]. Whereas there is a tendency for women to diet to achieve their ideal body, young men are more likely to be influenced by magazines to build up a bigger, more muscular body [31], [66]. So although the male ideal body is heavier, the additional weight is muscle rather than fat. As discussed above, BMI is a measure of body weight scaled for height and not a direct measure of percentage body fat. Its use in epidemiological studies is due to its ease of administration. The ideal male body set by both male and female participants is lean with high muscle definition (which requires a percentage body fat below 9–12%). Our participants' male ideal is both muscular and low in body fat.

The male participant's ideal body shows an increase in chest circumference (relative to their actual body) and a reduction in the waist and hips to produce a V-shaped upper body. Previous studies have also suggested that men prefer a body that is more muscular than the one they actually possess [31], [67]–[69]. It is suggested that a v-shaped upper body is a key predictor of male attractiveness judgements because this indicates upper body strength [8], [10], [30], [31], [70]. By contrast, the ideal lower body is narrowed relative to the actual body making it less curvy and more straight-up and down. This is the opposite of what is found for the ideal female bodies.

Do Men and Women Share Body Ideals?

The preferences for the ideal female body are broadly similar between the two genders. They both prefer the same low BMI and a relatively curvaceous body with WCR and WHR with values around 0.7. There is also general agreement between the genders on the ideal male body; this male ideal has a relatively large body with a V-shape upper torso and a narrow waist and hips. This is consistent with attractiveness rating studies which to show a strong correlation between male and female attractiveness ratings of male and female bodies (i.e. both genders seem to rate bodies of both genders the same way) [8], [17], [46], [48].This can be explained by mate selection theory which suggests that individuals will not only be able to judge the attractiveness of members of the opposite sex, but will also know their own attractiveness relative to other members of the same sex (i.e. their competitors) [3]. This information allows an individual to concentrate on potential partners of the same attractiveness as themselves, thus avoiding both unsuccessful courtship of a more attractive partner (potentially wasteful in time and resources) and accepting a less attractive partner (with a potentially negative impact on future reproductive success). Thus an individual must be able to assess bodies of their own gender using the same attractiveness criteria as the opposite sex, and by extension, must therefore have a good idea of the opposite gender's ideal partner. So our female and male participants should share the same ideals for both male and female bodies.

An alternative explanation would be that the ideals are influenced by a common media environment which pushes them towards the same concept of the ideal body. However, there are subtle gender-specific differences in the media images seen in the magazines targeted at men and women. For the male body, magazines aimed at a male audience contain male models which are more muscular than those aimed at a female audience [31], [66]. For the female body, female models in women's magazines are slimmer and have a smaller bust than female models in men's magazines [53], [54]. This would suggest that there should be systematic differences between the ideals favoured by the two genders.

This is partially what we find here. The male body selected by the male participants is indeed more muscular than the ideal male body chosen by the female participants. However, in the case of the ideal female body both men and women prefer a female body with the same low BMI, but the female participants prefer a larger bust size than the male participants. This directly contradicts what would be expected from the size and shape of the female models in their respective gender-specific media; the men should prefer a heavier female body than the women and a larger bust.

Previous studies have focussed on body size in women's bodies. These suggest that although women overestimate the level of female thinness desired by men (e.g. [36], [37], [71], [72]), when asked to simply rate images without reference to what they think men would find most attractive, women and men have the same ideal BMI for female attractiveness [4], [12], [22], [46], [48]. Our study asked the female participants what they thought was the ideal body size and shape, and if we had asked them to choose what they thought a man would choose we might have got a difference between this body and the male judgement of female ideal body size.

That still leaves the question of why the difference exists in male and female preferences for upper body shape; female participants prefer a larger bust in their ideal female body than men, and male participants prefer a larger chest in their ideal male body than women. This may be linked to within gender competition for status and prestige [31], [70]. Many forms of prestige and status competition are between members of the same gender. Such a competition could produce a runaway process in which a physical feature becomes increasingly exaggerated over time due to competition between same-gender individuals. As this is a within gender competition the possibility exists that these processes will lead to divergence between preferences of the two genders for a specific feature, such as muscularity in men or bust size in women [10], [31], [70].

An alternative socio-cultural explanation would emphasise how a culture-specific female ideal body size and shape potentially exerts a particularly strong influence on women's concept of what they should aspire to[73], [74]. This ideal, which is impossible for most women to achieve, is suggested to lead to body dissatisfaction and potentially in some cases to eating disorders [73], [75]. Women who do not conform to this ideal are more likely to receive negative comments and discrimination [76], [77], which serves to condition the importance of physical appearance as part of their estimation of self-worth [78], [79]. In this context, the importance of physical appearance is potentially clearest to young women (such as our participants) who are more likely to be actively involved in the mate selection process. This reinforcement of the perfect female ideal could potentially lead to an exaggeration of the internal representation of some of the ideal physical features [79]–[81], such as bust size in our female participants relative to the males. A similar process may also explain the exaggeration of the upper body musculature of the male ideals by the male participants. The propagation of the highly musculature male ideal through gender specific magazines [31],[66] and its reinforcement in young men by experience of mate competition with other men [1], may promote an exaggerated idea of the ideal male body shape.

CONCLUSIONS

The combination of the 3D morphing software and the regression analysis shows that the ideals for both genders have a specific body size (as indexed by BMI) and shape. For both sexes, the primary predictor of female beauty is a relatively low BMI combined with a relatively curvaceous body, whereas the features important for the male ideal are a slightly heavier, muscled body with a specific V-shaped upper body. Although, the results suggest a largely consistent preference for an ideal male and female body size and shape across both genders, but with subtle differences based on an own gender exaggeration of upper body shape.

ACKNOWLEDGMENTS

Thanks to Ellen Roberts and Laura Moxon for help in collecting some of the data.

REFERENCES

1. Buss DM (1987) Sex differences in human mate selection criteria: An evolutionary perspective. In: Crawford C, Smith M, Krebs D, editors. Sociobiology and psychology: Ideas, issues and application. Hillsdale, NJ Erlbaum. 335–351.
2. Thornhill R, Gangestad SW (1999) Facial attractiveness. Trends Cogn Sci 3: 452–460
3. Buss DM (2003) The evolution of desire: Strategies of human mating (rev. ed.). New York: Basic Books.
4. Fan JT, Liu F, Wu J, Dai W (2004) Visual perception of female physical attractiveness. Proc R Soc Lond B Biol Sci 271: 347–352
5. Fan JW, Dai W, Liu F, Wu J (2005) Visual perception of male body attractiveness. Proc R Soc Lond B Biol Sci 272: 219–226
6. Henss R (1995) Waist-to-hip ratio and attractiveness. A replication and extension. Pers Individ Dif 19: 479–488
7. Henss R (2000) Waist-to-hip ratio and attractiveness of the female figure. Evidence from photographic stimuli and methodological considerations. Pers Individ Dif 28: 501–513
8. Maisey DS, Vale ELE, Cornelissen PL, Tovée MJ (1999) Characteristics of male attractiveness for women. Lancet 353: 1500.
9. Puhl RM, Boland FJ (2001) Predicting female physical Attractiveness: Waist-to-hip ratio versus thinness. Psychol Evol Gender 3: 27–46
10. Sell A, Cosmides L, Tooby J, Sznycer D, von Rueden C, et al. (2008) Human adaptations for the visual assessment of strength and fighting ability from the body and face. Proc R Soc Lond B Biol Sci 276: 575–584
11. Singh D (1993) Body shape and women's attractiveness - the critical role of waist-to-hip ratio. Human Nature 4: 297–321

12. Smith KL, Cornelissen PL, Tovée MJ (2007) Color 3D bodies and judgments of human female attractiveness. Evol Hum Behav 28: 48–54

13. Streeter SA, McBurney DH (2003) Waist-hip ratio and attractiveness. New evidence and a critique of 'a critical test.'. Evol Hum Behav 24: 88–98

14. Swami V, Caprario C, Tovée MJ, Furnham A (2006) Female physical attractiveness in Britain and Japan: A cross-cultural study. Eur J Pers 20: 69–81

15. Thornhill R, Grammer K (1999) The body and face of woman: One ornament that signals quality? Evol Hum Behav 20: 105–120

16. Tovée MJ, Reinhardt S, Emery JL, Cornelissen PL (1998) Optimum body-mass index and maximum sexual attractiveness. Lancet 352: 548.

17. Tovée MJ, Hancock BJ, Mahmoodi S, Singleton BR, Cornelisson PL (2002) Human female attractiveness: Waveform analysis of body shape. Proc R Soc Lond B Biol Sci 269: 2205–2213

18. Tovée MJ, Edmonds L, Vuong QC (2012) Categorical perception of human female physical attractiveness. Evol Hum Behav 33: 85–93

19. Tovée MJ, Benson PJ, Emery JL, Mason SM, Cohen-Tovée EM (2003) Measurement of body size and shape perception in eating-disordered and control observers using body shape software. Brit J Psychol 94: 501–516

20. Prantl L, Grundl M (2010) Males prefer a larger bust size in women than females themselves: An experimental study on female bodily attractiveness with varying weight, bust size, waist width, hip width, and leg length independently. Aesth Plast Surg DOI 10.1007/s00266–011–9669–0.

21. Courtiol A, Picq S, Godelle B, Raymond M, Ferdy J-B (2010) From preferred to actual mate characteristics: The case of human body shape. PLoS ONE 5: e13010.

22. Tovée MJ, Cornelissen PL (2001) Female and male perceptions of female physical attractiveness in front-view and profile. Brit J Psychol 92: 391–402

23. Jasienska G, Ziomkiewicz A, Lipson SF, Ellison PT, Thune I (2004) Large breasts and narrow waist indicate high reproductive potential in women. Proc R Soc Lond B Biol Sci 271: 1213–1217

24. Manson JE, Willet WC, Stampfer MJ, Colditz GA, Hunter DJ, et al. (1995) Body weight and mortality among women. N Engl J Med 333: 677–685

25. Willet WC, Manson JE, Stampfer MJ, Colditz GA, Rosner B, et al. (1995) Weight, weight change and coronary heart disease in women: Risk within the 'normal' weight range. J Am Med Ass 273: 461–465

26. Frisch RE (1988) Fatness and fertility. Sci Am 258: 88–95

27. Lake JK, Power C, Cole TJ (1997) Women's reproductive health: the role of body mass index in early and adult life. Int J Obes 21: 432–438

28. Reid RL, Van Vugt DA (1987) Weight related changes in reproductive function. Fertil Steril 48: 905–913

29. Zaadstra BM, Seidell JC, van Noord PAH, Velde ER, Habbema JDF, et al. (1995) Fat and fecundity: Prospective study of effect of body fat distribution on conception rates. BMJ 306: 484–487

30. Hönekopp J, Rudolph U, Beier L, Liebert A, Constanze Müller C (2007) Physical attractiveness of face and body as indicators of physical fitness in men. Evol Hum Behav 28: 106–111

31. Frederick DA, Fessler DMT, Haselton MG (2005) Do representations of male muscularity differ in men's and women's magazines? Body Image 2: 81–86

32. Prospective Studies Collaboration (2009) Body-mass index and cause-specific mortality in 900,000 adults: collaborative analyses of 57 prospective studies. Lancet 373: 1083–1096

33. Emerging Risk Factors Collaboration (2011) Separate and combined associations ofbody-mass index and abdominal adiposity with cardiovascular disease: collaborative analysis of 58 prospective studies. Lancet377: 1085–95

34. Lean MEJ, Han TS, Morrison CE (1995) Waist circumference as a measure for indicating need for weight management. BMJ 311: 158–61

35. Zhu S, Wang Z, Heshka S, Heo M, Faith MS, Heymsfield SB (2002) Waist circumference and obesity-associated risk factors among whites in the third National Health and Nutrition Examination Survey: clinical action thresholds. Am J Clin Nutr 76, 743–749.

36. Fallon AE, Rozin P (1985) Sex differences in perceptions of desirable body shape. J Abnorm Psychol94: 102–105

37. Rozin P, Fallon AE (1988) Body-image, attitudes to weight, and misperceptions of figure preferences of the opposite sex – a comparison of men and women in two generations. J Abnorm Psychol 97: 342–345

38. Oakes ME, Slotterback C, Mecca EK (2003) Gender differences in perceptions of normal body weights.Curr Psych 22: 93–99

39. Cornelissen PL, Hancock PJB, Kiviniemi VW, George HR, Tovée MJ (2009) Patterns of eye-movements when Male and Female observers judge female attractiveness, body fat and waist-to-hip ratio.Evol Hum Behav 30: 417–428

40. Tovée MJ, Cornelissen PL (1999) The mystery of female beauty. Nature 399: 215–216

41. Tovée MJ, Swami V, Furnham A, Mangalparsad R (2006) Changing perceptions of attractiveness as observers are exposed to a different culture. Evol Hum Behav 27: 443–456

42. Health Survey for England (2008) National Centre for Social Research and University College London. Department of Epidemiology and Public Health. UK Data Archive. Colchester: Essex.

43. Pollock ML, Laughridge EE, Coleman B, Linnerud AC, Jackson A (1975) Prediction of body density in young and middle-aged women. J Appl Physiol 38: 745–749

44. Kryzwicki HJ, Chinn KSK (1967) Human body density and fat of an adult male population as measured by water displacement. Am J Clin Nutr 20: 305–310

45. Rosehthal R, Rosnow RL, Rubin DB (2000) Contrasts and effect sizes in behavioural research: a correlational approach. Cambridge: Cambridge University Press.

46. Tovée MJ, Maisey DS, Emery JL, Cornelissen PL (1999) Visual cues to female physical attractiveness. Proc R Soc Lond B Biol Sci 266: 211–218

47. Cornelissen PL, Tovée MJ, Bateson M (2009) Patterns of subcutaneous fat deposition and the relationship between body mass index and waist-to-hip ratio: Implications for models of physical attractiveness. J Theor Biol 256: 343–350

48. Swami V, Tovée MJ (2005) Male physical attractiveness in Britain and Malaysia: A cross-cultural study.Body Image 2: 383–393

49. Wells JCK, Treleaven P, Cole T J (2007) BMI compared with 3-dimensional body shape: The UK National Sizing Survey. Am J Clin Nutr 85: 419–425

50. Wells JCK, Cole TJ, Bruner D, Treleaven P (2008) Body shape in American and British adults: Between-country and inter-ethnic comparisons. Int J Obes 32: 152–159

51. World Health Organization (1995) Physical status: the use and interpretation of anthropometry. Report of a WHO Expert Committee. WHO Technical Report Series 854. Geneva: World Health Organization.

52. World Health Organization (2000) Obesity: preventing and managing the global epidemic. Report of a WHO Consultation. WHO Technical Report Series 894. Geneva: World Health Organization.

53. Tovée MJ, Mason SM, Emery JL, McCluskey SE, Cohen-Tovee EM (1997) Supermodels: stick insects or hourglasses? Lancet 350: 1474–1475

54. Voracek M, Fisher ML (2002) Shapely centrefolds? Temporal change in body measures: trend analysis. BMJ 325: 1447–1448

55. Stice E, Spangler D, Agras WS (2001) Exposure to media-portrayed thin-ideal images adversely affects vulnerable girls: A longitudinal experiment. J Soc Clin Psychol 20: 271–289

56. Spettigue W, Henderson KA (2004) Eating disorders and the role of the media. Can Child Adolesc Psychiatr Rev 13: 16–9

57. Field AE, Austin SB, Camargo CA, Taylor CB, Striegel-Moore RH, et al. (2005) Improve weight and shape among male and female adolescents. Pediatrics 116: e214–e220

58. Grabe S, Ward LM, Hyde JS (2008) The role of the media in body image concerns among women: A meta-analysis of experimental and correlational studies. Psychol Bull 134: 460–476

59. Andersen AE, DiDomenico L (1992) Diet vs. shape content of popular male and female magazines: A dose-response relationship to the incidence of eating disorders? Int J Eat Disord 11: 283–287

60. Gruber AJ, Pope HG, Lalonde JK, Hudson JI (2001) Why do young women diet? The roles of body fat, body perception, and body ideal. J Clin Psychiat 62: 609–611

61. Malinauskas BM, Raedeke TD, Aeby VG, Smith JL, Dallas MB (2006) Dieting practices, weight perceptions, and body composition: a comparison of normal weight, overweight, and obese college females.Nutr J 5: 1–8

62. Stice E, Shaw HE (1994) Adverse effects of the media portrayed thin-ideal on women and linkages to bulimic symptomatology. J Soc Clin Psychol 13: 288–308

63. Yamamiya Y, Cash TF, Melnyk SE, Posavac HD, Posavac SS (2005) Women's exposure to thin-and-beautiful media images: body image effects of media-ideal internalization and impact-reduction interventions. Body Image 2: 74–80

64. Fisher ML, Voracek M (2006) The shape of beauty: determinants of female physical attractiveness. J Cosmet Dermatol 5: 190–194

65. American Society of Plastic Surgeons (2010) Report of the 2010 plastic surgery statistics. APSP National Clearinghouse of Plastic Surgery Procedural Statistics.

66. Cafri G, Thompson JK, Ricciardelli L, McCabe M, Smolak L, Yesalis C (2005) Pursuit of the muscular ideal: Physical and psychological consequences and putative risk factors. Clin Psychol Rev 25: 215–239

67. Lynch SM, Zellner DA (1999) Figure preferences in two generations of men: The use of figure drawings illustrating differences in muscle mass. Sex Roles 40: 833–843

68. Olivardia R, Pope HG, Borowiecki JJ, Cohane GH (2004) Biceps and body image: The relationship between muscularity and self-esteem, depression, and eating disorder symptoms. Psychol Men Masc 5: 112–120

69. Pope HG, Gruber AJ, Mangweth B, Bureau B, deCol C, Jouvent R, et al. (2000) Body image perception among men in three countries. Am J Psychiatry 157: 1297–1301

70. Frederick DA, Haselton MG (2007) Why is male muscularity sexy? Tests of the fitness indicator hypothesis. Pers Soc Psychol Bull 33: 1167–1183

71. Cohn LD, Adler NE (1992) Female and male perceptions of ideal body shapes: Distorted views among Caucasian college students. Psychol Women Q 16: 66–79

72. Jacobi L, Cash TF (1994) In pursuit of the perfect appearance: discrepancies among self-ideal percepts of multiple physical attributes. J Appl Soc Psychol 24: 379–396

73. Cash TF (2003) Body image: Past, present, future. Body Image 1: 1–5

74. Markey CN (2004) Culture and the development of eating disorders: A tripartite model. Eat Disord: J Treat Prev 12: 139–156

75. Thompson JK, Stice E (2001) Thin-ideal internalization: mounting evidence for a new risk factor for body-image disturbance and eating pathology. Curr Dir Psychol 10: 181–183

76. Puhl RM, Heuer CA (2009) The stigma of obesity: A review and update. Obesity 17: 1–24

77. Swami V, Pietschnig J, Stieger S, Tovee MJ, Voracek M (2010) An investigation of weight bias against women and its associations with individual difference factors. Body Image 7: 194–199

78. Grover VP, Keel PK, Mitchell JP (2003) Gender differences in implicit weight identity. Int J Eating Disord 34: 125–135

79. Markey CN, Markey PM (2006) Romantic Relationships and Body Satisfaction Among Young Women.J Youth Adolesc 35: 271–279

80. Bergstrom RL, Neighbors C, Lewis MA (2004) Do men find "bony" women attractive?: Consequences of misperceiving opposite sex perceptions of attractive body image. Body Image 1: 183–191

81. Markey CN, Markey PM, Birch LL (2004) Understanding women's body satisfaction: The role of husbands. Sex Roles 51: 209–216

CHAPTER 7

MEVA AN INTERACTIVE VISUALIZATION APPLICATION FOR VALIDATION OF MULTIFACETED METEOROLOGICAL DATA WITH MULTIPLE 3D DEVICES

Carolin Helbig[1,2], Lars Bilke[1], Hans-Stefan Bauer[3], Michael Böttinger[4] , Olaf Kolditz[1,2]

[1]Department of Environmental Informatics, Helmholtz Centre for Environmental Research (UFZ), Leipzig, Germany,

[2]Faculty of Environmental Sciences, Technical University Dresden, Dresden, Germany,

[3]Institute of Physics and Meteorology, University of Hohenheim, Stuttgart, Germany,

[4]German Climate Computing Center (DKRZ), Hamburg, Germany

ABSTRACT

Background

To achieve more realistic simulations, meteorologists develop and use models with increasing spatial and temporal resolution. The analyzing, comparing, and visualizing of resulting simulations becomes more and more challenging due to the growing amounts and multifaceted character of the data. Various data sources, numerous variables and multiple simulations lead to a complex database. Although a variety of software exists suited for the visualization of meteorological data, none of them fulfills all of the typical domain-specific requirements: support for quasi-standard data formats and different grid types, standard visualization techniques for scalar and vector data, visualization of the context (e.g., topography) and other static data, support for multiple presentation devices used in modern sciences (e.g., virtual reality), a user-friendly interface, and suitability for cooperative work.

Methods and Results

Instead of attempting to develop yet another new visualization system to fulfill all possible needs in this application domain, our approach is to provide a flexible workflow that combines different existing state-of-the-art visualization software components in order to hide the complexity of 3D data visualization tools from the end user. To complete the workflow and to enable the domain scientists to interactively visualize their data without advanced skills in 3D visualization systems, we developed a lightweight custom visualization application (MEVA - multifaceted environmental data visualization application) that supports the most relevant visualization and interaction techniques and can be easily deployed. Specifically, our workflow combines a variety of different data abstraction methods provided by a state-of-the-art 3D visualization application with the interaction and presentation features of a computer-games engine. Our customized application includes solutions for the analysis of multirun data, specifically with respect to data uncertainty

Citation: Helbig C, Bilke L, Bauer H-S, Böttinger M, Kolditz O. MEVA - An Interactive Visualization Application for Validation of Multifaceted Meteorological Data with Multiple 3D Devices. Bohrer G, ed. PLoS ONE. 2015;10(4):e0123811. doi:10.1371/journal.pone.0123811.

and differences between simulation runs. In an iterative development process, our easy-to-use application was developed in close cooperation with meteorologists and visualization experts. The usability of the application has been validated with user tests. We report on how this application supports the users to prove and disprove existing hypotheses and discover new insights. In addition, the application has been used at public events to communicate research results.

INTRODUCTION

For many decades researchers have been working on the development of numerical weather and climate models with the aim to predict future developments. Climate modeling aims to simulate mean values and statistical properties of the climate system for a long time span. In contrast, weather modeling aims to predict the state of the atmosphere for a short time span [1]. Regional weather forecast simulations are based on the latest observations such as station, radar, and satellite data as well as on interpolated results of global models and older model predictions that serve to provide as accurate as possible initial and boundary conditions for the simulation [2]. To perform these predictions, modelers define the atmosphere as a fluid and simulate its state [1].

Meteorologists proceed in developing weather and climate models with ever increasing resolution and complexity. This results in the generation of larger and larger data sets. In order to reduce uncertainty in the predictions, it is common today to perform ensembles of simulations: multiple model runs are done with slightly varying initial conditions. With a statistical analysis of the resulting ensemble data, the probability of specific model results can be determined. The ensemble data is another source of the strong data growth in this discipline.

For the visual analysis of weather simulations, it is mandatory to display the data together with its geographical context (e.g., time-independent data such as topography, cities, and water bodies), since some meteorological phenomena can be correlated with these data. Furthermore, the integration of relevant observational data is necessary to visually compare this information with the simulation results and to identify relationships and inconsistencies. Consequently, meteorologists are faced with exploring and analyzing data sets from different sources *(multimodal)* that are very complex. Those data sets consist of *spatiotemporal* data and include numerous variables *(multivariate)*. In addition, data sets with multiple model runs *(multirun)*have to be analyzed [3]. To explore, analyze and present such data, scientific visualization methods for multifaceted data have to be developed [4–6].

Together with experts in meteorology and visualization, we develop a workflow and an application that can cooperatively be used to verify, falsify and generate hypotheses, study phenomena and their evolution, discover inconsistencies and unexpected features in the evolution and representation of processes in the atmosphere. We combine standard 3D visualization methods for time-dependent meteorological data and implement a visualization application with a graphical user interface (GUI) that supports users in analyzing these data sets. To tackle this challenge, we pursue an approach that combines the power and variety of data abstraction methods of a visualization tool (ParaView [7]) with the interaction and presentation capabilities offered by computer-games engines (Unity [8]).

To import the data sets into the visualization software, the data has to be converted (e.g., WRF netCDF data has to be converted to fit the netCDF conventions supported by ParaView) so that it can be incorporated by the visualization software. In addition, inconsistencies (e.g., differences in scale, resolution, coordinate system) between the data sets have to be identified and removed with the help of the visualization software or by prior processing steps [9]. To enable the user to analyze the data by using state-of-the-art visualization techniques (e.g., isosurfaces, streamlines with custom seeding, path lines, individual glyphs or calculation of variables before adding a visualization filter) and to be able to cooperatively analyze the data, the application needs to support multiple presentation devices that are used in modern sciences (e.g., virtual reality environments, head-mounted displays). Usually meteorologists are not using complex visualization methods in their daily work. Hence, the application should have an easy-to-use interface based on the needs of the user to enable him to quickly orient himself.

There are many visualization tools specifically developed for spatiotemporal, multivariate weather and climate data [10], but they usually offer only very limited support for multirun and multimodal data [3, 11]. Furthermore, most tools support just basic visualization methods (e.g., simple streamlines, isolines). To analyze phenomena like turbulence in high resolution simulations, it is necessary to provide a variety of more complex 3D visualization methods. Therefore representations such as 3D streamlines with high density, trajectories or path lines can be used.

In our case study, the input data sets are quite large (e.g., one time step with a size of 358 MB) which can cause problems with the performance in the visualization application. When applying visualization methods (e.g., isosurfaces) the original data is filtered. Usually, the data size of the resulting geometries is very small in comparison to the input data (e.g., geometries of one time step with eight variables has a size of 1.3 MB). Based on the visualization methods and parameters selected, 3D objects are generated.

Existing visualization applications include very sophisticated methods and parameters. For Example, the calculation of wind fields requires the application of a streamline algorithm including the integration of appropriate seed points. These methods are cumbersome for people who do not use visualization software in their daily work. To enhance the usability for end users, it is necessary to provide a simplified user interface that is well adapted to their needs and skills.

The weather simulation results consist of 3D data including 3D scalar and 3D flow data. Therefore, it is necessary to support advanced 3D visualization devices in order to deal with the high dimensionality of multifaceted data. As an example, virtual reality environments are a promising tool to visualize such complex data sets [12, 13]. A visualization application should support cooperative work because nowadays it is common for scientists to work closely together on the same project while they are spatially separated from each other.

We present an application, which is developed in close cooperation with meteorologists and visualization experts, that fulfills the described needs. Therefore, we developed a workflow based on the visualization software ParaView and the computer-games engine Unity to deliver a portable and easy-to-use visualization application together with the pre-computed geometries for the case study. We offer methods for displaying and comparing different simulation runs side-by-side or with outlines (marking the borders of geometries) and change settings of 3D objects (e.g., opacity). The application is portable to different platforms; hence, it can easily be used in discussions, presentations, and even for educational purposes. The use of the application is not restricted to weather data. It is usable for all kind of environmental data such as groundwater simulation results and geological models.

Finally, the usability and benefits of the application are evaluated with user tests.

RELATED WORK

The visualization of research results has a growing importance in nearly all scientific domains (e.g., environmental research [9, 14, 15], genetics [16, 17], biomedicine [18], and animal movement [19, 20]). We provide an overview of related work with a focus on visualization applications for earth system sciences, visualization applications in general, using movie applications for presenting, and approaches for using computer-games engines for scientific visualization.

Domain Specific Visualization Applications for Earth System Sciences

Since data from weather and climate simulations are quite large and complex, many visualization applications have been developed especially for this application domain. An overview is given by Nocke et al. [10]. One of the first applications specifically developed for meteorological model data is Vis5D [21] in 1988. Unfortunately, the development ended in 2002, but Vis5D is still used by many meteorologists. A more recent example is VAPOR [22], a visualization application developed especially for *Weather Research and Forecast model* (WRF) simulation data, which is the model used in our case study (see Meteorological background). Another approach is to integrate climate and weather data into web-based applications like Google Maps and Google Earth [23, 24].

All these applications allow for the visualization of weather simulations based on single model runs, and are mainly restricted to only a few data types common in weather and climate research (as NetCDF, HDF or GRIB). But, because of its huge relevance in climate and weather research, the visualization of multirun and scenario-based simulations is also an important challenge. Domain specific visualization applications such as Noodles [25] support multirun data, but unfortunately are restricted to 2D data and just take one or two different variables into account [26]. In general, the visualization methods used in these applications are very basic. The data types that could be imported are restricted to those common in weather and climate simulation research (mostly netCDF).

General Purpose Visualization Applications

Scientific visualization is applied in a broad range of application fields, ranging from molecular to medical to outer space research. There are some open source (e.g., ParaView [7], VisIt [27], OpenDX [28]) and commercial (e.g., Avizo [29]) general purpose 3D visualization systems available that meet domain specific requirements and provide some of the newest visualization methods and algorithms. However, due to the variety of operations and parameters offered by the systems, their user interfaces are quite complex and it is quite challenging and time-consuming for non-visualization experts to familiarize themselves with the different systems and produce meaningful visualizations of their data.

Movie Applications

Video technology is used to overcome lacking interactivity in the visualization process in particular for large, time-dependent data. Using movies to publish visualizations especially designed with respect to a broader public is another approach which we also used in our previous work [30]. The production of complex videos can be very time-consuming because of the large development effort [31]. A screenplay has to be developed, camera paths have to be generated in synchronization with the time animation of the data, and the resulting sequences need to be captured and finally edited. The visualization developer has to identify the most effective camera positions and settings, the most appropriate parameters for the visualization and for how long a specific view should be available. This is necessary because besides the player controls, no interaction methods can be provided to the user. An advantage of this approach is the platform independence: movies can be played back on various different devices and over the Internet. Depending on the rendering, it is also possible to export stereoscopic videos meeting the requirements of 3D players.

Game Applications

The reasons for using a computer-games engine for scientific visualization tasks are manifold. For example, they allow for rapid prototyping: the desired application functionality can be easily integrated with the help of visual editing and scripting, and a strong background in software engineering is not necessarily required. Their support of a variety of presentation devices such as PCs, game consoles and mobile devices [32] is advantageous. Furthermore, they enable access to one of the most sophisticated rendering pipelines that can be found [33]. But up to now only a few examples demonstrate the use of computer-games engines for scientific visualization (e.g. molecular structures using the Unity engine [31], cave reconstructions with the Quake3 engine, terrain shaping and vegetation placement with the Far engine [34] and prediction of protein structures with Foldit [35]).

WORKFLOW AND APPLICATION

At the beginning of our work, we specified a set of requirements together with domain experts. We summarized them in the use case diagram shown in Fig 1. These requirements relate to the gaps we defined in the introduction and will be presented in the following. For the development of our application, we followed an iterative process consisting of designing, testing and prototyping/implementation phases. We defined three workflow steps: (1) data integration, (2) abstraction and representation, and (3) interaction and presentation. As a result, we developed MEVA, an easy-to-use application that provides interaction methods for a large variety of data. We provide MEVA and the scripts used to build our application on our GitHub website [36]. Data of the simulation results and of the generated geometries are available through this repository, too.

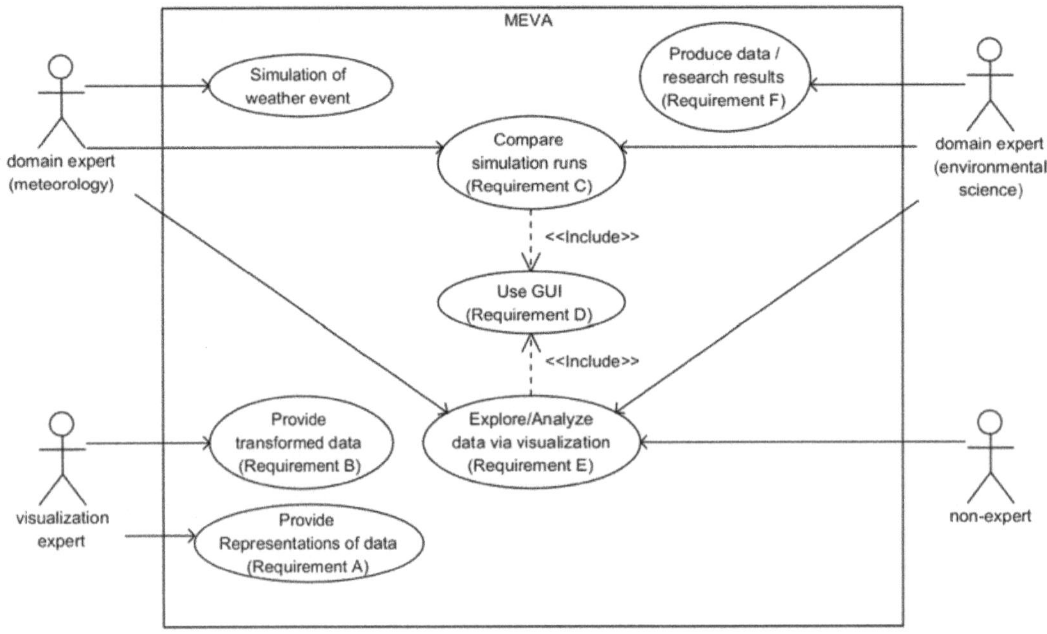

Figure 1: Use case diagram.

Requirements

A: Apply Advanced 3D Visualization Methods

The first requirement we defined is to support most standard 3D visualization methods as well as advanced methods and algorithms such as streamlines or trajectories for vector quantities. The multivariate character of the data requires these methods to find representations for all variables that are intuitive, capable of being differentiated, avoid clutter and occlusion, and are aware of outliers and important patterns [3, 37]. In terms of glyphs, attributes like shape, size, and color can be used to represent different variables and their values. It is also necessary to utilize characteristics of spatiotemporal data like "near things are more related than distant things" [38] and even use space between representations to encode additional information [3]. The fact that the data is not only multivariate, but also multimodal and multirun underscores the importance of using these possibilities. Therefore, it is crucial to place careful consideration into the representation methods of the data, such as side-by-side display.

B: Integrate Static Data, Observations and Simulation Data

The data we take into account is spatiotemporal (x,y,z,t) and consist of the following types:

- High resolution simulation data that is the result of different model runs Sim_a (a = 1 … n model runs) with time step size Δt_a (dependent on the simulation run) and including numerous variables Var_i (i = 1 … n variables, e.g., temperature, wind vector components, mass fraction).
- Observation data *Obs* with time step size Δt and including numerous variables Var_j (j = 1 … n variables, e.g., measured surface temperature, amount of precipitation).
- Static data *Stat* (time-independent) that includes numerous variables Var_k (k = 1 … n data set consisting of cities, rivers, borders, orography, land use characteristics, etc.).

We aim to combine these data in an effective way.

C: Compare experiments (multirun data)

Our objective is to provide visualization and interaction methods to deal with different simulation runs. Therefore, we implement split-screen views and the possibility to distinguish the different runs by marking them with colored outlines. Uncertainty can be displayed by overlaying the runs in one view and manipulating the opacity of each individually. Also, the comparative visualization of data with different temporal and spatial resolution needs to be supported.

D: Deploy visualization with easy-to-use GUI

We aim to deploy a graphical user interface (GUI) that is intuitive, self-explanatory and as slim as possible (limited functionality) to avoid overburdening the user. On the other hand, the user needs to be able to interact with the visualization: to move the camera, to select subsets of variables for display, to change the opacity of their visual representations, to control the time animation, or to additionally blend in a 2D overview map. To get an impression of values changing over time, we need to generate animations at constant frame rates.

E: Support for 3D presentation devices

In order to reduce the technical restrictions for using the visualization application to a minimum, it should work on a variety of different platforms, including different 3D presentation devices such as virtual reality environments, head-mounted displays, 3D projectors, and desktop PCs. The scientists can use it for analysis of their results in order to gain more insight and even arrange for interactive collaborative walkthroughs with colleagues and project partners [31]. In addition, one aim addressed by this requirement is to inform the public by providing an application for edutainment [34, 39].

F: Support for various environmental data

The application should also be usable with other environmental data, different spatial and temporal resolutions or coordinate reference systems. By using modular code and providing scripts for displaying variables, time controls, camera controls and an overview map, a basis application for visualization projects in environmental sciences can be provided.

Development and Implementation of the Visualization Workflow

Data Integration

According to requirement B, it is necessary to provide data integration methods to transform input data into a format readable by the visualization application. We used open source tools like NCO and the OpenGeoSys (OGS) Data Explorer [9, 40] (cf. Helbig et al 2014 [30]) for this preprocessing step (see Fig 2).

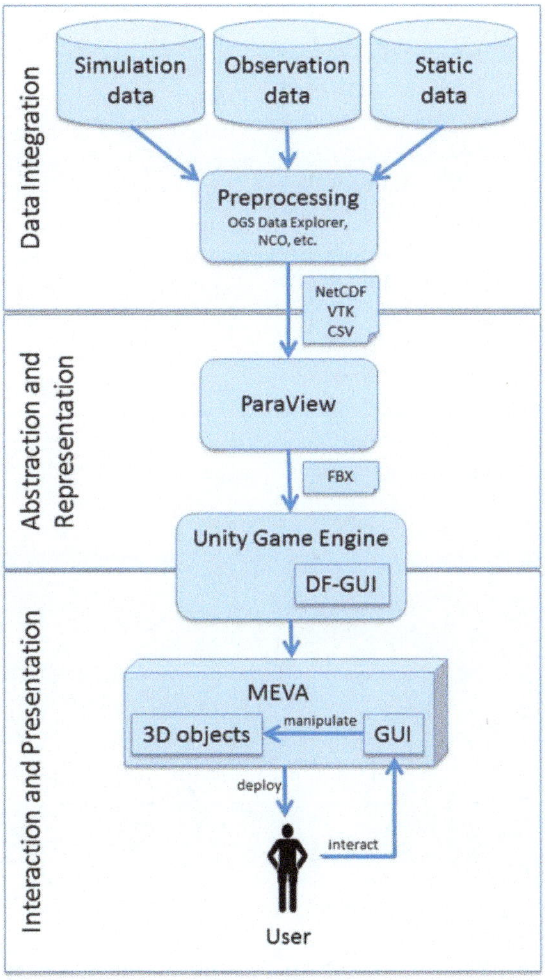

Figure 2: Workflow leading from input data to visualization.

Abstraction and Representation

The data sets in environmental research are often very large, especially if 3D simulations are involved. To deal with such large data sets (e.g., with regard to the performance of the visualization), it is obligatory to reduce the data size in a first step (abstraction): e.g. by downsampling (spatially, temporally), selection of subsets, or selection of variables. We worked together with our domain experts to find appropriate abstraction methods.

At first, a minimal subset of the variables required for the analysis is defined. The selection of the subset is done by the domain experts and based on the specific phenomena they want to analyze. It includes variables such as the three wind components, various mass fraction components (e.g., ice, graupel), and temperature. The further abstraction is realized by applying advanced visualization methods (cf. requirement A) to the data sets. The visualized variables for all time steps are then exported in the form of time series of geometries for each variable.

We used the open source visualization software ParaView, which provides a wide range of visualization methods. It also gives us the opportunity to implement our own methods, e.g., using Python scripts. ParaView is well documented and supported by a large, substantial community. In order to create an interface from ParaView to the Unity game engine, we implemented a module for exporting the resulting geometries (e.g., streamlines, isosurfaces) in the widely-used FBX format [8], including meta-data like color scale and range of values [41].

Interaction and Presentation

In order to fulfill requirements D and E, we decided to use a computer-games engine as a technical basis for the final visualization presentation application. The idea of using a computer-games engine was already discussed by Rhyne in 2000 [42]. It provides several advantages like state-of-the-art graphics, automatic support of major platforms, and low costs (only the developers need to license the software, while the end users can get a free executable to run the application) [31, 33]. We decided to use the computer-games engine Unity because it is freely available and well documented. There are many additional*assets* available that can be integrated into the application. Thus, implemented functionality (e.g., for the GUI, for camera) can be used without investing significant effort into its implementation. In addition, the computer-games engine Unity has a large support community. Besides these advantages, one challenge of using a computer-games engine such as Unity for scientific visualization is the high development effort in the beginning including implementation of basic functions (e.g., for data import).

We implemented an import interface that can handle the metadata using C#. For the presentation of the visualization in a virtual reality environment, it is required to use a GUI that supports multiple displays. The standard GUI included in Unity does not support that. We decided to use the Daikon Forge GUI [43] and implemented elements for objects, time, cameras and overview map. To manage the different data sets, we developed and implemented a hierarchy that can handle multiple data sets that differ in their different temporal resolution and display them in parallel.

Application Interface

We implemented our interface using the computer-games engine Unity and provided the resulting executable to the users. The application shows the current scene and GUI elements, which are collapsible in order to save screen space.

During the development process, we discovered the demand of the user to show and hide single variables and groups of variables. Therefore, we implemented a GUI element (see Fig 3) in which all variables are assigned to a tab according to their type of data (static data, observation data, simulation run a, simulation run b, etc.). Every variable has an assigned radio button to toggle the according 3D objects and a box showing the color code of the 3D objects of the variable. In addition, users have the option to change the opacity of the 3D objects (e.g., reduce the opacity of a cloud to examine the wind circulation inside of it). The combination of variables of various types that are displayed can be defined individually by the user. Furthermore, we also provide the option to add outlines to the 3D objects to distinguish various simulation runs.

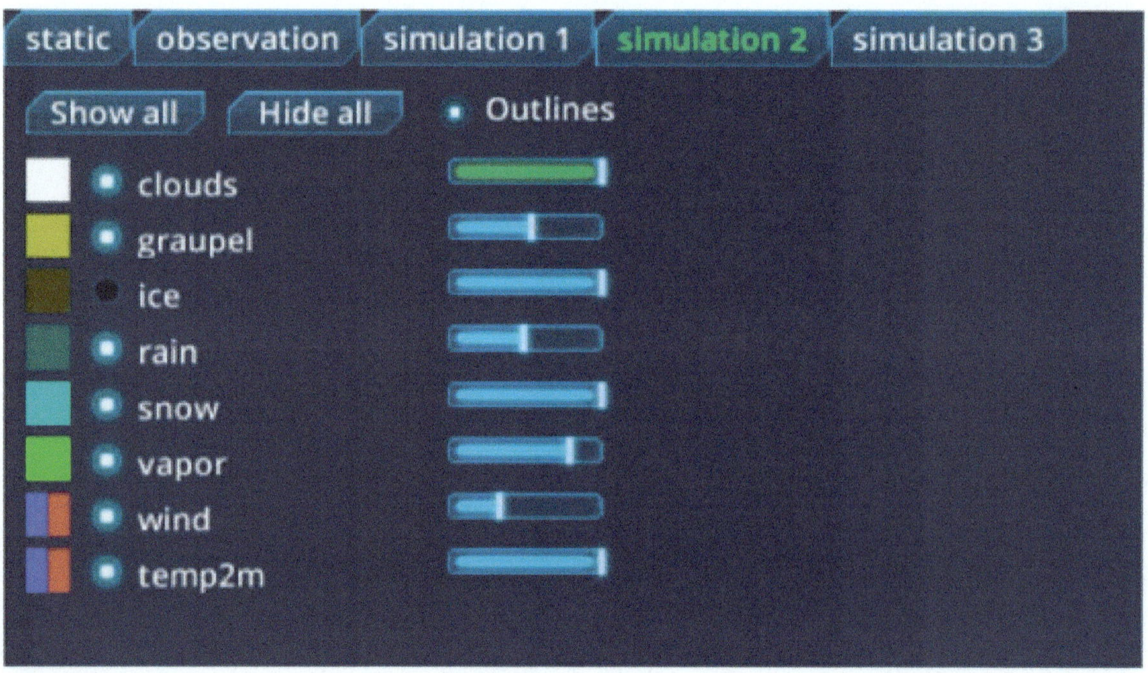

Figure 3: MEVA GUI objects element.

We implemented several camera modes that differ in their behavior: orbit (rotating around the center) and split-screen (side-by-side screens for displaying different simulation runs). The camera can be changed using a defined key (spacebar). For the orbit camera, we implemented a script that enables the user to change the position of the camera while rotating it around a predefined point (in our case, the case study area's center). This functionality is very common in computer games, and there are several functions provided by Unity to implement it (e.g., "RotateAround" function). We decided on this camera type because it allows the user to easily examine the visualization scene. The experiences with other camera types (e.g., first person controller camera that provides ego perspective) showed us that the user runs the risk of getting lost in the scene without the capability to reset to an appropriate perspective. For the split-screen camera we employ two cameras that use the same position but show different layers. A layer in Unity is commonly used to render only a part of a scene. In our case study, each simulation run is assigned to a separate layer.

The user can control the animation of the time steps with the GUI element time (see Fig 4). This component is implemented to handle groups of variables with different time step increments (e.g., simulation 1 with 15 minutes, simulation 2 with 5 minutes). The GUI includes the standard controls (play, pause, one step forward/back, go to first/last time step), and a slider to regulate the speed.

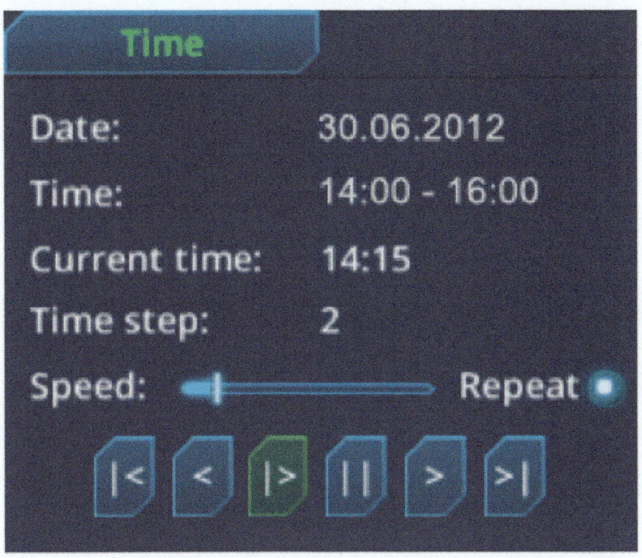

Figure 4: MEVA GUI time element.

In addition, there is a small overview map for orientation that shows the top view of the current scene and includes a polyhedron showing the position and angle of the camera perspective (see Fig 5). This script should be included especially for case studies with large areas where the orientation is challenging.

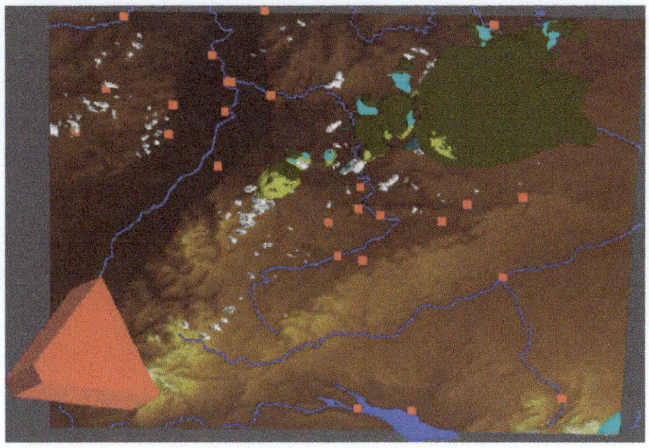

Figure 5: MEVA GUI overview element.

CASE STUDY

For our case study, we worked in a team of visualization experts and experts in meteorology. The case study area is Baden-Württemberg, a federal state of Germany, where we selected a subset area of about 40 x 35 km. This is the area where the specific meteorological phenomenon, which is described in the next section, appears within the simulation. We took into account a set of simulation data at different horizontal resolutions, observation data and static data, and defined a number of tasks that have to be solved with our approach. Finally, we validated the results together with the experts.

Meteorological Background

In our case study, we focus on simulation data of the limited area model (LAM) *Weather Research and Forecast model* (WRF) [1, 44]. WRF is a state-of-the-art numerical weather prediction model. Supported by a large community, new features and applications are added quickly. The model can be applied for research purposes as well as for operational numerical weather prediction on a wide range of spatial and temporal scales. A large variety of physical parameterizations of different complexity and capabilities to ingest observations by data assimilation are provided by the model. Hence, the model can be applied for numerical weather forecasts, for downscaling of global climate simulations as well as and for detailed studies of small scale meteorological processes, so-called large-eddy-simulations (LES) with grid increments on the order of 100 m.

MEVA is tested with the data from an interesting high-impact weather situation. In the area of interest, the 30th of June, 2012, was among the most significant summertime severe weather days of recent years. To the east of a trough over the British Isles, warm, moist air from the Mediterranean was transported to southern Germany. There had been almost no cloudiness in the morning hours and strong irradiance heated the ground surface to temperatures of 30 degrees C and higher. High temperatures and high moisture destabilized the atmosphere and prepared the environment for the development of severe convection.

Starting from 12 UTC, along a line from the northern Black Forest to the northeast, isolated severe thunderstorms, so-called supercells, developed. A supercell is the most rare of the four types of thunderstorms (single-cell, multi-cell, squall-line and supercell). At the same time, it has the largest potential to produce severe weather ranging from severe precipitation and hail to tornadoes. A deep and constantly rotating updraft, the so called mesocyclone, characterizes supercells. Thereby, a constant provision of energy to the system is guaranteed (with updraft and subsequent condensation). In contrast to single-cell thunderstorms the supercell can be active for hours.

One cell that developed near Pforzheim rapidly increased in strength, reaching radar reflectivity values of up to 60 dBZ within 30 minutes. Near Heilbronn, a cell-splitting—typical for long-lasting rotating thunderstorms—took place. While the left part of the cell died out quickly, the so-called right-mover strengthened and deviated to the right from the main wind direction. Later on, the cell produced hail stones with diameters up to 9 cm near the city of Feuchtwangen in Bavaria.

The situation was selected since the correct representation of especially isolated severe convection is a challenging task in numerical weather prediction. To be able to investigate the capability of the WRF model to represent the development, a downscaling from a global meteorological analysis was set up. The analysis forced the outer WRF domain with 3 km horizontal resolution. Two more domains with 1 km and 333 m horizontal resolution were nested into the outer domain and allow the investigation of the representation of the cell with different horizontal resolutions.

Tasks

To test and evaluate our application, we defined four tasks at two levels: analysis (task 1 and 2) and presentation (task 3 and 4). In the following, we present them and their results.

Task 1: Analyzing the simulated circulation and mass fraction in a supercell with different horizontal resolutions

The presentation of multirun data is challenging because of its complex character including space (3D), time, and a set of variables for each simulation run [3] which have to be represented avoiding clutter and occlusion [37]. Different simulation runs are important in the process of developing models and help to identify systematic errors in the model as well as in the parameterization schemes [45]. Therefore, the modelers want to analyze differences between the runs. As described in Meteorological background, we want to analyze the simulated circulation in a supercell as well as the distribution of different hydrometeor mass fractions of three simulations with different spatial and temporal resolutions. Fig 6 shows the topography of the three model runs with the following resolutions:

- Domain 1 with 3 km horizontal resolution and 15 min output interval
- Domain 2 with 1 km horizontal resolution and 5 min output interval
- Domain 3 with 333 m horizontal resolution and 5 min output interval

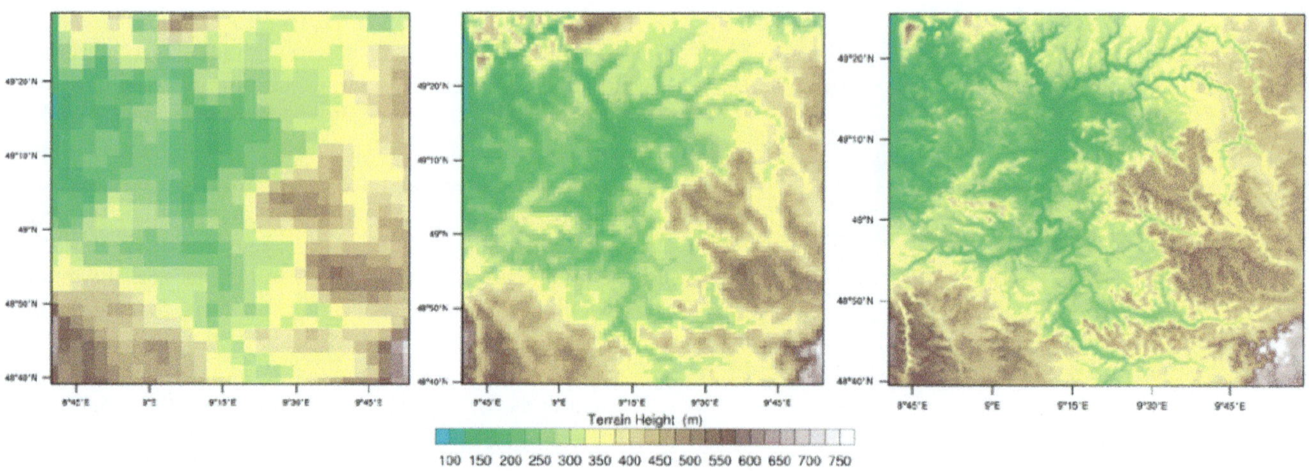

Figure 6: Topography of the subset.

For our case study, we define subsets of the original data sets in order to focus on the region of the supercell. Here, we visualize the mass fractions by isovolumes and the circulation with streamlines. For the comparison of the multirun data, we provide two approaches: side-by-side and overlay. For the first one, the screen can be split horizontally or vertically to visualize the different runs in separate canvases, but with the same viewing settings. The overlaid comparison uses outlines with a predefined color for every run to distinguish between the different runs. For the streamlines, the color is adapted according to the outline color. The user can adjust the opacity of each variable to create visualizations that suit his needs. At the same time, the overlapping of the isovolumes represents the uncertainty in the variable.

Fig 7 shows the overlaid view of the streamlines for all domains: domain 1 (blue), domain 2 (green), and domain 2 (red). The cone glyphs show the wind direction and their colors represent the wind speed. The simulation runs of domain 2 and 3 generated similar wind fields, whereas in domain 1 the updraft is shifted to the north (in the image right). Furthermore, the updrafts in the 1 km and 333 m simulations penetrate higher into the troposphere.

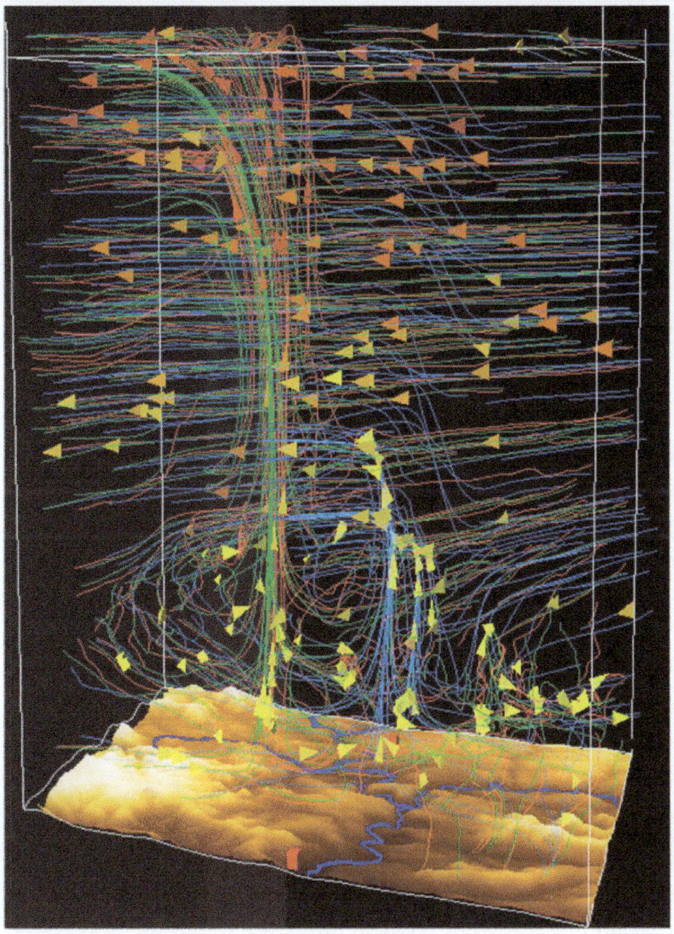

Figure 7: Overlaid view of all domains showing the wind.

As Fig 7 illustrates, a resolution of 3 km is too coarse to realistically represent the circulation details in the supercell. The broad and more diffusively represented updraft core splits into several branches and does not penetrate through the whole troposphere as expected by conceptual models of developing supercells [46, 47]. This suggests that other internal processes (e.g., stronger lateral inflow into the updraft at higher levels) are acting at that resolution. The representation of the updraft region is much more realistic in the two higher-resolution simulations. In both, the updraft is localized and penetrates to the top of the troposphere. Also, the leeward bending in the upper-tropospheric anvil region of developing thunderstorms is well represented.

In Fig 8, the mass fraction of rain is shown in the overlaid view using outlines to distinguish the domains. It can be seen that the domains with higher resolution produce finer structures in the mass fraction of the rain. The rain in domain 2 (green) and domain 3 (red) appears in small cells west of the region (in the image on the left-hand side). In the east, a large ground area is covered by rain in domain 1 (blue). On top of that, rain in domain 2 and 3 are overlapping. The simulation of finer structures with higher resolution is expected to a certain degree. However, it has to be investigated in detail whether the spatial scales of the simulated structures are not underestimated by the higher resolutions. If this is the case, a tuning of the cloud microphysics parameterization might be necessary to lead to an even better simulation.

Figure 8: Overlaid view of all domains showing the rain.

The large cell is captured by all three simulations. Whereas the 3 km simulation shows a flat top of the precipitation region, the influence of the stronger updraft cores in the finer-scale simulations is illustrated by an overshooting cloud top, reaching higher in the 333 m simulation. This also corresponds to conceptual models of developing thunderstorms.

To look at several variables of all domains at the same time, the side-by-side view is used as shown in Fig 9. Here it becomes apparent that the finer the resolution of the simulation, the finer are the appearing structures in the mass fraction, here graupel and rain. The upstream wind field in the western part of the domain is similar in all domains, but differs a little in its location. Especially in the lower troposphere, differences between the resolutions are clearly visible. In contrast to domain 2 and 3, where the updraft is (as expected from conceptual models) one concentrated column in the two higher-resolution simulations, the stream in domain 1 includes a curve. This leads to the hypothesis that different processes are occurring.

Figure 9: Side-by-side view of all domains.

An interesting difference between the three visualizations is the separation of the central updraft region and the area of main precipitation. This is an expected feature in a supercell thunderstorm and contributes to its long lifetime by avoiding evaporation and cooling in the central updraft. In the 3 km simulation, the updraft is embedded into the precipitation region. Therefore, evaporation of precipitation might influence the development of the updraft.

The representation improves in the 1 km simulation (domain 2). The central updraft and the region of precipitation are separated. But the resolution is too coarse to simulate internal structure in the precipitation and anvil regions.

In the 333 m simulation (domain 3) more structure is seen: turbulence in the region of the anvil base and in the swirling of the rain area around the precipitation-free updraft region (mesocyclone).

The visualization setup demonstrated its capabilitiy to support the verification of the representation and evolution of atmospheric processes in model simulations. Furthermore, the better view on the relations of different atmospheric variables support the improvement of process understanding, the necessary first step to improve the model physics. The simulations show the importance to operate the model with a fine horizontal resolution for detailed process studies of interesting atmospheric phenomena.

Task 2: Comparing observation and simulation data

An important task for the evaluation of model simulations is the comparison with observation data and static data, where users can examine correlations and inconsistencies between them. To manage that, we developed methods for the conversion of the different data sets with their different spatial and temporal resolutions, and implemented export and import interfaces. The integrated data includes observation data from stations of the German Weather Service (DWD) inside the focused region, providing data, e.g., temperature (four stations) and wind measurements (two stations) every hour. In addition, static data like the river network and the location of cities are included. Unfortunately, the observational station data is spatially and temporally very sparse compared to the model data.

We compare the data using visual data fusion: the different data types are overlaid in the visualization. The static data is represented using intuitive representations (e.g., tubes for rivers, boxes for cities) positioned on top of the digital elevation model (DEM). The observation data is represented by glyphs, where the color represents the values. For example, the temperature at a station is represented by the color of a sphere, and the wind is represented by an arrow showing the local wind direction and colorized according to the speed. For quick comparison, the color ranges of the observation data matches the color ranges of the simulation data.

To get an impression to which extent the simulated wind corresponds to the observed wind, we compare them in an overlaid view. Fig 10 shows the scene at two points in time: 1:00 pm (left) and 2:20 pm (right). In the first one, the wind direction at the stations (marked with circles) point to the opposite direction of the simulated wind in domain 3. On the other hand in the scene at 2:20 pm, the observed and the simulated wind match quite well, pointing to the upstream of the air in the supercell. The observations are only valid at one point in space (the location of the measurement). The simulated value is an averaged value in the model grid box. Furthermore, the static data (orography, land use, soil characteristics) are also interpolated to the grid resolution. Therefore, one cannot expect an exact match of observed and simulated values. This is the more the case, the more the comparison is done in orographic terrain. In general, the comparison of surface station data (especially wind) is a difficult issue since the observations are strongly influenced by the local characteristics of the measurement location (e.g. orography, exposition, land use, soil type). With an increasing resolution of the simulation, the consistencies increase.

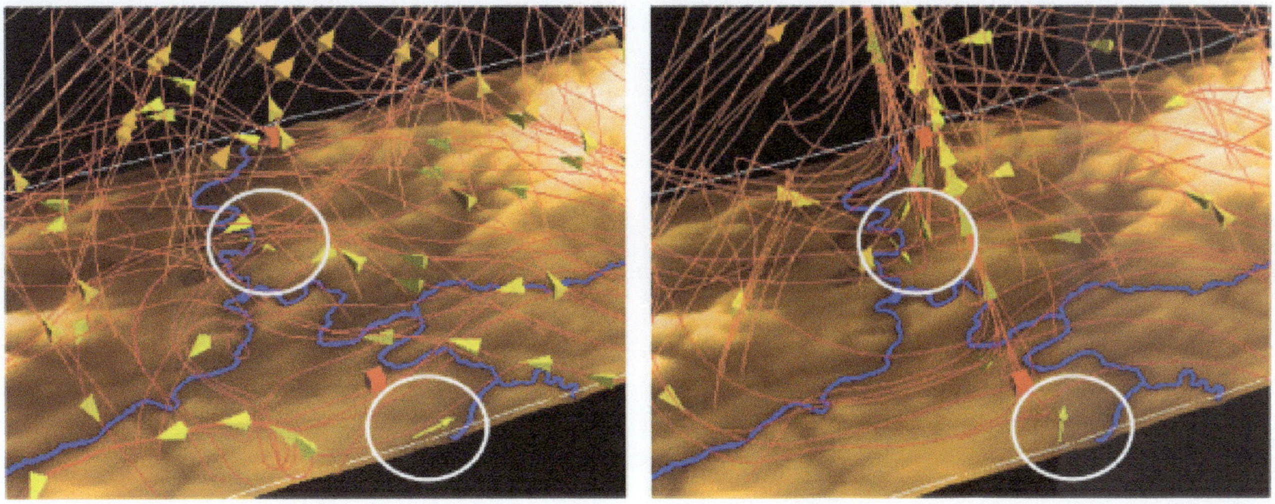

Figure 10: Overlaid view of simulated, observed and static data.

In Fig 11, we visually compare the simulated (left to right: 3km, 1km, 333m) and observed temperature at 2 m above ground. It is clearly visible that the river is occluded in the scene with domain 1. The reason is the finer resolution of the river network (90 m) compared to the resolution of the simulation (3 km). Using the same color map for observed and simulated temperature, we can see at first sight that in all domains and at the observation stations, the temperature around the river is higher than in the other regions.

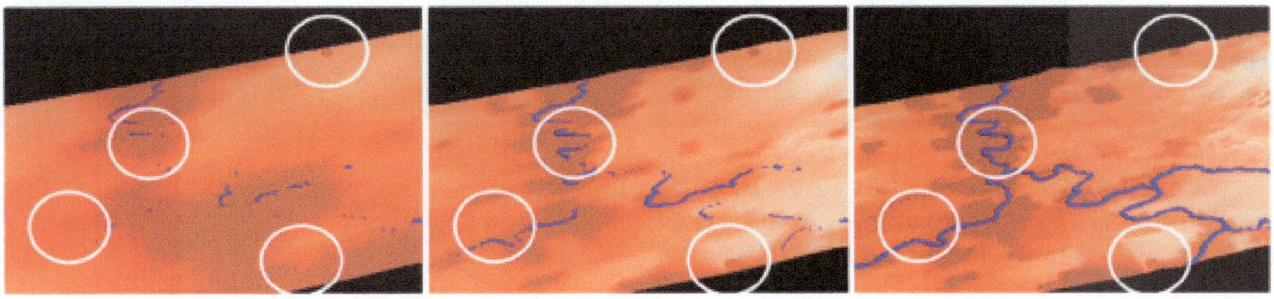

Figure 11: Side-by-side view of simulated, observed data, and static data.

The sequence of images shows the refinement of the resolution and therefore the better representation of orographic features of the Neckar valley and its confluences. Since the valley is deeper in the 333 m simulation, higher surface temperatures are simulated. The temperatures measured in the river valleys are well represented by the model whereas steep slopes in valleys are underestimated even with 333 m resolution. Apart from the warmer river valleys, the model simulation shows warm squared patches. This unrealistic near-surface temperature distribution can be attributed to the coarse grid of the land use data that was used as a boundary condition for the simulation.

Task 3: Using the interface by different user groups

To manage task 1 and 2, it is necessary to provide a set of interaction methods to handle time and multirun data as well as to navigate through the case study region (zooming, rotating, and panning). We focus on two different usage scenarios. In the first one, users are scientists, who want to explore and analyze their research results, evaluate their hypotheses, and have discussions with colleagues based on the visualization. In the second scenario, presentation to the public and other scientists is the aim as well as to use it for education. With MEVA, we provide an application, where users are able to easily discover recent research results and playfully explore them. Even inexperienced users find their way to use the application and interpret the data quickly because of the use of intuitive representations of variables (e.g., streamlines for wind, white isosurfaces for clouds) and the limited functionality of the provided GUI.

Our development process is iterative, including the following steps: designing, testing, and prototyping/ implementing. Together with experts in visualization, meteorologists, scientists from other domains, and based on our previous work [30], we identified a set of required interaction functionalities, designed the GUI, implemented the functionalities, and finally tested the prototypes with the users. We kept the functionalities to a minimum as described in requirement D.

With our approach, the domain experts were able to analyze the simulation runs and gain new insight as described in task 1 and 2. The interactivity of the visualization helped them to evaluate their results and explore the data to gain new perspectives. To prove our statements concerning our application, we conducted several user studies with different levels of formality. We did interviews with domain experts and presentations for the public in the context of events like open houses (see Fig 12), where we collected feedback and included it in the development process. The test environment was the TESSIN (Terrestrial Environmental System Simulation and Integration Network) VisLab, which was the visualization environment for several visualization projects in the past (see Bilke et al. 2014 [48]). The application was also tested on PC, 3D projector and head-mounted display. After a brief introduction, the users were able to navigate through the visualization and use the given interaction methods to control and manipulate the scene.

Figure 12: Presentation of the visualization.

To evaluate the final version of MEVA, we conducted a user study. Therefore, we organized a series of workshops in the VisLab. The 16 participants were from various scientific disciplines ranging from meteorology to mathematics to computer science. Because of the different backgrounds of the participants, not all of them had experience in meteorology (44%), with VR environments (50%), and with other 3D devices (38%). The users got a short introduction about the background of the underlying data and the interaction concept of the VisLab. They were told to solve two tasks with the help of the application: (1) compare the observed and simulated temperature; (2) find differences in the simulation runs with varying horizontal resolutions. With the help of a questionnaire, their feedback was collected. In the following, we present the outcomes of the study. As we gave our participants the freedom to skip questions, the number of answers for each question varies between 11 and 16.

We asked the participants if they would use the visualization application for their own data, and they stated that they would find it useful for presentation (94%), exploration (81%), and analysis (88%). In addition, some participants proposed some more fields of application such as teaching and discussions.

For 83% of the participants, it was easy to solve the given tasks after a short introduction, which was described as appropriate by all of the participants. They evaluated the usability as good or very good. The menu is intuitive (79%), gives the user control over the visualization scenes (86%), and the participants were able to correct erroneous input (92%). We also asked if there was a demand to further customize options for the menu (e.g., change font size, assignment of keys), but just 23% saw a need to have these options.

We provided a gamepad as the interaction device. In former tests, the participants used a flystick, which is a common device in VR environments. We observed that it is hard for most participants to become familiar with this device, which most of them never used before. The result of the user test with the gamepad was that 91% had no problems with the navigation. This result confirmed our impressions and led to the decision to support this device in future applications.

The representations of the variables were evaluated as intuitive by 80% of the participants. The majority, 93% of the participants was able to compare different simulations easily and could discover correlations between variables. Also, 69% of the participants state that they learned something new with the help of the visualization (e.g., about the circulation in a supercell and its structure).

With respect to the task of comparing the simulation data with observations, just 63% state that they were able to compare them. The reason was not the application itself, but the lack of observation data that was provided. For a better comparison, more observation data should be included (e.g., rain from radar data).

To sum up, the user study showed us that there is a large interest in MEVA. The participants want to use MEVA for their own work in the future. The provided data is a very important basis for a meaningful visualization. The participants from all scientific domains and with various experiences with 3D devices were able to get along with the application very quickly. A big advantage of using a VR environment as presentation device is that it attracts people and offers good preconditions to stimulate interdisciplinary discussions and productive work sessions (cf. Bohrer et al. 2008 [49]).

Task 4: Viewing the Presentation on Different Devices

In scientific projects it is common that cooperation partners are regionally distributed. In our case study, the domain experts are from different areas in Germany. Therefore, it is necessary to provide the visualization for different devices, ranging from common PCs to 3D devices like virtual reality environments, 3D projectors, and head-mounted displays. The use of high-resolution virtual reality environments enables the users to conduct collaborative, interactive walkthroughs, where they can explore their data in detail and navigate through the visualization while discussing and sharing their discoveries and thoughts.

We implemented an input device setup that is usable with a computer mouse, joystick, flystick (commonly used in VR environments), and gamepad. To reach a wide audience, it was necessary to support multiple platforms. The game engine Unity does that out of the box. We successfully tested our visualization with the virtual reality environment TESSIN VisLab (see Fig 12), a head-mounted display, a 3D projector, and several PCs with different hardware.

CONCLUSIONS AND FUTURE WORK

We have presented a workflow and application to explore, analyze, and present multifaceted meteorological data, at different spatial and temporal resolutions, from multiple sources (various simulation runs, observations, and static data) and including numerous variables. The workflow leads from input data to an interactive 3D visualization using advanced representations for the variables. To provide interaction for the user, we designed and implemented modular and reusable interfaces for the game engine Unity. We presented two different solutions for multirun data: side-by-side and overlaid comparison. Furthermore, we developed an easy-to-use interface in an iterative development process. We defined two usage scenarios for the application. The first one addresses domain experts and is used to analyze and explore their data as well as to evaluate hypotheses and facilitate discussions. The second one addresses presentation issues and is used to present research results to project partners, decision makers, and the interested public. To prove our statements concerning our application, we conducted several user studies with different levels of formality. Besides interviews with domain experts and presentations for the public, we conducted a user study with 16 participants to evaluate the final version of MEVA.

The novelty of our approach is to provide a flexible workflow that consists of software components and supports researchers of various disciplines to generate meaningful visualizations. Our aim was not to develop new algorithms and techniques, but to provide a toolbox that combines these existing techniques in a useful way (cf. Sedlmair et al 2012 [50]). By utilizing the general purpose visualization system Paraview as initial part of our workflow, MEVA can easily be used also in other application areas.

The analysis and comparison of the simulated circulations and mass fraction in a supercell with different horizontal resolutions (task 1) provided an interesting insight into the process chain in a developing supercell and allowed detailed verification of the model's representation of this hazardous meteorological phenomenon. This allows the further improvement of the model through the tuning and refinement of the model physics. Moreover, the visualization approach is predestined to improve the understanding of the processes taking place in nature.

With the help of the user study we could find out that the comparison of observation and simulation data (task 2) using MEVA was an easy task for the participants. But we also got the feedback that the participants found the observational data too sparse for a meaningful visual comparison. To solve that, additional observational data from other sources might be included, such as rain radar or satellite images. This could be done using the flexible workflow we developed: after preprocessing, the data could be visualized with ParaView and then added to MEVA, where it would appear as an additional item in the objects menu.

The user study showed us that the participants from various scientific domains were able to learn using the application very quickly (task 3). Even participants with little know-how in meteorology and no experience with VR environments were enabled to explore the data. Most of the participants could learn something new with the help of the visualization and want to use MEVA for their own work in the future.

MEVA was developed to be usable with different devices (task 4). The option of the presentation on a PC was very important during the development process because the project partners are situated in different parts of Germany. Thereby the current versions of MEVA could be tested regularly. The option to run it on a PC is also required for education purposes to make it usable by lecturers and students. For detailed analysis of the data, the use of MEVA in a VR environment was necessary to examine complex structures especially in the simulated wind circulation. During our user study in the VisLab we could observe that the participant's attention was caught very quickly and discussions have been stimulated while using MEVA.

The growing importance of visualization in many scientific domains caused by an increasing generation of complex, large data sets is apparent. With the development of a flexible workflow, the basis for the generation of visualizations for other application domains is provided. The use of existing software applications, which are enhanced and updated regularly, saves own development effort. Thereby even new and modified data types can be handled easily. The user study showed us that there is a large interest in MEVA and, with our approach to use freely available software as components of the workflow and provide the implemented scripts and example data, it is accessible by many users.

We believe that MEVA will be the basis for other visualization projects in environmental science, for example to compare climate scenarios, simulation results of groundwater and geological models, and for multimodal scenarios, which include multiple models with interacting parts. In the future, we would like to develop an automated parallelized workflow, which would then allow in-situ visualizations in high performance computing environments (cf. Olbrich 2000 [51]).

ACKNOWLEDGMENTS

The first author would like to express her gratitude to the European Social Fund (ESF) as part of the program "Europa fördert Sachsen" for the funding of the scholarship. We thank HIGRADE, the graduate school of UFZ, and CompeTE+ the school for doctoral students at the HTWK. The SRTM data necessary for the initialization of the high-resolution simulation was provided by the Centre for Tropical Agriculture (CIAT) from the CIAT-CSI SRTM webpage (http://srtm.csi.cigar.org).

REFERENCES

1. Lynch P (2008) The origins of computer weather prediction and climate modeling. J Comput Phys 227: 3431–3444. doi: 10.1016/j.jcp.2007.02.034
2. Lynch P (2006) Weather Prediction by Numerical Process In: The Emergence of Numerical Weather Prediction: Richardson's Dream, Cambridge: Cambridge University Press, volume 11 pp. 1–27.
3. Kehrer J, Hauser H (2013) Visualization and visual analysis of multifaceted scientific data: A survey.IEEE Trans Vis Comput Graph 19: 495–513. doi: 10.1109/TVCG.2012.110
4. Fuchs R, Hauser H (2009) Visualization of Multi-Variate Scientific Data. Comput Graph Forum 28: 1670–1690. doi: 10.1111/j.1467-8659.2009.01429.x
5. Thomas J, Cook K (2006) A visual analytics agenda. IEEE Comput Graph Appl 26: 10–13. doi:10.1109/MCG.2006.5
6. Aigner W, Miksch S, Müller W, Schumann H, Tominski C (2007) Visualizing time-oriented data - a systematic view. Comput Graph 31: 401–409. doi: 10.1016/j.cag.2007.01.030
7. Ahrens J, Geveci B, Law C (2005) ParaView: an end-user tool for large data visualization In: Hansen C, Johnson C, editors, The visualization handbook, London: Oxford; pp. 717–732.
8. Goldstone W (2011) Unity 3.x game development essentials Birmingham: Packt Publishing, 2 edition, 1–488 pp.
9. Rink K, Fischer T, Selle B, Kolditz O (2013) A data exploration framework for validation and setup of hydrological models. Environ Earth Sci 69: 469–477. doi: 10.1007/s12665-012-2030-3
10. Nocke T, Sterzel T, Böttinger M, Wrobel M (2008) Visualization of climate and climate change data: An overview In: Ehlers, editor, Digital Earth Summit on Geoinformatics 2008: Tools for Global Change Research, Heidelberg: Wichmann, volume 94 pp. 226–232.
11. Phadke MN, Pinto L, Alabi F, Harter J, Taylor RM, Wu X, et al. (2012) Exploring Ensemble Visualization. In: Proceedings of SPIE. [PMC free article]
12. van Dam A, Laidlaw DH, Simpson RM (2002) Experiments in Immersive Virtual Reality for Scientific Visualization. Comput Graph 26: 535–555. doi: 10.1016/S0097-8493(02)00113-9
13. Simpson R, LaViola J, Laidlaw D, Forsberg A, van Dam A (2000) Immersive VR for scientific visualization: a progress report. IEEE Comput Graph Appl Mag 20: 26–52. doi: 10.1109/38.888006
14. Castrillón M, Jorge P, López I, Macías a, Martín D, Nebot RJ, et al. (2011) Forecasting and visualization of wildfires in a 3D geographical information system. Comput Geosci 37: 390–396. doi:10.1016/j.cageo.2010.04.011
15. Lidal E, Patel D, Bendiksen M (2013) Rapid Sketch-based 3D Modeling of Geology. In: Workshop on Visualisation in Environmental Science. pp. 1–5.
16. Cain Aa, Kosara R, Gibas CJ (2012) GenoSets: visual analytic methods for comparative genomics. PloS one 7: e46401 doi: 10.1371/journal.pone.0046401 [PMC free article]
17. Meyer M, Wong B, Styczynski M, Munzner T, Pfister H (2010) Pathline: A Tool For Comparative Functional Genomics. Comput Graph Forum 29: 1043–1052. doi: 10.1111/j.1467-8659.2009.01710.x
18. Gerasch A, Faber D, Küntzer J, Niermann P, Kohlbacher O, Lenhof HP, et al. (2014) BiNA: a visual analytics tool for biological network data. PloS one 9: e87397 doi: 10.1371/journal.pone.0087397[PMC free article]
19. Shamoun-Baranes J, van Loon EE, Purves RS, Speckmann B, Weiskopf D, Camphuysen CJ (2012) Analysis and visualization of animal movement. Biol Lett 8: 6–9. doi: 10.1098/rsbl.2011.0764[PMC free article]
20. Grundy E, Jones MW, Laramee RS, Wilson RP, Shepard EL (2009) Visualisation of Sensor Data from Animal Movement. Comput Graph Forum 28: 815–822. doi: 10.1111/j.1467-8659.2009.01469.x
21. Hibbard WL, Anderson J, Foster I, Paul BE, Jacob R, Schafer C, et al. (1996) Exploring Coupled Atmosphere-Ocean Models Using Vis5D. Int J High Perform Comput Appl 10: 211–222. doi:10.1177/109434209601000208
22. Clyne J, Mininni P, Norton A, Rast M (2007) Interactive desktop analysis of high resolution simulations: application to turbulent plume dynamics and current sheet formation. New J Phys 9: 301–301. doi:10.1088/1367-2630/9/8/301
23. Wang Y, Huynh G, Williamson C (2013) Integration of Google Maps/Earth with microscale meteorology models and data visualization. Comput Geosci 61: 23–31. doi: 10.1016/j.cageo.2013.07.016
24. Sun X, Shen S, Leptoukh G, Wang P (2012) Development of a web-based visualization platform for climate research using google earth. Comput Geosci 47: 160–168. doi: 10.1016/j.cageo.2011.09.010
25. Sanyal J, Zhang S, Dyer J, Mercer A, Amburn P, Moorhead RJ (2010) Noodles: a tool for visualization of numerical weather model ensemble uncertainty. IEEE Trans Vis Comput Graph 16: 1421–30. doi:10.1109/TVCG.2010.181
26. Nocke T, Flechsig M, Bohm U (2007) Visual exploration and evaluation of climate-related simulation data. In: 2007 Winter Simulation Conference. IEEE, pp. 703–711.
27. Lawrence Livermore National Laboratory (2014). Visit: an interactive parallel visualization and graphical analysis tool. https://wci.llnl.gov/codes/visit/. Accessed 27 July 2014.
28. Braun J, Ford R, Thompson D (2001) Open DX Paths to Visualization. Austin, TX, U.S.A.: VIS Inc., 1–208 pp.
29. Visualization Sciences Group (2014). Avizo - 3d analysis software for scientific and industrial data.http://www.vsg3d.com/avizo/. Accessed 27 July 2014.
30. Helbig C, Bauer HS, Rink K, Wulfmeyer V, Frank M, Kolditz O (2014) Concept and workflow for 3D visualization of atmospheric data in a virtual reality

environment for analytical approaches. Environ Earth Sci72: 3881–3899. doi: 10.1007/s12665-014-3136-6

31. Lv Z, Tek A, Da Silva F, Empereur-mot C, Chavent M, et al. (2013) Game on, science - how video game technology may help biologists tackle visualization challenges. PloS one 8: e57990 doi:10.1371/journal.pone.0057990 [PMC free article]
32. Rhyne TM (2002) Computer games and scientific visualization. Commun ACM 45: 40–44. doi:10.1145/514236.514261
33. Lewis M, Jacobson J (2002) Game Engines in Scientific Research. Commun ACM 45: 27–31.
34. Friese K, Herrlich M, Wolter F (2008) Using game engines for visualization in scientific applications In:New Frontiers for Entertainment Computing, New York: Springer Publishing Company, Incorporated,volume 279 pp. 11–22.
35. Cooper S, Khatib F, Treuille A, Barbero J, Lee J, Beenen M, et al. (2010) Predicting protein structures with a multiplayer online game. Nature 466: 756–60. doi: 10.1038/nature09304 [PMC free article]
36. Helbig C, Bilke L (2014). Meva - multifaceted environmental data visualization application.http://carolinh.github.io/unityatmos/. Accessed 12 December 2014.
37. Roberts JC (2007) State of the art: Coordinated & multiple views in exploratory visualization In: Proc. Coordinated and Multiple Views in Exploratory Visualization. Ieee, Cmv, pp. 61–71.
38. Tobler W (1970) A computer movie simulating urban growth in the Detroit region. Econ Geogr 46: 234–240. doi: 10.2307/143141
39. Mayo Merrilea J (2009) Video Games: A Route to Large-Scale. Science 323: 79–82. doi:10.1126/science.1166900
40. Rink K, Bilke L, Kolditz O (2014) Visualisation strategies for environmental modelling data. Environ Earth Sci 72: 3881–3899. doi: 10.1007/s12665-013-2970-2
41. Bilke L (2014). VtkFbxConverter. https://github.com/ufz-vislab/. Accessed 29 July 2014.
42. Rhyne TM (2000) Computer games' influence on scientific and information visualization. Computer 33: 154–159. doi: 10.1109/2.889099
43. Daikon Forge (2014). Daikon forge gui library. www.daikonforge.com/. Accessed 27 July 2014.
44. Skamarock WC, Klemp JB, Gill DO, Barker DM, Duda MG, Wang W, et al. (2008) A Description of the Advanced Research WRF Version 3. Technical report. http://www2.mmm.ucar.edu/wrf/users/docs/arwv3.pdf. Accessed 29 July 2014.
45. Hamby DM (1994) A review of techniques for parameter sensitivity analysis of environmental models.Environ Monit Assess 32: 135–54. doi: 10.1007/BF00547132
46. Doswell CA, Burgess DW (1993) Tornadoes and toraadic storms: A review of conceptual models. In: Church C, et al. , editors, The Tornado: Its Structure, Amer. Geophys. Union; pp. 161–172.
47. Moller AR, Doswell CA (1994) The operational recognition of supercell thunderstorm environments and storm structures. Weather Forecast 9: 327–347. doi: 10.1175/1520-0434(1994)009%3C0327:TOROST%3E2.0.CO;2
48. Bilke L, Fischer T, Helbig C, Krawczyk C, Nagel T, Naumov D, et al. (2014) TESSIN VISLab - laboratory for scientific visualization. Environ Earth Sci 72: 3881–3899. doi: 10.1007/s12665-014-3785-5
49. Bohrer G, Longo M, Zielinski D, Brady R (2008) VR Visualisation as an Interdisciplinary Collaborative Data Exploration Tool for Large Eddy Simulations of Biosphere-Atmosphere Interactions In: Bebis G, Boyle R, Parvin B, Koracin D, Remagnino P, et al. , editors, Advances in Visual Computing, Springer Berlin Heidelberg, volume 5358 of Lecture Notes in Computer Science; pp. 856–866.
50. Sedlmair M, Meyer M, Munzner T (2012) Design Study Methodology: Reflections from the Trenches and the Stacks. IEEE Transactions on Visualization and Computer Graphics 18: 2431–2440. doi:10.1109/TVCG.2012.213
51. Olbrich S (2000) Ein leistungsfähiges System zur Online-Präsentation von Sequenzen komplexer virtueller 3D-Szenen. Ph.D. thesis. https://www.deutsche-digitale-bibliothek.de/binary/FLXJOX4WGW6S7ECJTKK5RLLA7NWYLUEP/full/1.pdf. Accessed 27 July 2014.

CHAPTER 8

VIRTUAL REALITY BASED SUPPORT SYSTEM FOR LAYOUT PLANNING AND PROGRAMMING OF AN INDUSTRIAL ROBOTIC WORK CELL

Hwa Jen Yap[1], Zahari Taha[2], Siti Zawiah Md Dawal[1], Siow-Wee Chang[3]

[1]Department of Mechanical Engineering, Faculty of Engineering, University of Malaya, Kuala Lumpur, Malaysia,

[2]Faculty of Mechanical Engineering, Universiti Malaysia Pahang, Pekan, Pahang, Malaysia,

[3]Bioinformatics Program, Institute of Biological Sciences, Faculty of Science, University of Malaya, Kuala Lumpur, Malaysia

ABSTRACT

Traditional robotic work cell design and programming are considered inefficient and outdated in current industrial and market demands. In this research, virtual reality (VR) technology is used to improve human-robot interface, whereby complicated commands or programming knowledge is not required. The proposed solution, known as VR-based Programming of a Robotic Work Cell (VR-Rocell), consists of two sub-programmes, which are VR-Robotic Work Cell Layout (VR-RoWL) and VR-based Robot Teaching System (VR-RoT). VR-RoWL is developed to assign the layout design for an industrial robotic work cell, whereby VR-RoT is developed to overcome safety issues and lack of trained personnel in robot programming. Simple and user-friendly interfaces are designed for inexperienced users to generate robot commands without damaging the robot or interrupting the production line. The user is able to attempt numerous times to attain an optimum solution. A case study is conducted in the Robotics Laboratory to assemble an electronics casing and it is found that the output models are compatible with commercial software without loss of information. Furthermore, the generated KUKA commands are workable when loaded into a commercial simulator. The operation of the actual robotic work cell shows that the errors may be due to the dynamics of the KUKA robot rather than the accuracy of the generated programme. Therefore, it is concluded that the virtual reality based solution approach can be implemented in an industrial robotic work cell.

INTRODUCTION

According to a survey from the UN Economic Commission for Europe published in "World Robotics 2001"[1], there were approximately 750,000 robots running in industries worldwide. After a decade, the statistics, market analysis, forecasts, case studies and profitability of robot investment published by the International Federation of Robotics (IFR) in "World Robotics 2011" [2] revealed that the annual robot sales in 2010 was 118,337 units. The sales almost doubled compared to that in 2009, during the worldwide economic and financial crisis. The forecasted figures for year 2011 and 2014 were 139,300 and 166,700 respectively. On the other hand, the sale for robots in 2010 was US$ 5.7 billion. It was estimated that the sales value for robot systems in 2010 was US$ 17.5 billion, which includes software, peripherals and systems engineering. On the other hand, over 1,300 labours were struck by moving objects in their workplace. This appears to be the most common accident which causes injuries, which is due to the complexity

Citation: Yap HJ, Taha Z, Md Dawal SZ, Chang S-W. Virtual Reality Based Support System for Layout Planning and Programming of an Industrial Robotic Work Cell. Slater M, ed. PLoS ONE. 2014;9(10):e109692. doi:10.1371/journal.pone.0109692.

of hardware and software. In Sweden and Japan [3], these incidents have been associated with robot operations, such as malfunction to stop, rapid motion of the robot arm, injuries in the workspace during repair and maintenance, illegal entrance to the workspace as well as unpredictable robot work functions to the operator.

In this research, a virtual reality based programming of a robotic work cell (VR-Rocell) is proposed and developed. In general, robot programming methods can be classified as online and offline programming, which is dependent on the presence of the actual robot while the programme is being created. In online programming, the actual robot must be present and used in the teaching process. Offline programming is the most popular method for robot programming as it does not involve real robots. This reduces robot downtime and the production line will be unaffected [4]. The system is designed to improve human-machine interface and is divided into two major sub-systems, namely, VR-Robotic Work Cell Layout (VR-RoWL) and VR-Robot Teaching System (VR-RoT). Open platform architecture is used along with a universal file format. The system uses the teach-points concept for offline programming to generate robot commands, which prevents users from potential uncertain manufacturing environments. The objective of this research is to develop a VR-based system for cell layout planning and design, path teaching, programming and simulation for an industrial robot which will achieve zero robot downtime during path teaching using a generic approach.

Solution Approaches for an Industrial Robotic Work Cell

The synergy between virtual reality and robotics in various areas was reviewed in 1999 [5]. It is known that traditional approaches in robot programming are tedious and require knowledge of a specific robot language. Hence, a new virtual reality based system is needed to provide users with a more user-friendly interface by means of high-level programming language. Various solution approaches for industrial robots have been proposed in past studies and are presented in the following sections.

Simulation Solution Approaches

There are various commercial software available which provide solutions for facility layout planning involving robotic work cells. In addition, commercial software have also been specifically developed for industrial robot simulation and visualization. Simulation and visualization of robotic work cells are necessary prior to procurement of hardware. The simulation software allow users to check the reachability, workspace, safety issues as well as other aspects of industrial robots.

In general, facility layout planning software is based on the discrete-event simulation concept, in which the operation of the system corresponds accordingly to the chronological order of events. Discrete-event simulation software provides an overall picture of the simulated robotic work cell, whereby the industrial robot is considered as an event within the cell. Hence, detailed simulation and information of the industrial robot (such as path planning and robot programming) may be unavailable in this type of simulation since robot commands cannot be generated automatically from the simulation results. In addition, the simulation environment does not consider spatial constraints and ergonomic issues. Although this technique is most popular compared to other techniques (e.g. gaming), it has a relatively low stakeholder commitment due to modelling lead time and purpose [6].

Geometric or continuous simulation provides a geometric graphical representation with constant time interval. Unlike discrete-event simulation with irregular time intervals, geometric simulation is more suitable for 3D visualization, offline programming of robots and collision detection during manufacturing processes [7]. Geometric simulation is also called virtual manufacturing [8].

On the other hand, various robotic simulation software have been developed by robot manufacturers such as ABB, KUKA, Fanuc and Mitsubishi. In this research, industrial robots manufactured by KUKA are used. KUKA SimPro is the official software developed specifically for offline programming of KUKA robots. The product is connected to KUKA OfficeLite, which is a virtual KUKA controller and enables real-time cycle time analyses and generation of robot programmes. However, the software requires adequate knowledge of robotic systems and special training. The cost of the software may not be within the financial means of many industries.

There are several generic robotic software available in the market such as Delmia from Dassault Systems, RobCAD from Technomatrix, Robsim from Camelot and Cosimir from Festo. These packages are flexible for various industrial robots produced by different manufacturers and are fully coupled with product life management (PLM). However, the disadvantages of these simulation software are the absence of direct outputs to the actual robots and require strong knowledge of robotic systems. The fundamental criteria which enable the adoption of such software in SMEs and academic sectors are affordability, ease of use and short learning curves [9], in which generic robot software are unable to do so. Furthermore, system integration is particularly a challenging task, whereby compatibility issues between some robot manufacturers and data structure of different software vendors may occur. Thus, simulation software with open architecture are actively developed by many researchers, in which the codes written for the simulation are portable to real robotic platforms [10].

Virtual Manufacturing Solution Approaches

Virtual Manufacturing (VM) is the use of virtual reality technology in real-time manufacturing based simulations in order to optimize product design and processes for a specific manufacturing goal such as design for assembly, quality, lean operations, and/or flexibility. VM is used to generate 3D models and real-time simulations of manufacturing processes to aid the design and production of products. VM is known as the next revolution of global manufacturing [11].

Lawrence Associates [12] introduced three virtual manufacturing paradigms which have been applied by Kim & Choi [13] to solve problems in flexible manufacturing systems. These three paradigms are:

a. Design-centred VM: Provides designers/engineers with tools to design products which fulfil the design criteria. Design-centred VM can be performed early during the conceptual design stage, which is easily modified and tested. This type of VM also uses simulation technology to optimize design specifications such as stress, strain and failure analyses.

b. Production-centred VM: Provides the means to develop and analyse alternative production and process plans. Factory models and layout designs can be created to simulate the production process, assembly process, material flow and human factors. Production simulation is capable of simulating manufacturing process models which provides a means for inexpensive, fast evaluation of various processing alternatives.

c. Control-centred VM: Evaluates product designs, production plans and control strategies. This VM interactively improves these three aspects through simulations of the control process. It is an additional tool to control models and actual processes, enabling seamless simulation for optimization during the actual production cycle.

Mujber et al. [14] classified the applications of VR in manufacturing into three areas, which are design, operations management and manufacturing process. In the design area, the VM technique has been used to design the die geometry for sheet metal forming in a virtual space. It is also widely used to generate virtual prototypes in order to replace physical prototypes [15]. The time required for design and development can be shortened and costs can be reduced when producing physical prototypes. Furthermore, virtual prototypes can be used to support design and development through simulation and analysis results [16], [17]. The integration of virtual reality with rapid prototyping technology is capable of creating digital prototypes to assist product development.

Moreover, VM technology is widely employed in the planning phase of manufacturing processes. This technology is used to estimate, optimize and minimize the processing time prior to the actual production [18]. In addition, this technology can be used to investigate material and information flow through the virtual operation control concept. The factors affecting the operation of a manufacturing system can be tracked to evaluate design and operational performance. In-house virtual training can be developed for machine tools[19], milling machines [20], turn-milling centres [21], maintenance [22] and so on.

VR application in manufacturing processes covers machining, assembly and inspection. Virtual machining is commonly used to solve cutting processes such as milling, turning and drilling. Collision detection can be checked to prevent accidental collisions which will damage machine tools during actual processes [23]. In addition, VM can be used in assembly planning to investigate the feasibility of assembly processes. The mechanical and physical characteristics of equipment and tooling can be modelled and simulated virtually. Furthermore, the quality, quantity, product cycle and cost can be determined in order to obtain optimum results. Virtual inspection is used to check the physical and mechanical properties of inspection equipment such as collision checks [24].

The effectiveness of VR-based training systems has also been tested. A VR-based training system for Metal Active Gas Welding (MAG) has been developed to integrate virtual robot path teaching system with a practical training system, in which VR-based offline robot programming techniques are implemented. The process paths are triggered, recorded and converted into robot commands for robotic welding. A survey has been carried out on controlled and uncontrolled groups of students in Faculty of Engineering, University of Malaya, in order to evaluate their motivation towards virtual welding learning and conventional welding learning. The results revealed that more than 85% of the students supported the use of VR integrated training system compared to traditional welding training. The respondents agreed that VR technology is a valuable tool and the prototype with VR-based system is feasible to be implemented as a supplement in the welding training programme as well as in path planning for robotic welding [4].

MATERIALS AND METHODS

Due to limited accessibility, high cost and strong demand for technical knowledge for commercial simulation software, a number of research projects have been conducted to develop alternative solution approaches for industrial robots. Several virtual robots have been created in the works of [25]–[27]. However, these systems still require strong technical knowledge. Furthermore, the interactions between humans and virtual robots are very poor, and are limited to conventional input/output devices.

VR-based industrial robotic systems are able to overcome the above mentioned problems [28]. VR technology was used to develop a virtual gripper for robotic inspections such that information of the inspected workpiece (position and orientation) can be defined interactively. The synergy between virtual reality and robots was reviewed and it was concluded that VR in robotics can be attained from haptic interfaces and human factor know-how [5]. Xu et al. [29] developed VR-based robot graphics simulation with built-in collision detection and coordinate translation functions.

Santis and Siciliano [30] developed a physical human-robot interaction model to measure the safety and dependability of robots. This model was implemented to evaluate the perceived safety during robot motion, which is dependent on shape, speed and posture. VR technology was also used to model a complex robotics system to study the perception of safe robot idle time [31]. They found that robot speed and size has an effect on idle time, whereby high robot speed and large-sized robot increases the waiting time of the participants.

Figure 1 shows the proposed model of this research, which is VR-Rocell. The model is designed to integrate the advantages of various simulation systems into a robotic work cell. The model proposes a VR-based layout planning and design for robotic work cell, which is similar to the functions provided by discrete-event simulation software. Furthermore, robot programming is conducted based on the teach-by-demonstrate concept for the VR-based robot teaching system. Robot paths are taught through an intuitive user interface compared to geometric simulation systems. The proposed model is based on open platform architecture, which includes programming language (C++), graphic libraries (OpenGL) and 3D formats (e.g. VRML).

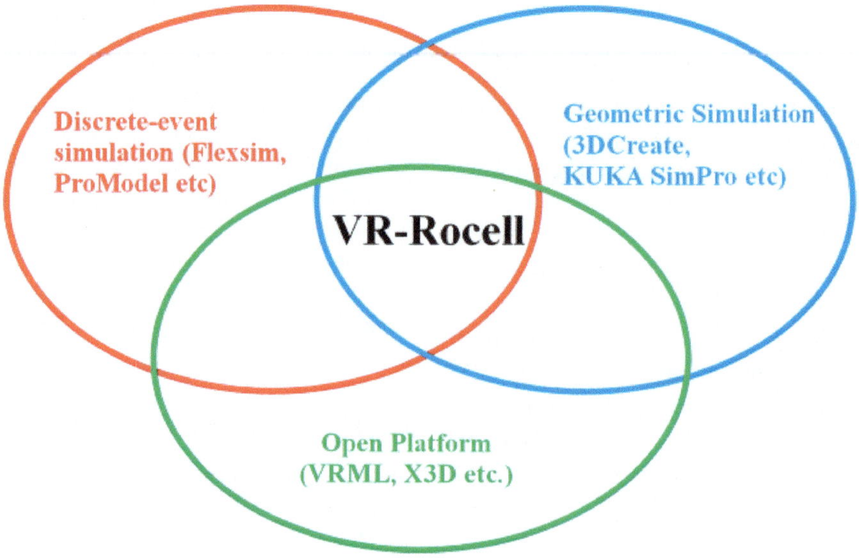

Figure 1: Proposed model of VR-Rocell.

CAD format can be converted into virtual reality (VR) by using either a commercial modelling software, library or database approach [32]. 3D models can be integrated in virtual manufacturing (VM) through Lex lexical analyser generator and Open Graphic Library [33]. However, there are no standard methods or algorithms which convert 3D models from CAD modelling software into VR and/or VM, which is monitored and controlled by the related society. In addition, there are no standard methods or algorithms which convert 3D models from CAD modelling software to virtual manufacturing systems. Therefore, a generic algorithm is proposed to import 3D models into the VR-based system based on an open platform concept. The system can be used in all operating systems such as Windows, Mac and Linux. In addition, the system supports all types of browsers such as Internet Explorer, Mozilla and Google Chrome.

Most of the commercial VR packages use native file formats to create their virtual environment. These formats consist of algorithms and data structures for a particular software, which may not be readable or visualized by others. The universal formats which are loaded into the CAD software are not editable in the native system. The assembly of universal formats cannot be saved back into the universal format. For these reasons, the proposed model consists of four universal file formats, which are supported by all related software. An algorithm is developed to extract and save 3D meshing information into four file formats. File format conversion is possible among the above formats. Thus, the output models from one VM can be transferred or imported to another VM using a universal file format. A database can be built to re-configure a modular-based VM, which further reduces the development time of the new system [33], [34]. The developed VR-based system is designed with simple and user-friendly interfaces to reduce the software training period and is therefore suitable for inexperienced users. Furthermore, the system is created based on an open architecture concept which allows the generated data to be transferred to/from different platforms without loss of information. In this manner, the output data can be viewed and configured by other systems.

The issues on human-computer interface design parameters in virtual environment have also been studied. A systematic axiomatic design method was used to design the virtual environment in order to minimize visual symptoms. The axiomatic design parameters were identified such as colour of background, brightness of virtual lighting, field of view (FOV), speed of virtual objects and display resolution. This method was developed and tested on a virtual robot manufacturing system [35].

Hardware Architecture

In this research, the hardware of the system is divided into two major parts, which are computer workstation and virtual reality hardware. The computer workstation consists of the computer CPU with a dedicated graphics card and is connected to two monitors, video splitters, two projectors, display extension devices and other peripherals (e.g. mouse, keyboard and speaker).

A Virtual Wall system has been built for the virtual reality hardware. The frame is made of an aluminium profile (40 mm×40 mm), which can be easily assembled and disassembled. At the same time, a mirror is used to reduce the projection distance and thus reduces the footprint of the Virtual Wall into half. Five major virtual reality hardware are integrated into the Virtual Wall, namely, 6-degree-of-freedom (6-DOF) tracker system, non-depolarized projection screen, polarized filter, polarized glasses and active-stereo glasses with emitter. The magnetic-type sensor tracker (Polhemus Patriot) is used in this research, which consists of a system electronics unit, a transmitter, a receiver (for head tracking) and a digitizer (user's manipulator). The maximum distance of the sensor tracker at which the resolution and noise performance of the system can be realized is 900 mm at default settings. Even though the useful range can be extended up to 1520 mm, this causes reduction in the accuracy of the data. These hardware are used to visualize and manipulate the virtual objects created in the virtual environment. The ergonomic dimensions (visual and arm reach) of the Virtual Wall needs to be decided. In this research, the visual limit can be neglected for the following reasons:

a. Visual position limits: The user can simply walk to the position of his/her choice within the range of the sensor. This means that the user can walk to the left, right, near to or far from the screen if he/she intends to see the virtual objects clearly.

b. Visual orientation limits: The 6-DOF sensor is attached to the glasses and virtual environment which will be updated according to his/her viewing angle for all attitudes.

The arm reach is the primary concern for screen position. The horizontal position has no effects on the arm reach since the user can walk either to the left or right if he/she is unable to reach the virtual objects. The vertical position of the projection screen is calculated based on the standing position of the user. The anthropometrics data for the Malaysian population are considered since the system is designed for a manufacturing environment [34]. The vertical position is designed to accommodate 95% of the population from short users (maximum height) to tall users (minimum height) in standing position. The height dimension is within the range of 2.5–97.5 percentiles. The reachable height depends on the shoulder height (H), arm reach forward length (L) and arm reach angle (θ) of the workers. Table 1 shows the parameters for standing position based on the anthropometrics data of Malaysian workers. From Table 1, the minimum reachable height can be calculated using the following equation:

$$reach.H = H - L\cos\theta \tag{1}$$

where:

$reach.H$ = reachable height

H = shoulder height

L = arm reach forward length

θ = arm reach angle

Table 1: Parameters for standing position

Specifications	Male (mm)		Female (mm)	
	Min	Max	Min	Max
Shoulder height, *H*	1,223.00	1,462.56	1,157.00	1,356.96
Arm reach forward length, *L*	724.00	920.00	690.56	843.44
Minimum reachable height – assumed the user's hand is vertically pointed down, with the arm reach angle, $\theta = 30°$)	596.00	665.82	558.96	626.52
Maximum reachable height – assumed the user's hand is vertically pointed up, with the arm reach angle, $\theta = 150°$)	1,850.00	2,259.30	1,755.04	2,087.40

Therefore, the minimum vertical reachable height is 665.82 mm and the maximum vertical reachable height is 1,755.04 mm. The design accommodates 95% of Malaysian workers.

Sensor Configuration

The Virtual Wall used by VR-RoWL involves four coordinate systems, namely, user, sensor transmitter, projection screen and virtual environment (VM). A standard reference coordinate system needs to be selected or defined, whereby other systems can be transformed into this coordinate frame. Thus, the projection screen coordinate system is used as the reference frame, as shown in Figure 2. The transformation between frames can be simplified into the following configurations:

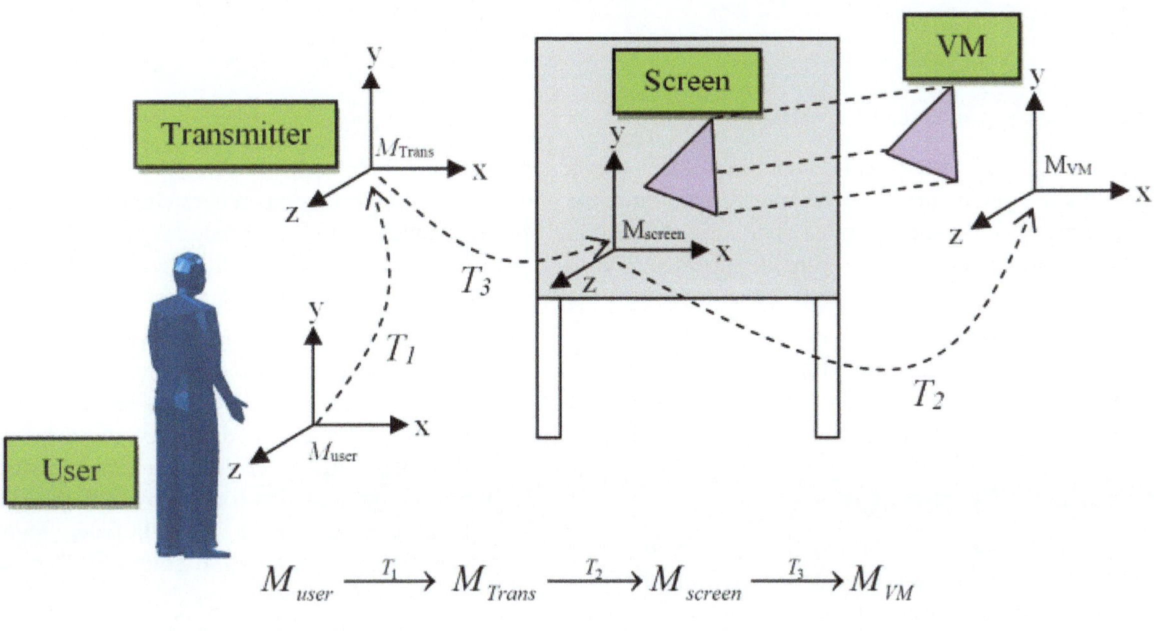

Figure 2: Reference coordinate system for Virtual Wall.

- First, the user's coordinate system needs to be transformed into the transmitter coordinate frame. In this research, the position and orien tation of the head and hand are tracked using sensors and the sensors are referred to the transmitter coordinate frame. Furthermore, the default location of the transmitter is programmed at the origin without the rotation angle. Therefore, the transformation matrix T_1 can be obtained through the position and orientation of the user's head/hand which is referenced to the transmitter, and these data can be directly tracked by the 6-DOF sensor and digitizer.
- The virtual scene is projected (clipped) onto the screen. The actual dimensions of the screen are used to set up the OpenGL projection, and the virtual objects will appear on a 1-to-1 scale (full-scale), which is the same as the actual size. The centroid of the projection screen is fixed as the origin of the virtual environment. Hence, the transformation matrix T_2 represents the position and orientation of the virtual objects referenced to the screen, and these data are referred to the global coordinate system of the virtual environment.

Only two coordinate frames need to be considered, which are transmitter and screen. Both coordinate frames are configured as the origin in their respective frame. Consequently, the transformation matrix T_3 can be obtained by measuring the distance and rotation angle of the transmitter which is referenced to the centroid of the screen. The user will feel as though he/she is located within the virtual manufacturing system and uses the same coordinate system through these transformation functions.

ErgoVR station is another configuration used in the robot teaching system (VR-RoT). The ErgoVR station is designed to make the user feel, see and interact with objects in the most natural and ergonomic position. The user will be able to see and interact with virtual objects in the same place, as shown in Figure 3. Thus, the screen coordinate system can be replaced by a new coordinate system, which is termed as the virtual-monitor coordinate system. The transformation of the frame for the VR-RoT can be simplified into the following configurations:

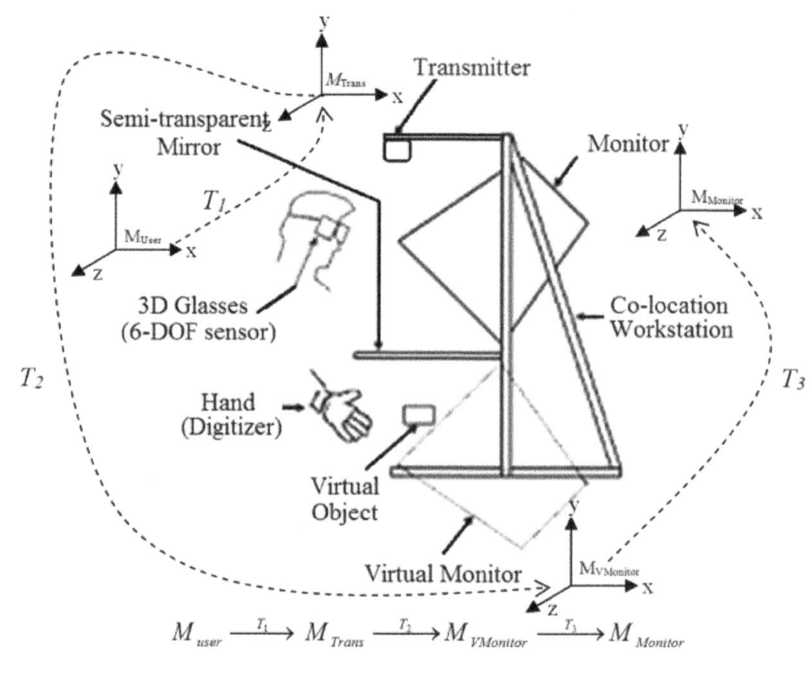

Figure 3: Coordinate system for ErgoVR workstation.

- Similar to the VR-RoWL condition, the user's coordinate system needs to be transformed into the transmitter coordinate frame. The transformation matrix T_1 is obtained directly from the recorded data.
- Based on the unique design of the co-location workstation, the virtual monitor will appear as the reflection of the actual monitor. Therefore, a mirror transformation matrix (T_3) at x-axis is applied to achieve the desired effect.

Robot Command Generator

The process path generated by the VR-based robot teaching system is recorded in a text file. Following this, the data acquisition and generator are designed to generate the robot commands, in which the generator is programmed specifically for KUKA industrial robots. The command generator is capable of converting the generated path into KUKA readable and executable formats. Hence, the robot commands can be tested by the KUKA OfficeLite simulator before being loaded into the real physical robot.

The software architecture of the KUKA robot system must be identified before conversion between VR-RoT and KUKA robot system can be established. Additionally, KUKA programming language and syntax must be analysed and studied in depth before conversion can be carried out. Each KUKA robot command is formed by two files, i.e. source file (*.*src*) and data file (*.*dat*). The data file contains data of the coordinate points whereas the source file consists of robot motion data such as velocity etc.

In this research, the content of each source file is similar to one another, whereby the motion of the robot will be determined during the actual robot testing. However, the source file is somewhat dependent on the data file. For example, the motion loop controls the dynamics and type of movement from one point to another, whereby the number of loops (counter) must tally with the number of points in the data file. The data file consists of all points in the path. Figure 4 shows the flow chart of the robot command generator.

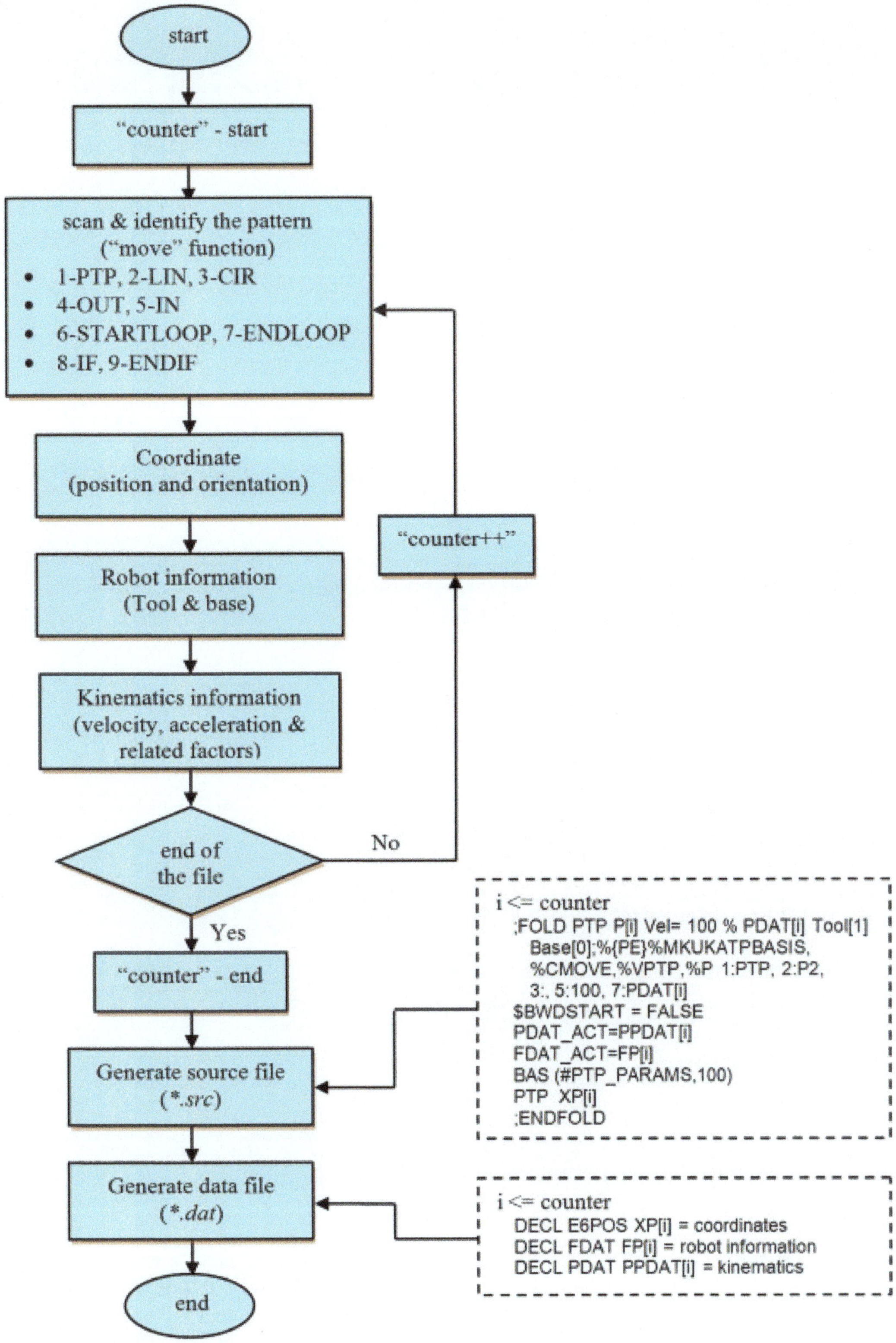

Figure 4: Flow chart of robot command generator.

In general, there are six parameters used to define a point in the KUKA robot command, namely, *x*, *y*, *z*,*pitch*, *yaw* and *roll*. Thus, the coordinates tracked by VR-RoT can be inserted into the data files accordingly. Figure 5 and Figure 6 shows an example of the source file and data file, respectively.

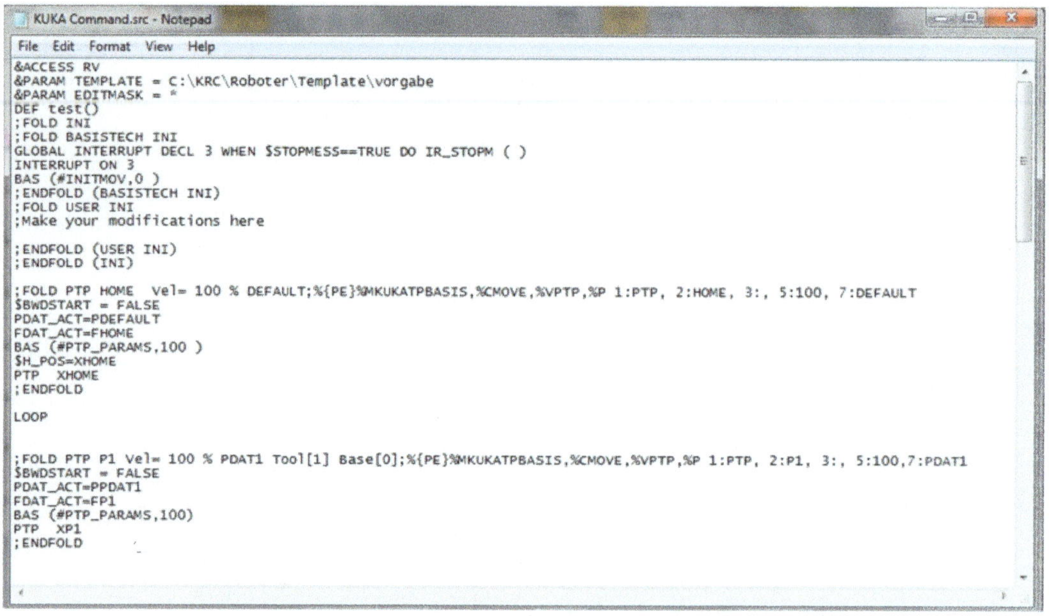

```
KUKA Command.src - Notepad
File Edit Format View Help
&ACCESS RV
&PARAM TEMPLATE = C:\KRC\Roboter\Template\vorgabe
&PARAM EDITMASK = *
DEF test()
;FOLD INI
;FOLD BASISTECH INI
GLOBAL INTERRUPT DECL 3 WHEN $STOPMESS==TRUE DO IR_STOPM ( )
INTERRUPT ON 3
BAS (#INITMOV,0 )
;ENDFOLD (BASISTECH INI)
;FOLD USER INI
;Make your modifications here

;ENDFOLD (USER INI)
;ENDFOLD (INI)

;FOLD PTP HOME  Vel= 100 % DEFAULT;%{PE}%MKUKATPBASIS,%CMOVE,%VPTP,%P 1:PTP, 2:HOME, 3:, 5:100, 7:DEFAULT
$BWDSTART = FALSE
PDAT_ACT=PDEFAULT
FDAT_ACT=FHOME
BAS (#PTP_PARAMS,100 )
$H_POS=XHOME
PTP  XHOME
;ENDFOLD

LOOP


;FOLD PTP P1 Vel= 100 % PDAT1 Tool[1] Base[0];%{PE}%MKUKATPBASIS,%CMOVE,%VPTP,%P 1:PTP, 2:P1, 3:, 5:100,7:PDAT1
$BWDSTART = FALSE
PDAT_ACT=PPDAT1
FDAT_ACT=FP1
BAS (#PTP_PARAMS,100)
PTP  XP1
;ENDFOLD
```

Figure 5: Example of source file (*.src) generated by the VR-based system.

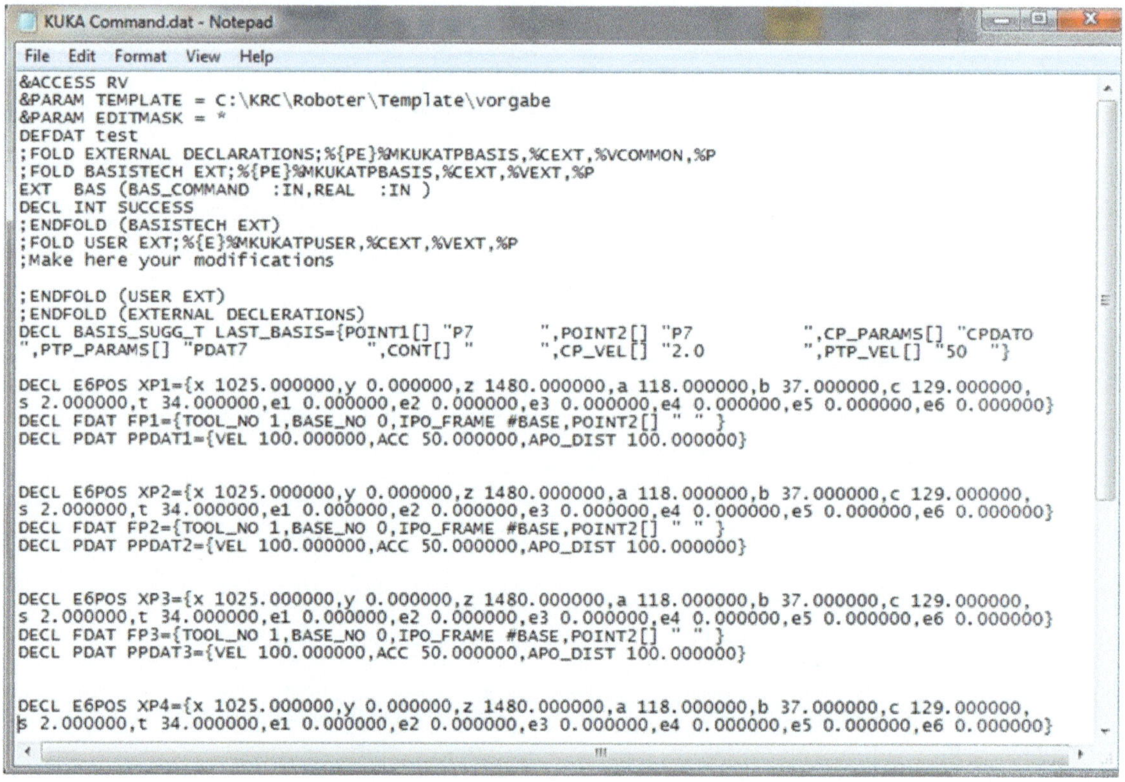

```
KUKA Command.dat - Notepad
File Edit Format View Help
&ACCESS RV
&PARAM TEMPLATE = C:\KRC\Roboter\Template\vorgabe
&PARAM EDITMASK = *
DEFDAT test
;FOLD EXTERNAL DECLARATIONS;%{PE}%MKUKATPBASIS,%CEXT,%VCOMMON,%P
;FOLD BASISTECH EXT;%{PE}%MKUKATPBASIS,%CEXT,%VEXT,%P
EXT  BAS (BAS_COMMAND  :IN,REAL  :IN )
DECL INT SUCCESS
;ENDFOLD (BASISTECH EXT)
;FOLD USER EXT;%{E}%MKUKATPUSER,%CEXT,%VEXT,%P
;Make here your modifications

;ENDFOLD (USER EXT)
;ENDFOLD (EXTERNAL DECLERATIONS)
DECL BASIS_SUGG_T LAST_BASIS={POINT1[] "P7      ",POINT2[] "P7          ",CP_PARAMS[] "CPDATO
",PTP_PARAMS[] "PDAT7            ",CONT[] "      ",CP_VEL[] "2.0         ",PTP_VEL[] "50  "}

DECL E6POS XP1={x 1025.000000,y 0.000000,z 1480.000000,a 118.000000,b 37.000000,c 129.000000,
s 2.000000,t 34.000000,e1 0.000000,e2 0.000000,e3 0.000000,e4 0.000000,e5 0.000000,e6 0.000000}
DECL FDAT FP1={TOOL_NO 1,BASE_NO 0,IPO_FRAME #BASE,POINT2[] " " }
DECL PDAT PPDAT1={VEL 100.000000,ACC 50.000000,APO_DIST 100.000000}


DECL E6POS XP2={x 1025.000000,y 0.000000,z 1480.000000,a 118.000000,b 37.000000,c 129.000000,
s 2.000000,t 34.000000,e1 0.000000,e2 0.000000,e3 0.000000,e4 0.000000,e5 0.000000,e6 0.000000}
DECL FDAT FP2={TOOL_NO 1,BASE_NO 0,IPO_FRAME #BASE,POINT2[] " " }
DECL PDAT PPDAT2={VEL 100.000000,ACC 50.000000,APO_DIST 100.000000}


DECL E6POS XP3={x 1025.000000,y 0.000000,z 1480.000000,a 118.000000,b 37.000000,c 129.000000,
s 2.000000,t 34.000000,e1 0.000000,e2 0.000000,e3 0.000000,e4 0.000000,e5 0.000000,e6 0.000000}
DECL FDAT FP3={TOOL_NO 1,BASE_NO 0,IPO_FRAME #BASE,POINT2[] " " }
DECL PDAT PPDAT3={VEL 100.000000,ACC 50.000000,APO_DIST 100.000000}


DECL E6POS XP4={x 1025.000000,y 0.000000,z 1480.000000,a 118.000000,b 37.000000,c 129.000000,
s 2.000000,t 34.000000,e1 0.000000,e2 0.000000,e3 0.000000,e4 0.000000,e5 0.000000,e6 0.000000}
```

Figure 6: Example of data file (*.dat) generated by the VR-based system.

Software Architecture

The virtual reality-based programming of a robotic work cell (VR-Rocell) is developed by solid modelling using commercial CAD software. The 3D models can be saved in a universal file format, either in STL, VRML, XML or OBJ. Following this, these models are integrated into the VR-Robotic Work Cell Layout (VR-RoWL) using a generic approach, in which the assembly models are saved in the VRML format. Virtual models can be transformed and saved in new files. The assembly models can be designed, modified and re-arranged accordingly. The models will be used to teach the robot path via the VR-based Robot Teaching system (VR-RoT). The robot commands and layout information can be loaded into the actual robotic work cell to execute the desired tasks. In addition, the VM models can be saved in different universal file formats. Therefore, the output models from one VM can be transferred or imported to another VM in the universal file format. Figure 7 shows the overview of the research.

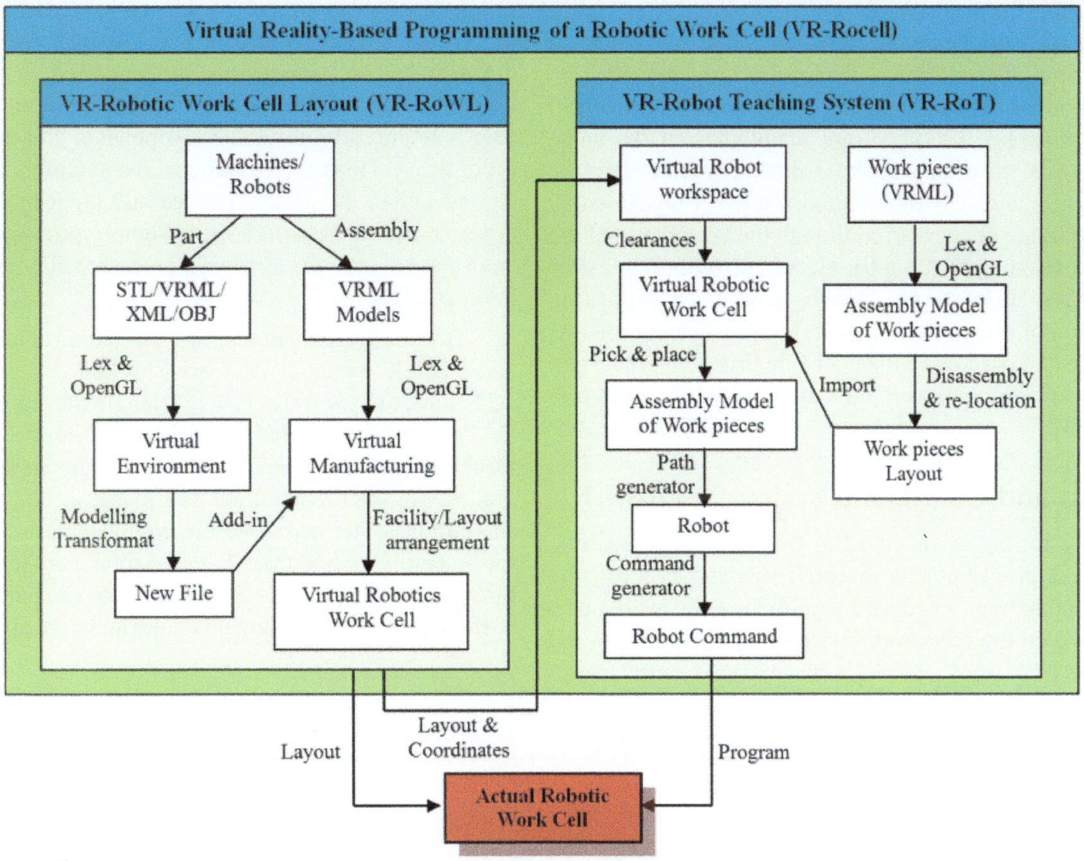

Figure 7: Overview of the project.

3D Models and Coordinate System in VR-RoWL

In the robot layout planning system (VR-RoWL), the layout is re-arranged according to the user's input (via digitizer) in the VRML format. In this case, the models inside the assembly file solely consist of 3-DOF, which allows translation in the horizontal plane (xz-plane) and rotation around the y-axis. This is due to the constraints of the physical (real) environment, such that the models must be placed on the floor with the y-axis always equal to zero. Thus, the machines/robots are unable to rotate around the x- and z-axes.

The global coordinate system in the virtual environment is the fixed reference coordinate system for all virtual objects. The global coordinate system maintains the position and orientation of the models such that they are unaffected by modelling transformation. The coordinate system is located on the virtual floor rather than on any virtual object since the virtual objects are allowed to move and hence, only the coordinates of the transformed models will be updated. The relative distance between the virtual objects can be calculated using vector analysis. As an example, the material handling of a milling machine is handled by an industrial robot while an operator is assigned to the lathe machine. The new layout involves switching the role (location) of the robot and operator. The global coordinate is fixed at the front-centre of the grid and the transformed coordinates (set-A) of the virtual objects are tabulated in Table 2. An additional set of coordinates (set-B) is shown if the industrial robot is selected as the reference frame (local coordinate) for comparison purposes.

Table 2: Global coordinate versus local coordinate in layout design

	Virtual Object	Original Layout	New Layout
Set A	Forklift	(−900, 0, 400), 135°CCW	(−900, 0, 400), 135°CCW
	Milling	(−200, 0, 1500)	(−200, 0, 1500)
	Lathe	(1000, 0, 1000), 315°CCW	(1000, 0, 1000), 315°CCW
	Robot	(−200, 0, 1000), 270°CCW	*(550, 0, 650), 45°CCW*
	Operator	(550, 0, 650), 135°CCW	*(−200, 0, 1000), 180°CCW*
Set B	Forklift	(600, 0, −700), 225°CCW	*(−120.2, 0, 848.5), 90°CCW*
	Milling	(−500, 0, 0), 90°CCW	*(70.7, 0, 1131.4), 315°CCW*
	Lathe	(0, 0, 1200), 45°CCW	*(565.7, 0, −70.7), 270°CCW*
	Robot	(0, 0, 0)	*(0, 0, 0)*
	Operator	(350, 0, 750), 225°CCW	*(−282.8, 0, 777.8), 135°CCW*

In general, the coordinates of the virtual objects in set-A can be measured directly from the reference point and are independent of each other. Therefore, only the position and orientation of the moving objects (operator and robot) are updated whereas the others are maintained at the same position. In set-B, the robot (local coordinate) is selected as the reference frame, and it is moved and rotated. Consequently, the reference frame changes even though the coordinate of the robot always remains at (0, 0, 0). Hence, the coordinates for other objects need to be re-calculated based on the new position and orientation of the industrial robot. For this reason, set-A is more suitable and practical for use in a flexible manufacturing environment. The global coordinate system is then applied in the VR-RoWL.

3D Models and Coordinate System in VR-RoT

The assembly format is also used in the robot path teaching system. Before proceeding to the path teaching process, the user is required to plan the robot assembly process in advance, which is based on the reverse engineering concept. Firstly, the original assembly model of the workpieces is imported into the virtual environment at the output buffer. Following this, the user has to disassemble the imported assembly models and place the related parts in the assembly area. Finally, the user is required to move the components into the input buffer accordingly. Hence, validation of the assembly models can be carried out in the assembly area and input buffer. Only then the user can demonstrate the assembly process in the virtual environment and this assembly process will be recorded as the robot path.

The imported model is validated, and the vertices and triangle meshing data are checked and compared. The coordinates of the vertices and triangle meshing sequence are also confirmed. Validation of the disassembled model in the assembly area is carried out (Table 3). The imported assembly models are then dismantled and placed in the assembly area accordingly. The layout of the components is used to design the assembly jig and the VR-based robot path teaching system. Finally, the components are moved into the pallet at the input buffer and the output model is validated accordingly.

Table 3: Example of validation in VR-RoT

Parameter		Original (Solidworks)	Imported (VR-RoT)	Dismantled (VR-RoT)	Exported (Internet browser)
Geometric Information	Number of components	3	3	3	3
	Total vertices	3708	3708	3708	3708
	Total triangles	1236	1236	1236	1236
Vertices' coordinates checking		Yes (same)	Yes (same)	Yes (same)	Yes (same)
Triangle meshing sequence checking		Yes (All CCW)	Yes (All CCW)	Yes (All CCW)	Yes (All CCW)

The origin of the industrial robot coordinate frame is located at its base. However, the local coordinate system is a convenient and common method of robot programming in a virtual environment, and is located at the end effector of the robot arm at the "home" position. If the user likes to change the end effector, the relative distance between the new end effector and work-pieces need to be updated although the origin (robot base) is unaltered. Therefore, the local coordinate system is preferable for use in VR-RoT.

Frame Delay

Real-time updating simulation is an important visual feedback in order to provide a sense of presence in the virtual environment. Moreover, simulation delays will create flickering images, which will cause cyber sickness. In general, delays are caused by insufficient computational load for handling large quantities of polygons in the virtual environment. The situation may become more critical if the user likes to interact with virtual objects in real-time. Furthermore, the active stereo is set at 120 Hz (120 frames-per-second) to create flicker-free stereoscopic images. Therefore, frame delays should be validated by checking the refreshing rate (frames-per-second) in real-time simulations. The validation process is checked with examples of virtual environment and the results are tabulated in Table 4.

Table 4: Validation of frame delay

Virtual environment	Passive stereo (60 Hz)		Active stereo (120 Hz)	
	without sensor	with sensor	without sensor	with sensor
Five (5) Equipment with Vertices (21,040) and Triangles (38,145)	√	√	√*	√*+
Five (5) Equipment with Vertices (54,107) and Triangles (97,275)	√	√	√*	√*+
Nine (9) Equipment with Vertices (140,116) and Triangles (142,608)	√	√	√*	√*+

The validation results show that the virtual environment created matches the computational load for the selected computer workstation. There are no frame delays when the refreshing rate is tested at 60 frames-per-second (FPS or Hz). The first two frames fall below the rated FPS (marked by "*" in Table 4) when the refreshing rate is doubled. This is attributed to the initial processing of the rendering process, whereby the OpenGL technique "Display List" is used for compiling for later use. This technique evaluates all rendering information (vertices and triangles) and stores the information into the server machine at the beginning of the process. The information created can be re-used without re-evaluating and re-transmitting the data during simulations.

Furthermore, the frame delay remains constant (120 Hz) when the sensors are activated in the active stereo mode. However, the frequencies of the sensors are limited to 60 Hz. Thus, the following methods are used to overcome frame delay:

The following frame will be updated accordingly when the data is triggered and sent from the sensors. A total of 60 frames will be updated based on the input information from the sensors in an event-mode input model.

The previous information will be used to update the frames if no data is received. The information is applied to the remaining frames, where the OpenGL is able to simulate the virtual environment until the next input is received (marked by "+" in Table 4).

RESULTS AND DISCUSSION

A case study is conducted at the Robotics Laboratory, Department of Mechanical Engineering, Faculty of Engineering, University of Malaya. The industrial robotic work cell is required to assemble an electronics casing which consists of three components: a main casing and two identical heat sink assemblies. The main casing is made of aluminium and the heat sink assemblies are attached onto the main casing. Each heat sink assembly is a combination of a heat sink and a side plate, which has been assembled prior to the robotic work cell. The work cell initially consists of two industrial robots (KUKA KR6/2). However, due to the high maintenance and performance of the robots (payload: 6 kg only), the existing robots are replaced with KUKA KR16 KS (payload: 16 kg). Therefore, a new layout design is required since both robots have different workspace and reach requirements. The machines and robots are listed in Table 5.

Table 5: Industrial robotic work cell

No.	Machine/Robot	Task
1	KUKA KR 16 (A)	Pick-and-place
2	KUKA KR 16 (B)	Fastening
3	Assembly jig with pneumatic system	Assembly area for the casing
4	PLCs	Synchronize and control assembly process
5	Pallet with fibre optic sensor	Input buffer and use sensor to trigger process

The assembly process of the electronics casing is identified during the design stage, as shown in Figure 8. Firstly, the main body (*O*-1) is placed at its position in the jig. Following this, the left heat sink assembly (*O*-2) is fixed onto the main body using two screws. Finally, the right heat sink assembly (*O*-2) is fixed onto the main body using two other screws.

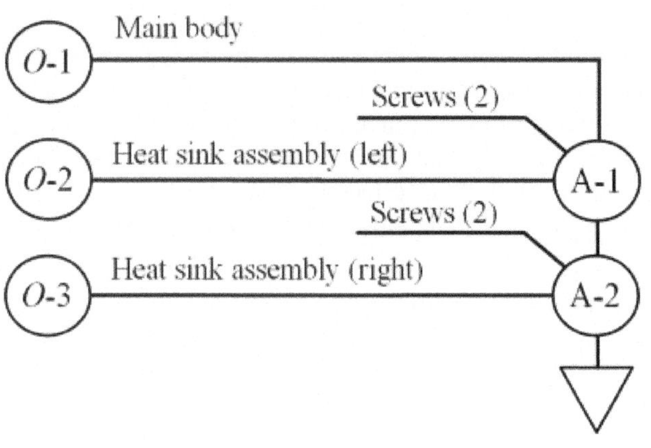

Figure 8: Assembly process chart for case study.

VR-Robotic Work Cell Layout (VR-RoWL)

The 3D models of the robots and related components are developed using commercial CAD modelling software. Following this, these models are imported into the virtual work cell for layout planning, in which the assembly model is saved in the VRML format. The layout of the laboratory needs to be studied in advance. The location of the power supply, entrance (door), windows (if any), materials supply and others should be confirmed before re-arranging the layout of the robotic work cell. In this case study, the upper-right corner of the laboratory is assigned as the global coordinate system. The dimensions of the robotic zone will be used to set up the layout for the VR-RoWL. All components and machines must be located within this zone, with the exception of materials supply mechanisms/machines.

The original assembly model is then imported into the VR-RoWL. The models are free to move within the fixed zone (grid). In general, the robots should be placed in parallel and facing towards the positive *y*-axis direction. The assembly zone is located between the robots and within the workspace of both robots. This layout design can be implemented for various assembly processes involving two robots, whereby the materials supply is located in front of the robots. Therefore, the new layout can be obtained by moving the related machines/robots in the virtual layout. The new layout can be saved in the VRML format, which can be shared between various Internet browsers. The proposed robotic work cell created by the VR-RoWL and the actual laboratory setup are shown in Figure 9.

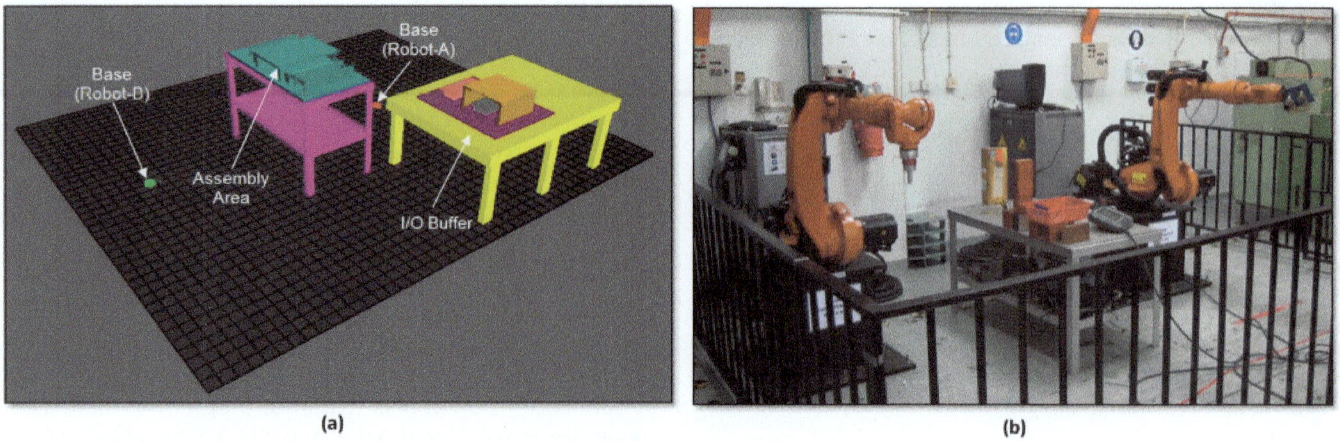

Figure 9: (a) Proposed robotic work cell (b) Actual robotic work cell setup.

The laboratory is set up according to the above mentioned criteria and the proposed new layout design. The actual dimensions of the equipment are measured based on the proposed global coordinate system. Comparisons between the actual coordinates and virtual layout coordinates for the robotic work cell are tabulated in Table 6. There are some constraints while installing the robots and equipment in the actual laboratory setup; one of them being the accuracy of the measurement tools. Thus, the unit used in the actual setup is centimetres (cm) and the tolerance is ±0.5 cm. The orientation of the equipment in the actual installation is difficult to control and therefore the tolerance is set as ± 3° (for all axes). The digital coordinates (virtual layout) do not impose limitations on the environmental conditions (e.g. floor) and instrumentations (e.g. measuring tape).

Table 6: Comparisons between virtual layout and laboratory setup

No	Equipments	Virtual Layout (x, y) (mm)	Laboratory (x, y) (mm)	Absolute Variation (x, y) (mm)
1	KUKA KR 16 (A)	(1459, 1716)	(1450, 1700)	(9, 12)
2	KUKA KR 16 (B)	(3965, 1713)	(3950, 1700)	(15, 13)
3	Input buffer (Conveyor)	(1556, 3203)	(1570, 3230)	(14, 27)
4	Output buffer (Conveyor)	(1555, 3209)	(1570, 3230)	(15, 21)
5	Assembly jig (Table)	(2515, 2041)	(2530, 2025)	(15, 16)

There are differences between the virtual layout and actual laboratory setup, and the virtual layout provides an overview of the arrangement. There are no measurement tools used in the virtual environment. The user locates the equipment using his/her visual feedback only. For example, the user is unable to "see" a difference of 5 mm in a virtual environment. The user is also unable to "see" whether two robots are exactly located at the same axis (y=1700 mm). This can be improved by applying the "snap-fit" feature, in which each grid is 10 mm.

VR-Robot Teaching System (VR-RoT)

The imported assembly model is located on the input buffer by default. The user is requested to dismantle the assembly model and move the components to the pallet and arrange them accordingly. In this mode, only the final position and orientation of the components will be stored, neglecting the path of the movement. The path planning mode is then activated after assembly planning and the user is requested to either plan the robot path by moving the components (material handling, e.g. suction cup) or demonstrate the process path (material processing, e.g. fastening). The path will be recorded and translated into robot commands. The loaded information will not be substituted by the final location of the components. In addition, there are two basic ways of describing the movement of industrial robots in a robot control system, i.e. angle of the manipulator's joints or location of the end effector. If all joint variables are known, then forward kinematics can be used to determine the position and orientation of the robot's end (tool centre point). However, inverse kinematics is used to calculate the value of each joint variable by knowing the position and orientation of the end effector. Thus, the inverse kinematic model is developed and used in this research [28].

Robot-A is programmed to pick-and-place (PnP) the components between the input buffer and assembly area. There is no force or tactile feedback provided to the developed robot teaching system. The collision detection algorithms are applied to detect whether the end effector touches the components, in which a bounding box will be drawn at the end effector when collision is detected. Furthermore, the colour of the bounding box will change at different stages of the PnP process, as follows:

- No bounding box: No collision is detected (Figure 10 (a)).

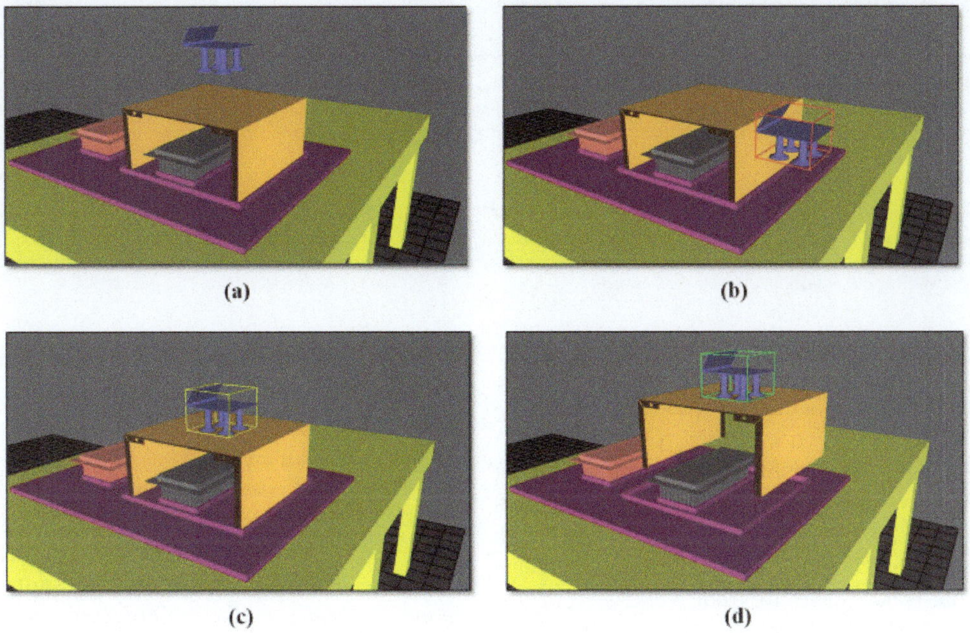

Figure 10: Bounding box for end effector.

- Red (blinking): Collision is detected between the end effector and workpiece, but the PnP process cannot be continued (Figure 10 (b)).
- Yellow (blinking): Collision is detected between the end effector and workpiece, and ready for the PnP process (Figure 10 (c)).
- Green (blinking): The colour of the workpiece will stop blinking when the end effector is activated. The workpiece is now grasped by the end effector and moves together with it (Figure 10 (d)).

In the virtual teaching system, collision detection is enabled when the end effector touches the components or the components collide with the assembly jig. Thus, the end effector always pick-and-place at the right height (*z*-axis). Only the horizontal positions (*x*-axis and *y*-axis) may differ slightly from the exact location, as shown in Table 7.

Table 7: Data of PnP process for Robot-A

Step	Task	VR generated Coordinate (x, y, z) (mm)	Predefined Coordinate (x, y, z) (mm)	Absolute Variation (x, y, z) (mm)
a1	Pick: Main Body	(1677, −126, 867)	(1670, −120, 867)	(7, 6, 0)
a2	Place: Main Body	(327, −1076, 1200)	(325, −1080, 1200)	(2, 4, 0)
b1	Pick: Heat Sink Assembly (left)	(1283, −115, 750.5)	(1290, −120, 750.5)	(7, 5, 0)
b2	Place: Heat Sink Assembly (left)	(−91, −1078, 994)	(−94, −1080, 994)	(3, 2, 0)
c1	Pick: Heat Sink Assembly (right)	(1635, −116, 750.5)	(1640, −120, 750.5)	(5, 4, 0)
c2	Place: Heat Sink Assembly (right)	(740, −1076, 994)	(744, −1080, 994)	(4, 4, 0)

On the other hand, the fastening paths and coordinates need to be controlled, as the fastening coordinates are fixed. These coordinates can be tracked by the screw holes available in the heat sink assembly, which enables the user to change the sequence of the fastening process. Figure 11 shows the fastening sequence defined in the virtual environment and Table 8 shows the coordinates of the fastening process.

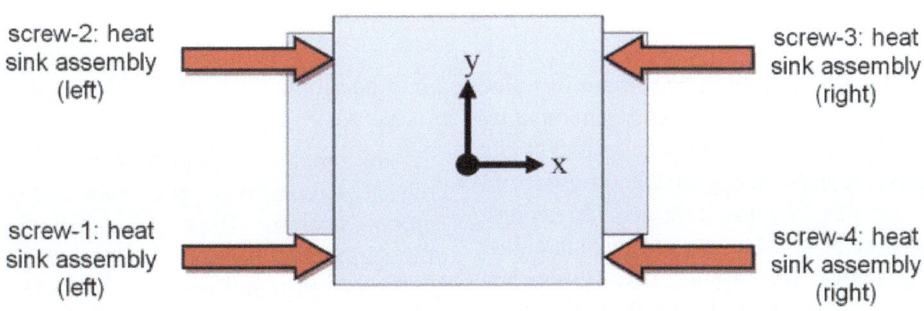

Figure 11: Fastening sequence.

Table 8: Data of fastening process for Robot-B

Screw	Predefined Coordinate (x, y, z) (mm)	Fastening direction (Robot's Coordinate system)	Fastening Depth (mm)
1	(113.4, 1270, 1190)	positive *x*-axis	10
2	(113.4, 1570, 1190)	positive *x*-axis	10
3	(536.6, 1570, 1190)	negative *x*-axis	10
4	(536.6, 1270,1190)	negative *x*-axis	10

Laboratory Testing

The actual testing is carried out in the Robotics Laboratory. The original work cell consists of two industrial robots of KUKA KR6/2 (payload: 6 kg). The existing robots require higher maintenance and the low payload is unable to fulfil the current requirements. Therefore, the existing robots are replaced with KUKA KR16 KS (payload: 16 kg). Thus, a new layout design is needed since both robots have different workspace, reach, speed and other requirements. Following this, the VR-RoWL model is used to determine and evaluate the layout prior to the actual hardware installation. The robot teaching system (VR-RoT) is used to plan and determine the robot path in advance. The robot process path is taught and the robot commands are generated offline. The layout and process path planned in the VR-RoWL are validated using KUKA SimPro Simulation environment to determine clearance, reach, safety etc. KUKA OfficeLite is used to validate the robot commands generated by the VR-RoT. Figure 12 shows the simulation and validation using KUKA software before being loaded into the actual system.

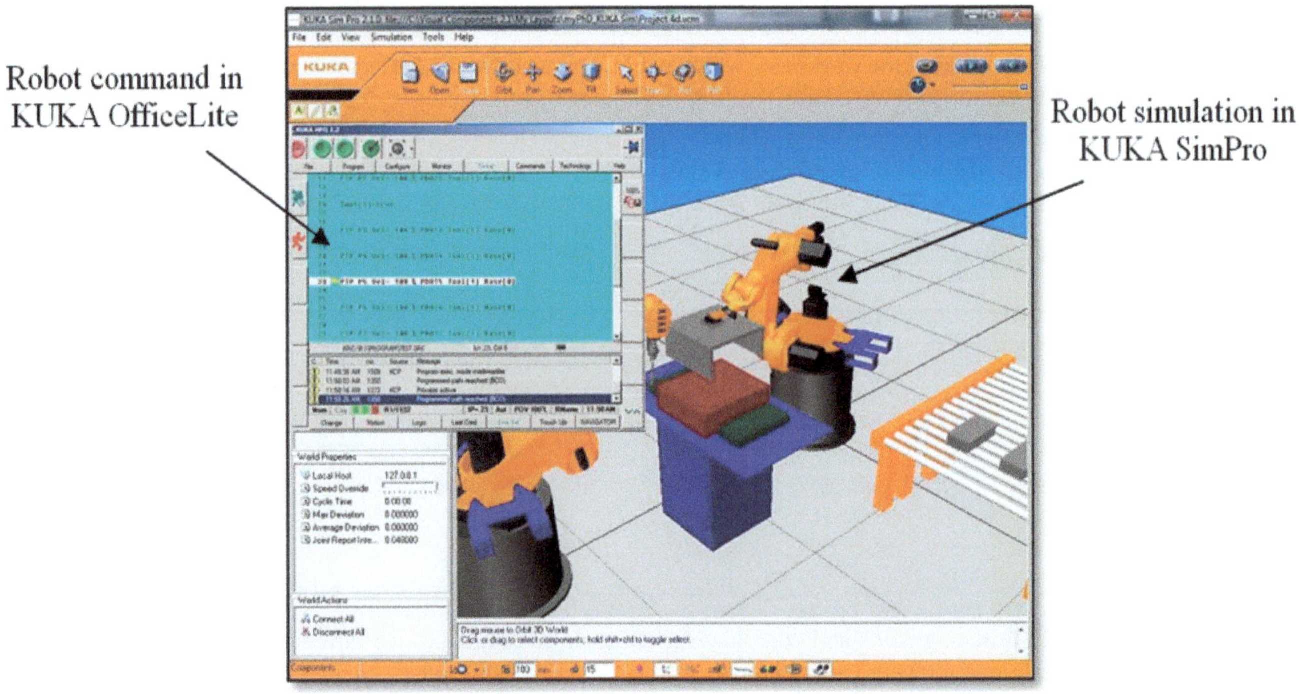

Figure 12: Simulation and validation using KUKA software.

Finally, the actual robots and all equipment are installed at their respective locations based on the layout generated by the VR-system. The KUKA robot commands generated by the VR-based system are saved into a floppy disk and transferred into Robot-A (pick-and-place) and Robot-B (fastening), respectively. The end effector is then attached to each robot. The assembly jig is assembled with a pneumatic system and the input buffer is equipped with a fibre optic sensor. The PLCs are programmed to synchronize and control the assembly process for all equipment.

The robots are operated at 10% of their maximum speed under two conditions (i.e. with and without load) during laboratory testing. The speed is then increased to 30 and 50% of the maximum speed for both conditions. These changes are made using the control panel of the KUKA robot. The average cycle time of the test (five tests for each condition and speed) is recorded and tabulated in Table 9.

Table 9: Average cycle time for various speeds

Process	10% of Speed		30% of Speed		50% of Speed	
	with load (sec)	without load (sec)	With load (sec)	without load (sec)	with load (sec)	without load (sec)
1. Trigger the sensor	1.0	1.0	1.0	1.0	-	1.0
2. Robot-A: Pick the components from input buffer and place at assembly area	103.12	102.88	36.56	36.28	-	23.08
3. Position the jig	2.0	2.0	2.0	2.0	-	2.0
4. Robot-B: Fastening the four (4) screws.	49.38	49.12	17.41	17.14	-	11.08
5. Position the jig	2.0	2.0	2.0	2.0	-	2.0
6. Robot-A: Pick the completed assembly at assembly jig and place at output buffer. Return to home position.	38.56	38.36	14.16	13.92	-	8.68
Cycle Time	**196.06**	**195.36**	**73.13**	**72.34**	-	**47.84**

Overall, it can be seen that the cycle time is reduced when the speed of the robots is increased, whereby the relationship between the cycle time and percentage speed of the robot is fairly linear. It is also found that the cycle time without load is slightly faster than that with load, which is about 0.70 and 0.79 sec for 10% and 30% speed, respectively. The suction cup is unable to hold the components firmly when the robot arm travels at high speed (50% speed) and thus the data for 50% speed is unsuitable for load-carrying use in this case study.

CONCLUSIONS

A virtual reality based programming for an industrial robotic work cell has been developed in this research, and is given the name "VR-Rocell". The system is divided into two major sub-systems, namely, VR-Robotic Work Cell Layout (VR-RoWL) and VR-Robot Teaching System (VR-RoT). The system is designed to improve the human-machine interface (human-robot interface), whereby off-line programming is used to generate robot commands which will reduce robot downtime and prevents the programmer from exposure to potential hazardous environment.

The system has been validated and tested using a case study. Several settings and assumptions have been made during the validation process. The generic approach of integrating and converting file formats is validated for both part and assembly files and the frame delay has also been tested. It is found that the algorithms and approaches used are able to generate a flicker-free virtual environment for both passive and active stereoscopic visualization systems. Moreover, a case study has been conducted for an electronics casing assembly. The results reveal that there is an absolute variation of 9–15 mm (x-axis) and 12–27 mm (y-axis) between the virtual layout and actual setup. The virtual path teaching shows an absolute variation of 2–7 mm (x-axis) and 2–6 mm (y-axis). It is found that the variations in dimensions have no effects on the performance of the robotic work cell. Thus, it is concluded that the VR-Rocell can be used in an industrial robotic work cell based on the accuracy and effectiveness achieved in the case study. Future work can be focused on layout and sequence optimization, analysis of human factors in workstation design and multi-criteria decision making for sustainable manufacturing in order to further improve the proposed system.

REFERENCES

1. IFR (2001) World Robotics: Statistics, market analysis, forecasts, case studies and profitability of robot investment: United Nations Economic Commission for Europe (UNECE), International Federation of Robotics.
2. IFR (2011) World Robotics: Statistics, market analysis, forecasts, case studies and profitability of robot investment: United Nations Economic Commission for Europe (UNECE), International Federation of Robotics.
3. OSHA Japan (1995) Study on accidents Involving industrial robot report no. 5, Labor Standard Bureau: Occupational Safety and Health Department, Japan Ministry of Labour.
4. Choo HK (2014) Virtual Reality-based Training System for Metal Active Gas Welding. Department of Mechanical Engineering: University of Malaya, Malaysia.
5. Burdea GC (1999) Invited review: The synergy between virtual reality and robotics. IEEE Transactions on Robotics and Automation 15: 400–410
6. Jahangirian M, Eldabi T, Naseer A, Stergioulas LK, Young T (2010) Simulation in manufacturing and business: A review. European Journal of Operational Research 203: 1–13
7. Li E, Du TC (2005) Advances in electronic business, volume I Idea Group Publishing.
8. Klingstam P, Gullander P (1999) Overview of simulation tools for computer-aided production engineering Computers in Industry. 38: 173–186
9. Neto P, Pires JN, Moreira AP (2010) Robot path simulation: a low cost solution based on CAD; Singapore. pp.333–338.
10. Staranowicz A, Mariottini GL (2011) A survey and comparison of commercial and open-source robotic simulator software; Greece.
11. Radhakrishnan R, Balasubramanian S (2008) Business process re-engineering: text and cases. New Delhi: Prentice-Hall of India.
12. Lawrence Associates (1994) Technical Report. Virtual Manufacturing User Workshop.
13. Kim SC, Choi KH (2000) Development of Flexible Manufacturing System using Virtual Manufacturing

Paradigm. International Journal of the Korean Society of Precision Engineering 1: 84–90

14. Mujber TS, Szecsi T, Hashmi MSJ (2004) Virtual reality applications in manufacturing process simulation. Journal of Materials Processing Technology 155–156: 1834–1838

15. Raneda A, Pessi P, Siuko M, Handroos H, Palmer J, et al. . (2003) Utilization of virtual prototyping in development of CMM. Fusion Engineering and Design 69: 183–186

16. Gökdere LU, Benlyazid K, Dougal RA, Santi E, Brice CW (2002) A virtual prototype for a hybrid electric vehicle. Mechatronics 12: 575–593

17. Guerra D, Ramírez R, Donato L, Molina A (2006) Digital Manufacturing for Aerospace industry: Experimental Aircraft; Mexico.

18. Lee WB, Cheung CF, Li JG (2001) Applications of virtual manufacturing in materials processing.Journal of Materials Processing Technology 113: 416–423

19. Iqbal M, Hashmi MSJ (1999) Application of virtual reality for the training of engineering personnel; Dublin, Ireland. pp.1959–1966.

20. Crison F, Lécuyer A, d'Huart DM, Burkhardt J-M, Michel G, et al. Virtual technical trainer: learning how to use milling machines with multi-sensory feedback in virtual reality; 2005; Bonn, Germany. pp.139–145.

21. Zhu L, Wang J, Chen E, Yang J, Wang W (2008) Applications of virtual reality in turn-milling centre; Qingdao, China. pp.2302–2305.

22. Wang QH, Li JR (2004) A desktop VR prototype for industrial training applications. Virtual Reality 7: 187–197

23. Kadir AA, Xu X, Hämmerle E (2011) Virtual machine tools and virtual machining-A technological review. Robotics and Computer-Integrated Manufacturing 27: 494–508

24. Lee WB, Li JG, Cheung CF (2002) Development of a virtual training workshop in ultra-precision machining. International Journal of Engineering Education 18: 584–596

25. Mitsi S, Bouzakis K-D, Mansour G, Sagris D, Maliaris G (2005) Off-line programming of an industrial robot for manufacturing. International Journal of Advanced Manufacturing Technology 26: 262–267

26. Fauadi HF, Jumali MS (2008) Modeling and simulation of programmable universal machine for assembly (PUMA) industrial robot for automotive-related assembly process; Kuala Lumpur.

27. Kumar R, Kalra P, Prakash NR (2011) Virtual RV-M1 robot system. Robotics and Computer-Integrated Manufacturing 27: 994–1000

28. Yap HJ, Taha Z, Lee JV (2008) VR-based Robot Programming and Simulation System for an Industrial Robot. International Journal of Industrial Engineering – Theory, Application and Practice 15: 314–322

29. Xu ZJ, Zhao ZX, Wu MH, Liao JB, Tian GY (1997) Virtual reality based robot graphic simulation and virtual manufacturing system; Manchester.

30. Santis AD, Siciliano B (2008) Safety issues for human-robot cooperation in manufacturing systems; Napoli, Italy. pp.1–12.

31. Or CKL, Duffy VG, Cheung CC (2009) Perception of safe robot idle time in virtual reality and real industrial environments. International Journal of Industrial Ergonomics 39: 807–812

32. Whyte J, Bouchlaghem N, Thorpe A, McCaffer R (2000) From CAD to virtual reality: modelling approaches, data exchange and interactive 3D building design tools. Automation in Construction 10: 43–55

33. Yap HJ, Taha Z, Dawal SZ (2012) A Generic Approach of Integrating 3D Models into Virtual Manufacturing. Journal of Zhejiang University-SCIENCE C (Computers & Electronics) 13: 20–28

34. Malaysian Anthropometry (2008) Anthropometric data for the Malaysian adult population. Unpublished data, Centre for Product Design & Manufacture, University of Malaya, Kuala Lumpur.

35. Taha Z, Soewardi H, Dawal SZM (2014) Axiomatic design principles in analysing the ergonomics design parameter of a virtual environment. International Journal of Industrial Ergonomics 44: 368–373

CHAPTER 9

MODELING 3D FACIAL SHAPE FROM DNA

Peter Claes[1] , Denise K. Liberton[2] , Katleen Daniels[1] , Kerri Matthes Rosana[2] , Ellen E. Quillen[2] , Laurel N. Pearson[2] , Brian McEvoy[3] , Marc Bauchet[2] , Arslan A. Zaidi[2] , Wei Yao[2] , Hua Tang[4] , Gregory S. Barsh4,[5], Devin M. Absher[5] , David A. Puts[2] , Jorge Rocha[6,7], Sandra Beleza[4,8], Rinaldo W. Pereira[9] , Gareth Baynam[10,11,12], Paul Suetens[1] , Dirk Vandermeulen[1] , Jennifer K. Wagner[13], James S. Boster[14], Mark D. Shriver[2]

[1]Medical Image Computing, ESAT/PSI, Department of Electrical Engineering, KU Leuven, Medical Imaging Research Center, KU Leuven & UZ Leuven, iMinds-KU Leuven Future Health Department, Leuven, Belgium,

[2]Department of Anthropology, Penn State University, University Park, Pennsylvania, United States of America,

[3]Smurfit Institute of Genetics, Dublin, Ireland,

[4]Department of Genetics, Stanford University, Palo Alto, California, United States of America,

[5]HudsonAlpha Institute for Biotechnology, Huntsville, Alabama, United States of America,

[6]CIBIO: Centro de Investigac¸a˜o em Biodiversidade e Recursos Gene´ticos, Universidade do Porto, Porto, Portugal,

[7]Departamento de Biologia, Faculdade de Cie^ncias, Universidade do Porto, Porto, Portugal,

[8]IPATIMUP: Instituto de Patologia e Imunologia Molecular da Universidade do Porto, Porto, Portugal,

[9]Programa de Po´s-Graduac¸a˜o em Cie^ncias Geno^micas e Biotecnologia, Universidade Cato´lica de Brası´lia, Brasilia, Brasil,

[10]School of Paediatrics and Child Health, University of Western Australia, Perth, Australia,

[11]Institute for Immunology and Infectious Diseases, Murdoch University, Perth, Australia,

[12]Genetic Services of Western Australia, King Edward Memorial Hospital, Perth, Australia,

[13]Center for the Integration of Genetic Healthcare Technologies, University of Pennsylvania, Philadelphia, Pennsylvania, United States of America,

[14]Department of Anthropology, University of Connecticut, Storrs, Connecticut, United States of America

ABSTRACT

Human facial diversity is substantial, complex, and largely scientifically unexplained. We used spatially dense quasilandmarks to measure face shape in population samples with mixed West African and European ancestry from three locations (United States, Brazil, and Cape Verde). Using bootstrapped response-based imputation modeling (BRIM), we uncover the relationships between facial variation and the effects of sex, genomic ancestry, and a subset of craniofacial candidate genes. The facial effects of these variables are summarized as response-based imputed predictor (RIP) variables, which are validated using self-reported sex, genomic ancestry, and observer-based facial ratings (femininity

Citation: Claes P, Liberton DK, Daniels K, et al. Modeling 3D Facial Shape from DNA. Luquetti D, ed. PLoS Genetics. 2014;10(3):e1004224. doi:10.1371/journal.pgen.1004224.

and proportional ancestry) and judgments (sex and population group). By jointly modeling sex, genomic ancestry, and genotype, the independent effects of particular alleles on facial features can be uncovered. Results on a set of 20 genes showing significant effects on facial features provide support for this approach as a novel means to identify genes affecting normal-range facial features and for approximating the appearance of a face from genetic markers.

SUMMARY

The face is perhaps the most inherently fascinating and aesthetic feature of the human body. It is a principle subject of art throughout human history and across cultures and populations. It provides the most significant means by which we communicate our emotions and intentions in addition to health, sex, and age. And yet features such as the strength of the brow ridge, the spacing between the eyes, the width of the nose, and the shape of the philtrum are largely scientifically unexplained. Here, we use a novel method to measure face shape in population samples with mixed West African and European ancestry from three locations (United States, Brazil, and Cape Verde). We show that facial variation with regard to sex, ancestry, and genes can be systematically studied with our methods, allowing us to lay the foundation for predictive modeling of faces. Such predictive modeling could be forensically useful; for example, DNA left at crime scenes could be tested and faces predicted in order to help to narrow the pool of potential suspects. Further, our methods could be used to predict the facial features of descendants, deceased ancestors, and even extinct human species. In addition, these methods could prove to be useful diagnostic tools.

INTRODUCTION

The craniofacial complex is initially modulated by precisely-timed embryonic gene expression and molecular interactions mediated through complex pathways [1]. As humans grow, hormones and biomechanical factors also affect many parts of the face [2], [3]. The inability to systematically summarize facial variation has impeded the discovery of the determinants and correlates of face shape. In contrast to genomic technologies, systematic and comprehensive phenotyping has lagged. This is especially so in the context of multipartite traits such as the human face. In typical genome-wide association studies (GWAS) today phenotypes are summarized as univariate variables, which is inherently limiting for multivariate traits, which, by definition cannot be expressed with single variables. Current state-of-the-art genetic association studies for facial traits are limited in their description of facial morphology [4]–[7]. These analyses start from a sparse set of anatomical landmarks (these being defined as "a point of correspondence on an object that matches between and within populations"), which overlooks salient features of facial shape. Subsequently, either a set of conventional morphometric measurements such as distances and angles are extracted, which drastically oversimplify facial shape, or a set of principal components (PCs) are extracted using principal components analysis (PCA) on the shape-space obtained with superimposition techniques, where each PC is assumed to represent a distinct morphological trait. Here we describe a novel method that facilitates the compounding of all PCs into a single scalar variable customized to relevant independent variables including, sex, genomic ancestry, and genes. Our approach combines placing spatially dense quasi-landmarks on 3D images [8], [9], principal component analysis (PCA), and a new partial least squares regression (PLSR, [10]) derived method we call "bootstrapped response-based imputation modeling" (BRIM) to measure and model facial shape variation.

Given the multivariate nature of the face and the large number of genes likely affecting variation in the face, we chose to focus attention on the between-population variation with a genetic admixture approach using research participants from three West African/European admixed populations. Ancestry informative markers (AIMs) can be used to estimate individual genomic ancestry from DNA [11], which can be used to investigate population differences and map genes for genetically determined traits that vary between populations. Non-random mating and continuous gene flow in admixed populations results in admixture stratification or variation in individual ancestry [12], [13]. The process of admixture also results in admixture linkage disequilibrium or the non-random association among both AIMs and traits that vary between the parental populations. These characteristics make admixed populations uniquely suited to investigations into the genetics of such traits [14]–[16]. By simultaneously modeling facial shape variation as a function of sex and genomic ancestry along with genetic markers in craniofacial candidate genes, the effects of sex and ancestry can be removed from the model thereby providing the ability to extract the effects of individual genes.

RESULTS/DISCUSSION

A spatially dense mesh of 7,150 quasi-landmarks was used to map 3D images of participants' faces onto a common coordinate system (Figure 1). Quasi-landmarks are defined here as largely homologous vertices in this mapped mesh. The mesh is applied automatically, eliminating the difficult and error-prone procedure of manually indicating facial landmarks [8], [9], [17]. Deviations from bilateral symmetry were removed by averaging each face with its mirror image [18], [19]. PCA on the symmetrized 21,450 quasi-landmark 3D coordinates (X, Y, and Z for each of the 7,150 quasi-landmarks) using all 592 participants produces 44 principal components (PCs) that together summarize 98% of the variation in face shape and define a multidimensional face space. The effects of the first 10 PCs are illustrated in Figure 2. Some of these PCs (*e.g.*, PC4, PC5) capture the effects of changes in only particular parts of the face. However, many PCs (*e.g.*, PC1, PC2, PC3) capture effects in multiple parts of the face. Moreover, although the PCs are statistically independent, any particular part of the face is affected by several PCs. As such, it is likely incorrect to assume that each PC represents a distinct morphological trait resulting from the action of specific genes. Our use of BRIM to combine the independent effects of PCs is agnostic about their biological meaning, if any, and provides for the compounding of the information from any or all of the PCs together into a single variable that is customized to the predictor variable being modeled. In this way, BRIM also overcomes the problem of multiple testing inherent to other methods for summarizing facial variation. In other words, the hypothesis, *does this gene have significant effects on facial shape*, can be addressed with a single statistical test.

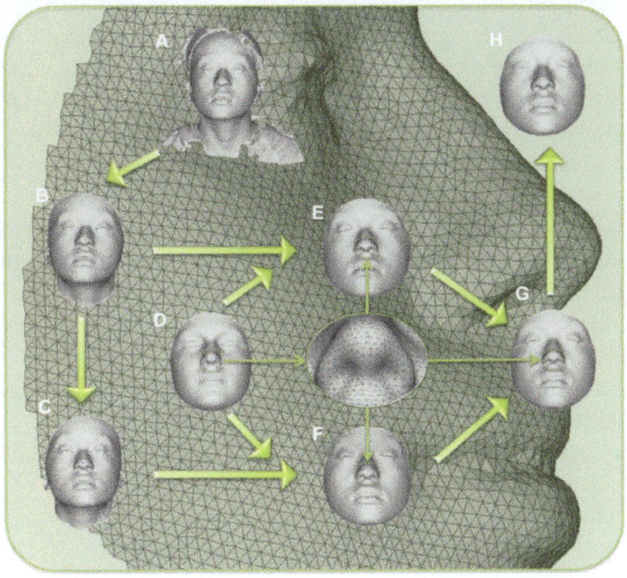

Figure 1: Workflow for 3D face scan processing.

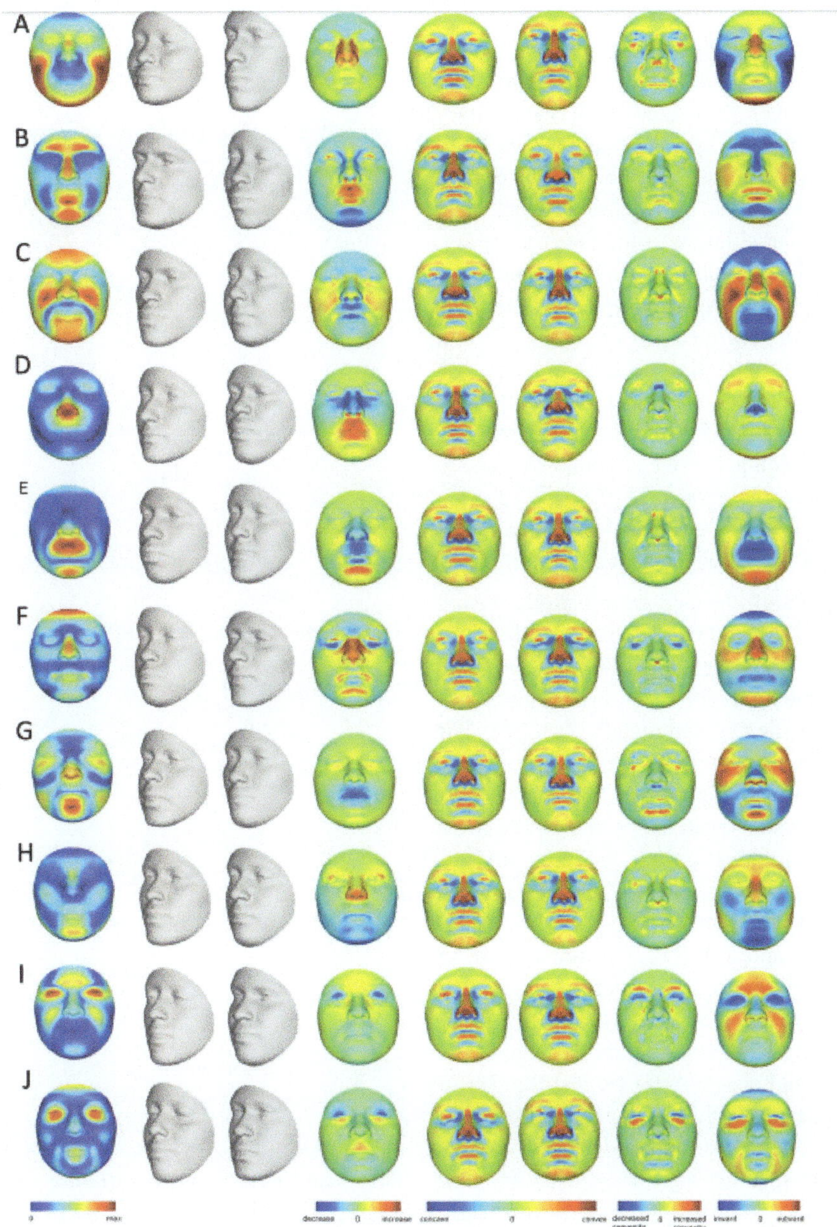

Figure 2: PCA effects on facial morphology.

BRIM is an extension of existing relationship modeling techniques that uses response variables to refine and, in some cases, to transform one or more initial predictor variables. In other words and in contrast to alternate techniques, BRIM uses a multivariate matrix of response variables in a leave-one-out forced imputation setup to update the initial predictor variable values, creating a new type of variable – the response-based imputed predictor (RIP) variable (Figure S2). The BRIM process is bootstrapped, and estimator improvement over successive iterations can be monitored (Figures S5, S6, S7, S8, S9). BRIM also functions to correct observation error, misspecification of predictor values, and other sources of statistical confounding. Within the iterative bootstrapping scheme, a nested leave-one-out approach is used to avoid model over-fitting and to allow hypothesis testing using standard statistical techniques, such as correlation analysis, ANOVA, and receiver operating characteristic (ROC) curve analysis [20], to test the significance of the association between the predictors and RIP variables. Likewise, the relationships between the RIP variables and the response variables, *e.g.*, the 21,450 facial parameters, allows for the visualization and quantitation of their effects on face shape.

RIP variables modeling sex (RIP-S) and genomic ancestry (RIP-A), as well as those modeling the effects of particular genetic markers (RIP-Gs), can be visualized using two primary methods – shape transformations and heat maps. We used three summary statistics (area ratio, normal displacement, and curvature difference), which can be illustrated using heat maps, to quantify the particular changes to the face that result. These measures of facial change, along with particular inter-landmark distances, angles, and spatial relationships, can together be termed *face shape change parameters* (FSCPs). FSCPs provide a means of translating face shape changes from the abstract face space into both visual representations into words. Such terms are used in clinical and anthropological descriptions of faces and by doing so we can compare these to the BRIM results (*e.g.*, Figures S28, S29, S30, S31, S32, S33, S36, S37, S38, and Table S1). The statistical significance of these and related FSCPs can be tested using permutation.

As expected, many parts of the face are affected by both ancestry and sex. Figure 3 illustrates the partial effects of RIP-A and RIP-S on facial shape using transformations and heat maps for effect size (R^2) and the three primary FSCPs. Facial regions that are statistically significant ($p<0.001$) for effect size and the FSCPs are shown in Figure 3 as the yellow (not green regions in the bottom panels). The RIP-A and RIP-S shape transformations shown are set to the points three standard deviations plus and minus the mean RIP-A and RIP-S levels in these samples. As seen in the effect-size (R^2) panels in Figure 3, the proportion of the total variance in particular facial features explained by RIP-A and RIP-S can be substantial. In general, up to a third of the variance in several parts of the face is explained by these two variables. RIP-A primarily affects the nose and lips and, to lesser extents, the roundness of the face, the mandible, and supraorbital ridges. Sex has a much larger effect than ancestry on the supraorbital ridges and cheeks, and smaller effects on the nose and under the eyes. The FSCPs help to illustrate the specific ways in which particular RIP variables affect the face. For example, the area ratio shows increased surface area for the medial canthus, sides of the nose, and front of the chin on the European end of RIP-A and a greater surface area for the nostrils and lips on the West African end of RIP-A. The curvature difference highlights the top of the philtrum as a facial feature that is highly convex on the European end and highly concave on the West African end of RIP-A. Regions showing curvature differences for RIP-A are also seen in the nasal bridge, supraorbital ridges, and chin. RIP-S shows greatest effects on the supraorbital ridges, nasal bridge, nasal ridge, zygomatics, and cheeks. The nose, lips, medial canthus, and mandible are also affected by RIP-S. The largest differences in facial curvature related to changes in RIP-S are on the supraorbital ridges and the nasal bridge.

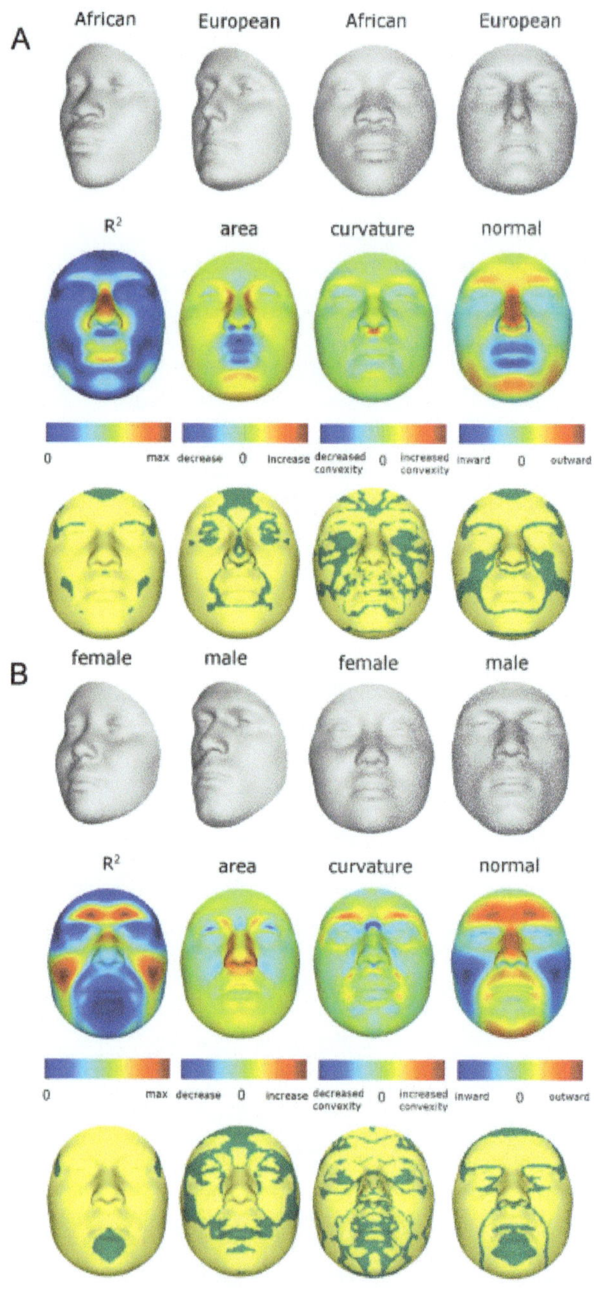

Figure 3: Transformations and heat maps showing how face shape is affected by (A) RIP-A and (B) RIP-S.

Despite the complex ways in which faces are affected by RIP-A and RIP-S, these variables are useful summaries of the degree to which particular faces are more or less ancestry-typical and sex-typical, respectively. This is evident in the strong relationship observed between RIP-A and genomic ancestry as measured with a panel of 68 AIMs ($r=0.81$, $p<0.001$; Figure 4A). Approximately two thirds of the variation in RIP-A across these three West African/European admixed populations is explained by genomic ancestry. Likewise, as seen in Figure 4B, RIP-S is very distinctive between the sexes. ROC analyses (Figure S32) show that the AUC for RIP-S on sex is 0.994 ($p<0.001$). Genomic ancestry, independently from sex, explains 9.6% of the total facial variation, while sex independently from ancestry explains 12.9% of the total facial variation (Table S3). Most facial variation, like human genetic variation in general, is shared among different human populations and by members of both sexes.

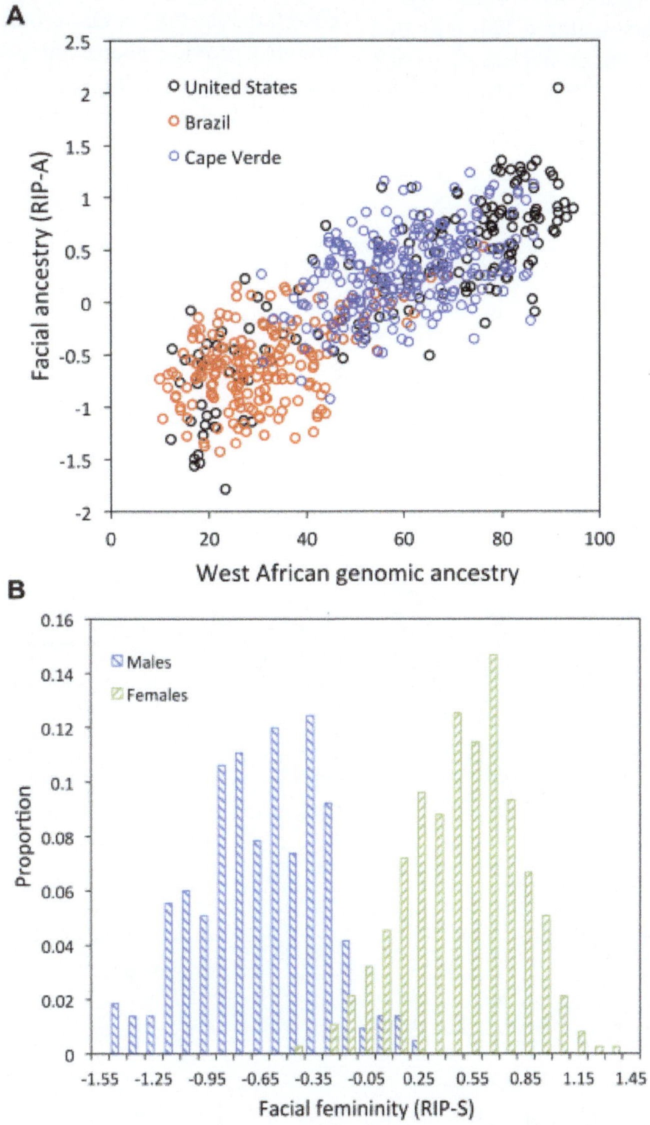

Figure 4: Relationships between the ancestry and sex RIP variables and their initial predictor variables.

We used alternate subsets of AIMs and alternate population samples to test the robustness of the facial ancestry (RIP-A) estimation. RIP-A values were derived using different initial predictor variables and compared. The pairwise correlations of RIP-A estimates are high ($R^2>0.99$), showing that very similar estimates of facial ancestry result from different panels of AIMs (Figure S9) and alternate population samples (Figures S10, S11). The robustness of RIP-A estimates to both marker panel and population sample substantiates the generality and, thus, practical usefulness of these models.

We also see that RIP-A estimates generated using AIMs panels with lower ancestry-information content show stronger correlations with more accurate genomic ancestry estimates than with the genomic ancestry estimates that were used to generate them (Figure S9). To further evaluate the performance of BRIM when less information is available, we performed noise injection experiments by adding or subtracting randomly defined quantities from the estimates of genomic ancestry and misclassifying the sex of persons in the sample (Figures S4, S5, S6, S7, S8 and Figures S12, S13, S14, respectively). These experiments demonstrate the same patterns noted above using alternate panels of AIMs: Accurate RIP variables for these two traits are possible with incorrect coding of sex and imprecise estimates of genomic ancestry. The initial predictor variable values of both sex and ancestry can be reduced in precision by as much as 30% (*i.e.*, $r^2=0.7$ between the original predictor variable and the noise predictor injected variable) and still show correlation coefficients of about $r=0.95$ between the RIP measures generated with these noisy estimates and RIP measures generated with the original estimates (Figure S8 and Figure S14). BRIM is efficient in using the latent covariance structure of the facial PCs to discover the paths through face space that reflect sex and ancestry and can accurately summarize the relative positions of individual faces on these paths as RIP-S and RIP-A, respectively.

Humans are also very adept at observing faces and can infer many aspects of the variability among faces[21], [22]. Given this, we attempted to test whether the human observer might provide a means of validating the RIP-A and RIP-S variables. Observers were shown false-colored 3D animated GIF images of research participants' faces and asked to rate the proportion of West African ancestry (from 0% to 100%) and the femininity (using a Likert scale from 1 to 7). Observers were also asked to judge the sex and the population group. As shown in Figures 5A and 5B, the correlations between RIP-A and observer ratings of proportional facial ancestry and judgments of facial population are strong (all $r>0.85$ and $p<0.0001$). Similarly, RIP-S and observer ratings of facial femininity and judgments of facial sex are also highly correlated ($r>0.85$ and $p<0.0001$; Figures 5C and 5D). These findings provide additional validation that RIP-A and RIP-S are informative summary statistics representing the relative levels of facial ancestry and facial femininity.

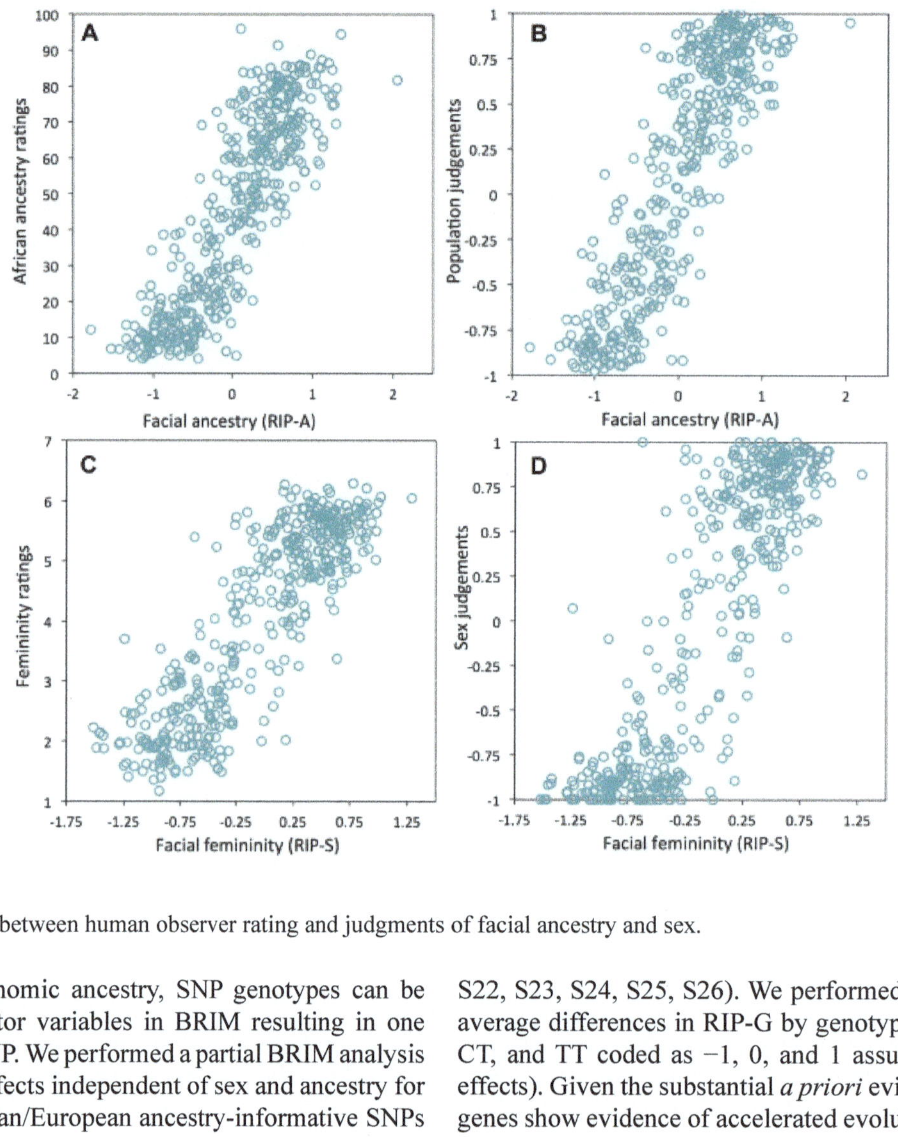

Figure 5: Relationships between human observer rating and judgments of facial ancestry and sex.

Like sex and genomic ancestry, SNP genotypes can be used as initial predictor variables in BRIM resulting in one RIP-G variable per SNP. We performed a partial BRIM analysis modeling genotype effects independent of sex and ancestry for each of 76 West African/European ancestry-informative SNPs located in 46 craniofacial candidate genes. These 46 genes were selected primarily from a set of 50 craniofacial genes that also showed genomic signatures of accelerated evolution in a survey of 199 genes (Table S2). Since properly conditioned tests of genetic association in admixed populations are an efficient approach to discover genes affecting traits that differ between populations and since RIP-A is an efficient means of summarizing overall facial ancestry, it is perhaps somewhat counterintuitive that RIP-A conditioning is superior to genomic ancestry conditioning in our partial BRIM modeling (Figures S15, S16, S17, S18, S19, S20 andS27). Likewise, RIP-S proved to be a better conditioning variable than sex in the partial BRIM analyses to estimate RIP-G (Figures S21, S22, S23, S24, S25, S26). We performed ANOVAs to test for average differences in RIP-G by genotype category (*e.g.*, CC, CT, and TT coded as −1, 0, and 1 assuming additive allelic effects). Given the substantial *a priori* evidence, *viz.*, that these genes show evidence of accelerated evolution in one or both of the parental populations and that mutations in these genes can cause overt murine or human craniofacial dysmorphology, we consider our analysis of each gene to be a separate statistical test and, as such, do not require adjustments for multiple testing. Twenty-four of 76 RIP-G variables (in 20 different genes) show $p<0.1$ (Table S2). The relatively low threshold was motivated by the strong *a priori*evidence for each gene noted above, the single trait summary provided by RIP-G, and an expected small effect of single genes on normal-range variation across the whole face. Additionally, given the general finding that clinically relevant genes can also affect subclinical and normal-range variation (e.g., [23]), we performed detailed *post hoc* descriptions of the effects of these RIP-Gs using

FSCPs (Figures S34, S35,Figures S39, S40, S41, S42, S43, S44 and Table S4).

Summaries of the effects of three of these 24 RIP-G variables (rs1074265 in *SLC35D1*, rs13267109 in*FGFR1* and rs2724626 in *LRP6*) presented in Figures 6A, 6B, and 6C illustrate these results. A detailed analysis and description of each of the 24 SNP effects using FSCPs is given in the supporting material. The gene solute carrier family 35 member D1 gene (*SLC35D1*; OMIM#610804) is located on human chromosome 1p31.3 [24]. Mutations in *SLC35D1* have been shown to result in Schneckenbecken dysplasia (OMIM#269250), which affects the face causing the characteristic feature of "superiorly oriented orbits." The normal-range results of the SNP in rs1074265 in *SLC35D1* (Figure 6A) indicate strong effects at the eyes and periorbital regions, including notable differences at the supraorbital region, as well as at the midface and the chin. Mutations in the human fibroblast growth factor receptor 1 (*FGFR1*;OMIM#136350) gene located on chromosome 8p21.23-p21.22 can result in four autosomal dominant craniofacial disorders: Jackson-Weiss syndrome (OMIM#123150), which is characterized by craniosynostosis and midfacial hypoplasia; trigonocephaly (OMIM#190440), which is characterized by a keel-shaped forehead resulting in a triangle-shaped cranium when viewed from above; osteoglophonic dysplasia (OMIM#166250), which is characterized by craniosynostosis prominent supraorbital ridge and depressed nasal root; and Pfeiffer syndrome (OMIM#101600), which is characterized by midface hypoplasia and, depending on the subtype, ocular proptosis, short cranial base, and cloverleaf skull. The normal-range results of the SNP rs13267109 in*FGFR1* depicted in Figure 6B indicate the strongest effects in the supraorbital ridges, the eyes, the midface, the nose, and the corners of the mouth. The strongest differences in the shape transformations are indeed the forehead, supraorbital ridges and nasal bridge. The mouse homologue of the human low-density lipoprotein receptor-related protein 6 (*LRP6*; OMIM#603507) gene is known to be critical for the development of lips in the mouse resulting in bilateral cleft lips in the knockout *LRP6* mouse model [25]. As yet, no human craniofacial diseases have been linked to the *LRP6* gene or to the gene region on human chromosome 12p13.2 although the gene product is known to interact on a molecular level with WNT signaling. Observing the shape transformation in Figure 6C, a change from a prominent lip region, including the appearance of a thick and convex vermilion, to a less prominent lip region, including an apparently thinner and less convex (more concave) vermilion, is noted. This is confirmed by inspecting the normal displacement results and the significance maps, in which the lips are clearly delineated (Figure S43).

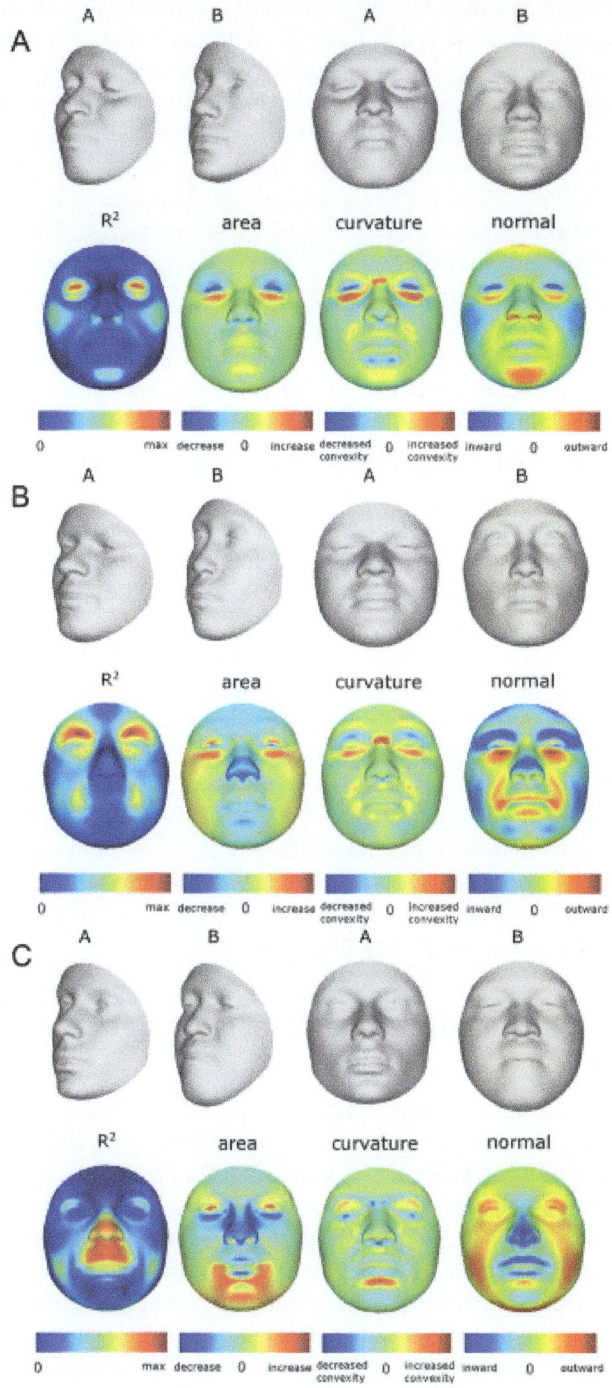

Figure 6: Transformations and heat maps showing how face shape is affected by three particular RIP-G variables.

In general, some RIP-G variables show localized effects (*e.g.*, rs1074265 in *SLC35D1*), changing only certain aspects in facial shape, while others display changes in several facial regions (*e.g.*, rs13267109 in*FGFR1*). Summary statistics for the underlying distributions of effect sizes across the quasi-landmarks are presented in Table S3. In the case where multiple SNPs in the same gene are modeled, overlapping and similar effects are seen across the different SNPs for the same gene (*e.g.*, *DNMT3B* and *SATB2*) and different SNPs from genes within the same biological pathway (*e.g.*, *WNT3*, *FGFR1*, and *FGFR2*). We present a graphical user interface (GUI) so that effects of changes in these 24 RIP-G variables, RIP-A, RIP-S, or any of the top 44 PC variables can be visualized in more detail. These transformations can be visualized with the texture map as well as shape only, and the GUI (http://tinyurl.com/DNA2FACEIN3D) allows for the illustration of the comparison of transformed faces to the consensus face using the three primary FSCPs.

Since both categorical and continuous variables can be modeled using BRIM, this approach might be used to test for relationships between facial features and other factors, *e.g.*, age, adiposity, and temperament. The methods illustrated here also provide for the development of diagnostic tools by modeling validated cases of overt craniofacial dysmorphology. Most directly, our methods provide the means of identifying the genes that affect facial shape and for modeling the effects of these genes to generate a predicted face. Although much more work is needed before we can know how many genes will be required to estimate the shape of a face in some useful way and many more populations need to be studied before we can know how generalizable the results are, these results provide both the impetus and analytical framework for these studies.

MATERIALS AND METHODS

Population Samples and Participant Recruitment

Population samples were collected in the United States (State College, PA, Williamsport, PA, and The Bronx, NY); Brasilia, Brazil; and Cape Verde (São Vicente, and Santiago), all under a Penn State University Internal Review Board (IRB) approved research protocol titled, "Genetics of Human Pigmentation, Ancestry and Facial Features." Skin pigmentation was measured using narrow-band reflectometry with the DermaSpectrometer (Cortrex Technology, Hadsund, Denmark) in the United States and Brazil and the DSMII (Cortrex Technology, Hadsund, Denmark) in Cape Verde. DermaSpectrometer readings were rescaled to the DSMII scale by multiplying by 1.19, the slope derived from a comparison of readings with both instruments on the same set of participants (data not shown). Height, weight, age, self-reported ancestry, and sex were collected by survey. DNA was collected both with buccal cell brushes and using finger-stick blood on four-circle Whatman FTA cards (Whatman, Florham Park, NJ).

To minimize age-related variation in facial morphology, we only recruited participants between the ages of 18 and 40. From these recruits, we selected individuals with >10% West African ancestry and <15% combined Native American and East-Asian ancestry as measured with the 176 ancestry informative marker (AIM) panel. We assigned these cutoff points to reduce admixture from parental populations other than West African and European. Ancestry-based exclusion criteria were not applied to Cape Verdeans given the largely dihybrid nature of this population. Finally, we excluded participants whose 3D images were obstructed by facial or head hair. After excluding participants by these criteria, we were left with 592 participants (154 from the US, 191 from Brazil, and 247 from Cape Verde).

SNP Genotyping and Genomic Ancestry Estimates

Genotyping of 176 AIMs for the US and Brazilian samples was performed on the 25 K SNPstream ultra-high-throughput genotyping system (Beckman Coulter, Fullerton, CA) as previously described [11]. Ancestry was estimated using the various panels of AIMs by one of two methods. Ancestry using full set of 176 AIMs was estimated in the US and Brazilian subsample using maximum likelihood on a four-population model; European, West African, Native American, and East Asian [11].The 68-AIM ancestry estimates were generated using the full sample (U.S., Brazilian, and Cape Verdean) using ADMIXMAP as these markers were available on all 592 participants. One marker (rs917502) from the original 176 had a call rate of less than 30% and was omitted from the ADMIXMAP analyses.

The Cape Verdean sample was assayed for the Illumina Infinium HD Human1M-Duo Beadarray (Illumina, San Diego, CA) following the manufacturer's recommendations. A total of 537,895 autosomal SNPs that passed quality controls were used to estimate ancestry using the program FRAPPE [26], assuming two ancestral populations (West African and European). HapMap genotype data, including 60 unrelated European-Americans (CEU) and 60 unrelated West Africans (YRI), were incorporated in the analysis as reference panels (phase 2, release 22, The HapMap Project; [27]).

We identified a list of selection-nominated candidate genes for testing against normal-range facial variation in admixed individuals of European and West African descent. Ancestry information and tests for accelerated evolution [28] were used to prioritize among a larger set of craniofacial genes. Since most genomic regions show low levels of allele frequency change across human populations, genes affecting traits that vary across populations are usually distinctive in showing large differences in frequency and other features of local variation and allele frequency spectra consistent with rapid local evolution. A preliminary set of craniofacial candidate genes was developed by searching the Online Mendelian Inheritance in Man (OMIM) database[24]. The keywords "craniofacial" and "facial" were searched to determine a set of genes known to affect craniofacial development. The OMIM entries for each gene included in the search output were then scanned manually to remove genes where the term appeared as a result of phrases such as "no craniofacial associations found" and other similar negative results. OMIM searching resulted in a list of 199 unique craniofacial candidate genes. Because this work focused on admixed populations of West African and European descent, the statistical power to detect linkage with craniofacial variation is greatest for SNPs that show large allele frequency differences between West African and European parental populations. Therefore, allele frequency differences among parental groups were further used to prioritize among the candidate genes. SNP frequency data in putative parental population (CEPH Europeans (CEU) and Yoruban (YRI) West Africans) for all SNPs within the 199 OMIM candidate genes were pulled from the HapMap database. This reduced subset of genes was then tested for signatures of non-neutral evolution in a 200 kb window surrounding each gene using a combination of three statistical tests: Locus-Specific Branch Length (LSBL)[29], the log of the ratio of the heterozygosities (lnRH) [30], and Tajima's D [31]. Because these tests are inferring different concepts regarding population history, we considered as significant any gene with statistical evidence of selection for all three measures or strong evidence of non-neutral evolution for two measures in either West African and/or European parental populations as a Selection-nominated candidate gene. It is notable that these steps were taken to increase the likelihood that a functional SNP would be available to test the ability of methods like BRIM to model individual gene effects on the human face. We are making no strong claims in this analysis that craniofacial genes generally or this subset in particular have been subject to greater than average levels of non-neutral evolution or that these genes do in fact have genetic variation that is affecting normal range facial variation in this sample. A total of 50 autosomal genes were thus selected (*SKI, LMNA, SIL1, EDN1, RSPO2, TRPS1, POLR1D, MAP2K1, ADAMTS10, TBX1, PEX14,HSPG2, CAV3, CTNND2, TFAP2A, PEX6, PEX3, MEOX2, RELN, ROR2, NEBL, CHUK, FGFR2, WT1,PEX16, BMP4, FANCA, RAI1, FOXA2, ECE1, DPYD, ZEB2, SATB2, FGFR3, NIPBL, NSD1, ENPP1,GLI3, COL1A2, BRAF, ASPH, FREM2, SNRPN, FBN1, MAP2K2, RPS19, DNMT3B, GDF5,* and*UFD1L*) and a set of SNPs with high allele frequency differences (delta >0.4) in these 50 craniofacial Selection-nominated candidate genes to test for associations with facial shape variation.

3D Facial Images and Phenotyping

3D images composed of surface and texture maps were taken using the 3dMDface system (3dMD, Atlanta, GA). Participants were asked to close their mouths and hold their faces with a neutral expression for the picture. Images were then exported from the 3dMD Patient software in OBJ file format and imported into a scan cleaning program for cropping and trimming, removing hair, ears, and any dissociated polygons. The complete work flow involved in processing face scans is depicted in Figure 1. Five positioning landmarks were placed on the face to establish a rough facial orientation. Subsequently, an anthropometric mask (7,150 quasi-landmarks) was non-rigidly mapped onto the original 3D images and their reflections [8], [9], [17], which were constructed by changing the sign of the x-coordinate [18], [32]. This established homologous spatially-dense quasi-landmark (Q-L) configurations for all original and reflected 3D images (*8*). Note that, by homologous, we mean that each quasi-landmark occupies the same position on each face relative to all other quasi-landmarks. Subsequently, a generalized Procrustes superimposition [18], [33] is used to eliminate differences in position, orientation, and scale of both original and reflected configurations combined was performed. This constructed a tangent space of the Kendall shape-space centered on the overall consensus configuration [25]. Procrustes shape coordinates, representing the shape of an object [34], were obtained for all 3D faces and their reflections. After Procrustes superimposition, the overall consensus configuration is perfectly symmetrical and a single shape can be decomposed into its asymmetric and its bilaterally symmetric part [18]. The average of an original and its reflected configuration constitutes the symmetric component while the difference between the two configurations constitutes the asymmetric component [19], [35]. The analyses in this report were all based on facial shape as represented using the component of symmetry only. Although deviations from bilateral symmetry are thought to be the effects of developmental noise and/or environmental factors [36], it is likely there are genetic effects on asymmetry, which would compel independent investigation.

Principal components analysis (PCA) [9] on the superimposed and symmetrized quasi-landmark configurations of the panel of 592 participants resulted in 44 PCs that together summarize 98% of the total variation in face space. To examine the effect of excluding lower PCs, we first reconstructed actual quasi-landmark configuration from the 44 PCs only and compared these to the original remapped face. We found that the average root mean squared error (RMSE) is as small as 0.2 mm per quasi-landmark. The localized differences between the original faces and the faces as represented by the first 44 PCs are largest around the iris, eyelids, under the nose, and the corners and opening of the mouth and are at most about 0.45 mm. How a PC or any other independent variable affects the face can be shown with heat maps and shape transformations: heat maps use contrasting colors to highlight the specific parts of the face that are affected, while shape transformations illustrate the changes in overall face shape with two or more images of the face at set intervals. Shape transformations are obtained from the average face in the direction of each PC at −3 and +3 times the accompanying standard deviation (square-root of the eigenvalue). Figure 2 shows how the first 10 PCs affect the face. Some of these PCs (*e.g.*, PC1, PC2, PC3) summarize effects on many parts of the face, while other PCs (*e.g.*, PC4, PC5) summarize the effects of changes in only particular parts of the face. The effects of each of the 44 PCs as well as the RIP variables can be visualized with a GUI software tool that we have written called DNA2FACEIN3D.EXE. The program and instruction manual can be downloaded here: http://tinyurl.com/DNA2FACEIN3D.

We have used three methods to visualize and quantify facial difference so that we can systematically express the effects of particular response-based imputed predictor (RIP) variables on the face into anatomically interpretable results. These are based on comparing faces pairwise, such as comparing the most feminine RIP-S to the most masculine RIP-S transformed consensus faces using three fundamental measures: area ratio, normal displacement, and curvature ratio. These two ratios and one displacement along with particular inter-landmark distances and angles can together be termed "face shape change parameters" (FSCPs) and are a means of translating face shape changes from the abstract face space into language of facial characteristics such that comparisons between clinical or anthropological descriptions of faces can be compared to bootstrapped response-based imputation modeling (BRIM) results. The statistical significance of these FSCPs can be estimated using permutation. A more detailed description on how this is done is given in the supplementary online material.

Human Observer Ratings and Judgments

Ancestry and Sex Observations

Given the dexterity humans have for discerning numerous traits, features, and expressions, it is reasonable to expect the observer would provide a useful reference point for studies of the genetics of facial traits. We accessed observer ratings and judgments of sex and ancestry in order to test the informativeness of RIP-A and RIP-S.

Selection of Stimuli

A total of 500 participant faces were selected and divided into twenty-five panels of twenty faces, with each panel including faces of research participants across the range of genomic ancestry levels and similar numbers of male and female faces. We used false colored grey GIF animations so that ancestry and sex ratings and judgments would be based on face-shape cues but not cues of skin, iris, or hair pigmentation or hair texture. Animation order was randomized.

Administration of Instruments

We administered the instruments containing the animated, false-colored GIFs with accompanying questions using Survey Monkey (SurveyMonkey.com LLC; Palo Alto, CA). Four survey questions were asked for each of 20 faces participants observed:

- "What proportion (from 0% to 100%) of this person's ancestry appears to be West African?" *(Ratings made with a number between 0 and 100.)*
- Which single categorical group best describes this person? *(Judged with Black African, or African-American; White, European or European-American; or Mixed)*
- Does this person appear to be male or female? *(Judged with "male" or "female")*
- "How feminine does this person's face appear to you?" *(Ratings made with a choice from a 7-point Likert scale ranging from 1 "extremely feminine" to 7 "extremely masculine".)*

Observers were randomly assigned to one of the 20 panels through a link on the Anthropology Department homepage. Observers were recruited from students enrolled at Penn State University. Of the 1,156 participants, 938 (81.1%) completed the surveys. The number of observers for the 20 alternative surveys ranged from 27 to 70, with a mean of 47. Observers who completed fewer than half of the survey as well as three whose discrepancies were more than three standard deviations from the mean were excluded from the analysis. Observers were not trained and were not familiar with the research participants whose faces were shown as stimuli.

SUPPORTING INFORMATION

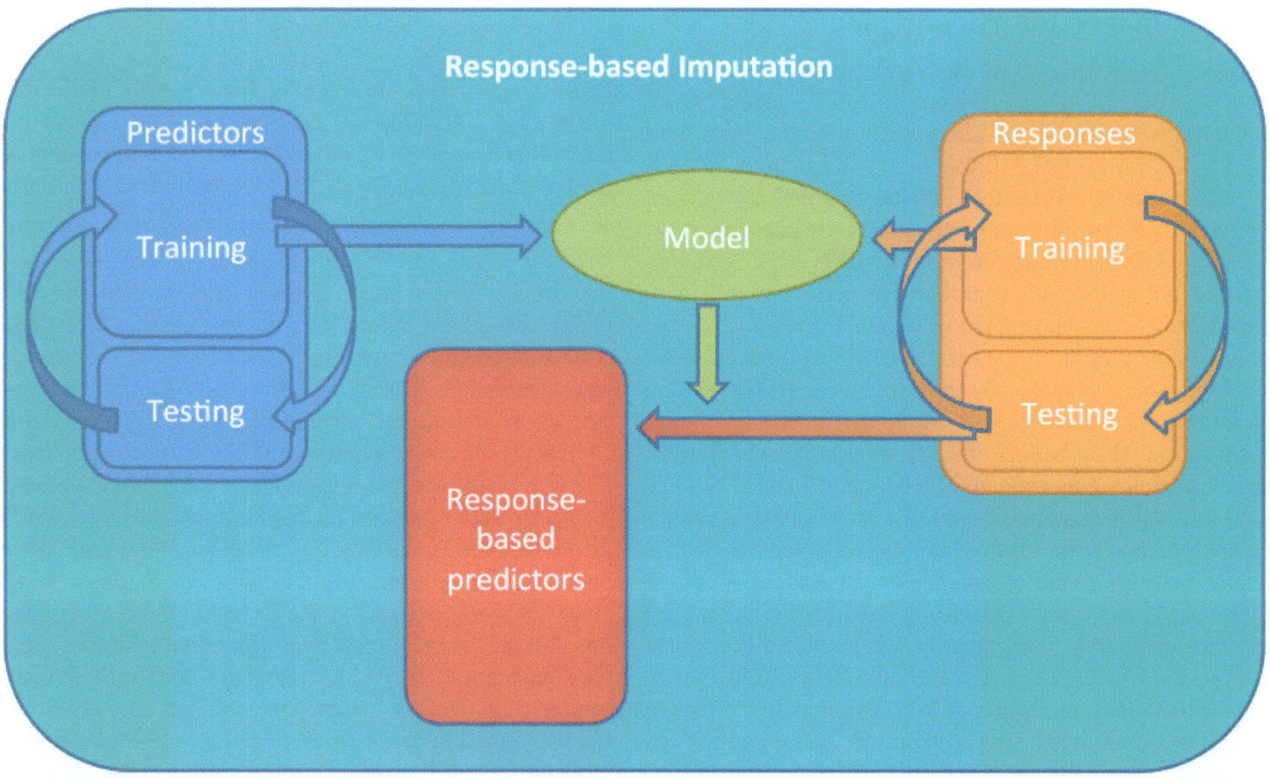

Figure S1: Response-based imputation based on Leave-OneOut.

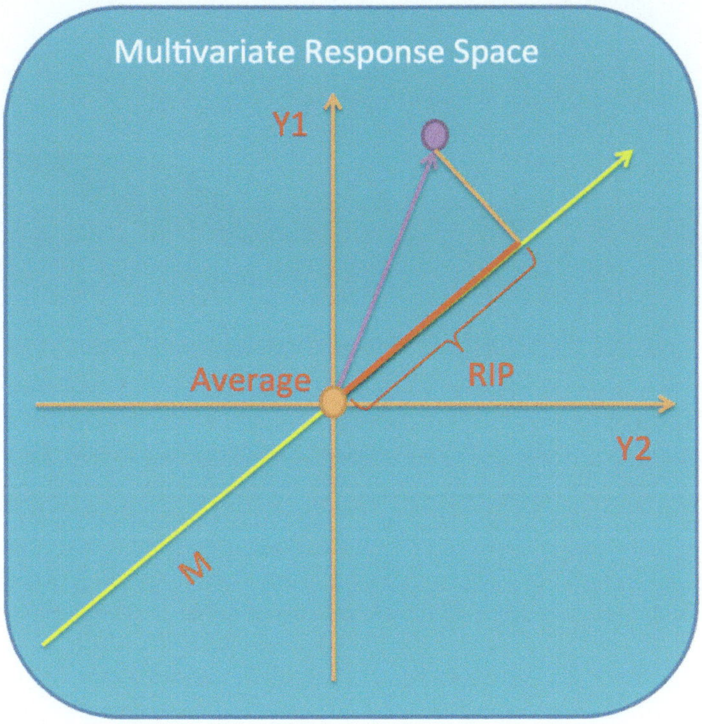

Figure S2: Response-based imputation using distance decomposition.

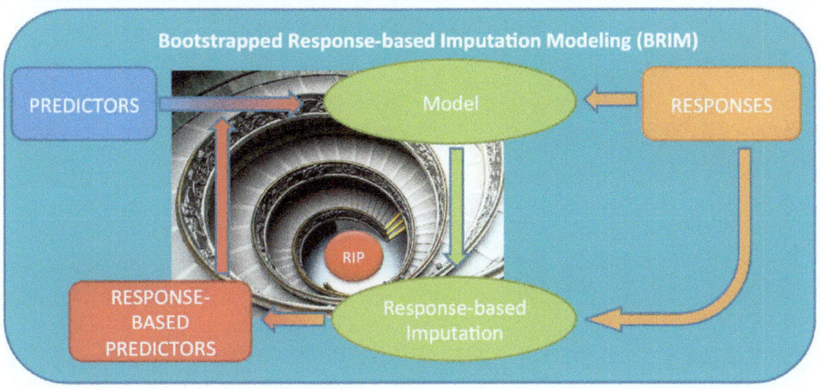

Figure S3: Bootstrapping.

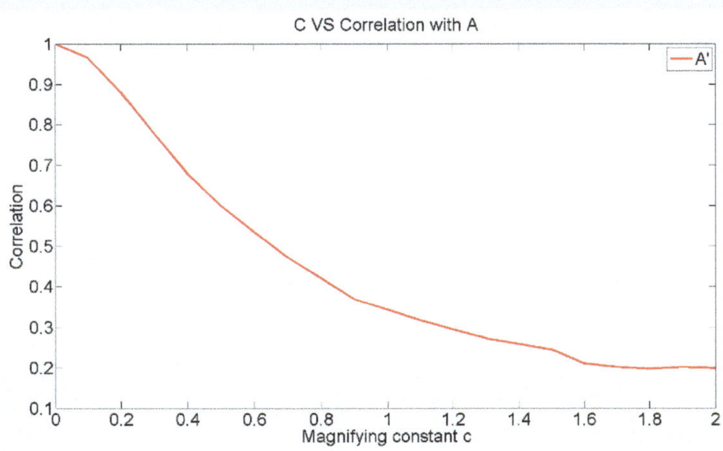

Figure S4: The correlation of A9 with A in function of the magnification constant c. The higher the constant, higher the level of injected noise, and hence, the lower the correlation.

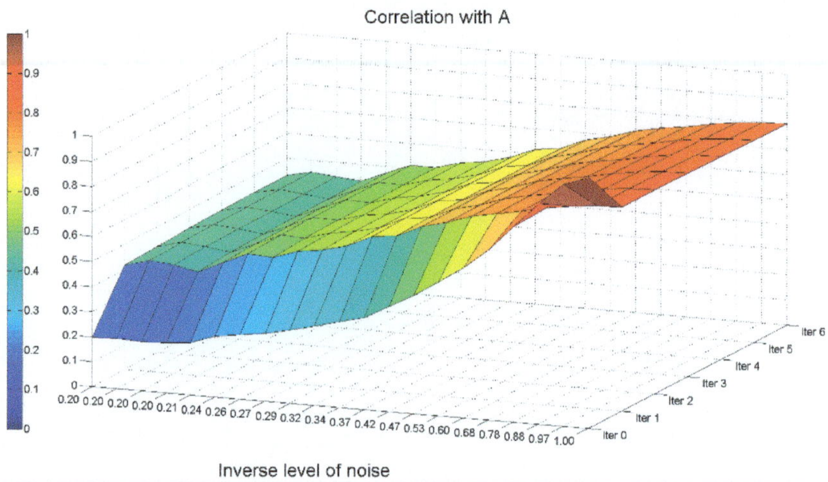

Figure S5: Absolute correlations of retrieved RIP-A9 variables for each iteration and for each noise level with the genomic ancestry variable A. Color bar ranges from 0 to 1. Note the peak correlation of 1 in the situation of no noise injection and no iteration, this correlation is of A against itself.

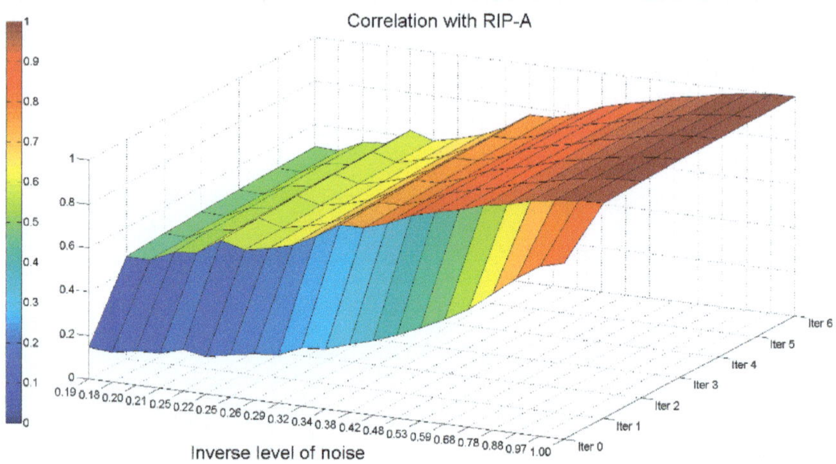

Figure S6: Correlations of retrieved RIP-A9 variables for each iteration and for each misclassification level with RIP-A. Color bar ranges from 0 to 1.

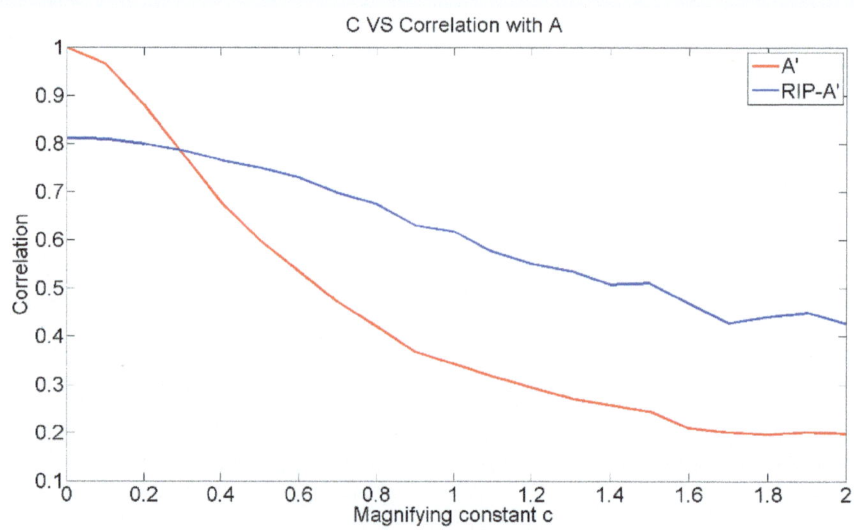

Figure S7: Correlation of A9 and RIP-A9 with A for different levels of noise.

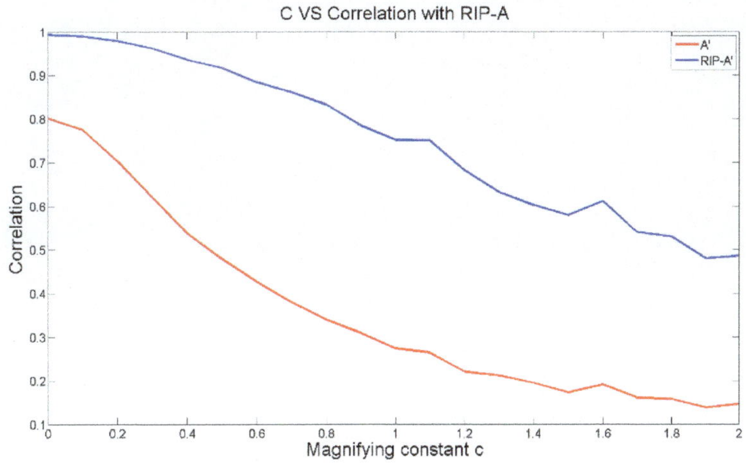

Figure S8: Correlation of A9 and RIP-A9 with RIP-A for different levels of noise.

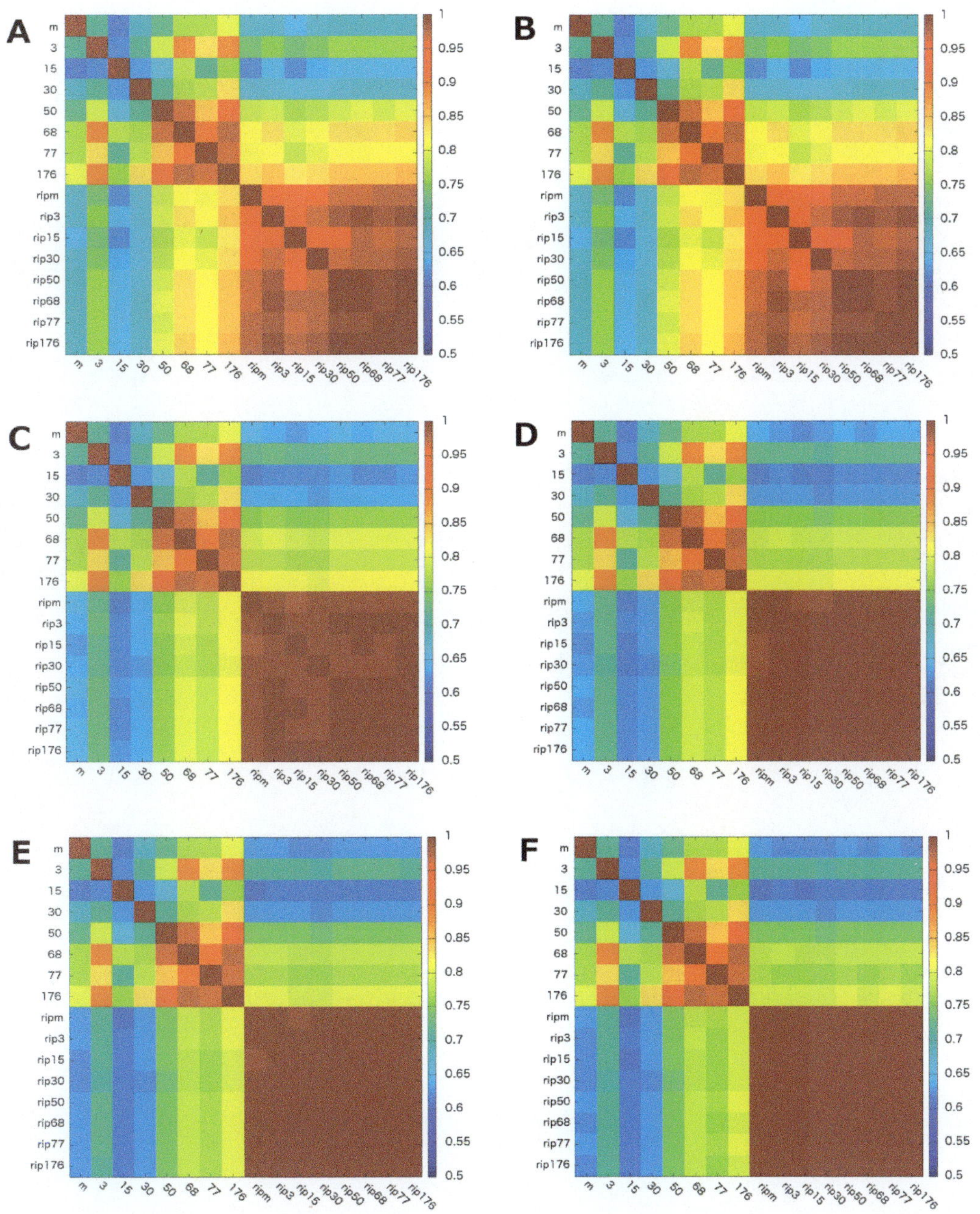

Figure S9: Correlation matrices for different AIMs subsets over each iteration 1–6 (A–F).

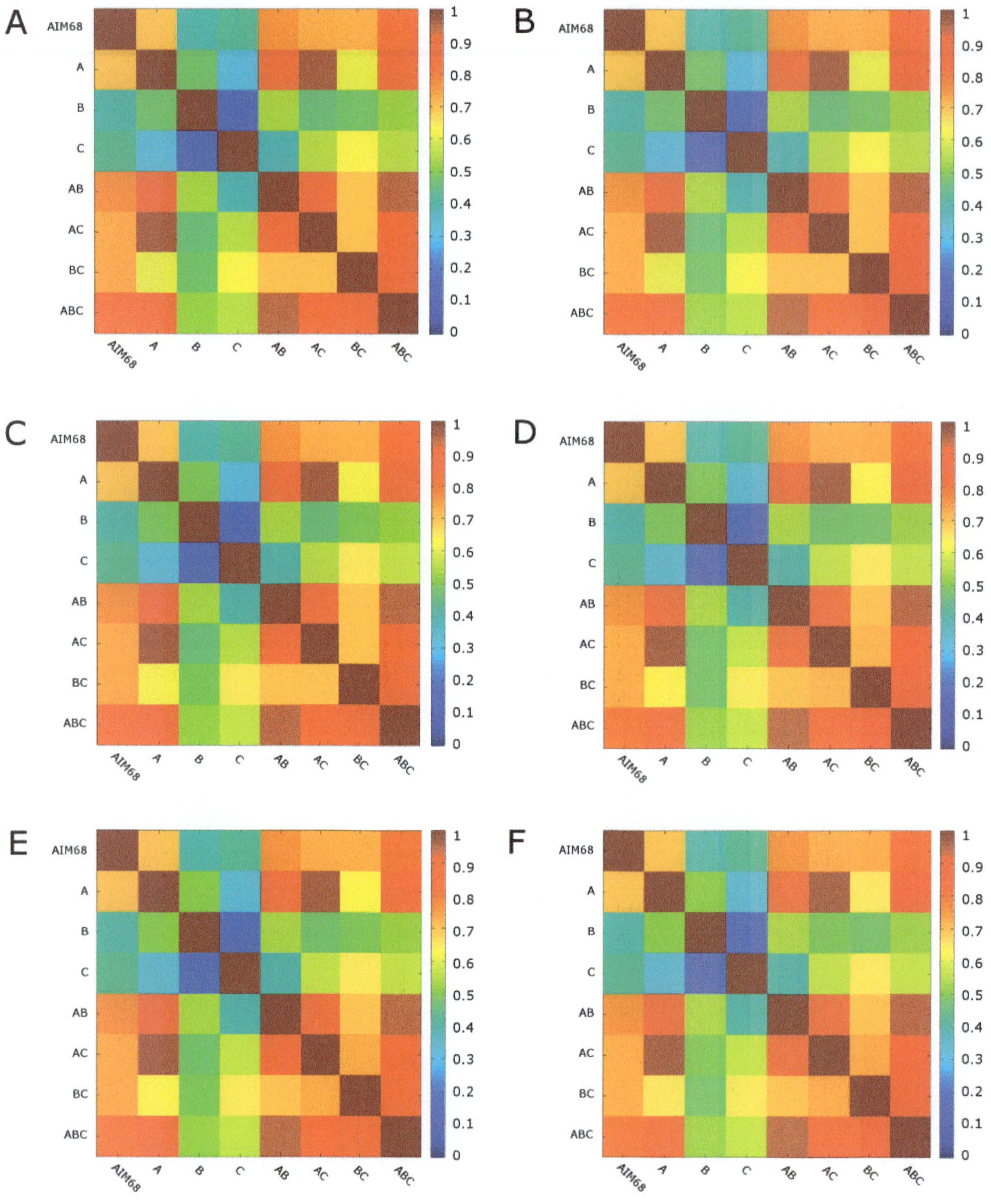

Figure S10: Correlation matrices of AIM 68 for different population subsamples over each iteration 0–6 (A–F) Note that iteration 0, implies correlations in between the original predictor variables. A=American, B=Brazilian, C=Cape Verdean.

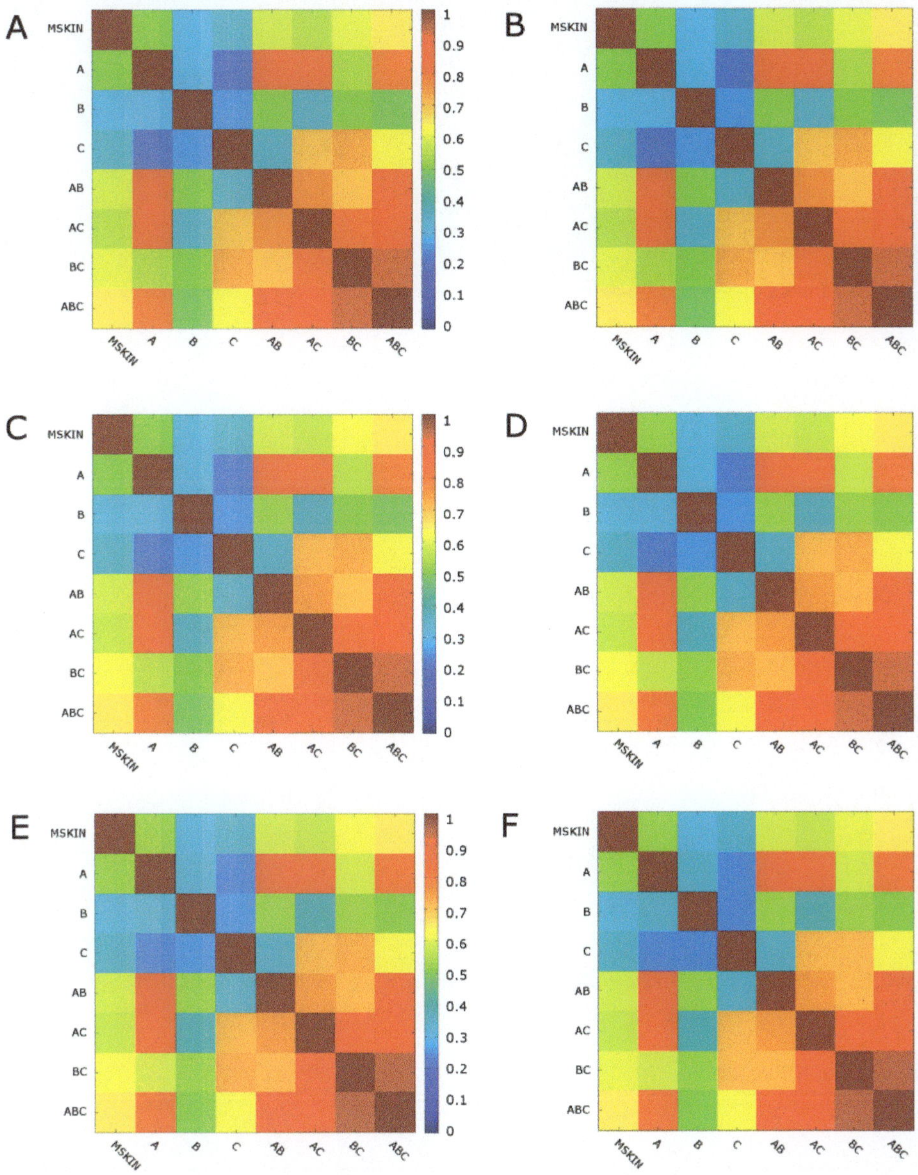

Figure S11: Correlation matrices of M-index for different population subsamples over each iteration 0–6 (A–F) Note that iteration 0, implies correlations in between the original predictor variables. A=American, B=Brazilian, C=Cape Verdean.

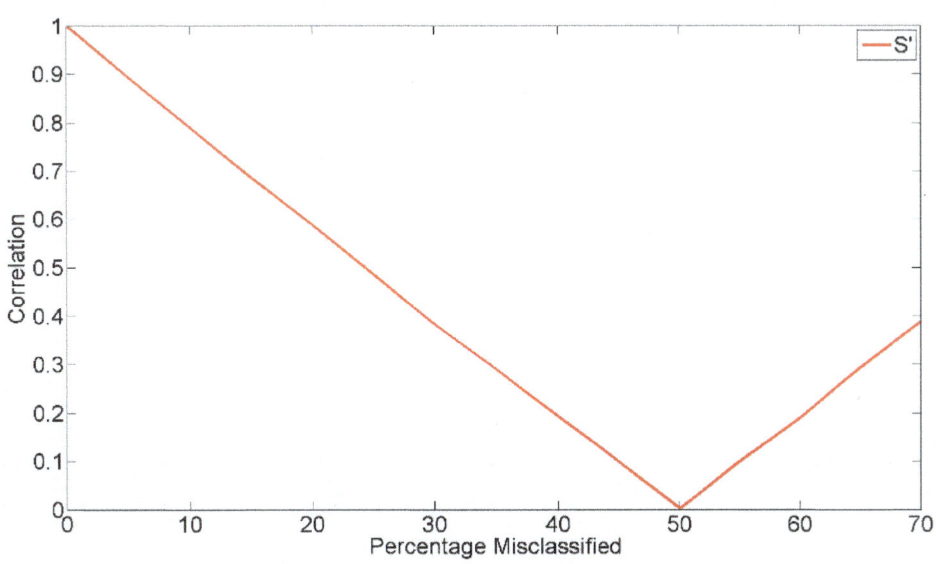

Figure S12: AUC in function of percentage misclassification.

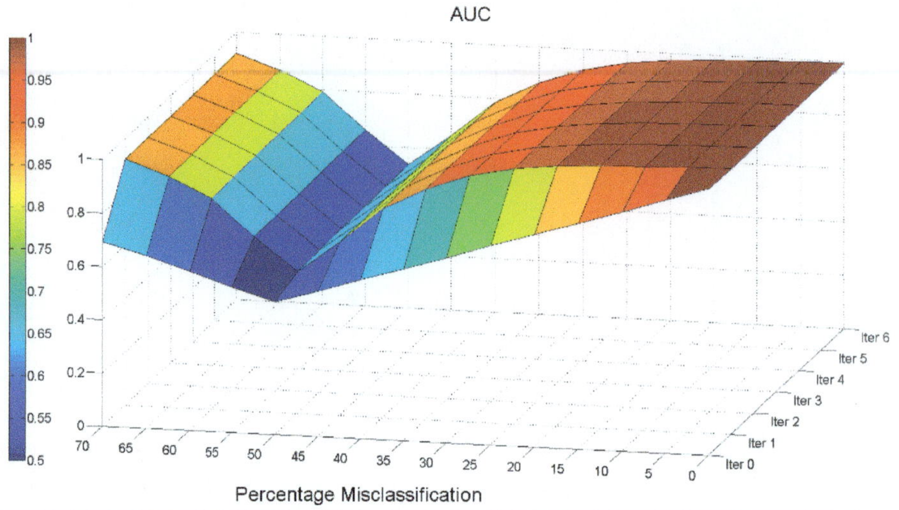

Figure S13: Average AUC values of ROC analyses for each iteration and for each level of misclassification.

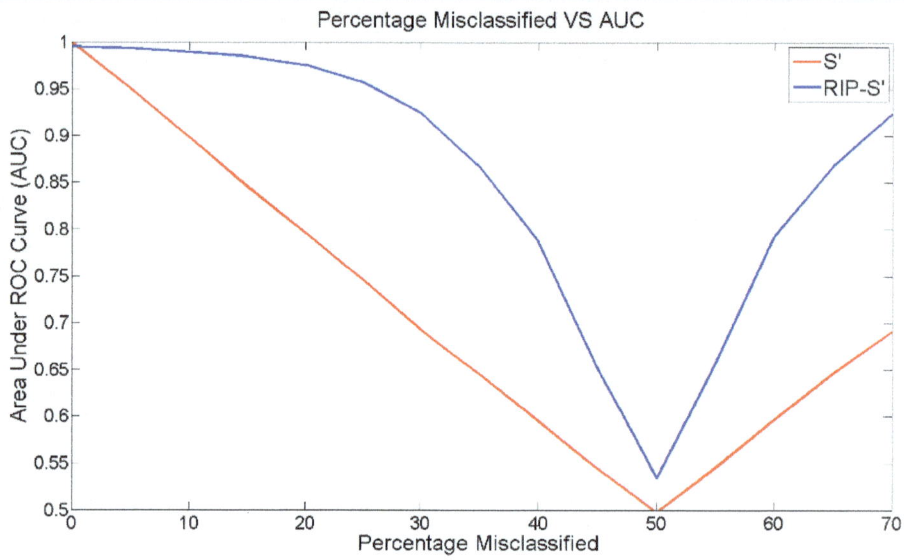

Figure S14: AUC of S′ and RIP-S′ with S as grouping variable in an ROC analysis for different levels of misclassification.

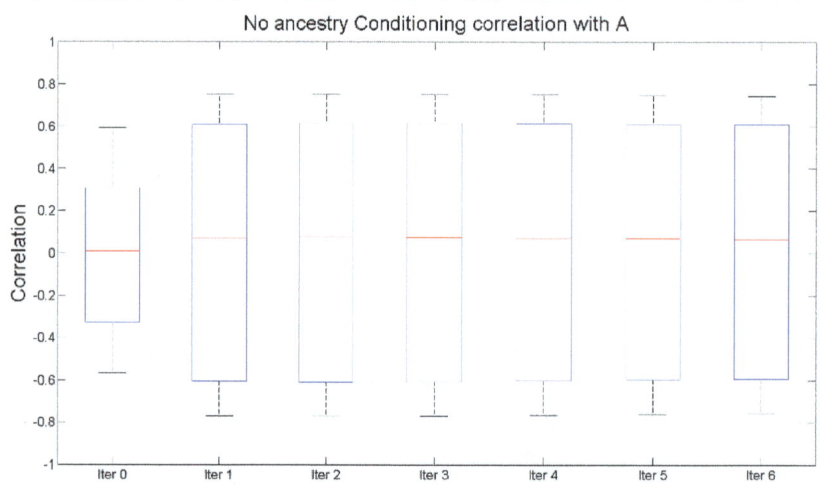

Figure S15: Correlation boxplots of RIP-G values for each iteration with genomic ancestry A, without ancestry conditioning.

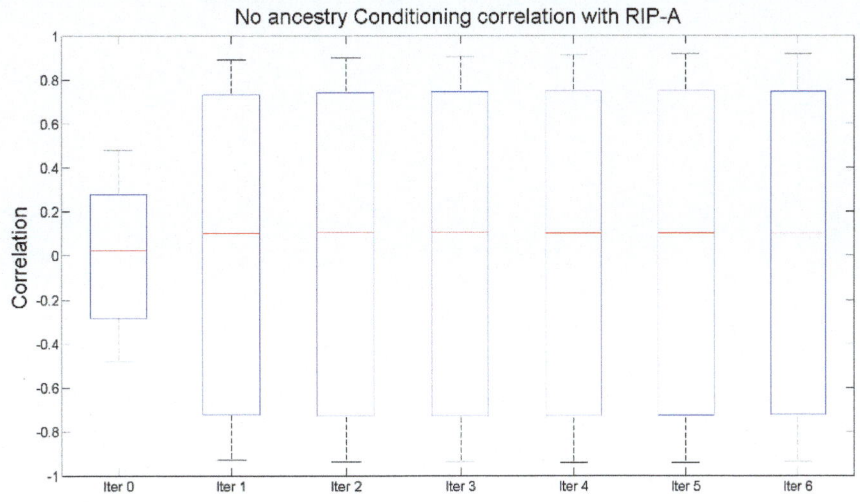

Figure S16: Correlation boxplots of RIP-G values for each iteration with RIP-A, without ancestry conditioning.

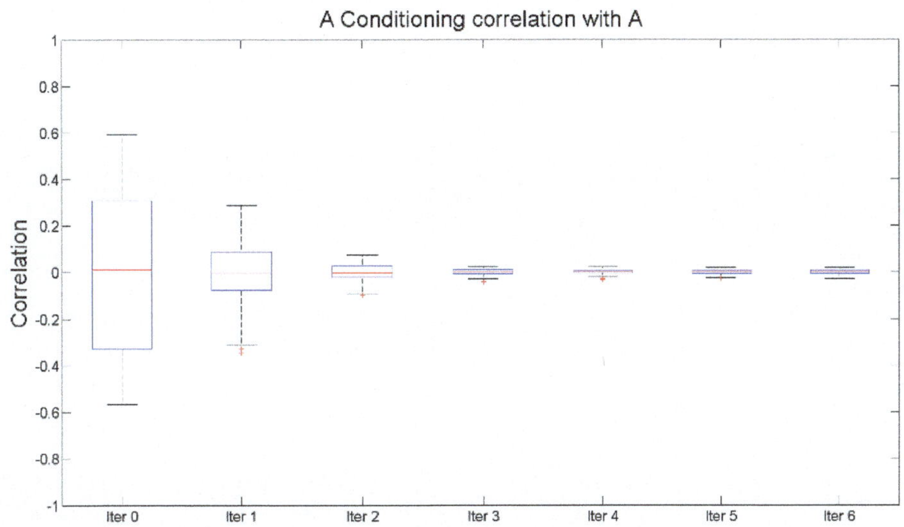

Figure S17: Correlation boxplots of RIP-G values for each iteration with A, conditioned on A.

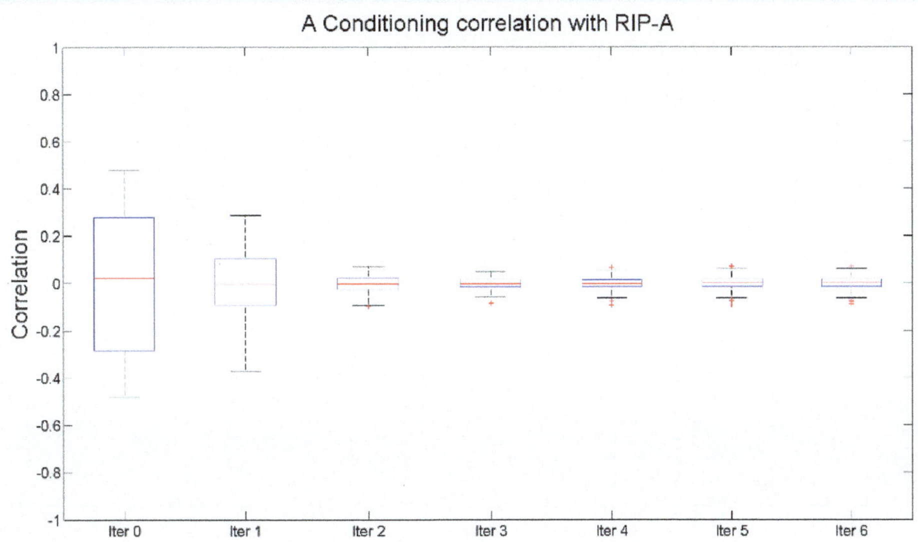

Figure S18: Correlation boxplots of RIP-G values for each iteration with RIP-A, conditioned on A.

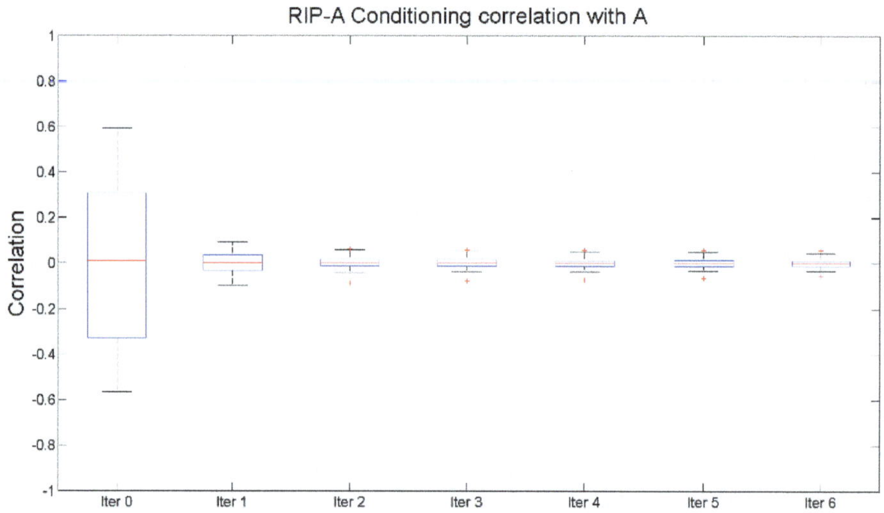

Figure S19: Correlation boxplots of RIP-G values for each iteration with A, conditioned on RIP-A.

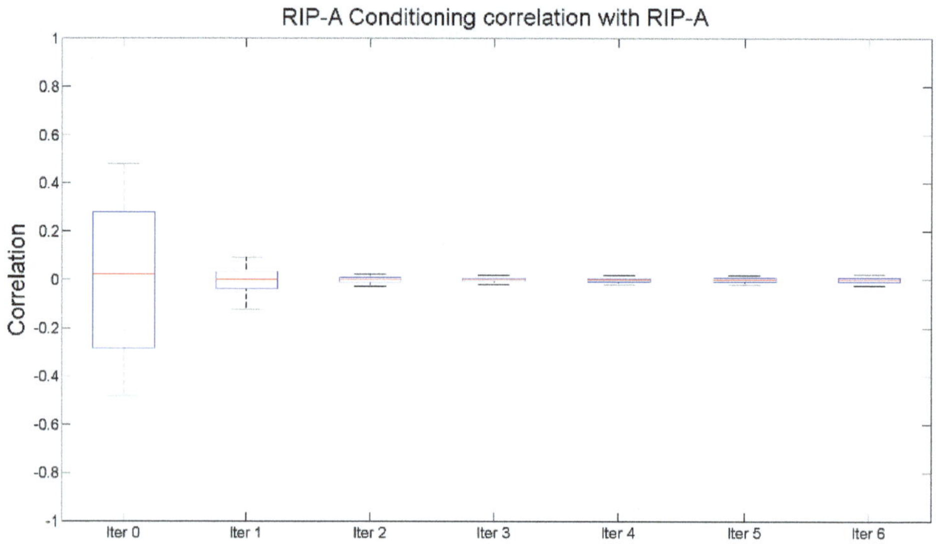

Figure S20: Correlation boxplots of RIP-G values for each iteration with RIP-A, conditioned on RIP-A.

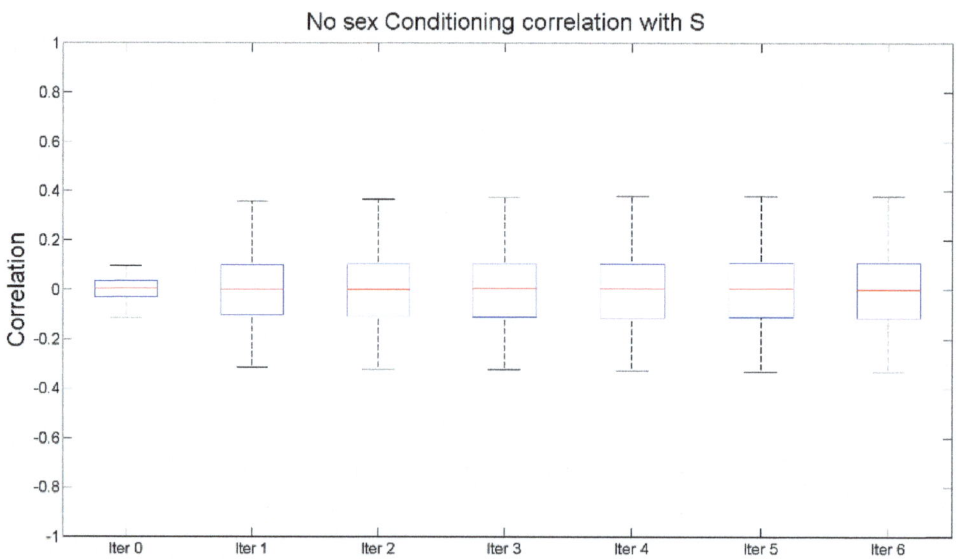

Figure S21: Correlation boxplots of RIP-G values for each iteration with S, without sex conditioning.

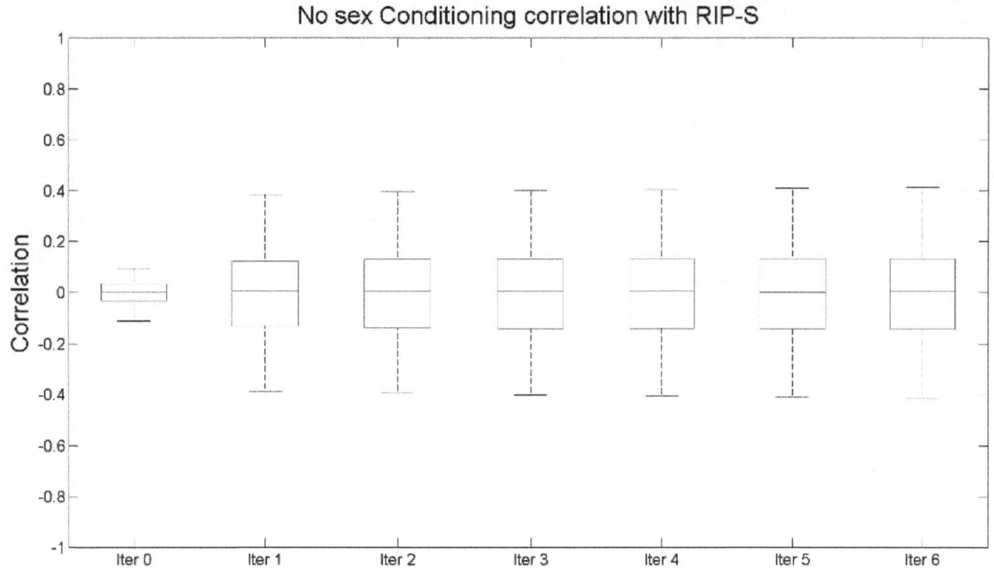

Figure S22: Correlation boxplots of RIP-G values for each iteration with RIP-S, without sex conditioning.

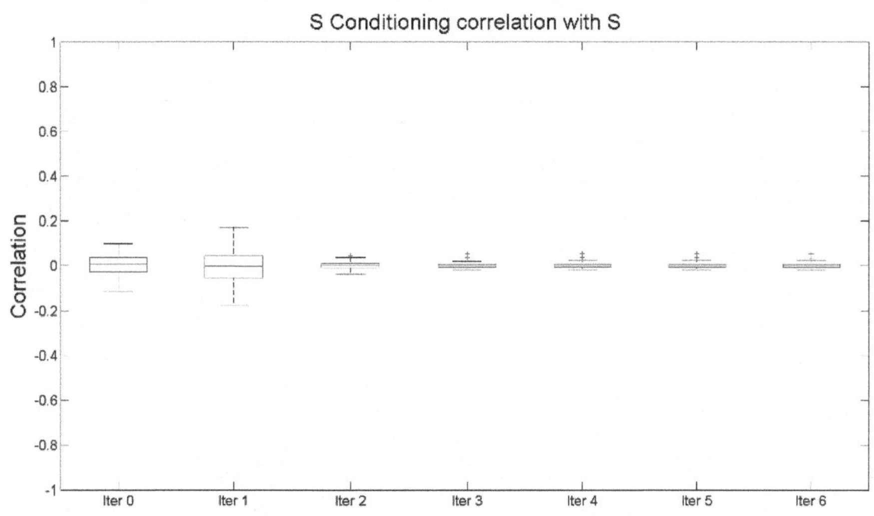

Figure S23: Correlation boxplots of RIP-G values for each iteration with S, conditioned on S.

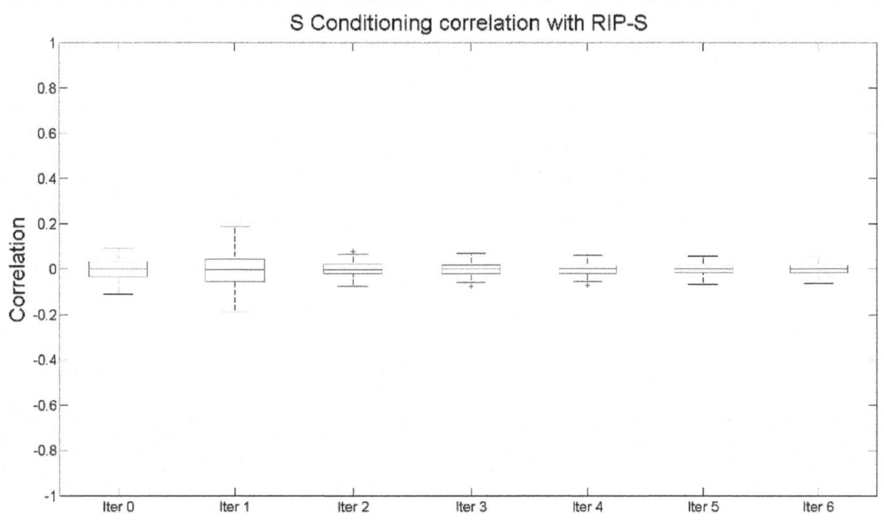

Figure S24: Correlation boxplots of RIP-G values for each iteration with RIP-S, conditioned on S.

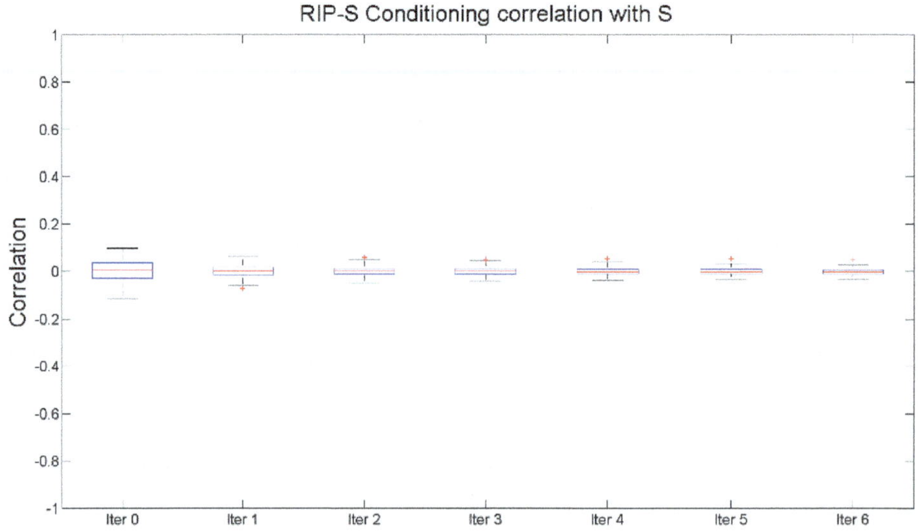

Figure S25: Correlation boxplots of RIP-G values for each iteration with A, conditioned on RIP-A.

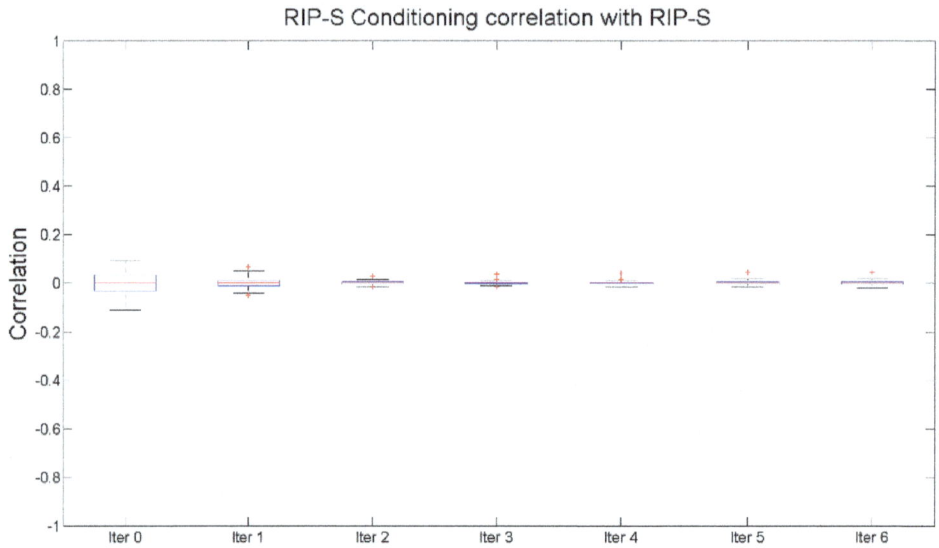

Figure S26: Correlation boxplots of RIP-G values for each iteration with RIP-A, conditioned on RIP-A.

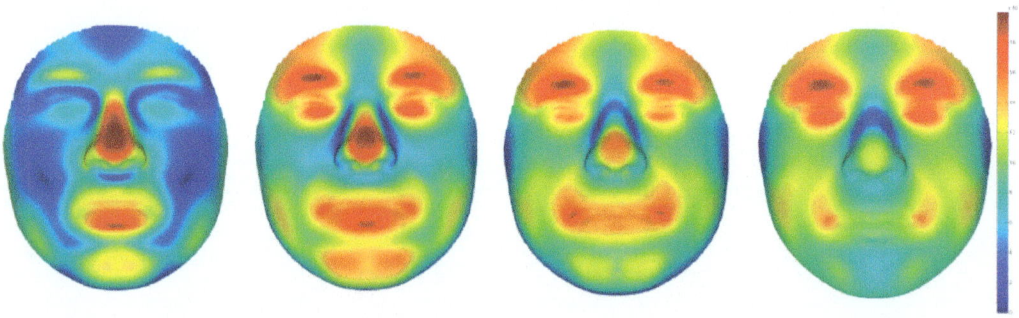

Figure S27: The effect of the rs13267109 SNP in *FGFR1* using approach 1 (no conditioning for ancestry), left, approach 2 (conditioning for genomic ancestry), second from left, approach 3 (conditioning for RIP-A), second from right and approach 4 (conditioning for RIP-A using BRIM), right.

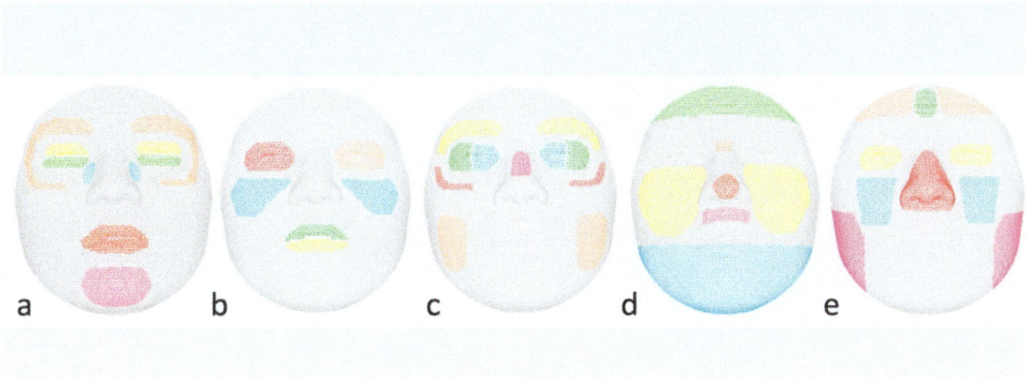

Figure S28: Facial Regions: (A) Orange, Orbital Ridges; Red, Lips; Yellow, Eyes Superior; Green. Eyes Iinferior; Blue, Paranasal Tissues; Pink, Chin. (B) Orange, Left Eye; Red, Right Eye; Yellow, Lower Lip; Green, Upper Lip; Blue Cheek Bones. (C) Orange, Cheeks; Pink, Nose ridge; Yellow, Browridges; Green, Lateral half of eyes; Blue; Medial half of eyes; Red, Cheekbones. (D) Green, Forehead; Yellow, Midface; Red, Nasal tip; Pink, Philtrum; Orange, Nasal bridge; Blue, Lower face. (E) Green, Metopic ridge; Orange, Forehead sides; Yellow, Eyes; Red, Nose; Blue, Malars; Pink, Lateral midface.

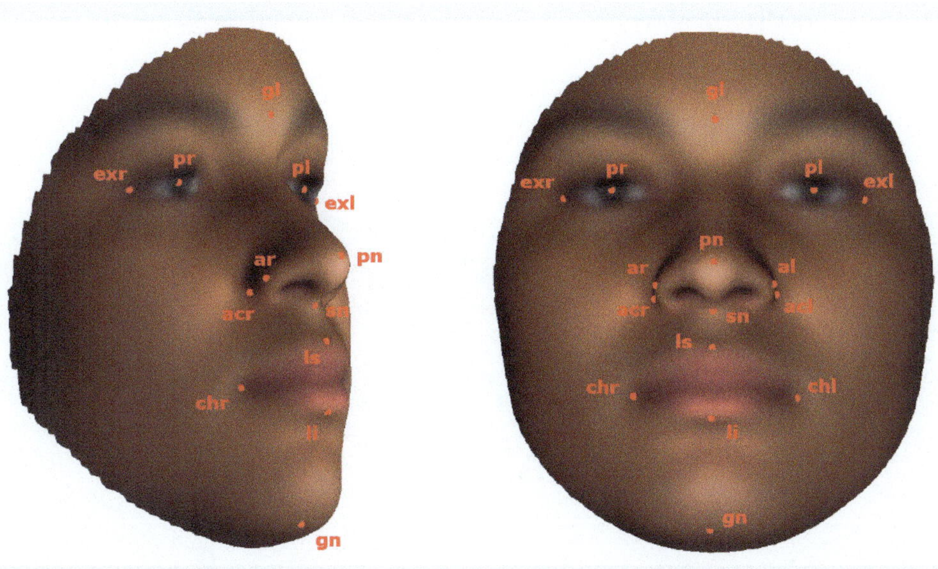

Figure S29: Manually annotated landmarks: gl=Glabella; pn=Pronasale; sn=Subnasale; ls=Labiale Superiusinferius; li=Labiale Iinferius; gn=Gnathion; exr=Right Eendocanthion; pr=Right Ppupil; pl=Left Ppupil; enl=Left Eendocanthion; ar=Right Alar; chr=Chelion right; chl=Chelion left.

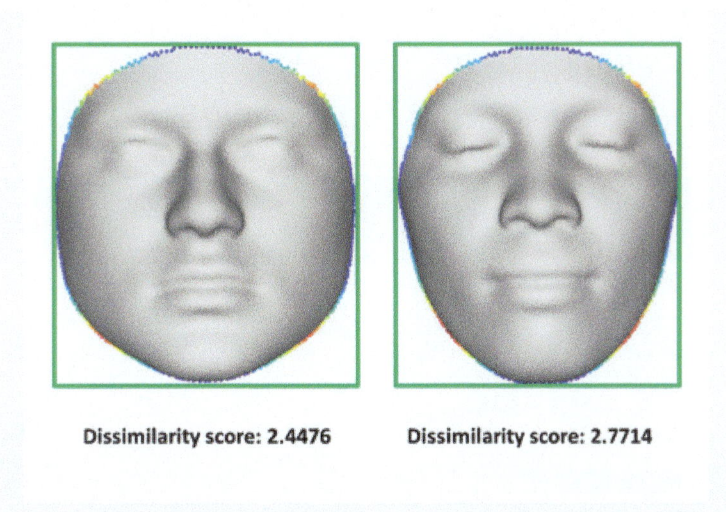

Figure S30: Measurement of facial squareness: The facial border is projected onto the XY plane and compared to a fitted square. The result is a dissimilarity score with a higher/lower score indicating a less/more square face. The colors in the border points indicate their distance to the square.

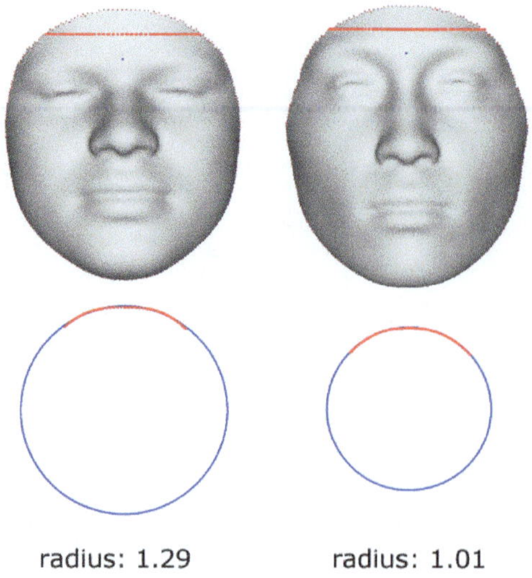

Figure S31: A proxy for head circumference: The forehead intersection with a plane halfway the Glabella and the top of the forehead is determined. This generates and arc segment through which a circle is fitted. The radius of the fitted circle serves as a proxy for head circumference.

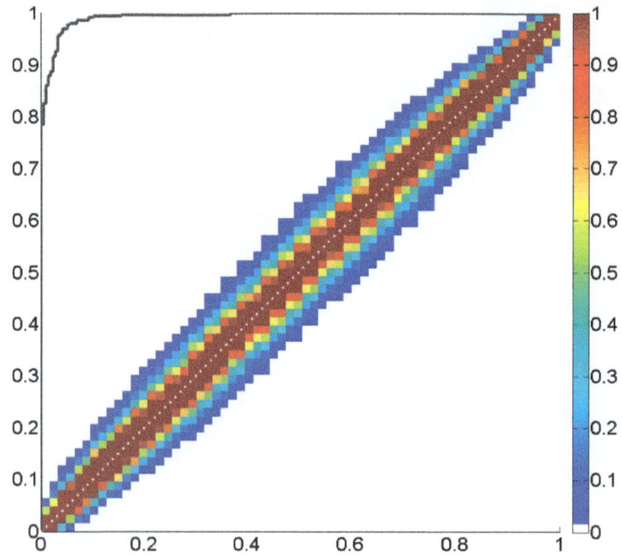

Figure S32: Receiver operator curve (ROC) showing the ability of facial femininity (RIP-S) to correctly classify faces by self-reported sex.

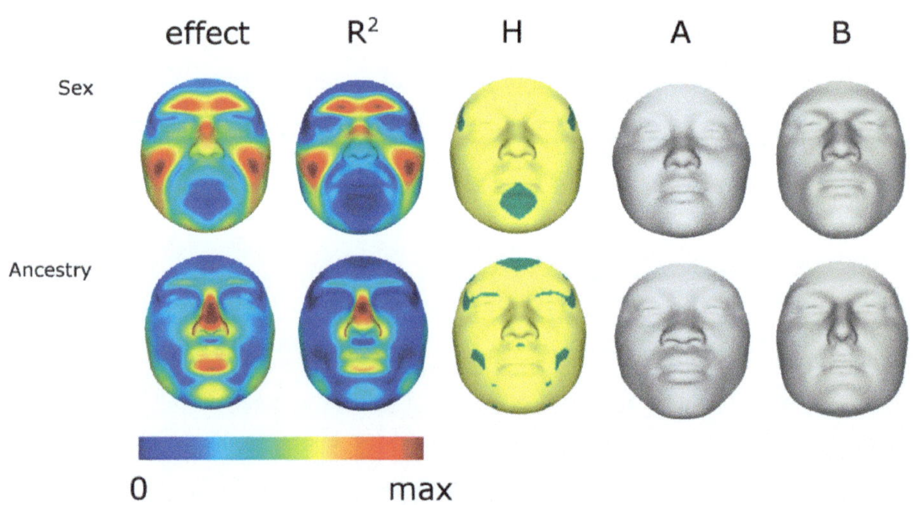

Figure S33: The effect, effect-size (r^2), significance of the effect (H), and two shape transformations at opposite sides (+3 and −3 times the standard deviation) of the RIP distribution for sex (top row) and ancestry (bottom row). The maximum values for r^2 **can be found in Table S3.**

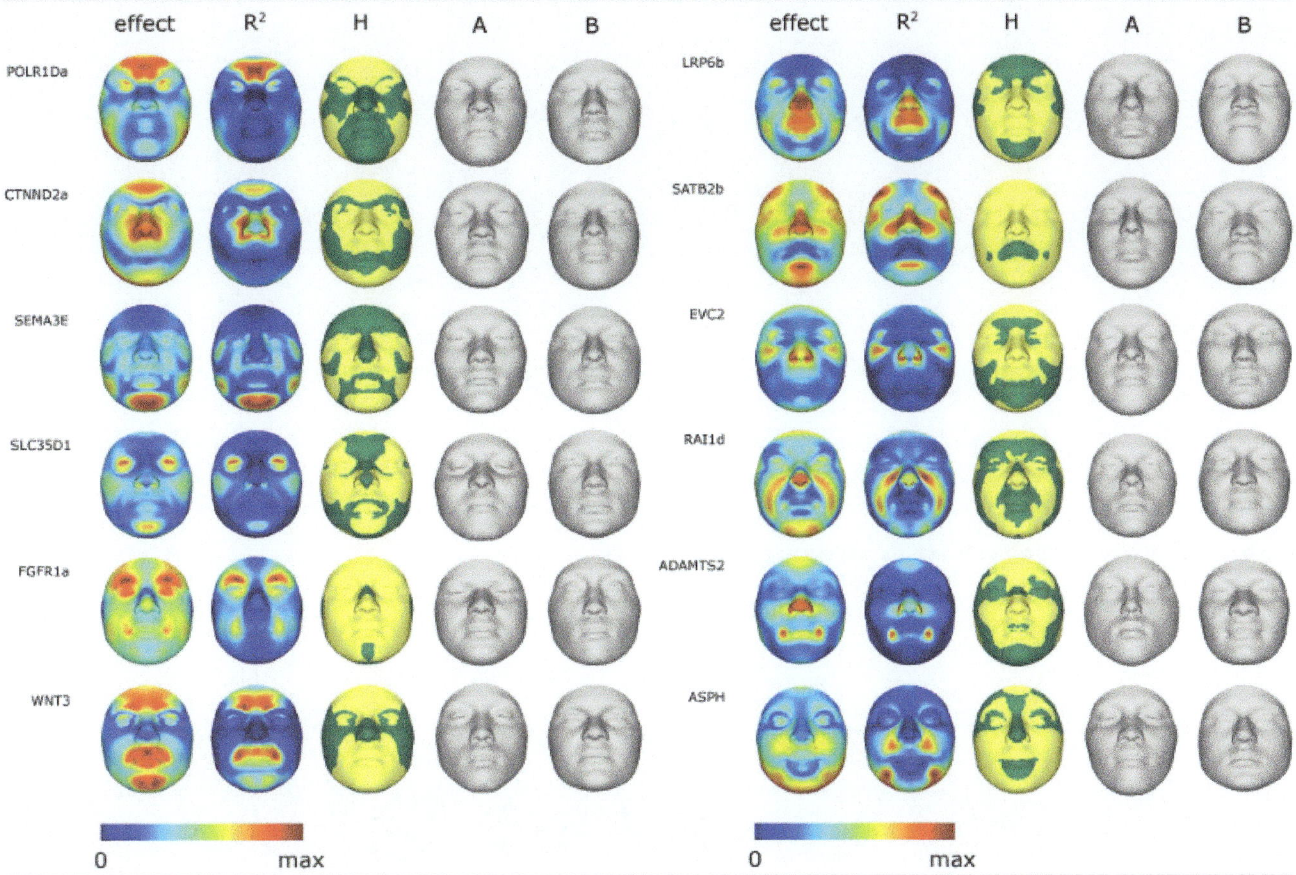

Figure S34: The effect, effect-size (r^2), significance of the effect (H), and two shape transformations at opposite sides (+X and −X times the standard deviation) of the RIP distribution. The maximum values for R^2 can be found in Table S3. PART 1.

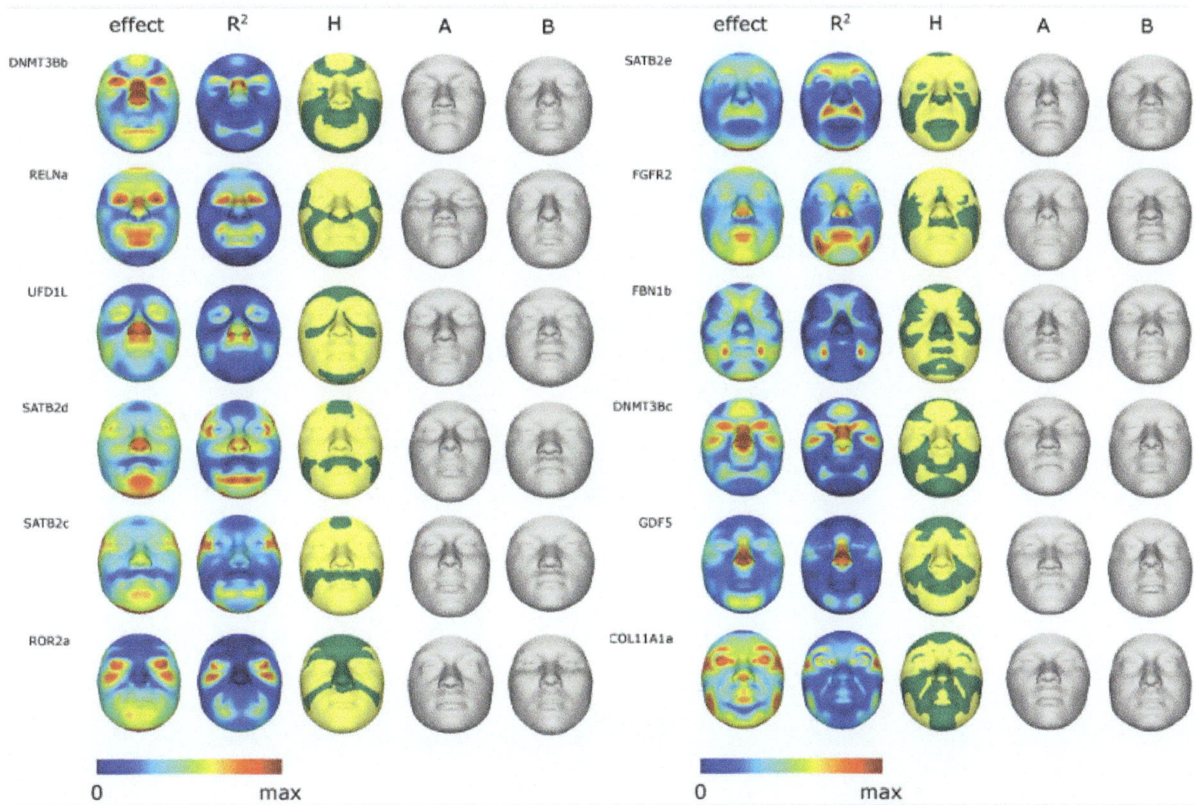

Figure S35: The effect, effect-size (r^2), significance of the effect (H), and two shape transformations at opposite sides (+X and −X times the standard deviation) of the RIP distribution. The maximum values for R^2 can be found in Table S3. PART 2.

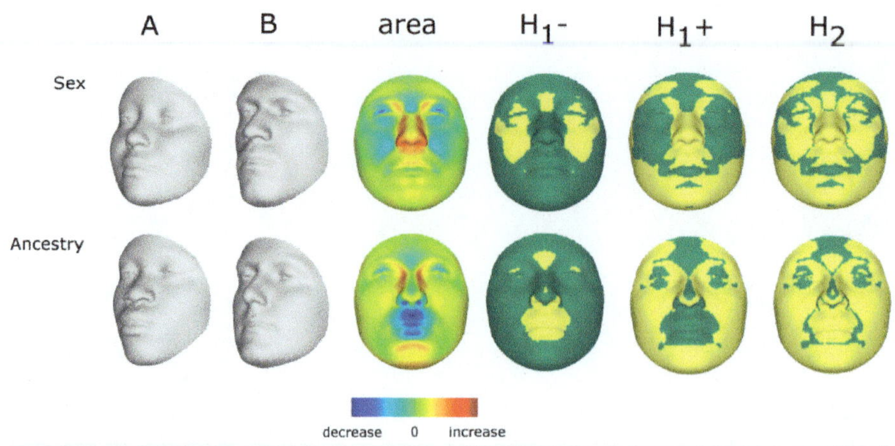

Figure S36: Facial area changes due to sex and ancestry.

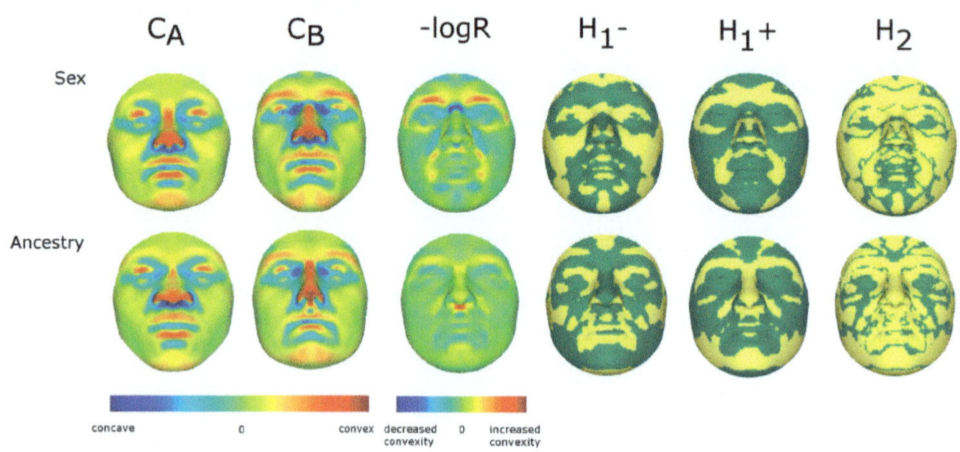

Figure S37: Facial curvature changes due to sex and ancestry.

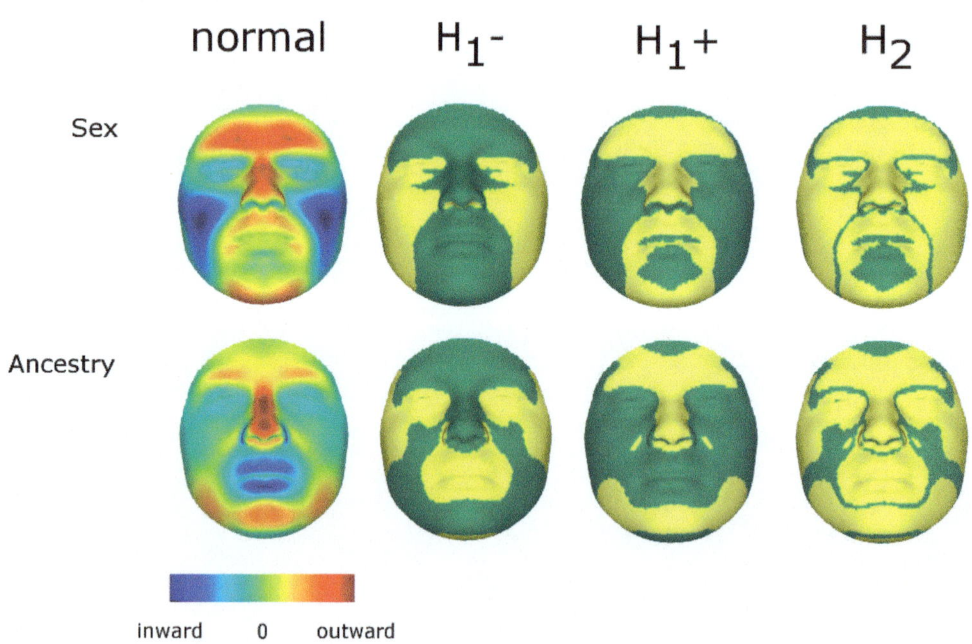

Figure S38: Normal displacements due to sex and ancestry.

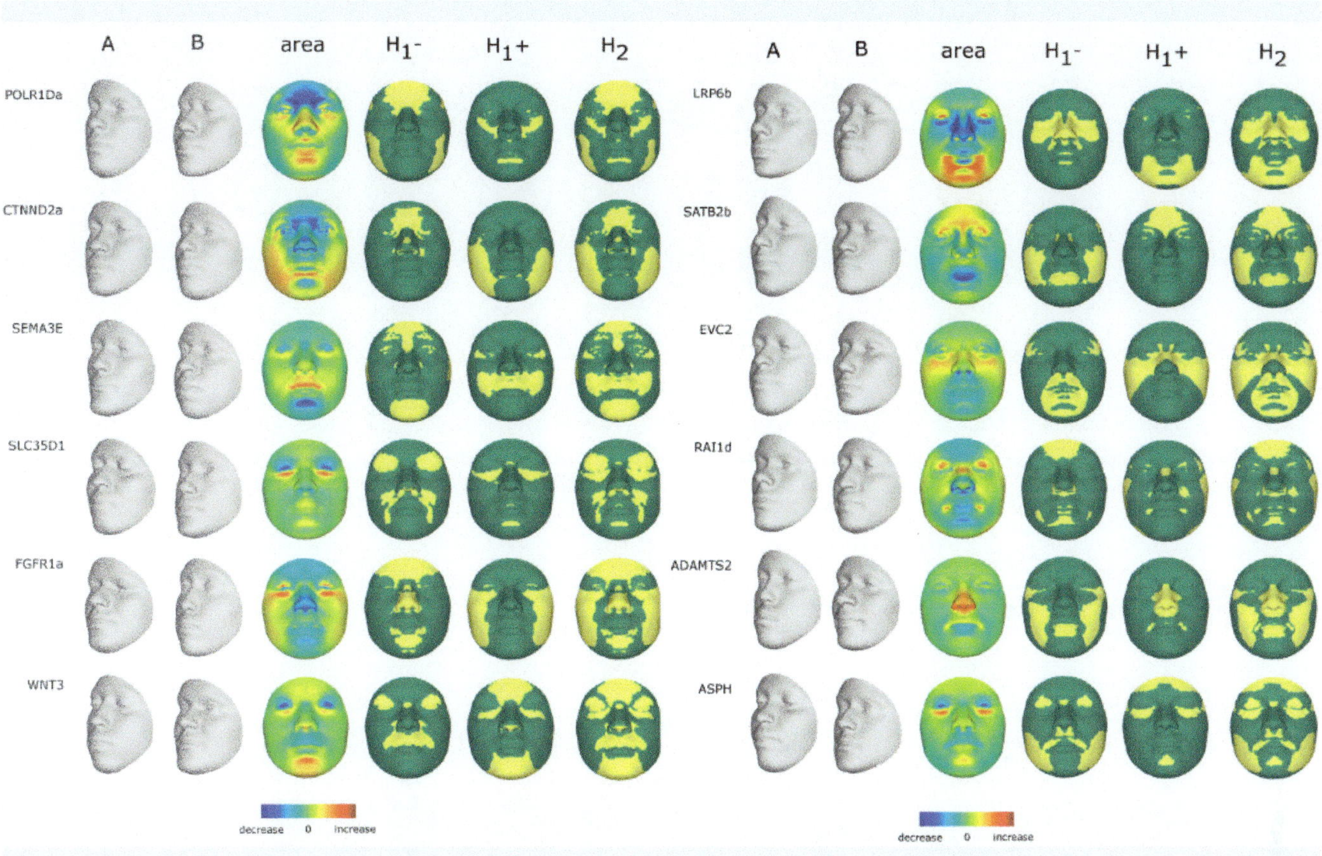

Figure S39: Facial area changes due to candidate genes. PART 1.

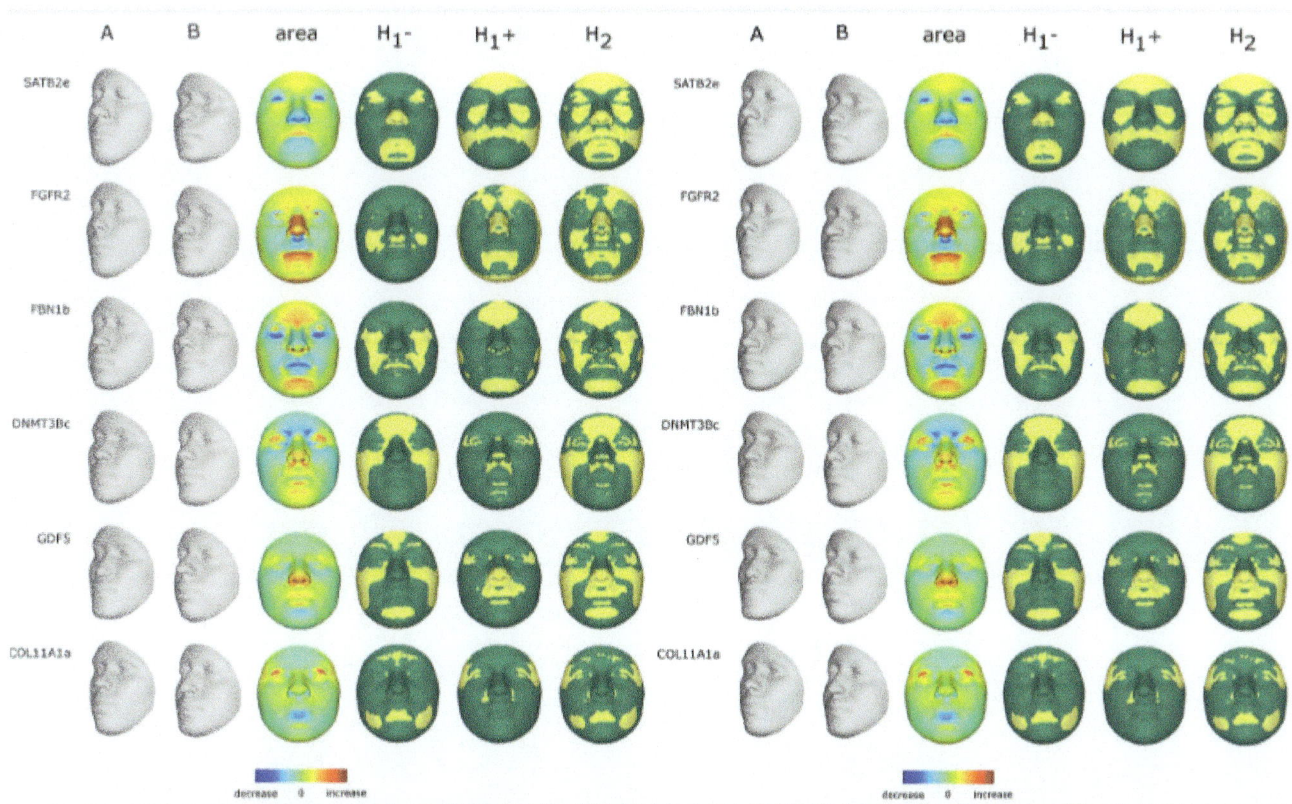

Figure S40: Facial area changes due to candidate genes. PART 2.

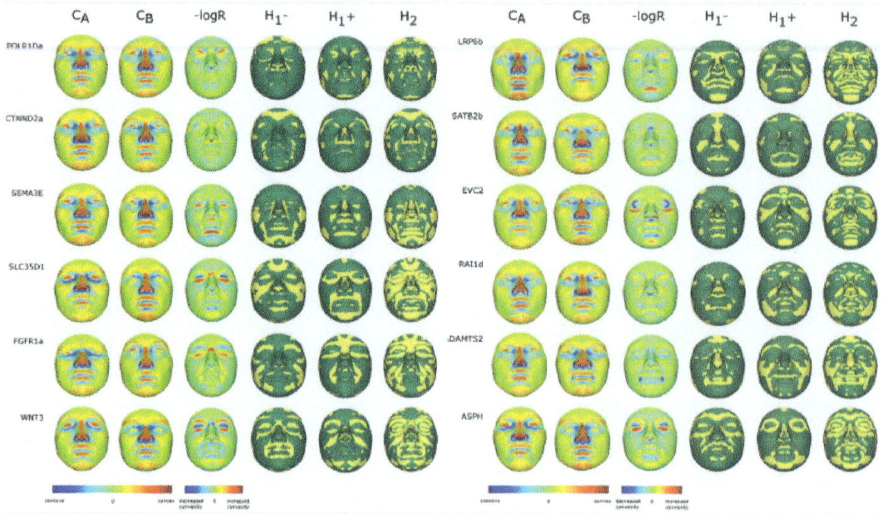

Figure S41: Facial curvature changes due to candidate genes. PART 1.

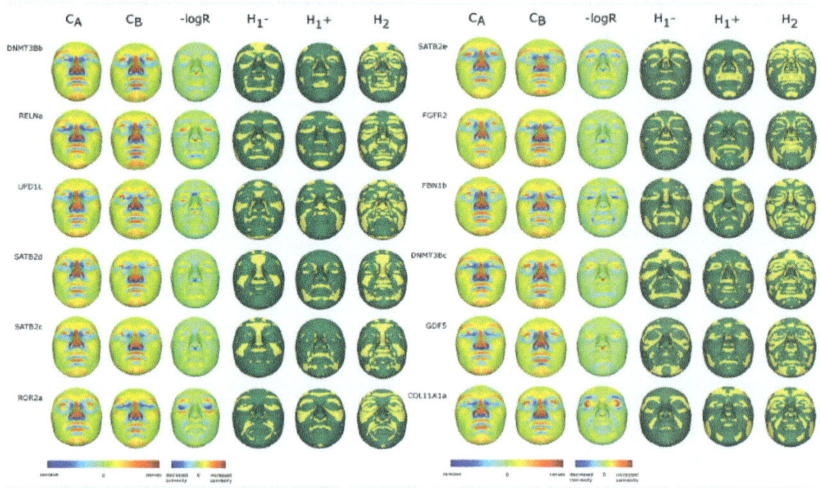

Figure S42: Facial curvature changes due to candidate genes. PART 2.

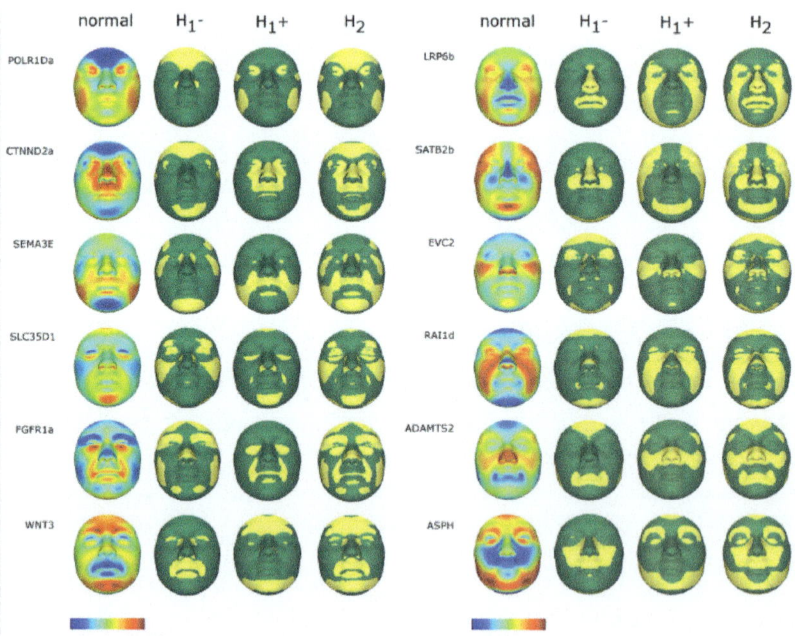

Figure S43: Normal displacements due to candidate genes. PART 1.

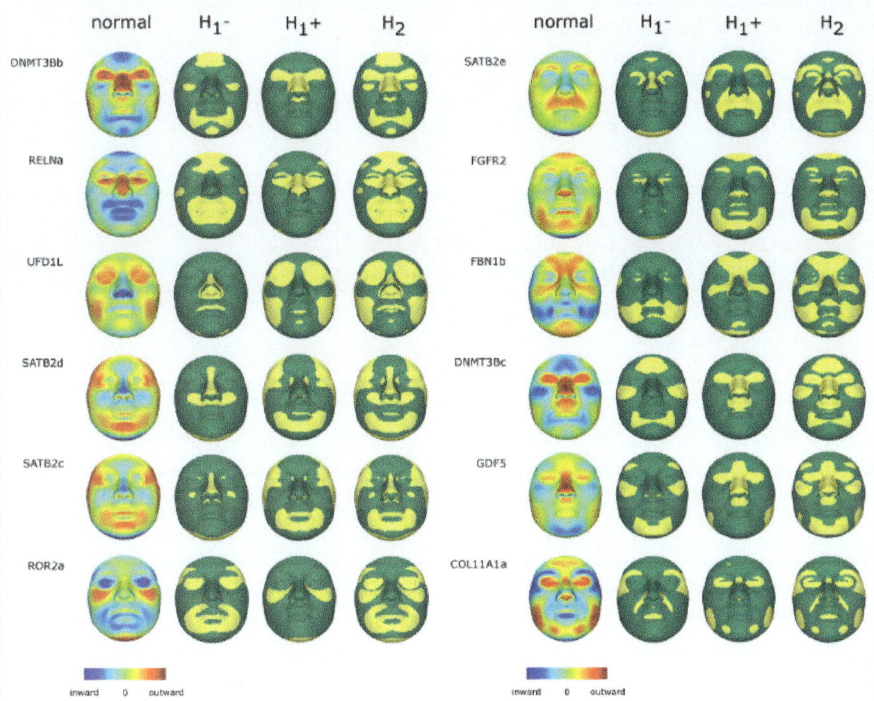

Figure S44: Normal displacements due to candidate genes. PART 2.

Table S1: List of local face shape change parameters (FSCPs) measured.

Facial Characteristic	*ID*	*Explanation*	*Technical Formulation*
Malar flattening	A	Change in surface curvature at the cheekbones	
	B	Change in variation or dispersion in position (standard deviation) along the anteroposterior (sagittal) axis of the cheekbones	
	C	Inward/outward movement of the medial part of the midface, in relation to the lateral part of the lower face	
Square/round face	A	Change in similarity between facial contours and a square (Fig S30)	*Take a border projection of each shape transformation and compare it with a fitted square:* *The square fitted to each of the morphs is defined by placing the 2 vertical lines at the most lateral positions of the facial border (left and right) and the 2 horizontal lines at the most superior and inferior positions of the facial border.*
Micrognathia	A	Inward/outward movement of the chin	
	B	Area increase/decrease of the size of the chin	
	C	Change in distance between labiale inferius and gnathion, along the vertical (longitudinal) axis	
Microcephaly	A	Change in size of the circle fitted to the surface of the forehead. The circle is placed in the transverse plane, halfway the glabella and the top of the face (Fig S31)	*Take the intersection plane with the facial mesh (through point halfway between glabella and top of head, parallel to the XZ plane), fit a circle through the intersection points, and take radius as approximate for head circumference*
	B	Inward/outward movement of the upper face	

Midface retrusion / Flat midface	A		Inward/outward movement of the midface in relation to the inward/outward movement of the upper and lower face	
	B		Change in surface curvature at the midface	
Frontal bossing	A		Inward or outward movement of the bilateral parts of the upper face, relative to the movement of the metopic ridge	
Metopic ridge prominent	A		Change in curvature at the surface over the metopic ridge	
Face long	A		Change in height-width ratio of the entire face	
Forehead short	A		Change in height of the upper face	
Supraorbital ridge underdeveloped	A		Inward/outward movement of the supraorbital ridge	
	B		Displacement of the supraorbital ridge along the antero-posterior (sagittal) axis.	
Forehead sloping	A		Inward/outward movement of the upper face	
	B		Change in angle that the antero-posterior (sagittal) axis makes with the anterior surface of the forehead, at the intersection with the sagittal (medial) plane, through the glabella	*where TopForehead is the most superior point of the intersection line between the ZY-plane through the glabella and the facial mesh.*
Forehead narrow/ broad	A		Change in width of the upper face	
Shallow orbits	A	AND	Inward/outward movement of the orbital ridges	
	B		Inward/outward movement of the supraorbital ridge	
	C		Inward/outward movement of the infraorbital ridge	
	D	AND	Change in surface curvature at the orbital ridges	
	E		Change in surface curvature at the supraorbital ridge	
	F		Change in surface curvature at the infraorbital ridge	
Superiorly oriented orbits	A		Change in surface curvature at the superior half of the eye, compared to the change in surface curvature at the inferior half of the eye.	
	B		Inward/outward movement of the superior half of the eye, compared to the inward/outward movement of the inferior half of the eye.	
	C		Area increase/decrease of the superior half of the eye, compared to the area increase/decrease of the inferior half of the eye.	

Palpebral fissures downslanted	A			Change in distance along the vertical (longitudinal) axis from the average position of the medial half of the eye, to the average position of the lateral half of the eye.	
	B			Change in angle between the principal axis of the left eye and the principal axis of the right eye	
Eyes widely spaced	A			Change in interpupillary distance	
	B			Change in outer canthal distance	
	C			Change in distance along the medio-lateral (horizontal) axis from the average position of the left eye, to the average position of the right eye.	
Proptosis	A	AND		Movement of the eyes along the antero-posterior (sagittal) axis	
	B	AND	AND	Movement of the lateral half of the eyes along the antero-posterior (sagittal) axis.	
	C	AND	AND	Movement of the medial half of the eyes along the antero-posterior (sagittal) axis.	
	D	AND	AND	Movement of the superior half of the eyes along the antero-posterior (sagittal) axis.	
	E	AND	AND	Movement of the inferior half of the eyes along the antero-posterior (sagittal) axis.	
	F	AND	NOT	Change in distance along the antero-posterior (sagittal) axis from the average position of the superior half of the eyes, to the average position of the inferior half of the eyes. (This indicates 'superiorly oriented orbits')	
Nasal ridge narrow	A			Change in width of the nasal ridge	
	B	AND		Change in curvature at the surface over the nasal ridge, compared to the change in surface curvature at the paranasal tissues	
	C	AND		Change in surface curvature at the nasal ridge	
	D	AND		Change in surface curvature at the paranasal tissues	
Nasal ridge retruded	A			Inward/outward movement of the nasal ridge	
Nasal bridge depressed	A			Inward/outward movement of the nasal bridge	
Nasal bridge wide	A			Change in surface curvature at the nasal bridge	
	B			Change in width of the nasal bridge	
Nose wide	A			Change in distance between alar curvature left and right	
	B			Change in distance between alare left and right	

Nose snubbed	A	Change in angle between columella (represented as a vector from subnasale to pronasale) and philtrum (represented as a vector from subnasale to labiale superius)	
Nares anteverted	A	Change in distance along the vertical (longitudinal) axis between subnasale and pronasale	
Cleft lip	A	Area increase/decrease of the philtrum	
	B	Inward/outward movement of the philtrum	
	C	Change in surface curvature of the philtrum	
	D	Area increase/decrease of the nose	
	E	Inward/outward movement of the nose	
	F	Change in surface curvature of the nose	
Mouth wide	A	Change in distance between left and right chelion	
Vermilion upper lip thick	A	Area increase/decrease of the upper lip (upper vermilion)	
Vermilion lower lip thick	A	Area increase/decrease of the lower lip (lower vermilion)	
Lips thick	A	Area increase/decrease of the lips	
Maxilla prominent	A	Movement of the midface along the antero-posterior (sagittal) axis	
	B	Inward/outward movement of the medial part of the midface, in relation to the lateral part of the lower face	
Philtrum long	A	Change in length of the philtrum (represented as a vector from subnasale to labiale superius)	
Sunken cheeks	A	Area increase/decrease of the cheeks	
Large eyes	A	Area increase/decrease of the eyes	
	B	Area increase/decrease of the eyes, projected onto the coronal (frontal) plane	
Large nose tip	A	Area increase/decrease of the nasal tip	
	B	Area increase/decrease of the nasal tip, projected onto the coronal (frontal) plane	
	D	Inward/outward movement of the nasal tip	

Notes: C = curvature, S = area, LM = landmark, N = normal displacement, P = point, B = shape transformation -X times the standard deviation of RIP values, A = shape transformation +X times the standard deviation of RIP values, $\sigma_{x,y,z}$ the standard deviation of quasi-landmarks in the x,y or z direction (used as surrogate for length and width measurements), θ = angle measurement, = distance between two points, = distance in the X, Y, or Z direction, = vector from point to point .

Table S2: ANOVA analysis of genotypes on facial morphology using RIP variables. The Table is sorted from low to high p-value on the three group ANOVA results.

Gene Symbol	SNP symbol in facial figures	reference SNP ID	chromosome	position	F statistic	p-value	A	B	YRI	CEU	DELTA
POLR1D	POLR1Da	rs507217	13	27106517	6.98	**0**	C	A	0.858	0.398	0.46
CTNND2	CTNND2a	rs2277054	5	11213043	5.84	**0.004**	G	A	0.013	0.535	0.522

SEMA3E	SEMA3E	rs2709922	7	82863254	6	**0.004**	G	A	0.119	0.707	0.588
SLC35D1	SLC35D1	rs1074265	1	67263852	5.36	**0.005**	T	A	0.625	0.1	0.525
FGFR1	FGFR1a	rs13267109	8	38496214	4.5	**0.01**	G	A	0.78	0.228	0.552
WNT3	WNT3	rs199501	17	42217772	4.62	**0.011**	G	A	0.252	0.788	0.536
LRP6	LRP6b	rs2724626	12	11721982	4.19	**0.015**	A	C	0.108	0.883	0.775
SATB2	SATB2b	rs1357582	2	200287112	4.27	**0.015**	G	A	0.143	0.761	0.618
EVC2	EVC2	rs10001971	4	5689700	4.26	**0.016**	A	G	0.796	0.042	0.754
RAI1	RAI1d	rs4925108	17	17590148	4	**0.017**	A	G	0.009	0.677	0.668
ADAMTS2	ADAMTS2	rs3822601	5	178554182	3.75	**0.021**	A	G	0.587	0.066	0.521
ASPH	ASPH	rs4738909	8	62650091	3.39	**0.033**	C	A	0.867	0.283	0.584
DNMT3B	DNMT3Bb	rs2424905	20	30816588	3.42	**0.037**	G	A	0.95	0.358	0.592
RELN	RELNa	rs471360	7	103168085	3.15	**0.044**	A	G	0.55	0.108	0.442
UFD1L	UFD1L	rs2073730	22	17817946	3.18	**0.045**	G	C	1	0.491	0.509
SATB2	SATB2d	rs6759018	2	200329008	3.12	**0.045**	G	A	0.115	0.695	0.58
SATB2	SATB2c	rs4530349	2	200329665	2.88	**0.054**	A	G	0.204	0.704	0.5
ROR2	ROR2a	rs7029814	9	93602502	2.74	**0.064**	G	A	0.978	0.389	0.589
SATB2	SATB2e	rs4673339	2	199981621	2.68	**0.068**	G	A	0.004	0.456	0.452
FGFR2	FGFR2	rs2278202	10	123233187	2.55	**0.081**	G	A	0.925	0.42	0.505
FBN1	FBN1b	rs6493315	15	46354306	2.33	**0.1**	G	A	0.774	0.196	0.578
DNMT3B	DNMT3Bc	rs2424928	20	30852297	2.28	**0.103**	G	A	0.929	0.403	0.526
GDF5	GDF5	rs143384	20	33489170	2.24	**0.106**	G	A	1	0.358	0.642
COL11A1	COL11A1a	rs11164669	1	103326755	2.25	**0.106**	G	A	0.064	0.629	0.565
DHCR7	DHCR7	rs11603330	11	70831107	2.06	0.122	C	A	0.889	0.279	0.61
FGFR1	FGFR1b	rs7818839	8	38607758	2.03	0.134	G	A	0.106	0.681	0.575
RPS19	RPS19	rs7254214	19	47060578	2.03	0.135	G	A	0.959	0.509	0.45
ROR2	ROR2b	rs7037255	9	93736775	1.98	0.136	G	A	0.052	0.619	0.567
DNMT3B	DNMT3Ba	rs1997797	20	30851615	2.02	0.138	G	C	0.864	0.331	0.533
RAI1	RAI1c	rs4315391	17	17270453	2	0.141	G	A	0.093	0.704	0.611
CHD7	CHD7	rs10092214	8	61889433	1.95	0.149	G	A	0.996	0.58	0.416
GNAS	GNAS	rs6123837	20	56898966	1.88	0.152	G	A	0.991	0.588	0.403
NIPBL	NIPBL	rs300063	5	37059655	1.68	0.185	A	C	0.903	0.414	0.489
SIL1	SIL1	rs10074485	5	138523427	1.63	0.192	G	A	0.31	0.903	0.593

CTNND2	CTNND2c	rs2561627	5	11708192	1.62	0.201	G	A	0.844	0.261	0.583
RAI1	RAI1a	rs2955382	17	17888435	1.6	0.201	A	G	0.137	0.642	0.505
RSPO2	RSPO2	rs2514838	8	109118075	1.63	0.203	G	A	0.996	0.518	0.478
WT1	WT1	rs5030317	11	32366913	1.63	0.204	G	C	0.849	0.248	0.601
SKI	SKI	rs2843159	1	2225532	1.6	0.204	A	G	0.562	0.137	0.425
DPYD	DPYDb	rs526645	1	97521968	1.56	0.209	G	A	0	0.417	0.417
GLI3	GLI3a	rs10951667	7	42118246	1.5	0.225	G	A	0.155	0.611	0.456
FBN1	FBN1a	rs16961205	15	46669609	1.44	0.239	C	A	0.491	1	0.509
FANCA	FANCA	rs10852623	16	88392743	1.41	0.241	G	A	0.841	0.416	0.425
RELN	RELNb	rs7799028	7	103133462	1.38	0.245	A	C	0.755	0.242	0.513
COL1A1	COL1A1	rs1934709	1	102740624	1.38	0.251	A	G	0.235	0.885	0.65
DPYD	DPYDa	rs12568335	1	98046720	1.32	0.27	G	A	0.119	0.58	0.461
CTNND2	CTNND2d	rs7733427	5	11896303	1.15	0.316	G	A	0	0.628	0.628
WT1	WT1b	rs5030320	11	32366578	1.1	0.333	G	A	0.823	0.288	0.535
FREM2	FREM2a	rs2496425	13	38162690	1.06	0.343	G	A	0.854	0.301	0.553
CCBE1	CCBE1	rs2564464	18	55387663	1.05	0.346	C	A	0.167	0.667	0.5
FLNB	FLNB	rs1127745	3	58487277	1.04	0.359	G	A	0.677	0.088	0.589
LRP6	LRP6c	rs3741800	12	12387226	0.99	0.364	A	G	0.035	0.535	0.5
FRAS1	FRAS1	rs345528	4	79448326	0.99	0.372	G	A	0.447	0.004	0.443
ECE1	ECE1	rs3026900	1	21432667	0.93	0.39	G	A	0.487	0.035	0.452
PEX3	PEX3b	rs9403540	6	144334280	0.92	0.4	C	A	0.004	0.536	0.532
PEX3	PEX3a	rs161062	6	143843526	0.82	0.441	G	A	0.177	0.69	0.513
SNRPN	SNRPN	rs12591149	15	22727610	0.8	0.445	G	A	0.947	0.478	0.469
POLR1D	POLR1Db	rs542610	13	27133069	0.76	0.471	A	G	0.398	0.805	0.407
HDAC4	HDAC4a	rs10207474	2	239720868	0.69	0.504	G	A	0.425	0.019	0.406
GLI3	GLI3b	rs11772482	7	42002148	0.64	0.536	G	A	0.85	0.434	0.416
FGFR2	FGFR2b	rs2912755	10	123252698	0.53	0.596	G	A	0.531	0.035	0.496
TBX1	TBX1	rs2301558	22	18131829	0.45	0.638	A	G	0.593	0.181	0.412
ADAMTS10	ADAMTS10	rs10401300	19	8579881	0.44	0.65	G	A	0.554	0.018	0.536
SATB2	SATB2a	rs1014497	2	199949474	0.3	0.741	A	G	0.264	0.708	0.444
BRAF	BRAF	rs10487888	7	140145576	0.28	0.76	A	G	0.004	0.522	0.518
EVC2	EVC2	rs7670299	4	5627041	0.23	0.795	G	A	0.808	0.307	0.501

FGFR1	FGFR1c	rs6474464	8	38617794	0.22	0.802	A	G	0.279	0.842	0.563
CTNND2	CTNND2b	rs249237	5	11494822	0.22	0.805	G	A	0.721	0.084	0.637
COL11A1	COL11A1b	rs6577351	1	103325508	0.13	0.878	G	A	0.606	0.088	0.518
RAI1	RAI1b	rs8079502	17	17262921	0.11	0.89	G	A	0.199	0.883	0.684
GLI3	GLI3c	rs6969239	7	42001356	0.11	0.895	G	A	0.933	0.526	0.407
HDAC4	HDAC4b	rs7573680	2	239834014	0.1	0.9	A	G	0.633	0.205	0.428
LMNA	LMNA	rs505058	1	154372809	0.1	0.908	G	A	0.73	0.066	0.664
FREM2	FREM2b	rs990909	13	38188569	0.08	0.921	A	G	0.142	0.642	0.5
LRP6	LRP6a	rs12823243	12	12220953	0.06	0.948	T	A	0.925	0.5	0.425
NSD1	NSD1	rs12660023	5	176633491	0.01	0.987	G	A	0.792	0.19	0.602

Table S3: RIP effect-size statistics.

	Overall partial R^2	mean	maximum	standard deviation	1st quartile	median	3rd quartile
POLR1Da	**1.24**	1.58	**7.06**	1.42	0.55	1.15	2.10
CTNND2a	**1.60**	2.07	**8.10**	1.58	0.90	1.60	2.76
SEMA3E	**1.42**	1.65	**6.39**	1.27	0.77	1.35	2.18
SLC35D1	**1.44**	1.88	**11.68**	1.55	0.84	1.42	2.45
FGFR1a	**2.82**	3.98	**15.16**	2.69	1.91	3.41	5.27
WNT3	**1.94**	2.49	**9.95**	2.32	0.74	1.47	3.81
LRP6b	**1.91**	2.38	**10.10**	2.14	0.81	1.71	3.15
SATB2b	**3.31**	4.37	**10.09**	2.30	2.61	4.01	5.96
EVC2	**1.87**	2.50	**13.70**	2.45	0.72	1.61	3.39
RAI1d	**1.14**	1.46	**5.34**	1.03	0.68	1.26	2.00
ADAMTS2	**1.60**	2.08	**15.28**	2.21	0.59	1.31	2.68
ASPH	**2.17**	2.71	**9.25**	2.08	1.04	2.12	3.89
DNMT3Bb	**1.35**	1.81	**9.91**	1.62	0.65	1.33	2.48
RELNa	**1.93**	2.59	**11.31**	2.21	0.94	1.89	3.79
UFD1L	**2.95**	3.71	**17.17**	2.78	1.63	3.23	4.89
SATB2d	**2.11**	2.62	**6.57**	1.39	1.49	2.45	3.67
SATB2c	**2.13**	2.69	**8.00**	1.60	1.58	2.33	3.50
ROR2a	**1.79**	2.34	**10.51**	1.90	1.07	1.91	3.25
SATB2e	**1.81**	2.30	**8.40**	1.80	0.91	1.69	3.22
FGFR2	**1.56**	1.90	**4.66**	0.93	1.15	1.76	2.48
FBN1b	**1.24**	1.63	**9.60**	1.26	0.67	1.43	2.24
DNMT3Bc	**1.18**	1.65	**6.83**	1.56	0.41	1.05	2.51
GDF5	**1.99**	2.48	**13.17**	2.46	0.84	1.69	3.31
COL11A1a	**0.93**	1.22	**4.64**	0.76	0.67	1.09	1.58
Sex	**12.92**	14.08	**38.21**	9.29	6.99	12.47	20.23
Ancestry	**9.55**	9.86	**40.83**	7.97	3.78	7.82	13.55

Table S4: Empirical *p*-values under 10,000 permutations for the local FSCPs listed in TableS1 tested for the 24 candidate genes, sex and ancestry.

		SEMA3E	WNT3	POLR1Da	CTNND2a	SLC35D1	LRP6b	FGFR1a	DNMT3Bb	DNMT3Bc	ADAMTS2	RAI1d	UFD1L	SATB2b	SATB2c	SATB2d	SATB2e	RELNa	ASPH	FGFR2	COL11A1a	EVC2	FBN1b	ROR2a	GDF5	Sex	Ancestry
Malar Flattening	A	0,216	0,511	0,850	0,493	0,000	0,003	0,001	0,332	0,000	0,000	0,000	0,002	0,001	0,000	0,000	0,538	0,001	0,000	0,628	0,611	0,000	0,000	0,000	0,000	0,000	0,048
	B	0,000	0,001	0,000	0,000	0,005	0,030	0,000	0,222	0,010	0,819	0,000	0,652	0,000	0,458	0,001	0,266	0,014	0,069	0,635	0,273	0,079	0,697	0,002	0,002	0,122	0,000
	C	0,027	0,538	0,001	0,002	0,501	0,981	0,000	0,000	0,000	0,000	0,001	0,002	0,000	0,000	0,000	0,684	0,841	0,000	0,343	0,000	0,000	0,000	0,000	0,000	0,370	0,004
Face square/round	A	0,046	0,001	0,057	0,021	0,090	0,000	0,271	0,556	0,002	0,000	0,727	0,000	0,075	0,000	0,830	0,000	0,000	0,000	0,001	0,000	0,048	0,003	0,008	0,050	0,146	0,000
Micrognathi	A	0,000	0,000	0,426	0,000	0,000	0,754	0,110	0,000	0,379	0,257	0,000	0,860	0,000	0,002	0,000	0,376	0,041	0,000	0,010	0,219	0,550	0,001	0,045	0,000	0,000	0,000
	B	0,000	0,000	0,173	0,839	0,113	0,000	0,000	0,490	0,004	0,701	0,017	0,000	0,640	0,103	0,030	0,002	0,014	0,277	0,005	0,648	0,001	0,000	0,000	0,000	0,000	0,000
	C	0,000	0,000	0,228	0,640	0,022	0,000	0,241	0,552	0,867	0,167	0,011	0,100	0,157	0,326	0,191	0,000	0,001	0,112	0,155	0,003	0,000	0,000	0,013	0,003	0,000	0,000
Microcephaly	A	0,000	0,095	0,136	0,135	0,482	0,083	0,000	0,000	0,000	0,000	0,213	0,930	0,000	0,000	0,000	0,001	0,000	0,862	0,130	0,000	0,000	0,786	0,016	0,001	0,000	0,470
	B	0,205	0,000	0,000	0,000	0,456	0,772	0,000	0,001	0,010	0,000	0,000	0,186	0,000	0,070	0,008	0,176	0,002	0,005	0,002	0,262	0,000	0,001	0,110	0,915	0,000	0,000
Midface retrusion	A	0,455	0,029	0,067	0,000	0,009	0,214	0,057	0,303	0,139	0,000	0,000	0,241	0,074	0,564	0,166	0,001	0,502	0,000	0,232	0,077	0,000	0,358	0,002	0,157	0,000	0,000
/ flat midface	B	0,000	0,676	0,050	0,285	0,460	0,704	0,002	0,000	0,000	0,003	0,001	0,000	0,242	0,123	0,815	0,000	0,000	0,000	0,019	0,000	0,000	0,000	0,000	0,000	0,000	0,000
Frontal bossing	A	0,000	0,298	0,028	0,016	0,073	0,468	0,001	0,000	0,000	0,000	0,439	0,105	0,000	0,000	0,000	0,001	0,000	0,145	0,010	0,014	0,006	0,262	0,064	0,000	0,031	0,291
Metopic ridge prominent	A	0,007	0,781	0,022	0,000	0,509	0,025	0,000	0,000	0,000	0,000	0,089	0,000	0,000	0,001	0,000	0,000	0,219	0,911	0,200	0,001	0,000	0,012	0,001	0,003	0,000	0,000
Face long	A	0,927	0,004	0,064	0,048	0,001	0,645	0,000	0,042	0,628	0,089	0,149	0,001	0,000	0,000	0,000	0,000	0,003	0,000	0,000	0,783	0,171	0,268	0,001	0,071	0,000	0,058
Forehead short	A	0,933	0,919	0,770	0,004	0,011	0,952	0,000	0,234	0,384	0,075	0,003	0,004	0,000	0,182	0,374	0,004	0,000	0,004	0,004	0,015	0,000	0,552	0,457	0,770	0,000	0,008
Supraorbital ridge	A	0,023	0,000	0,000	0,006	0,000	0,046	0,000	0,622	0,005	0,012	0,358	0,000	0,000	0,000	0,000	0,000	0,380	0,000	0,000	0,000	0,020	0,024	0,001	0,109	0,000	0,000
underdeveloped	B	0,111	0,000	0,002	0,246	0,002	0,000	0,000	0,217	0,428	0,964	0,191	0,000	0,000	0,188	0,002	0,234	0,752	0,000	0,055	0,000	0,000	0,010	0,001	0,239	0,000	0,000
Forehead sloping	A	0,205	0,000	0,000	0,000	0,456	0,772	0,000	0,001	0,010	0,000	0,000	0,186	0,000	0,070	0,008	0,176	0,002	0,005	0,002	0,262	0,000	0,001	0,110	0,915	0,000	0,000
	B	0,738	0,263	0,021	0,282	0,000	0,826	0,000	0,007	0,624	0,010	0,000	0,874	0,037	0,000	0,000	0,001	0,000	0,000	0,000	0,016	0,000	0,000	0,004	0,000	0,000	0,000
Forehead broad/narrow	A	0,001	0,000	0,000	0,000	0,029	0,004	0,000	0,203	0,520	0,006	0,687	0,280	0,000	0,000	0,000	0,000	0,001	0,024	0,014	0,059	0,967	0,006	0,082	0,163	0,959	0,120
Shallow orbits	A	0,001	0,000	0,042	0,029	0,000	0,007	0,000	0,574	0,000	0,000	0,993	0,000	0,000	0,000	0,000	0,000	0,225	0,004	0,001	0,000	0,022	0,040	0,558	0,000	0,000	0,000
	B	0,023	0,000	0,000	0,008	0,000	0,048	0,000	0,638	0,006	0,013	0,368	0,000	0,000	0,000	0,000	0,000	0,380	0,000	0,000	0,000	0,020	0,026	0,001	0,113	0,000	0,000
	C	0,003	0,825	0,048	0,992	0,000	0,001	0,380	0,021	0,000	0,001	0,014	0,794	0,000	0,296	0,212	0,086	0,035	0,004	0,595	0,000	0,000	0,022	0,000	0,000	0,000	0,000
	D	0,555	0,005	0,188	0,001	0,011	0,465	0,831	0,000	0,000	0,000	0,144	0,011	0,539	0,059	0,009	0,000	0,000	0,091	0,037	0,000	0,000	0,617	0,000	0,000	0,000	0,000
	E	0,354	0,000	0,011	0,004	0,757	0,161	0,503	0,000	0,000	0,000	0,503	0,337	0,510	0,279	0,020	0,000	0,002	0,001	0,067	0,000	0,000	0,439	0,001	0,000	0,000	0,000
	F	0,003	0,017	0,010	0,982	0,000	0,965	0,772	0,257	0,000	0,180	0,903	0,000	0,090	0,001	0,025	0,491	0,000	0,001	0,155	0,000	0,000	0,373	0,000	0,000	0,000	0,125
Superiorly oriented orbits	A	0,000	0,000	0,592	0,773	0,000	0,000	0,000	0,002	0,433	0,087	0,139	0,281	0,415	0,038	0,270	0,000	0,000	0,000	0,900	0,863	0,939	0,001	0,001	0,633	0,062	0,006
	B	0,002	0,000	0,287	0,171	0,000	0,024	0,000	0,000	0,001	0,145	0,537	0,852	0,020	0,287	0,372	0,038	0,000	0,000	0,432	0,541	0,001	0,273	0,566	0,027	0,800	0,148
	C	0,001	0,000	0,660	0,887	0,000	0,000	0,019	0,000	0,014	0,194	0,075	0,638	0,116	0,038	0,084	0,000	0,001	0,000	0,855	0,265	0,034	0,023	0,280	0,051	0,714	0,363
Palpebral fissures	A	0,009	0,000	0,068	0,001	0,052	0,000	0,966	0,000	0,000	0,026	0,000	0,284	0,006	0,087	0,248	0,000	0,000	0,001	0,004	0,000	0,000	0,540	0,139	0,000	0,000	0,007
downslanted	B	0,043	0,000	0,014	0,002	0,024	0,000	0,105	0,000	0,000	0,029	0,000	0,067	0,007	0,080	0,148	0,000	0,000	0,000	0,012	0,000	0,000	0,992	0,020	0,000	0,000	0,272
Eyes widely spaced	A	0,000	0,000	0,000	0,000	0,279	0,037	0,000	0,393	0,022	0,000	0,237	0,000	0,000	0,000	0,000	0,000	0,155	0,217	0,000	0,064	0,269	0,000	0,001	0,077	0,000	0,000
	B	0,000	0,815	0,014	0,000	0,061	0,251	0,000	0,385	0,049	0,000	0,404	0,003	0,000	0,000	0,000	0,000	0,420	0,131	0,001	0,432	0,257	0,009	0,000	0,560	0,000	0,000
	C	0,000	0,023	0,000	0,000	0,815	0,034	0,000	0,376	0,013	0,000	0,292	0,000	0,000	0,000	0,000	0,000	0,627	0,492	0,000	0,035	0,283	0,000	0,006	0,054	0,000	0,000
Proptosis	A	0,640	0,080	0,000	0,001	0,947	0,000	0,000	0,000	0,000	0,113	0,000	0,000	0,985	0,383	0,043	0,000	0,000	0,004	0,002	0,000	0,000	0,139	0,000	0,000	0,000	0,000
	B	0,589	0,101	0,000	0,495	0,894	0,000	0,000	0,000	0,000	0,288	0,000	0,000	0,119	0,007	0,000	0,000	0,000	0,097	0,026	0,000	0,000	0,438	0,000	0,000	0,000	0,000
	C	0,387	0,133	0,000	0,000	0,951	0,000	0,000	0,000	0,000	0,066	0,000	0,000	0,190	0,441	0,446	0,000	0,000	0,000	0,001	0,000	0,002	0,174	0,000	0,000	0,000	0,000
	D	0,468	0,002	0,000	0,010	0,000	0,000	0,000	0,000	0,000	0,214	0,000	0,000	0,522	0,611	0,039	0,000	0,000	0,069	0,001	0,000	0,000	0,209	0,000	0,000	0,000	0,000
	E	0,142	0,992	0,000	0,000	0,001	0,001	0,000	0,001	0,000	0,090	0,000	0,000	0,598	0,192	0,039	0,008	0,000	0,000	0,005	0,000	0,010	0,112	0,000	0,000	0,000	0,000
	F	0	0	0,091	0,009	0,000	0,002	0,000	0,000	0,000	0,264	0,793	0,702	0,041	0,123	0,626	0,000	0,000	0,000	0,835	0,950	0,000	0,359	0,276	0,238	0,000	0,097
Nasal ridge narrow	A	0,000	0,045	0,458	0,000	0,010	0,409	0,000	0,000	0,654	0,000	0,438	0,000	0,000	0,000	0,000	0,033	0,725	0,122	0,000	0,019	0,041	0,000	0,266	0,002	0,000	0,000
	B	0,000	0,000	0,278	0,034	0,000	0,002	0,042	0,000	0,066	0,000	0,000	0,001	0,000	0,000	0,000	0,000	0,005	0,000	0,413	0,266	0,042	0,000	0,509	0,003	0,219	0,000
	C	0,000	0,194	0,771	0,000	0,000	0,000	0,003	0,058	0,312	0,000	0,002	0,530	0,000	0,000	0,000	0,000	0,020	0,000	0,052	0,020	0,049	0,000	0,436	0,003	0,000	0,000
	D	0,000	0,000	0,113	0,401	0,000	0,645	0,602	0,000	0,000	0,147	0,000	0,000	0,002	0,000	0,000	0,000	0,014	0,000	0,647	0,764	0,132	0,000	0,052	0,029	0,079	0,000
Nasal ridge retruded	A	0,194	0,819	0,488	0,000	0,078	0,000	0,004	0,000	0,000	0,001	0,531	0,000	0,000	0,000	0,000	0,000	0,000	0,784	0,769	0,059	0,075	0,361	0,155	0,000	0,000	0,000
Nasal bridge depressed	A	0,088	0,857	0,000	0,040	0,728	0,000	0,514	0,000	0,000	0,024	0,033	0,012	0,000	0,001	0,000	0,001	0,001	0,939	0,128	0,001	0,006	0,000	0,097	0,000	0,000	0,000
Nasal bridge wide	A	0,024	0,000	0,033	0,179	0,000	0,000	0,000	0,411	0,080	0,000	0,023	0,000	0,000	0,000	0,000	0,000	0,035	0,043	0,001	0,006	0,032	0,000	0,351	0,035	0,000	0,000
	B	0,000	0,646	0,000	0,000	0,601	0,963	0,000	0,000	0,101	0,038	0,570	0,543	0,000	0,000	0,000	0,000	0,003	0,000	0,798	0,025	0,000	0,000	0,012	0,011	0,063	0,946
Nose wide	A	0,099	0,128	0,799	0,010	0,935	0,000	0,000	0,882	0,551	0,000	0,312	0,130	0,000	0,262	0,131	0,040	0,000	0,000	0,007	0,000	0,002	0,000	0,001	0,061	0,000	0,000
	B	0,347	0,018	0,574	0,000	0,546	0,001	0,000	0,933	0,389	0,000	0,321	0,248	0,000	0,756	0,012	0,990	0,100	0,001	0,000	0,000	0,000	0,000	0,000	0,998	0,000	0,000
Nose snubbed	A	0,199	0,013	0,000	0,000	0,149	0,002	0,117	0,000	0,000	0,000	0,000	0,125	0,000	0,000	0,000	0,000	0,000	0,083	0,000	0,378	0,023	0,000	0,171	0,000	0,221	0,000
Nares anteverted	A	0,552	0,028	0,069	0,000	0,846	0,000	0,553	0,000	0,000	0,018	0,000	0,168	0,000	0,000	0,000	0,000	0,000	0,204	0,000	0,515	0,170	0,000	0,016	0,000	0,015	0,000
Cleft lip	A	0,000	0,000	0,240	0,129	0,042	0,023	0,087	0,028	0,001	0,000	0,009	0,051	0,477	0,002	0,001	0,011	0,955	0,000	0,000	0,695	0,000	0,183	0,586	0,000	0,000	0,000
	B	0,000	0,000	0,601	0,000	0,486	0,001	0,105	0,002	0,000	0,001	0,312	0,588	0,002	0,309	0,000	0,000	0,000	0,000	0,061	0,031	0,029	0,004	0,005	0,000	0,000	0,000
	C	0,000	0,000	0,042	0,002	0,060	0,113	0,013	0,006	0,000	0,095	0,743	0,000	0,023	0,000	0,000	0,000	0,363	0,000	0,000	0,000	0,000	0,039	0,017	0,000	0,000	0,000
	D	0,618	0,396	0,170	0,013	0,511	0,000	0,000	0,027	0,015	0,000	0,070	0,000	0,568	0,291	0,757	0,000	0,000	0,248	0,000	0,757	0,358	0,673	0,680	0,000	0,000	0,091
	E	0,565	0,776	0,553	0,000	0,781	0,000	0,000	0,000	0,000	0,000	0,827	0,000	0,000	0,008	0,002	0,308	0,000	0,116	0,004	0,986	0,087	0,393	0,996	0,000	0,000	0,000
	F	0,191	0,000	0,000	0,000	0,737	0,000	0,111	0,000	0,000	0,126	0,000	0,537	0,000	0,000	0,000	0,000	0,007	0,000	0,000	0,980	0,681	0,230	0,005	0,000	0,000	0,000
Mouth wide	A	0,995	0,033	0,001	0,146	0,826	0,881	0,000	0,418	0,377	0,028	0,354	0,015	0,001	0,000	0,000	0,658	0,000	0,000	0,002	0,751	0,000	0,035	0,000	0,178	0,076	0,000
Upper vermilion thick	A	0,000	0,000	0,886	0,677	0,082	0,034	0,333	0,961	0,273	0,087	0,151	0,028	0,008	0,001	0,000	0,000	0,148	0,002	0,146	0,447	0,047	0,000	0,495	0,000	0,000	0,000
Lower vermilion thick	A	0,895	0,540	0,036	0,981	0,421	0,756	0,012	0,668	0,753	0,000	0,016	0,000	0,000	0,053	0,145	0,000	0,063	0,244	0,000	0,000	0,004	0,028	0,000	0,039	0,216	0,000
Lips thick	A	0,000	0,005	0,253	0,837	0,121	0,318	0,052	0,838	0,433	0,034	0,021	0,000	0,000	0,001	0,000	0,869	0,052	0,217	0,017	0,010	0,004	0,000	0,074	0,074	0,000	0,000
Maxilla prominent	A	0,005	0,079	0,307	0,000	0,269	0,003	0,000	0,054	0,014	0,004	0,000	0,023	0,000	0,011	0,000	0,086	0,972	0,000	0,055	0,441	0,000	0,506	0,004	0,001	0,000	0,294
	B	0,027	0,538	0,001	0,002	0,501	0,981	0,000	0,000	0,000	0,000	0,001	0,002	0,000	0,000	0,000	0,684	0,841	0,000	0,343	0,000	0,000	0,000	0,000	0,000	0,370	0,004
Philtrum long	A	0,000	0,000	0,105	0,976	0,832	0,000	0,015	0,147	0,243	0,591	0,003	0,676	0,018	0,000	0,000	0,037	0,841	0,002	0,000	0,810	0,052	0,746	0,000	0,000	0,000	0,000
Sunken cheeks	A	0,087	0,422	0,000	0,000	0,892	0,276	0,000	0,000	0,000	0,000	0,653	0,000	0,000	0,359	0,007	0,000	0,000	0,000	0,874	0,001	0,001	0,282	0,151	0,001	0,000	0,000
Eyes large	A	0,256	0,000	0,591	0,005	0,000	0,055	0,000	0,131	0,010	0,017	0,033	0,956	0,171	0,470	0,234	0,000	0,000	0,577	0,696	0,000	0,658	0,002	0,000	0,008	0,000	0,005
	B	0,700	0,166	0,469	0,018	0,001	0,000	0,000	0,281	0,363	0,147	0,568	0,004	0,100	0,150	0,951	0,723	0,000	0,000	0,020	0,000	0,448	0,000	0,000	0,055	0,000	0,200
Nasal tip large	A	0,206	0,074	0,291	0,018	0,247	0,000	0,000	0,006	0,004	0,000	0,439	0,000	0,363	0,482	0,481	0,000	0,000	0,101	0,000	0,318	0,478	0,332	0,849	0,000	0,000	0,019
	B	0,330	0,096	0,269	0,083	0,745	0,000	0,000	0,009	0,013	0,000	0,489	0,000	0,957	0,594	0,015	0,000	0,000	0,125	0,000	0,579	0,536	0,709	0,646	0,000	0,000	0,000
	C	0,110	0,233	0,542	0,000	0,738	0,000	0,000	0,000	0,000	0,000	0,082	0,000	0,000	0,001	0,000	0,014	0,000	0,880	0,001	0,465	0,011	0,308	0,992	0,000	0,000	0,000

ACKNOWLEDGMENTS

We thank the participants in this study, without whom none of this research would have been possible. We also thank the many colleagues who assisted us in our sampling trips, namely, Xianyun Mao, Kirk French, Abigail Bigham, Janet L. Markman, Breno Abreu, Joanna Abreu, Erika Horta Grandi Monteiro, Tulio Lins, Isabel Ine^s Arau´jo, Tovi M. Anderson, Joana Campos, Crisolita Gomes, and Jailson Lopes. We would like to acknowledge the advice and assistance of Rich Doyle, Yann Klimentidis, Andrea Hendershot, Sam Richards, Manolis Kellis, and Jose Fernandez.

REFERENCES

1. Carlson DS (2005) Theories of craniofacial growth in the postgenomic era. Seminars in Orthodontics 11: 172–183.
2. Sperber GH (2001) Craniofacial development. (Decker Inc., Ontario, B.C.).

3. Williams SE, Slice DE (2010) Regional shape change in adult facial bone curvature with age. American Journal of Physical Anthropology 143: 437–447.
4. Coussens AK, van Daal A (2005) Linkage disequilibrium analysis identifies an FGFR1 haplotype-tag SNP associated with normal variation in craniofacial shape. Genomics 85: 563–573.
5. Liu F, van der Lijn F, Schurmann C, Zhu G, Chakravarty MM, et al. (2012) A genome-wide association study identifies five loci influencing facial morphology in Europeans. PLoS Genetics 8: e1002932.
6. Boehringer S, van der Lijn F, Liu F, Günther M, Sinigerova S, et al. (2011) Genetic determination of human facial morphology: links between cleft-lips and normal variation. European Journal of Human Genetics 19: 1192–1197.
7. Paternoster L, Zhurov AI, Toma AM, Kemp JP, St Pourcain B, et al. (2012) Genome-wide association study of three-dimensional facial morphology identifies a variant in PAX3 associated with nasion position. American Journal of Human Genetics 90: 478–485.
8. Claes P, Walters M, Vandermeulen D, Clement JG (2011) Spatially-dense 3D facial asymmetry assessment in both typical and disordered growth. Journal of Anatomy 219: 444–455.
9. Claes P, Walters M, Clement J (2012) Improved facial outcome assessment using a 3D anthropometric mask. International Journal of Oral and Maxillofacial Surgery 41: 324–330.
10. Abdi H (2010) Partial least squares regression and projection on latent structure regression (PLS Regression). Wiley Interdisciplinary Reviews: Computational Statistics 2: 97–106.
11. Halder I, Shriver MD, Thomas M, Fernandez JR, Frudakis T (2008) A panel of ancestry informative markers for estimating individual biogeographical ancestry and admixture from four continents: utility and applications. Human Mutation 29: 648–658.
12. Long JC (1991) The genetic structure of admixed populations. Genetics 127: 417–428.
13. Pfaff CL, Parra EJ, Bonilla C, Hiester K, McKeigue PM, et al. (2001) Population structure in admixed populations: effect of admixture dynamics on the pattern of linkage disequilibrium. American Journal of Human Genetics 68: 198–207.
14. McKeigue PM (2005) Prospects for admixture mapping of complex traits. American Journal of Human Genetics 76: 1–7.
15. Hoggart CJ, Parra EJ, Shriver MD, Bonilla C, Kittles RA, et al. (2003) Control of confounding of genetic associations in stratified populations. American Journal of Human Genetics 72: 1492–1504.
16. Hoggart CJ, Shriver MD, Kittles RA, Clayton DG, McKeigue PM (2004) Design and analysis of admixture mapping studies. American Journal of Human Genetics 74: 965–978.
17. Claes P (2007) A robust statistical surface registration framework using implicit function representations-Application in craniofacial reconstruction. Ktholieke Universiteit Leuven. Available:https://lirias.kuleuven.be/handle/123456789/174028).
18. Mardia K, Bookstein FL, Moreton IJ (2000) Statistical assessment of bilateral symmetry of shapes.Biometrika 87: 285–300.
19. Klingenberg CP, Barluenga M, Meyer A (2002) Shape analysis of symmetric structures: quantifying variation among individuals and asymmetry. Evolution 56: 1909–1920.
20. Wray NR, Yang J, Goddard ME, Visscher PM (2010) The genetic interpretation of area under the ROC curve in genomic profiling. PLoS Genetics 6: e1000864.
21. Webster MA, Kaping D, Mizokami Y, Duhamel P (2004) Adaptation to natural facial categories. Nature428: 557–561.
22. Klimentidis YC, Shriver MD (2009) Estimating genetic ancestry proportions from faces. PLoS One 4: e4460.
23. Beleza S, Johnson NA, Candille SI, Absher DM, Coram MA, et al. (2013) Genetic architecture of skin and eye color in an African-European admixed population. PLoS Genetics 9: e1003372.
24. Hamosh A, Scott AF, Amberger JS, Bocchini CA, McKusick VA (2005) Online Mendelian Inheritance in Man (OMIM), a knowledgebase of human genes and genetic disorders. Nucleic Acids Research D514–517.
25. Song L, Li Y, Wang K, Wang YZ, Molotkov A, Gao L, et al. (2009) Lrp6-mediated canonical Wnt signaling is required for lip formation and fusion. Development 136: 3161–171.
26. Tang H, Peng J, Wang P, Risch NJ (2005) Estimation of individual admixture: analytical and study design considerations. Genetic Epidemiology 28: 289–301.
27. The International HapMap Consortium (2003) The International HapMap Project. Nature 426: 789–796.
28. Hawks J, Wang ET, Cochran GM, Harpending HC, Moyzis HK (2007) Recent acceleration of human adaptive evolution. Proceedings of the National Academy of Sciences of the United States of America 104: 20753–20758.
29. Shriver MD, Kennedy GC, Parra EJ, Lawson HA, Sonpar V, et al. (2004) The genomic distribution of population substructure in four populations using 8,525 autosomal SNPs. Human Genomics 1: 274–286.
30. Kauer MO, Dieringer D, Schlötterer C (2003) A microsatellite variability screen for positive selection associated with the “out of Africa” habitat expansion of Drosophila melanogaster. Genetics 165: 1137–1148.
31. Tajima F (1989) Statistical method for testing the neutral mutation hypothesis by DNA polymorphism.Genetics 123: 585–595.
32. Klingenberg CP, McIntyre GS (1998) Geometric morphometrics of developmental instability: analyzing patterns of fluctuating asymmetry with Procrustes methods. Evolution 52: 1363–1375.
33. Rohlf FJ, Slice D (1990) Extensions of the Procrustes Method for the Optimal Superimposition of Landmarks. Systematic Zoology 39: 40–59.

34. Mitteroecker P, Gunz P (2009) Advances in Geometric Morphometrics. Evolutionary Biology 36: 235–247.

35. Kimmerle EH, Jantz R (2005) Secular Trends in Craniofacial Asymmetry Studied by Geometric Morphometry and Generalized Procrustes Methods. In: Slice D, editor. Modern Morphometrics in Physical Anthropology. pp 247–263.

36. Palmer AR (1994) Waltzing with Asymmetry Is fluctuating asymmetry a powerful new tool for biologists or just an alluring new dance step?? BioScience 46: 518–532.

CHAPTER 10 A GENERALIZED APPROACH TO THE MODELING AND ANALYSIS OF 3D SURFACE MORPHOLOGY IN ORGANISMS

Janice L. Pappas, Daniel J. Miller

Museum of Paleontology, University of Michigan, Ann Arbor, Michigan, United States of America

ABSTRACT

The surface geometry of an organism represents the boundary of its three-dimensional (3D) form and can be used as a proxy for the phenotype. A mathematical approach is presented that describes surface morphology using parametric 3D equations with variables expressed as x, y, z in terms of parameters u, v. Partial differentiation of variables with respect to parameters yields elements of the Jacobian representing tangent lines and planes of every point on the surface. Jacobian elements provide a compact size-free summary of the entire surface, and can be used as variables in principal components analysis to produce a morphospace. Mollusk and echinoid models are generated to demonstrate that whole organisms can be represented in a common morphospace, regardless of differences in size, geometry, and taxonomic affinity. Models can be used to simulate theoretical forms, novel morphologies, and patterns of phenotypic variation, and can also be empirically-based by designing them with reference to actual forms using reverse engineering principles. Although this study uses the Jacobian to summarize models, they can also be analyzed with 3D methods such as eigensurface, spherical harmonics, wavelet analysis, and geometric morphometrics. This general approach should prove useful for exploring broad questions regarding morphological evolution and variation.

INTRODUCTION

The world is filled with organisms that display a broad diversity of surface morphologies. From the rippled texture of a tropical mollusk shell to the exquisitely delicate frustule of a diatom, surfaces are everywhere. It is surprising that so few studies have sought to rigorously describe the surface geometry of biological forms. The fact that the surface of an organism represents the boundaries of its three-dimensional (3D) morphology means that surface geometry can be used as a proxy for the phenotype. Studying surfaces may lead to a deeper understanding of the function and evolution of biological structures. As organisms evolve, so too may their surfaces, and studying the spatiotemporal distribution of surface morphology may lead to a greater understanding of evolutionary processes. Surfaces often change during growth, and their study may provide a different way of describing patterns of ontogeny.

While a great deal of effort has been devoted to the quantitative measurement of organisms, most studies have resorted to the measurement of distances between points or outlines. Often measurements are made from two-dimensional (2D) projections of 3D objects. Advances in 3D scanning

Citation: Pappas JL, Miller DJ. A Generalized Approach to the Modeling and Analysis of 3D Surface Morphology in Organisms. Evans AR, ed. PLoS ONE. 2013;8(10):e77551. doi:10.1371/journal.pone.0077551.

and imaging technologies have generated great interest in the digitization of biological form, and 3D data have been incorporated in studies using methods such as eigensurface analysis [1], [2], spherical harmonics [3], wavelet analysis [4], and geometric morphometrics [5]. Some studies in pattern recognition and computer visualization have described methods that emphasize using point clouds in shape matching of 3D surfaces [6], [7]; other studies are concerned with automating shape matching in 3D space [8].

We present a new approach to characterize 3D surfaces that differs in an important respect from existing methods. We begin with a parameterized 3D surface model rather than an empirically-derived array of landmarks or a point cloud from an actual specimen. Consequently, our approach permits the generation of theoretical or abstracted morphologies and this, in itself, presents some interesting opportunities for the study of biological form. However, while model-based, our method is fully capable of yielding empirical results, and we demonstrate how morphologically accurate models of real forms can be produced using reverse engineering principles [9]. That is, we can create a model from the empirical object, but numerical representation is extracted from equations of the model rather than directly from the empirical object itself. From our models, we calculate a matrix of values summarizing the entire surface, the elements of which can be used in conventional multivariate ordinations to generate surface morphospaces.

Our broader objective is to develop a generalized approach that can be used to study a wide variety of morphologies and morphological evolution in a more mathematically rigorous and flexible way. Our approach can be used alone or in conjunction with other analytical methods for the study of forms, both theoretical and empirical, and be applied to problems ranging from large scale macroevolution to small scale intrapopulation variation. Using examples from the Mollusca and Echinodermata, we show how forms with different morphologies and taxonomic affinities can be plotted meaningfully in a common morphospace. We also provide an example of how to study ontogenetic trajectories, real or theoretical, including allometric and isometric growth.

Mollusks and Echinoderms as Test Organisms

Both mollusks and echinoderms are taxonomically rich, well-studied, and have a well-documented fossil record. These phyla are suitable for studies of ecological and evolutionary patterns and processes over long time spans. The external morphology of many shelled mollusks has been the focus of numerous ecological and evolutionary studies, and the distinct spiral geometry of many forms has long attracted the interest of both mathematicians and biologists [10]–[11], providing an ideal test case for our approach. Others have analyzed echinoid form mathematically [12]–[13] and we show how our approach can be used on this group as well.

The study of spiral form in mollusks was codified by Thompson [14] early in the 20th century. However, the prominent spiral geometry of many mollusks was studied even earlier by Moseley [15]–[16], and his studies were the basis for Raup's [17]–[20] classic analyses of mollusk shells using a generating curve that moves along a logarithmic spiral with a fixed reference frame. Raup's model stimulated a number of additional studies [21]–[23], and several modifications were introduced, including the use of a moving reference frame[24]–[26], multiple helico-spirals [27]–[28], and the explicit inclusion of growth and morphogenesis[29]–[32].

For models with a generating curve, the shell surface is an epiphenomenon that reflects the cumulative history of the position of the generating curve as it moves along the spiral. These models, although treating the surface as a byproduct of growth, nevertheless succeed in simulating the broadest aspects of marginal accretion. However, describing mathematical properties such as the generating curve, type of spiral, and reference frame required to produce a coiled surface is not the same as describing the surface as an entity. Shell growth develops via marginal accretion, but the resulting form is often what interacts with the environment and can be the target of selective pressure. Thus, although we might be able to reduce a shell mathematically to marginally accreting spiral growth, morphological evolution involves the whole shell, and there is a need for methods that describe the surface phenotype as completely and directly as possible.

The objective of theoretical coiling models, in part, is to produce a low dimension morphospace with axes that represent particular parameters of the model. In contrast, our approach is to generate 3D theoretical models of whole surfaces that are then analyzed in an empirical n-dimensional space where the meaning of the axes is determined *a posteriori*.

Background on 3D Surface Geometry

In general, geometric forms can be parameterized and expressed in terms of particular basic trigonometric functions. Such expressions indicate how basic geometric forms are closely related, and any one of these forms can be a starting point to produce whole 3D forms (Fig. 1). Rather than use helicospiral curves as a starting geometry, as has been done in most previous studies, we instead use a torus as the initial surface and modify it to model a range of whole 3D surfaces. The general form of the organisms of interest ultimately determines which basic geometric form is most appropriate, and our choice of the torus does not preclude using other basic geometric forms as a starting point (Fig. 1). As a regular surface, the torus is expressed as a differentiable function. It can be represented by three parameterized variables given as $x(u, v) = (R + r \cos v) \cos u$, $y(u, v) = (R + r \cos v) \sin u$, $z = r \sin v$, where R is the major (outer) radius of the torus horizontally and r is the minor (inner) radius of the cross-section of the torus; the parameters are $u, v \in [0, 2\pi]$ with u defining the angle between R and the x-axis, and v defining the angle between R and r [33].

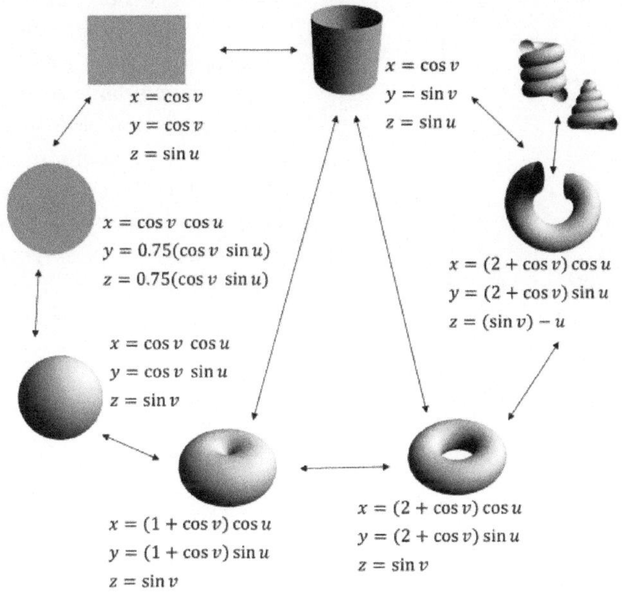

Figure 1: Basic geometric forms are depicted with their sets of parametric 3D equations.

For gastropod shells, an apical view (i.e., looking at a shell from its top) is generally circular with an offset end, and that end – the aperture – is a cross-section of a tube. The simplest closed-form tube is a torus. The simplest cut is a circular cross-section or meridian, and a cross-sectionally cut torus is used to expose one end to make a coil which can be viewed as a whorl. Elliptical cuts in the torus are produced via Villarceau [34] circles. One end of the cut torus is fixed and the other end is freely suspended, and if we pull downward on the free end, the curvature and torsion of what is now a spiral whorl will change. A coil of n-whorls can be produced by extending the length of the tube. The perpendicular of each whorl is also the radius of the cut torus tube, and every whorl has its own coiling axis. Different shapes and sizes of each whorl can be achieved as a result of changes in the geodesic properties (i.e., parallels and meridians) of the cut torus. Aperture and whorl diameters can change at any stage in the coiling process. As the whorls and apertures change in size, they will come in contact with each other as a result of their expansion or contraction. The cut torus may become a cylindrical or conical spiral (Fig. 1, upper right). In general, the whole 3D form is made by the interaction of all aspects of the geodesic properties of the torus.

The torus has geodesic properties on its surface [35]. Lines of curvature on a torus are meridians (lines that create cross-sectional, or poroidal circles) and parallels (lines of latitude with boundaries as outer circles, or toroidals) [36]. Parallels and meridians in x, y, z are orthogonal, and all meridians are of constant diameter, while all parallels are of variable diameter. All meridians and parallels, except for the maximum sized parallel, are geodesics. Tangent vectors of meridians and parallels are principal curvatures [36].

Curvature is the rate of change of tangents given as first derivatives that are perpendicular to each other on the surface [36]–[37]. The first derivatives are lines of curvature on a surface that are curves with tangential points along a principal curvature [38]. For a torus, the principal curvatures are $\kappa^1=-\cos\alpha/R+r\cos\alpha$ and $\kappa^2=-1/r$ of some angle α with major and minor radii R and r, respectively [35]. Torsion can be defined as a point along a curve that can move in different directions and be determined along any of the n-space curves of a given surface. Torsion depends on the rate at which a curve is turning along its arc length. On a surface, torsion is definable only for curves with $\kappa>0$.

Moving Reference Frames

At each point on the surface of a torus, an orthonormal frame consisting of three vectors can be defined. A moving reference frame with respect to velocity and acceleration of that frame occurs at each point along any curve on the surface where the frame is differentiable, and the result is a complete description of change in curvature and torsion. A moving reference frame on the whole torus can be a Darboux frame [39] which is a generalization of a Serret-Frenet [40]–[41] frame.

A Serret-Frenet frame consists of the unit tangent vector, t, the principal unit normal, **n**, and unit binormal, **b** at a point, p, along a curve. These vectors comprise an orthonormal frame, {**t**, **n**, **b**} [37]. In general, **b** is a multiple of **n**, and torsion is a factor in **b** and **n**, while curvature is a factor in **t** and **n** [38]. A Serret-Frenet frame parameterized by arc length, s, with velocity vector, **w**, is $\mathbf{t}(s)=\frac{d(\mathbf{w})}{ds}$, $\mathbf{n}(s)=\frac{dt/ds}{\|dt/ds\|}$, $\mathbf{b}(s)=\mathbf{t}\times\mathbf{n}$ with first derivatives $\frac{dt}{ds}=\kappa\mathbf{n}$, $\frac{dn}{ds}=-\kappa\mathbf{t}+\tau\mathbf{b}$, and $\frac{db}{ds}=-\tau\mathbf{n}$. For curvature with respect to s, $\kappa=\|\mathbf{t}'\|=\left|\frac{dt}{ds}\right|$ and for torsion, $\tau=-\mathbf{b}'\cdot\mathbf{n}$. If $\kappa\neq0$, $\mathbf{n}=\frac{1}{\kappa}\frac{dt}{ds}$ and $\mathbf{t}=\frac{1}{\kappa}$; **n** and **t** are perpendicular [37], [42].

A Darboux frame with respect to s has the vectors $\mathbf{t}(s)=\frac{d\mathbf{w}}{ds}$, **u**($s$) = **u**(**w**), and $\mathbf{v}(s)=\mathbf{u}(s)\times\mathbf{t}(s)$, where **w** is the velocity vector, **t**(s) is the unit tangent vector as it is in a Serret-Frenet frame, **u** is the unit normal vector, and **v** is tangent normal vector (Fig. 2). Essentially, a Darboux frame is an orthonormal frame where the unit tangent and tangent normal vectors are two principal directions and are linearly independent; **u** and **v** are in tangent planes that are perpendicular to **t** [43]. Since a Darboux frame is a moving reference frame $\frac{d\mathbf{t}}{ds}=\kappa_g\mathbf{u}+\kappa_n\mathbf{v}$,

$\frac{d\mathbf{u}}{ds} = -\kappa_g \mathbf{t} + \tau_g \mathbf{v}$, $\frac{d\mathbf{v}}{ds} = -\kappa_n \mathbf{t} - \tau_g \mathbf{u}$, where κ_g is the geodesic curvature, and τ_g is the geodesic torsion [36].

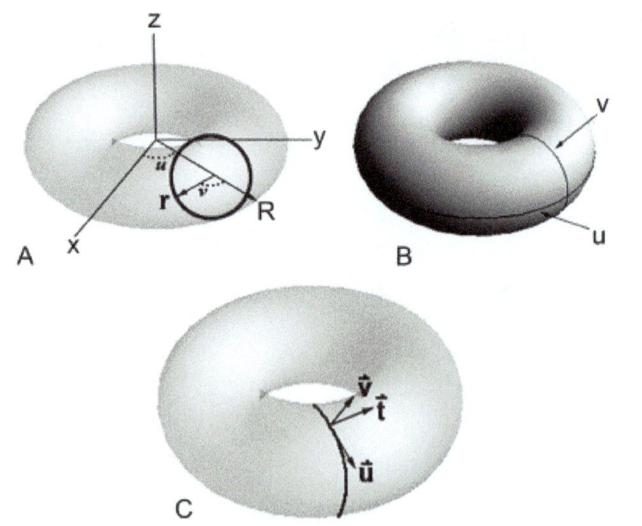

Figure 2: The torus as a model depicted with a Darboux frame.

A Darboux frame exists at any non-umbilic point on a surface. An umbilic point occurs on a surface where the principal curvatures are equal and every tangent direction is a principal direction [39]; i.e., curvature is the same in any direction. Since a torus has no umbilics, and lines of curvature are parallels and meridians [43], Darboux frames exist on all points on a torus. Moreover, with parallels and meridians as principal curves, *u*and *v* are parameters defining angles of the principal curves in terms of the major and minor radii of a torus, respectively [36].

From a Darboux vector field for a unit velocity vector [37], the relationship between Serret-Frenet and Darboux frames is determined by $\mathbf{t}' = \mathbf{D} \times \mathbf{t}$, $\mathbf{n}' = \mathbf{D} \times \mathbf{n}$, $\mathbf{b}' = \mathbf{D} \times \mathbf{b}$, where the Darboux vector field is $\mathbf{D} = \tau\mathbf{t} + \kappa\mathbf{b}$with κas curvature, τas torsion, and vectors {**t**, **n**, **b**} are a Serret-Frenet frame. For a unit velocity vector with $\kappa > 0$ [37], $\frac{d\mathbf{t}}{ds} = \kappa_n \mathbf{n} - \kappa_g \mathbf{b}$, $\frac{d\mathbf{n}}{ds} = -\kappa_n \mathbf{t} + \tau_g \mathbf{b}$, $\frac{d\mathbf{b}}{ds} = \kappa_g \mathbf{t} - \tau_g \mathbf{n}$, so that the relationship between Serret-Frenet and Darboux frames is evident in terms of curvature and torsion.

Parametric 3D Equations

Parametric 3D equations can be used to construct bounded surfaces [44]–[47]such as tori. At each point on the surface of a torus, two parameters, *u* and *v*, and a tangent vector define a Darboux frame (Fig. 2). The torus is a smooth manifold and differentiable everywhere, and parametric 3D equations in *x*, *y*, and *z* are differentiable with respect to parameters *u* and *v*. The derivatives of the vectors of a Darboux frame indicate the behavior of a curve locally on a surface. By modifying the parametric 3D equations for a torus, we construct various models of mollusks and echinoids.

First partial derivatives are tangent lines and planes of the whole surface. Numerically, tangent lines and planes at every point on the surface (Fig. 3) are solutions to Jacobian determinants that can also be the elements of the Jacobian matrices (i.e., **Jacobians**). These elements represent degrees of stretching from variables *x*, *y*, *z* to parameters *u*, *v* and represent angular attributes with respect to radii of models constructed. Solutions to the Jacobians are used as variables in principal components analysis (PCA) to create a morphospace. Euclidean distances can be found as the change in tangents of a curve between two points, and because tangents are slopes of the curve, a morphological *surface space* can be constructed based on the Jacobians. The Jacobians are a mathematical summary of the entire surface and present a powerful new way of comparing phenotypes.

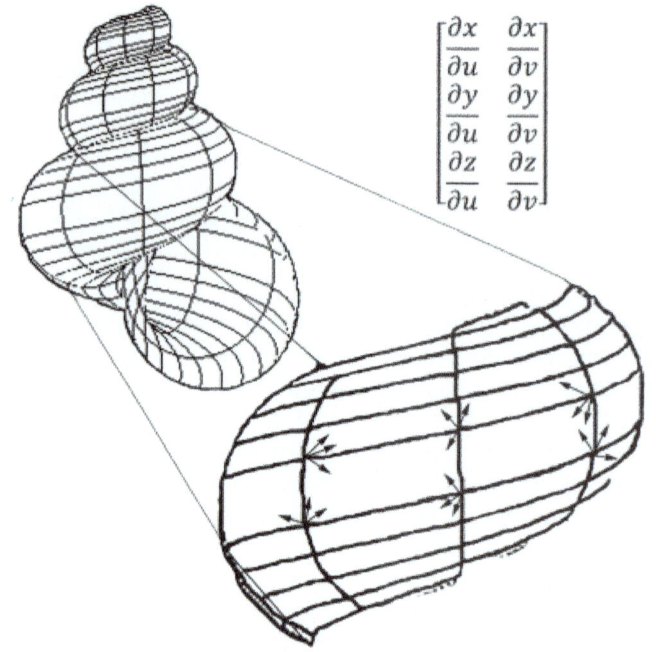

$$\begin{bmatrix} \frac{\partial x}{\partial u} & \frac{\partial x}{\partial v} \\ \frac{\partial y}{\partial u} & \frac{\partial y}{\partial v} \\ \frac{\partial z}{\partial u} & \frac{\partial z}{\partial v} \end{bmatrix}$$

Figure 3: Tangent lines and planes representing Jacobians on a shell model with a gridded surface.

METHODS

Parametric 3D Equations

All systems of parametric 3D equations are derived from the generalized set that is

$$x = ((\boldsymbol{R} + \boldsymbol{r}\,\boldsymbol{cos}\,\boldsymbol{v}*)[\pm \mathit{function}])([\mathit{function}] \times \boldsymbol{cosu}*)\left\{\binom{\pm}{\times}\right\}[\mathit{function}] \quad (2.1.1)$$

$$y = ((\boldsymbol{R} + \boldsymbol{r}\,\boldsymbol{cos}\,\boldsymbol{v}*)[\pm \mathit{function}])([\mathit{function}] \times \boldsymbol{sinu}*)\left\{\binom{\pm}{\times}\right\}[\mathit{function}] \quad (2.1.2)$$

$$z = ([\mathit{function}](\pm \boldsymbol{r}\,\boldsymbol{sinv}*))\left\{\binom{\pm}{\times}\right\}[\mathit{function}]\left\{\binom{\pm}{\times}\right\}[\mathit{function}] \quad (2.1.3)$$

where emboldened, italicized terms and operators represent a torus, asterisked variables can have coefficients, and multiple operators are possible with respect to other functions. The label, [*function*], can be a value multiplied by $1/2\pi$, an exponential with respect to $u,v \in [0,2\pi]$, a sine, cosine, hyperbolic tangent, hyperbolic secant, or some combination. Some of the models are depicted in Figure 4. Parametric 3D equations used to construct gastropod shells (with the exception of limpets) are identified with the term "System" and model number.

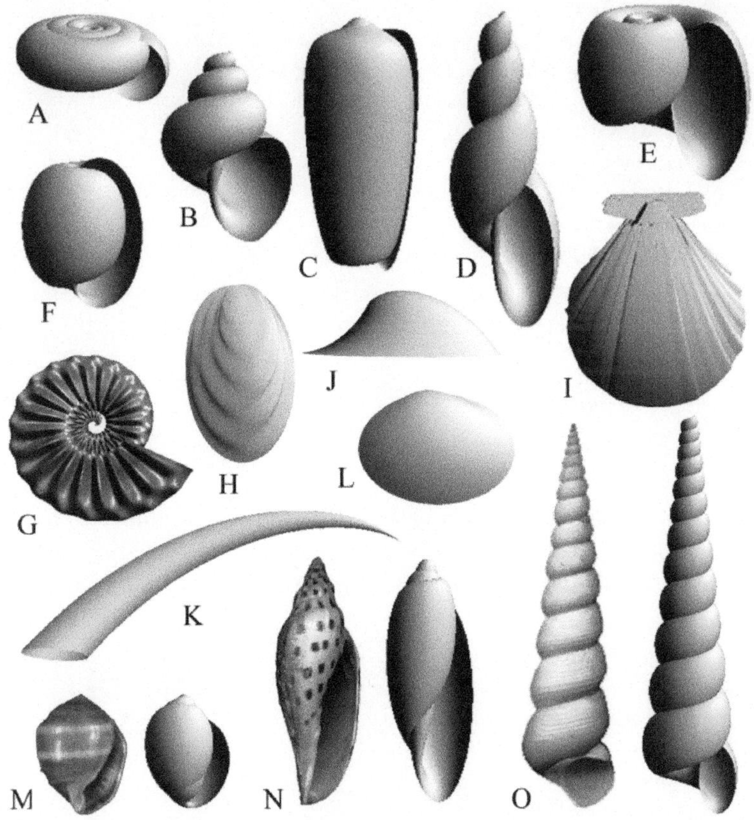

Figure 4: Illustrations of mollusk shell models created from parametric 3D equations.

Specimen pictures [48], [49] (e.g., Fig. 4M, N) used for comparison to gastropod shell models and measurements of R_{max} and r_{max} are listed in Table 1. Measurements were made from pictures [48], [49] and verified using specimens from the University of Michigan Museum of Zoology (UMMZ) (Tables 1 and and2).2). Scientific names are accepted in the World Register of Marine Species (WoRMS; www.marinespecies.org) and Integrated Taxonomic Information System (IT IS; www.itis.gov) databases. R_{max} and r_{max} are measured, and the nearest integer values are reduced to the lowest possible ratio.

Table 1: Measurement of maximum whorl and aperture radii ($\boldsymbol{R_{max}}$ and$\boldsymbol{rmax}$) of gastropod shells.

Mollusk shell example	Actual measurement (cm)		Actual proportion		Integer proportion	
	R_{max}	r_{max}	R_{max}	r_{max}	R_{max}	r_{max}
Scaphander lignarius	1	1.5	1	1.5	1	2
Sinum perspectivum	1.6	1.6	1	1	1	1
Melampus coffeus	0.7	0.85	1	1.214	1	1
Oliva sayana	0.8	1.35	1	1.687	1	2
Longchaeus candidus	0.55	0.45	1.222	1	1	1
Lithopoma phoebium	1.8	0.75	2.4	1	2	1
Scalenostoma subulatum	0.55	0.4	1.375	1	1	1
Architectonica nobilis	1.45	0.4	3.625	1	4	1
Heliacus variegatus	0.8	0.3	2.667	1	3	1
Calliostoma bairdi	1.2	0.45	2.7	1	3	1
Janthina globosa	1.35	1.25	1.08	1	1	1
Vitrinella oldroydi	0.7	0.25	2.8	1	3	1
Scaphella junonia	0.9	1.55	1	1.722	1	2
Buccinum glaciale	1.4	1.05	1.333	1	1	1

Table 2: Material examined from the University of Michigan Museum of Zoology (UMMZ).

Name of Taxon; UMMZ Number; Locality and Collector Information.
Scaphander lignarius (Linnaeus, 1758); UMMZ 9423; England.
Sinum perspectivum (Say 1831); UMMZ 19526; Bahamas.
Melampus coffeus (Linnaeus, 1758); UMMZ 236294; Bird Key, NR., St. Petersburg, Pinellas Co., Florida; coll.: J.B. Clark.
Oliva sayana Ravenel, 1834; UMMZ 251580; Big Gaspilla Island, Florida; coll: Carl Fellons, 3/62.
Longchaeus candidus (Mörch, 1875); UMMZ 165937; Grassy Key, Florida Keys; coll.: B.R. Bales.
Lithopoma phoebium (Röding, 1798); UMMZ 198132; McGinty's Ocean Ridge, Florida; coll: M.M. Solem.
Scalenostoma subulatum (Broderip, 1832); UMMZ 19940; Mauritius.
Architectonica nobilis Röding, 1798; UMMZ 250913; Florida, Gulf of Mexico; dredged.
Heliacus variegatus (Gmelin, 1791); UMMZ 250920; Hawaii.
Calliostoma bairdi Verrill & Smith, 1880; UMMZ 250829; Florida.
Janthina globosa Swainson, 1822; UMMZ 18272; Japan.
Vitrinella oldroydi Bartsch, 1907; UMMZ 174914; Monterey, Pacific Grove, California; coll: Chase 1941.
Scaphella junonia (Lamarck, 1804); UMMZ 251654; 20 fms., off Cameron, Louisiana.
Buccinum glaciale Linnaeus, 1761; UMMZ 17166; Greenland.

Jacobians

The following example using a set of parametric 3D equations for a ring torus illustrates how the surface is represented by a Jacobian. Let the equations for a ring torus be $x = (1+ \cos v) \cos u$, $y = (1+ \cos v) \sin u$, $z = \sin v$, where $R = r=1$ and $u, v \in [0, 2\pi]$. From Jacobian determinants of the mapping $x = f(u, v)$, $y = g(u, v)$, $z = h(u, v)$, the Jacobian (i.e., Jacobian matrix) evaluated at $u = v=0$ is

$$\begin{bmatrix} \frac{\partial x}{\partial u} & \frac{\partial x}{\partial v} \\ \frac{\partial y}{\partial u} & \frac{\partial y}{\partial v} \\ \frac{\partial z}{\partial u} & \frac{\partial z}{\partial v} \end{bmatrix} =$$

$$\begin{bmatrix} (-\cos v \sin u) - \sin u & (-\sin v \sin u) + \cos u \\ \cos v \cos u + \cos u & -\sin v \sin u \\ \cos u & 0 \end{bmatrix} = \begin{bmatrix} 0 & 1 \\ 2 & 0 \\ 1 & 0 \end{bmatrix}. \tag{2.1.4}$$

Relationship between Parametric 3D Equations and Jacobians

A surface is implicitly defined as $x = f(u,v)$, $y = g(u, v)$, $z = h(u, v)$ with differentials that form tangents [50]. The derivative of a parametric function is a velocity vector since tangents to a curve are the speeds at which a point on the curve moves.

Partial derivatives for x, y, and z with respect to u and v via the general chain rule [50] are

$\left(\frac{\partial x}{\partial u}\right)_v = \frac{\partial x}{\partial v}\frac{\partial v}{\partial u}$ and $\left(\frac{\partial x}{\partial v}\right)_u = \frac{\partial x}{\partial u}\frac{\partial u}{\partial v}$ for differential $dx = \frac{\partial x}{\partial u}du + \frac{\partial x}{\partial v}dv$, $\left(\frac{\partial y}{\partial u}\right)_v = \frac{\partial y}{\partial v}\frac{\partial v}{\partial u}$ and $\left(\frac{\partial y}{\partial v}\right)_u = \frac{\partial y}{\partial u}\frac{\partial u}{\partial v}$ for differential $dy = \frac{\partial y}{\partial u}du + \frac{\partial y}{\partial v}dv$, $\left(\frac{\partial z}{\partial u}\right)_v = \frac{\partial z}{\partial v}\frac{\partial v}{\partial u}$ and $\left(\frac{\partial z}{\partial v}\right)_u = \frac{\partial z}{\partial u}\frac{\partial u}{\partial v}$ for differential $dz = \frac{\partial z}{\partial u}du + \frac{\partial z}{\partial v}dv$.

Define three implicit functions [49,51] of the surface as $F(x, y, z) = F[f(u, v), g(u, v), h(u, v)]$,

$G(x, y, z) = G[f(u, v), g(u, v), h(u, v)]$, and $H(x, y, z) = H[f(u, v), g(u, v), h(u, v)]$ that describe the relation among x, y, z with respect to u and v. The differentials for F, G, and H in terms of x, y, z are $d[F(x,y,z)] = \frac{\partial F}{\partial x}dx + \frac{\partial F}{\partial y}dy + \frac{\partial F}{\partial z}dz$, $d[G(x,y,z)] = \frac{\partial G}{\partial x}dx + \frac{\partial G}{\partial y}dy + \frac{\partial G}{\partial z}dz$, $d[H(x,y,z)] = \frac{\partial H}{\partial x}dx + \frac{\partial H}{\partial y}dy + \frac{\partial H}{\partial z}dz$, respectively, where dx, dy, and dz in terms of u and v can be substituted into each equation in this set of linear equations, and F, G, H are tangent planes that define the intersecting surfaces of a model.

Let $F(u,v) = f[x(u, v), y(u, v), z(u, v)]$, $G(u,v) = g[x(u, v), y(u, v), z(u, v)]$, and $H(u,v) = h[x(u, v), y(u, v), z(u, v)]$ be defined as implicit functions. The differentials for F, G, and H in terms of u and v are $d[F(u,v)] = \frac{\partial[F(u)]}{\partial u}du + \frac{\partial[F(v)]}{\partial v}dv$, $d[G(u,v)] = \frac{\partial[G(u)]}{\partial u}du + \frac{\partial[G(v)]}{\partial v}dv$, $d[H(u,v)] = \frac{\partial[H(u)]}{\partial u}du + \frac{\partial[H(v)]}{\partial v}dv$. The relation between x, y, z and u, v with respect to differentials for F is $d[F(x,y,z)] = d[F(u,v)]$. For $d[F(x,y,z)] = \frac{\partial F}{\partial x}dx + \frac{\partial F}{\partial y}dy + \frac{\partial F}{\partial z}dz$, and substituting equations for dx, dy, dz in terms of u and v gives

$$d[F(x,y,z)] = \frac{\partial[F(u,v)]}{\partial x}\left[\frac{\partial x}{\partial u}du + \frac{\partial x}{\partial v}dv\right] + \frac{\partial[F(u,v)]}{\partial y}\left[\frac{\partial y}{\partial u}du + \frac{\partial y}{\partial v}dv\right] + \frac{\partial[F(u,v)]}{\partial z}\left[\frac{\partial z}{\partial u}du + \frac{\partial z}{\partial v}dv\right].$$

By gathering terms and using the chain rule,

$$d[F(x,y,z)] = \left[\frac{\partial[F(u)]}{\partial x}\frac{\partial x}{\partial u} + \frac{\partial[F(u)]}{\partial y}\frac{\partial y}{\partial u} + \frac{\partial[F(u)]}{\partial z}\frac{\partial z}{\partial u}\right]du + \left[\frac{\partial[F(v)]}{\partial x}\frac{\partial x}{\partial v} + \frac{\partial[F(v)]}{\partial y}\frac{\partial y}{\partial v} + \frac{\partial[F(v)]}{\partial z}\frac{\partial z}{\partial v}\right]dv,$$

therefore, $d[F(x,y,z)] = d[F(u,v)]$. For G and H, the same treatment will yield $d[G(x,y,z)] = d[G(u,v)]$ and $d[H(x,y,z)] = d[H(u,v)]$, respectively. The total differential indicates the relation among x, y, z with respect to u and v for intersecting tangent planes in terms of F, G, and H. In general, three dimensions (x, y, z) are mapped into two dimensions (u, v), and any coordinate system can be used [50,52]. Intersecting tangent planes represent the surface of a model. In vector notation, the normal planes of $\nabla F = \frac{\partial F}{\partial x}\mathbf{i} + \frac{\partial F}{\partial y}\mathbf{j} + \frac{\partial F}{\partial z}\mathbf{k}$, $\nabla G = \frac{\partial G}{\partial x}\mathbf{i} + \frac{\partial G}{\partial y}\mathbf{j} + \frac{\partial G}{\partial z}\mathbf{k}$, $\nabla H = \frac{\partial H}{\partial x}\mathbf{i} + \frac{\partial H}{\partial y}\mathbf{j} + \frac{\partial H}{\partial z}\mathbf{k}$, where ∇F, ∇G, and ∇H are gradients of the models are F, G, and H, respectively.

A surface is a system of linear functions in dx, dy, dz with respect to differentials of implicit functions F, G, H. The Jacobian determinants give the best linear approximations [49] to the differentiable functions dx, dy, dz, and they are the generalized gradients in terms of F, G, H. Now, we can solve Jacobian determinants of implicit functions, F, G, H with respect to differentials dx, dy, dz in terms of functions f, g, h with respect to differentials du, dv that represent the surfaces of models as tangent planes and tangent lines in a form that is amenable to simpler calculation.

Implicit equations for F, G, H where variables defined as x, y, z, u, v are $F(x,y,z,u,v) = f(u,v) - x = 0$, $G(x,y,z,u,v) = g(u,v) - y = 0$, $H(x,y,z,u,v) = h(u,v) - z = 0$,and the differentials are $F_x dx + F_y dy + F_z dz + F_u du + F_v dv = 0$, $G_x dx + G_y dy + G_z dz + G_u du + G_v dv = 0$, $H_x dx + H_y dy + H_z dz + H_u du + H_v dv = 0$. For differentials du and dv with respect to f, g, h, the vector

$$\left(\frac{\partial f}{\partial u}\mathbf{i} + \frac{\partial g}{\partial u}\mathbf{j} + \frac{\partial h}{\partial u}\mathbf{k}\right) \times \left(\frac{\partial f}{\partial v}\mathbf{i} + \frac{\partial g}{\partial v}\mathbf{j} + \frac{\partial h}{\partial v}\mathbf{k}\right)$$

$$\equiv \frac{\partial(g,h)}{\partial(u,v)}\mathbf{i} + \frac{\partial(h,f)}{\partial(u,v)}\mathbf{j} + \frac{\partial(f,g)}{\partial(u,v)}\mathbf{k}$$

is normal to a point, p, on the surface of the model [50]. From Cramer's Rule [50,52], solutions to the partial derivatives with respect to F, G, H are

$$\left(\frac{\partial x}{\partial u}\right)_v = -\frac{\frac{\partial(F,G,H)}{\partial(u,y,z)}}{\frac{\partial(F,G,H)}{\partial(x,y,z)}} \text{ and } \left(\frac{\partial x}{\partial v}\right)_u = -\frac{\frac{\partial(F,G,H)}{\partial(v,y,z)}}{\frac{\partial(F,G,H)}{\partial(x,y,z)}}$$

$$\left(\frac{\partial y}{\partial u}\right)_v = -\frac{\frac{\partial(F,G,H)}{\partial(x,u,z)}}{\frac{\partial(F,G,H)}{\partial(x,y,z)}} \text{ and } \left(\frac{\partial y}{\partial v}\right)_u = -\frac{\frac{\partial(F,G,H)}{\partial(x,v,z)}}{\frac{\partial(F,G,H)}{\partial(x,y,z)}}$$

$$\left(\frac{\partial z}{\partial u}\right)_v = -\frac{\frac{\partial(F,G,H)}{\partial(x,y,u)}}{\frac{\partial(F,G,H)}{\partial(x,y,z)}} \text{ and } \left(\frac{\partial z}{\partial v}\right)_u = -\frac{\frac{\partial(F,G,H)}{\partial(x,y,v)}}{\frac{\partial(F,G,H)}{\partial(x,y,z)}}.$$

The partial derivatives are expressed in terms of the Jacobian determinants with $\frac{\partial(F,G,H)}{\partial(x,y,z)} \neq 0.$ Each partial derivative of x, y, z with respect to u, v is an element of the Jacobian matrix used to represent the surface of a model.

RESULTS

Solutions to the Generalized System of Parametric 3D Equations

Analytically, parametric 3D equations allow for flexibility in producing a wide range of 3D forms. In our examples, the protocol has been to make generalized 3D models of planispiral and conispiral gastropods, bivalves (unwound forms), and rounded echinoids. The parametric 3D equations that describe surface models explicitly include the values for maximum whorl and aperture radii (R_{max} and r_{max}, respectively) for mollusk shells and maximum radii (R_{max} and r_{max} are approximately equal) for echinoids. Following the generalized set of parametric 3D equations (Section 2.1.), generalized models may be produced based on R_{max} and r_{max} obtained from pictures of actual shells or specimens. These generalized forms serve as morphological "blanks" that, when subjected to refinements via further measurements and reverse engineering, can be quite realistic and dimensionally accurate. Refinements entail experimenting with coefficient values and iteratively comparing the resulting model with measurements made from real shells until an acceptable degree of similarity has been obtained.

In all cases, $u, v \in [0, 2\pi]$. For gastropod morphologies, three systems of equations are developed, one of which is for limpets. The first gastropod shell generating system (System 1) (e.g., Figs. 4A, B, C, D, F) is

$$x = (R + r\cos v)(a^u \cos ju) \tag{3.1.1}$$

$$y = (R + r\cos v)(-a^u \sin ju) \tag{3.1.2}$$

$$z = -c(b + r\sin v)(a^u)(k\sin v) + (m\cos v \sin u) - \begin{Bmatrix} (l\sin u) \\ (l\cos cu) \end{Bmatrix} \tag{3.1.3}$$

where coefficients are $1.1 \le a \le 1.3$, $3 \le b \le 16$, $1 \le c \le 2$, $2 \le j \le 12$, $0 \le k \le 3$, $0 \le l \le 3$, $-3 \le m \le 3$, $1 \le R \le 2$, and $1 \le r \le 2$. Special cases highlighted in our study include normally coiled turritellids (e.g., Fig. 4O) and spirals as

$$x = (R + r\cos v)(a^u \cos ju) \tag{3.1.4}$$

$$y = (R + r\cos v)(-a^u \sin ju) \tag{3.1.5}$$

$$z = -c(b + r\sin v)(a^u)(k\sin v) \tag{3.1.6}$$

Vermicularia as

$$x = (R + r\cos v)(a^u \cos ju) \tag{3.1.7}$$

$$y = (R + r\cos v)(-a^u \sin ju) \tag{3.1.8}$$

$$z = -c(b + r\sin v)(a^u)(k\sin v) - 2.2u^n - 5\sin u - \cos 0.2u \tag{3.1.9}$$

where coefficients are defined by those in System 1 equations with the exceptions of $2 \le c \le 2.5$ and $0.2 \le r \le 1.4$, and an additional coefficient is $2 \le n \le 2.2$ for *Vermicularia*. For $R_{max} = r_{max} = 1$, high values of *b*, and $8 \le j \le 12$, normally coiled turritellid shells result. *Vermicularia* start out as a normal turritellid shell, but negative values of *b* and a squared term in the *z*-direction produce the unwound bottom half of the shells. Spirals require that r_{max} approaches but is not equal to zero, making a "curve" with the smallest possible cross-section.

The second gastropod shell generating system (System 2) (e.g., Fig. 4E) is

$$x = [(R + r\cos v) + \cos u](u\cos u) \tag{3.1.10}$$

$$y = [(R + r\cos v) + \sin u](u\sin u) \tag{3.1.11}$$

$$z = (2u + r\sin v) + 2u \tag{3.1.12}$$

where $1.5 \le R \le 2$ and $r = 1$.

A system of equations for limpets (e.g., Fig. 4J) is

$$x = (R + r\cos v)(6ub\cos 0.15u) - ju \tag{3.1.13}$$

$$y = (R + r\cos v)(6ub\sin 0.15u) + ku \tag{3.1.14}$$

$$z = ju(r\sin v) \tag{3.1.15}$$

where $0.8 \le b \le 1$, $3 \le j \le 4$, $1 \le k \le 1.1$, $R = 0.5$ and $r = 1$.

For Systems 1 and 2, values for coefficients and R_{max} and r_{max} vary within narrow ranges. Coefficients *a* and *c* each partially affect overall shell width and height with respect to the *x*-, *y*-, *z*-directions. Coefficients *b* and *j* each partially affect overall height with respect to number of whorls in the *z*-direction. Coefficients *k*, *l*, and *m* affect overall shape of the shell in terms of changes in whorl radius with respect to the *z*-direction.

Specific values for R_{max} and r_{max} and coefficients are necessary to produce specific shells. When k=0.7, l=m=3, c=1, R_{max}=1.5, and r_{max}=2, a theoretical cone shell is created. For l=1 or 2 with respect to a sine function, $k\sin v$=1, m=0, R_{max}=1, and r_{max}=2, theoretical bubble shells are represented.

Systems of parametric 3D equations were developed for exemplar forms of bivalves, cephalopods, and scaphopods. For bivalves, three systems are based on the following: for clams (e.g., Fig. 4L),

$$x = (R + r\cos 2v)(au\cos 0.3u) + b\cos u \tag{3.1.16}$$

$$y = (R + r\cos 2v)(au\sin 0.3u) + b\sin u \tag{3.1.17}$$

$$z = ju(r\sin 2v) + bu \tag{3.1.18}$$

where $0.4 \le a \le 0.5$, $0.1 \le b \le 0.2$, $0.4 \le j \le 2$, R = r=1; for oysters (e.g., Fig. 4H),

$$x = (R + r\cos 2v)(cu\cos 0.2u) \tag{3.1.19}$$

$$y = (R + r\cos 2v)(cu \sin 0.2u) \quad (3.1.20)$$

$$z = ju(r\sin 2v) - (\cos 2u \sin u) - 0.5u(\cos 2v)(\cos 5u \sin 7u) \quad (3.1.21)$$

where $4 \leq c \leq 5$ and $R = r = 1$; for scallops (e.g., Fig. 4I),

$$x = (R + r\cos 2v)(u\cos mu) + a\tan hv \quad (3.1.22)$$

$$y = (R + r\cos 2v)(u\sin mu) \quad (3.1.23)$$

$$z = a\left(\frac{u}{2\pi}\right)(r\sin 2v) + \left\{\begin{array}{c} 2\cos u^2 l \operatorname{sech} u^q \\ 2\operatorname{sech} u^2\, l \operatorname{sech} u^q \end{array}\right\} - 0.2\sin u^2 \cos u^3 \quad (3.1.24)$$

where $0.4 \leq a \leq 0.85$, $1.7 \leq l \leq 2$, $0.11 \leq m \leq 0.15$, $8 \leq q \leq 10$, $R = 1.2$, and $1 \leq r \leq 1.1$.

For bivalves, coefficient *a* controls the presence and shape of the auricles for scallops with respect to the *x*- and *y*-directions. Coefficient *b* defines the position of the umbo in clams with respect to the *z*-direction. Coefficient *c* is important for oysters in terms of length and shape (degree of concavity or roundness) of the shell with respect to the *x*- and *y*-directions, while for clams, length is controlled by coefficient *j* with respect to the *z*-direction. Shape for scallops is influenced by coefficient *l* in the *z*-direction and coefficient *m* in the *x*- and *y*-directions in terms of width and degree of flatness, respectively.

For cephalopods (e.g., Fig. 4G) a system of parametric 3D equations for ammonites is

$$x = (R + r\cos v)(\cos lu)\left(\frac{3u}{2\pi}\right) \quad (3.1.25)$$

$$y = (R + r\cos v)(\sin lu)\left(\frac{3u}{2\pi}\right) \quad (3.1.26)$$

$$z = 3u(r\sin v) + 2u\cos qu \sin qu \cos 2v \quad (3.1.27)$$

where $2 \leq l \leq 4$, $20 \leq q \leq 40$, $R = 1.5$, and $r = 1$. In the *x*- and *y*-directions, degree of coiling is controlled by coefficient *l*. The degree of folding or pleating is defined by coefficient *q* with respect to the *z*-direction.

The system of equations representing scaphopods (Fig. 4K) is

$$x = [(R + r\cos v) - u](au\cos 0.2u) - kv \quad (3.1.28)$$

$$y = [(R + r\cos v) + u](au\sin 0.2u) + kv \quad (3.1.29)$$

$$z = au(r\sin v) \quad (3.1.30)$$

where $0.1 \leq a \leq 0.5$, $1 \leq k \leq 2$, $1 \leq R \leq 6$, and $0.5 \leq r \leq 1$. Coefficient *a* controls length and degree of bending with respect to the *x*-, *y*-, *z*-directions, while coefficient *k* influences the overall width with respect to the *x*- and *y*- directions.

Aside from changes in coefficient values, changes in *R* and *r*, when considered in tandem, reveal the extent that each variable – *x*, *y*, *z* – influences changes in modeled shell morphologies. If *R* changes and *r* is held constant, modeled shells will widen (coiled gastropods), change in symmetry (limpets), lengthen and become rounder (bivalves), become narrower (ammonites), or become flatter and narrower (scaphopods) with respect to *x* and *y*. If *R* is held constant and *r* changes, modeled shells will become wider, rounder and elongated (coiled gastropods), flatter and elongated (limpets), wider (clams, oysters and scaphopods), wider and rounder (scallops), or wider and have whorls that merge (ammonites) with respect to the *z*-direction.

Modeled counterparts to actual specimens were constructed to show the efficacy of using R_{max} and r_{max} in representing these taxa. Along with these values, a small range of specific values for coefficients is used. For example, when $k=1$, $l = m = 0$, $R_{max} = 1.5$, and $r_{max} = 2$, the result is a volute (e.g., Fig. 4N). For $l=2$, $k=1$, or 2 with respect to a cosine function, $m=0$, $R_{max} = 1$, and $r_{max} = 2$, an olive shell is created. When $l=1.5$, $m=2.3$, $R_{max} = 0.7$, and $r_{max} = 2.4$, a *Melampus* is made (e.g., Fig. 4M). To make a sundial shell, $R_{max} = 1.3$, and $r_{max} = 1$, $l \sin u = l \cos cu = 0$ with $a = b = 2$.

Finally, as a contrast to mollusk shells, a system of parametric 3D equations for echinoids was devised. Just as was done with mollusks, we start with a torus and only use sine, cosine, hyperbolic secant and hyperbolic tangent functions to produce the desired forms. The result is the system of equations given as

$$x = [(R + r\cos v) + 0.1\cos 10u](\cos 2u) \quad (3.1.31)$$

$$y = [(R + r\cos v) + 0.1\sin 10u](\sin 2u) \quad (3.1.32)$$

$$z = (-1 - r\sin v) - 0.2\sin au \cos bv \sin 5u \sin bv \quad (3.1.33)$$

where $5 \leq a \leq 10$, $40 \leq b \leq 60$, $R = 1$, and $r=2$. Negative terms are used in the *z*-direction with the last term being a polynomial in sines and cosines, and the hole in the starting torus almost disappears so that the resultant form is approximately spheroid to ellipsoid (see inset-Fig. 5).

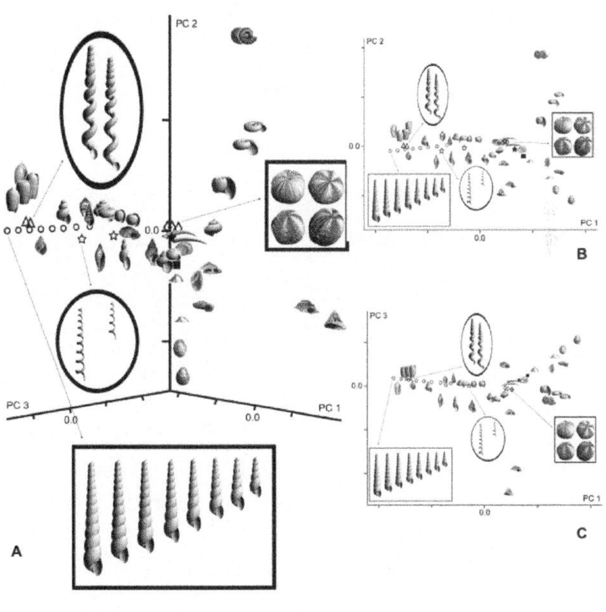

Figure 5: A, PCA ordination of modeled and actual mollusk shells as well as echinoid models and spiral curves in morphospace; ?= modeled clams; • = modeled scallops. Insets depicted are △ = ...

Jacobians and Interpretation of Mollusk and Echinoid Forms

The majority of the forms modeled are mollusk shells. In the Jacobians, the first column of elements is associated with whorl attributes. In the *x*-direction, small changes in whorl radii are generally associated with a high positive value for the first element, and in the *y*-direction, tightness of coiling is associated with a high negative value for the second element. In the *z*-direction, forms with a higher spire tend to have larger negative values for the third element and low spired to planispiral forms tend to have large positive values. The second column of the Jacobians is associated with aperture attributes. The highest positive values in the*x*-direction are associated with forms that have relatively large apertures. The highest negative values in the *y*- and *z*-directions are associated with irregular and elongated apertures, respectively. For echinoids, the Jacobians are mostly defined by a large positive value for the second element of the first column and a low negative value in the third element in the second column of values. These values reflect the roundness of echinoid forms.

Interpretation of Mollusk and Echinoid Surface Morphospace

The numerical solutions of the Jacobians were used as variables in a principal components analysis (PCA) to produce a morphospace (Fig. 5A– 3D PCA; Fig. 5B– PC2 vs. PC1; Fig. 5C– PC3 vs. PC1). Our use of Jacobians ordinated in an empirical morphospace differs from purely theoretical approaches in which the morphological axes are based directly on the underlying mathematical model. The empirical spaces generated with our method contain a combination of purely theoretical shells and empirically based forms. Whether using empirical or theoretical axes in morphospace construction, the representation of variation among whole 3D surfaces in our morphospace is depicted with independent axes.

From PCA, eigenvalues for the first six PCs are 2.042, 1.183, 1.037, 0.820, 0.566, and 0.352, and the percent variance explained is 34.029, 19.722, 17.275, 13.664, 9.435, and 5.874, respectively. The percent cumulative variance is 53.751 for the first two PCs, 71.027 for the first three PCs, 84.691 for the first four PCs, and 94.126 for the first five PCs.

The highest correlations of Jacobian elements with PCs are: PC1 with $\frac{\partial y}{\partial u}$(0.844); PC2 with $\frac{\partial x}{\partial u}$(0.840); PC3 with $\frac{\partial y}{\partial v}$ (0.896); PC 4 with $\frac{\partial z}{\partial v}$(0.647); PCs 5 and 6 have correlations that are <0.50. On PC1, the next highest correlation occurs with $\frac{\partial z}{\partial u}$(0.693) and $\frac{\partial z}{\partial v}$(0.694), while for PC2, the next highest correlation is associated with $\frac{\partial z}{\partial u}$(−0.608) as well.

For modeled mollusk shells, partial derivatives in the *z*-direction reflect changes in aperture radii as each whorl radius changes. Partial derivatives with respect to *u* (whorls) in the *x* and *y*-directions have more influence on overall modeled shell morphologies than partial derivatives with respect to *v* (apertures). Jacobian elements,$\frac{\partial y}{\partial u}$, $\frac{\partial z}{\partial u}$, and $\frac{\partial z}{\partial v}$, that are highly correlated to PC1, indicate tightness of coiling and relative aperture elongation. Whorl and aperture dimensions relative to each other are indicated by high correlation of Jacobian elements $\frac{\partial x}{\partial u}$, and $\frac{\partial z}{\partial u}$ with PC2. PC3 is highly correlated with Jacobian element $\frac{\partial y}{\partial v}$, which indicates aperture shape.

In general, the morphospace for modeled mollusk shells can be divided roughly in terms of the relationship between maximum whorl (R_{max}) and aperture radii (r_{max}). Small changes in whorl radii of high spired shells with small rounded to elongated apertures are represented by $R_{max} \leq r_{max}$ (Table 1). Large changes in whorl radii of low spired shells with irregularly rounded to elliptical apertures are represented by $R_{max} > r_{max}$ (Table 1). In the morphospace, high-spired mollusk models (i.e., those with spirals that are stretched in space) group together extending into gradients as do planispiral to lower spired mollusk forms (i.e., those with spirals in or nearly in the plane). Unwound and low-spired mollusk models produce gradients that converge near the rounded echinoid forms. The insets (middle left and lower center) in Figure 5 illustrate ontogenetic trajectories for simulated turitellid models. Isometric accretionary growth is depicted by a linear trajectory of a turritellid having from five to twelve whorls. The modeled *Vermicularia* have the same trajectory as a normally coiled turritellid until growth becomes allometric, and the whorls begin to separate [53]. Theoretical spiral elements also illustrate a separate "ontogenetic" trajectory (Fig. 5) and show how the method can be used to study basic spiral geometry.

The normally coiled turritellid series represents an ontogenetic trajectory illustrating isometric growth, with changes in number of whorls while the overall proportions of each whorl remain the same. The contrast between a normal turritellid and the *Vermicularia* illustrates allometry. The normal turritellid trajectory is a sequence of whorls numbering five (closest to the PC2 axis) to twelve, paralleling PC3 (Fig. 5). Modeled*Vermicularia* with nine whorls exhibit different degrees of uncoiling and occur at a slight angle from the normal turritellid with nine whorls (Fig. 5). In close proximity a parallel trajectory of spiral curves (with infinitesimal diameters) are plotted with five and nine coils, respectively, and occur below the turritellid sequence.

Each PC is correlated with Jacobian elements related to whorl or aperture characteristics for mollusk shells or spheroidal-ellipsoidal features when considering echinoids. For mollusk shells, PC1 was highly correlated with elements related to the tightness of coiling and relative elongation of the aperture. PC2 was most highly correlated with elements related to relative whorl and aperture dimensions. PC3 had a high correlation with elements related to aperture shape. For echinoids, PC1 defined degree of elongation, while PC2 defined breadth, and PC3 defined overall shape.

Jacobian elements with respect to *u* (PCs 1 and 2) for normal turritellids, *Vermicularia*, and spiral curves represent coiling attributes. In particular, $\frac{\partial y}{\partial u}$(PC1) indicates degree of coiling (i.e., number of coils), and $\frac{\partial z}{\partial u}$(PC1) indicates tightness of coiling. For $\frac{\partial x}{\partial u}$(PC2), height of spire is indicated. In addition, changes in $\frac{\partial z}{\partial v}$(PC1) indicate degree of change in aperture size from a normal turritellid or a *Vermicularia* form to the infinitesimal diameter of a spiral curve. As with modeled mollusk shells in general, $R_{max} \leq r_{max}$.

Modeled echinoid morphologies are spheroidal and ellipsoidal forms that occupy morphospace near PC2 (Fig. 5). Each element of the Jacobian recovers test shape as a closed form. The Jacobian element $\frac{\partial y}{\partial u}$ is the predominant influence (PC1) with $\frac{\partial z}{\partial v}$ a secondary influence (PC1 and PC4) as changes in the *z*-direction overtake changes in the *y*-direction with respect to a spheroidal or ellipsoidal shape. For echinoids, $R_{max} < r_{max}$.

For models where tangent lines and planes fall into the same plane, an interesting result emerged. The two ammonites that differ very slightly morphologically had identical Jacobians and plot in the same location in morphospace (Fig. 5; to allow for visibility, the ammonite models are jittered). One model exhibits a shift in a repeated sine wave (i.e., pleating) parallel to and shifted in the direction of v with respect to the other model (Figs. 4G and and5),5), producing indistinguishable Jacobians. In the second model, a shift in pleats parallel to and in the direction of u, produced Jacobians that were different but close in value. For the first case, a post-processing step related to Jacobians can be used numerically to produce morphological differences. More information and recommendations are given below.

Heuristic Tests of the Meaning of Surface Morphospace

The use of PCA in our method means that the axes of the resulting morphospace are independent. However, each axis represents both parameters u and v, and interpretation of the axes requires consideration of each parameter changing relative to the other, potentially making axis interpretation more complicated. To better understand the meaning of the axes and the properties of surface morphospaces based on Jacobians, 3D models of familiar basic geometric forms were created and their surfaces ordinated using PCA (Fig. 6). In addition, a limited number of bivalve and gastropod models, including slight variations, were generated to determine how these forms plot relative to one another. For basic geometric forms we used a torus, cylinder, cone, dome, sphere, and two ellipsoids in which the lengths of the major axis differ. For bivalves we generated models showing variation in the shape of the outline as well as some differences in the way the umbones coil. Gastropods with forms ranging from planispiral to conispiral and varying in whorl number were also included. With the exception of ellipsoids, we have not conducted controlled heuristic experiments in which one aspect of morphology is varied and all others held constant (but, see turritellid ontogenetic trajectory in Figure 5, where only whorl number is varied). Rather, our heuristic tests are intended simply to determine if ordination based on Jacobians succeeds in capturing the major differences in surface geometry among a smaller number of basic forms. The morphospaces generated using PCA are empirical, and therefore, not directly comparable. However, there are similarities in the models used in this heuristic test and the mollusk-echinoid morphospace shown earlier (Fig. 5). The tests should, therefore, be useful for interpretation of each space and for evaluating the method more generally.

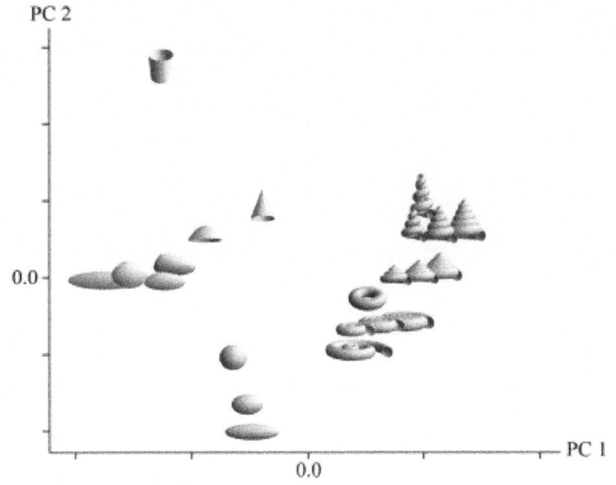

Figure 6: PCA of basic geometric forms and select mollusks to heuristically explore the properties of surface morphospaces based on Jacobians.

The resulting PCA (Fig. 6) very effectively ordinates forms based on differences in their surface geometry. Domed, conical, heliconical, toroidal/planar, and spherical/ellipsoidal forms all plot in different regions of the morphospace, and the relative positions of different morphologies reflect the degree of similarity in surface geometry. The dome plots relatively close to the cone and can be viewed as a rounded cone. Although bivalves are rapidly expanding helicones, their overall shape is similar to a dome, and their position near the dome in the morphospace confirms this. Among the coiled forms, there is a general progression from planispiral to more conispiral forms along PC2. Slight variations in the tightness of coiling and increases in the number of whorls within a morphology are captured and form morphoclines along PC1. The torus can be viewed as a planar uncut coil and is nearest the planispiral gastropods. Spheroidal and ellipsoidal forms occur as a morphocline reflecting elongation in the major axis and are roughly equidistant from coiled and uncoiled forms. The cylinder is an uncoiled tube that is located furthest from all other forms.

As was the case for the ordination of mollusks and echinoids, partial derivatives $\frac{\partial y}{\partial u}$, $\frac{\partial x}{\partial v}$, and $\frac{\partial z}{\partial v}$represent number of coils, whorl overlap (i.e., partial aperture shape), and aperture size but, because these are empirical spaces, the correlations with PC1 are slightly different (0.962, −0.679, and 0.899, respectively). Similarly, partial derivatives $\frac{\partial x}{\partial u}$, $\frac{\partial z}{\partial u}$, and $\frac{\partial y}{\partial v}$ represent height of form, tightness of coiling (or lack thereof) and aperture shape (and to some degree, size) and are correlated with PC2 (−0.669, 0.727, and 0.598, respectively). In contrast to other methods that have been used to model and plot shells, the partial derivatives used here include the number of coils/whorls (from undefined in rounded forms to zero in bivalves to some number for gastropods) and potentially allow for the parsing of other morphological attributes as well.

The results of these simple heuristic tests validate our earlier interpretations of the mollusk-echinoid space and demonstrate that this approach can be useful for analyzing a broad range of surface geometries. Many organisms and structures have overall forms that approximate some of the basic geometries used here and our analysis suggests that this method will effectively permit the ordination of a broad range of biological surfaces as well.

DISCUSSION

The great diversity of surface morphology that exists makes the study of surfaces a natural extension of the study of phenotypic form. Methods for capturing and manipulating 3D objects are becoming more accessible, and there is growing interest in approaches that make fuller use of 3D data, including the study of surfaces. For many questions, surface morphology is arguably a more complete description of the phenotype compared to methods that rely solely on points, distances between points, or outlines. Our general approach in which the entire surface is modeled and parameterized, summarized numerically with the Jacobian, and ordinated using conventional PCA presents a potentially powerful complement to existing methods for studying morphology more generally and biological surfaces explicitly.

Measurements obtained from actual specimens, pictures in monographs, digital images, or point cloud data can be used when making models. Accuracy and fine morphological differences can be obtained by testing variables, parameters and coefficients of parametric 3D equations in an ordered fashion to determine the effect that one change has on the form being produced. For example, one could change the value of one coefficient while holding all other coefficients constant to determine the effect. Extremely detailed models that may capture individual level variability are possible.

Jacobian solutions capture the behavior locally at each point on the surface. A property of Jacobians is size invariance, thus two identical forms that differ in size only will have identical Jacobians and occupy the same location in morphospace. Specimens with very different shapes and sizes can be jointly analyzed, but because 3D parametric models and their Jacobians exclude size, the resulting ordination will also be size-free and reflect differences in surface form only. Rotation and reflection are also important properties of Jacobians. Rotation is modeled as a switch in parameters in the *x*-, *y*-, or *z*- direction. Reflection (or change in symmetry) is modeled as a change in sign in the *x*-, *y*-, and/or *z*-direction. Because of the properties of size, rotation, and reflection associated with Jacobians, detecting large intragroup and/or intergroup variation is possible.

The Jacobian is capable of capturing morphological variation provided that tangent lines and planes differ among models. For 3D models starting with a torus, the degree to which Jacobians have similar numerical solutions and similar morphologies is affected by similarity in measured R_{max} and r_{max} if measurement is in approximately the same plane (i.e., u and v are approximately in the same plane). If measurement betweenR_{max} and r_{max} is farther apart, Jacobian solutions will be different, describing more disparate morphologies. In cases where 3D surfaces approach 2D planar figures, the Jacobian may not recover such differences numerically, or the differences may be quite small, despite the fact that positional shifts may have significance biologically. If the 3D model is reduced to 2D (i.e., changes in the *z*-direction are not measurable), the Jacobian determinant will be zero (i.e., the tangent lines and planes are equal to zero) and the surface is a plane. At this point, a 2D method (e.g., outline analysis) could be used.

In general, the Jacobian is best suited for curved surfaces. We recommend visually inspecting the models and the Jacobians (or the ordination) to ensure that perceived variation is captured. Alternatively, one could simulate relevant morphological differences and test if the Jacobian is sensitive to these. If a difference is not recordable in the Jacobian, then post-processing by using a Hessian matrix (Hessian) or some of the elements of a Hessian may be calculated and used in morphospace construction.

Parametric 3D Equations, Partial Differentiation, and Jacobians

In our analyses exemplifying the utility of the methods we describe, we use homogeneous equations to find numerical solutions to Jacobians. Boundary conditions on all sets of parametric 3D equations are $u, v \in [0, 2\pi]$. When constructing ammonites, we made incremental changes in one or two coefficients in sets of parametric 3D equations with the intent of producing very similar models. In general, small changes in coefficient and/or radii values can produce large changes in surface morphology from parametric 3D equations, and these differences are definable with respect to partial differential equations. However, if systems of equations used are almost identical to one another, the partial differential equations will have solutions, but those solutions are not necessarily or sufficiently smooth; therefore, Jacobians for very slightly different forms within a given set of parametric 3D equations might have (near-) identical solutions. That is, the tangent lines and planes of very similar surfaces may have solutions to Jacobians that are the same. We use sets of parametric 3D equations that are almost identical in the case of ammonite shells. For example, a difference in the spacing of the pleats (coefficient q) – a shift of a sine or cosine – can slightly affect the tightness of coiling (coefficient l) (Fig. 7). Using the same set of parametric 3D equations with very slight changes, the model differences just described occur only by slight stretching of the same basic form in one direction, or addition or subtraction of whorls overlapping in the same plane. In addition, there is a shift with respect to v from one model to the other. This is like a phase shift for the same tangent lines and planes. The result is the same solutions to the Jacobians occur for both models (Figs. 4G and and5)5) used in the morphospace.

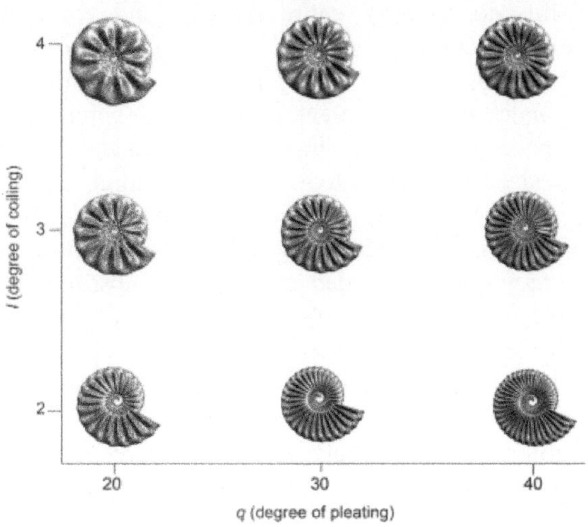

Figure 7: Ammonite shell models with ***R*** and ***r*** held constant, showing the result of combinations of coefficient ***l*** (degree of coiling) in the ***x***- and ***y***-directions and coefficient ***q*** (degree of pleating) in the *z*-direction.

With tangent lines and planes being the same in the *z*-direction with the *x*- and *y*- directions remaining the same, slight modifications of parametric 3D equations in the *z*-direction will produce different solutions to Jacobians for surface models that are nearly identical. In the case of the ammonites (Figs. 4G and and5),5), change from one model to the next can be used with respect to *v*, so this parameter is added to the *z*-direction in one of the models. That is, a new set of parametric 3D equations is created. The new model that results is the same as the previous one since the model adds a value of *v* that affects all of the *z*-direction. For one of the ammonite models, the addition is $v=-1$.

Alternatively, imposing the additional constraint of post-processing by calculating Hessian or partial Hessian matrices can be used to produce differences among forms, regardless of the behavior of the Jacobians for all models considered. For example, we did calculations for some of the elements of Hessians for some of the ammonite forms depicted in Figure 7. The elements of Hessians are second partial derivatives, including iterated and mixed partial derivatives. Jacobians have elements that are first partial derivatives. The same number of Hessian elements is calculated for each of the forms. Solutions to the Hessians produced different values for comparable elements in contrast to their corresponding Jacobian elements. Six Hessian elements were calculated, and differences were evident in second partial derivatives $\frac{\partial^2 y}{\partial u^2}$ and $\frac{\partial^2 z}{\partial u^2}$ for the ammonites.

Jacobians represent a linearization near critical points, and Hessians (or partial Hessians) represent critical points or extrema on surfaces, and therefore are descriptors of surface features (local maxima and minima). Hessian determinants represent local curvature at critical points of a surface. However, calculating entire Hessians for each form is not a trivial task. As stated above, post-processing by calculating enough elements of a Hessian can be used to detect differences in forms. Hessian solutions could be used as input in a PCA to devise a morphospace in addition to or instead of using Jacobian solutions. Alternatively, solutions to partial Hessians along with solutions to Jacobians can be used as the data to represent models in morphospace analysis.

Whether creating a new set of parametric 3D equations for a model or using solutions to Hessians (or partial Hessians), almost identical forms will plot very close to each other in morphospace. Differences in surface morphology for forms constructed with nearly identical parametric 3D equations may require a post-processing step. Since parametric 3D equations are not invariant, the behavior of the equations from form to form in a more comprehensive way has yet to be tested. Jacobians can be characterized generally as transformation matrices that are used in the linearization of non-linear systems such as the models we constructed using parametric 3D equations. Further testing is necessary and is beyond the purposes and scope of this current study. However, such testing may be necessary for additional analyses such as measuring disparity.

In the future it will be important to explore further how different geometries and surface features are summarized by the Jacobian. In addition, more analysis is needed on the mathematical properties of parametric 3D equations and Jacobians, including testing with regard to boundary conditions, evaluating homogeneous equations at degree *n*, linearization of a non-linear system, evaluating critical points and stability of the Jacobian. These would each be separate studies and are beyond the scope of this paper.

The Utility of Surface Morphospaces

We have demonstrated that parametric 3D models can be meaningfully ordinated using Jacobians, and forms with similar overall surface geometry in terms of tangent lines and planes plot close together in morphospace. The approach presented should be applicable to a broad range of questions where documenting and analyzing scale-invariant patterns of surface form is important. This could include questions related to morphological disparity, functional morphology, ontogeny, and morphological evolution more broadly. In principle, these questions could be explored at a range of taxonomic levels. As our simulations of turritellid ontogeny demonstrate (Fig. 5), even patterns of change at the level of individuals can be studied. The analysis of surfaces potentially provides a richer, more complete assessment of morphology because more of the phenotype may be incorporated when compared to methods based on curves, points, or other one or two dimensional representations of form.

The ability to include disparate forms in a common morphospace is of particular interest and should present new directions for research that have not been pursued before due to methodological limitations [54]. Many existing techniques require some degree of uniformity or similarity in form before analysis can proceed. Applying log spiral models, for example, requires that the phenotypes display spiral geometry; geometric morphometric methods require a common framework of homologous points that precludes the analysis of very different phenotypes. Our method, however, simply requires that the objects have bounded 3D surfaces that are amenable to modeling using parametric 3D equations.

The method shows promise for the study of patterns of ontogeny, but even here it is important to recognize what is being captured in the morphospace. The normally coiled turritellid ontogenetic trajectory (Fig. 5), for example, shows substantial changes as growth proceeds. This may seem counterintuitive as the growth is isometric, and there is no change in the underlying spiral geometry. However, from the perspective of change in surface form this pattern makes perfect sense as each growth stage includes the addition of an entire whorl of *new* shell surface. Compared to a typical bivalve, which does not add whorls during growth, a snail will display much more change in surface geometry during ontogeny. As an aside, it is worth noting that ontogenetic patterns could have been portrayed with each whorl (or growth increment) plotted separately rather than plotting the cumulative morphology. As patterns are scale-invariant, all increments that differ in size only (i.e., isometry) would plot at a single point in space, both for bivalves and gastropods.

While many biologists may be accustomed to studying surfaces or surface features, few have experience describing entire surfaces mathematically and intuiting what surface differences mean. The applicability of our method to a particular question will require a careful consideration of the appropriateness of surface data and the significance of differences in surface geometry (Fig. 6). When analyzing morphological patterns at higher taxonomic levels (e.g., [55]), for example, it may be important to determine if surface morphology is an effective proxy for higher taxa and that distances in morphospace are correlated with taxonomic/ phylogenetic rank.

The richness of information embedded in the elements of the Jacobian means that a surface morphospace is, in some ways, more complicated. We hope that this new method will promote the study of biological surfaces and stimulate thoughtful discussion of the concept and use of a morphospace [56]. We expect that comparing phenotypes in terms of their overall surface form in a mathematically rigorous framework will enrich the study of morphology in general, and as the method is applied, new ways in which to treat the data acquired will emerge.

CONCLUSIONS

Parametric 3D equations can be used to accurately model certain biological surfaces. As shown in Figure 1, our approach can be applied to a variety of basic geometric forms and used to model any bounded surface. Although we use mollusks and echinoids as examples, the same approach can be applied to a wide diversity of organismal forms. Our results clearly demonstrate the utility of the Jacobian in describing and analyzing surface morphology. The ability to summarize a broad range of surfaces with a six-element matrix without any loss of resolution or need for preprocessing is a potential advantage over other methods. However, we have also identified some potential limitations. As with any study, it is important to ensure that the method is appropriate for the question being addressed. The flexibility and mathematical rigor of our approach makes it usable alone or in conjunction with other 3D surface analytical methods to address a broad range of questions. Models constructed could be exported as point clouds and analyzed using methods such as eigensurface [1]–[2], spherical harmonics [3] wavelet analysis [4], or geometric morphometrics [5]. The models should permit the analysis of theoretical and empirical data in a common framework using any one of these 3D techniques.

The ability to quantify entire surfaces and construct multivariate surface spaces should generate new questions regarding morphological evolution. Patterns of phenotypic variation, changes in developmental trajectories, simulations of morphospace occupation, including unrealized forms or hypothetical patterns of ontogeny and evolution, can all be studied in the context of surface morphology. Although we emphasize whole surfaces, homologous or otherwise comparable parts can be treated as bounded substructures. The appearance of new parts on the surface does not preclude analysis, and in a sense, our method can be viewed as "homology-free [57]." Experiments involving the addition or removal of evolutionary novelties could be conducted to determine influences on morphological patterns. Our approach offers another avenue to study morphological variation, and more broadly, the potential to study evolutionary and non-evolutionary patterns and processes in new and fruitful ways.

ACKNOWLEDGMENTS

We wish to thank Takao Ubukata, David Raup, and Alistair Robert Evans for their thoughtful and thorough comments that improved our paper.

REFERENCES

1. Polly PD, MacLeod N (2008) Locomotion in fossil carnivore: an application of eigensurface analysis for morphometric comparison of 3D surfaces. Palaeontologia Electronica 11: 10A 13p.
2. MacLeod N (2008) Understanding morphology in systematic contexts: 3D specimen ordination and 3D specimen recognition. In: Wheeler Q, editor. The New Taxonomy. London: CRC Press, Taylor & Francis Group. 143–210.
3. Shen L, Farid H, McPeek MA (2009) Modeling three-dimensional morphological structures using spherical harmonics. Evolution 63: 1003–1016doi:10.1111/j.1558-5646.2008.00557.x [PMC free article]
4. Laga H, Takahashi H, Nakajima M (2006) Spherical wavelet descriptors for content-based 3D model retrieval. In: Proceedings of the IEEE International Conference on Shape Modeling and Applications, SMI2006: 15–25doi:10.1109/SMI.2006.39
5. Mitteroecker P, Gunz P (2009) Advances in geometric morphometrics. Evol Biol 36: 235–247doi:10.1007/s11692-009-9055-x
6. Memoli F, Sapiro G (2005) Theoretical and computational framework for isometry invariant recognition of point cloud data. J Foundations Comp Math 5: 313–347doi:10.1007/s10208-004-0145-y
7. Bronstein AM, Bronstein MM, Mahmoudi M, Kimmel R, Sapiro G (2010) A Gromov-Hausdorff framework with diffusion geometry for topologically-robust non-rigid shape matching. Internat J Computer Vision 89: 266–286doi:10.1007/s11263-009-0301-6
8. Boyer DM, Lipman Y, St Clair E, Puente J, Patel BA, et al. (2011) Algorithms to automatically quantify the geometric similarity of anatomical surfaces. Proc Natl Acad Sci 108: 18221–18226doi:10.1073/pnas.1112822108 [PMC free article]

9. Varady T, Martin R, Cox J (1997) Reverse engineering of geometric models – an introduction. Computer-Aided Design 29: 255–268doi:10.1016/S0010-4485(96)00054-1

10. Dera G, Eble G J, Neige P, David B (2008) The flourishing diversity of models in theoretical morphology: from current practices to future macroevolutionary and bioenvironmental challenges. Paleobiol34: 301–317doi:10.1666/07070.1

11. McGhee Jr G R (1999) Theoretical Morphology, The Concept and its Applications. New York: Columbia University Press.

12. Zachos LG, Sprinkle J (2011) Computational model of growth and development. In: Paleozoic echinoids. In: Elewa AMT, editor. Computational Paleontology, Berlin Heidelberg: Springer-Verlag. 75–94.

13. Chakra MA, Stone JR (2011) Holotestoid: a computational model for testing hypotheses about echinoid skeleton form and growth. J Theor, Bio 285: 113–125doi:10.1016/j.jtbi.2011.06.019

14. Thompson DW (1942) On Growth and Form. New York: Macmillan.

15. Moseley H (1838) On the geometrical form of turbinated and discoid shells. Philos Trans R Soc 128: 351–370

16. Moseley H (1842) On conchiliometry. Philos Mag 21: 300–305

17. Raup DM (1961) The geometry of coiling in gastropods. Proc Natl Acad Sci 47: 602–609doi:10.1073/pnas.47.4.602 [PMC free article]

18. Raup DM (1962) Computer as aid in describing form in gastropod shells. Science 138: 150–152

19. Raup DM (1966) Geometric analysis of shell coiling: general problems. J Paleontol 40: 1178–1190

20. Raup DM, Michelson A (1965) Theoretical morphology of the coiled shell. Science 147: 1294–1295

21. Cortie M (1993) Digital seashells, Computer & Graphics. 17:79–84doi:10.1016/0097-8493(93)90054-D

22. Prusinkiewicz P, Fowler DR (1995) Shells models in three dimensions. In: Meinhardt H, editor. The Algorithmic Beauty of Sea Shells. Berlin: Springer. 163–181.

23. Ubukata T (2005) Theoretical morphology of bivalve shell sculpture. Paleobiol 31: 643–655doi:10.1666/0094-8373

24. Okamoto T (1988) Analysis of heteromorph ammonoids by differential geometry. Palaeontol 31: 35–52

25. Okamoto T (1988) Developmental regulation and morphological saltation in the heteromorphy ammonite*Nipponites*. Paleobiol 14: 272–286

26. Ackerly SC (1989) Kinematics of accretionary shell growth, with examples from brachiopods and molluscs. Paleobiol 15: 147–164

27. Savazzi E (1990) Biological aspects of theoretical shell morphology. Lethaia 23: 195–212doi:10.1111/j.1502-3931.1990.tb01360.x

28. Checa A (1991) Sectorial expansion and shell morphogenesis in mollusks. Lethaia 24: 93–114doi:10.1111/j.1502-3931.1991.tb01183.x

29. Bayer U (1978) Morphogenetic programs, instabilities, and evolution—A theoretical study, Neues Jahrbuch für Geologie und Pal?ontologie, Abhandlungen. 156: 226–261

30. Hammer Ø, Bucher H (2005) Models for the morphogenesis of the molluscan shell. Lethaia 38: 111–122doi:10.1080/00241160510013222

31. Moulton DE, Goriely A, Chirat R (2012) Mechanical growth and morphogenesis of seashells. J Theor Biol 311: 69–79doi:10.1016/j.jtbi.2012.07.009

32. Boettiger A, Ermentrout B, Oster G (2009) The neural origins of shell structure and pattern in aquatic mollusks. Proc Natl Acad Sci 106: 6837–6842doi:10.1073/pnas.0810311106 [PMC free article]

33. do Carmo MP (1976) Differential Geometry of Curves and Surfaces. Englewood Cliffs, New Jersey: Prentice-Hall.

34. Villarceau M (1848) Théorème sur le tore, Nouv Ann Math. 7: 345–347

35. Pressley A (2001) Elementary Differential Geometry. United Kingdom: Springer.

36. Spivak M (1979) A Comprehensive Introduction to Differential Geometry 2nd edition, Vol. 3. Berkeley, California: Publish or Perish, Inc.

37. Gray A (1998) Modern Differential Geometry of Curves and Surfaces with Mathematica 2nd edition. Boca Raton, Florida: CRC Press.

38. Davies A (1996) Samuels P, An Introduction to Computational Geometry for Curves and Surfaces. Oxford: Clarendon Press.

39. Darboux G (1887, 1889, 1894, 1896) Leçons sur la théorie génerale des surfaces: Volume I, Volume II, Volume III, Volume IV. Gauthier-Villars.

40. Serret JA (1851) Sur quelques formules relatives à la théorie des courbes à double courbure. J Math Pures Appl 16: 193–207

41. Frenet F (1852) Sur les courbes à double courbure, Thèse, Toulouse, Abstract in J Math Pures Appl. 17: 437–447

42. Spivak M (1979) A Comprehensive Introduction to Differential Geometry 2nd edition, Vol. 2. Berkeley, California: Publish or Perish, Inc.

43. Struik DJ (1961) Lectures on Classical Differential Geometry. Reading,Massachusetts: Addison-Wesley.

44. Pappas JL (2005) Geometry and topology of diatom shape and surface morphology for use in applications of nanotechnology. J Nanosci Nanotechnol 5: 120–130doi:10.1166/jnn.2005.009

45. Pappas JL (2005) Theoretical morphospace and its relation to freshwater gomphonemoid-cymbelloid diatom (Bacillariophyta) lineages. J Biol Systems 13: 385–398doi:10.1142/S0218339005001586

46. Pappas JL (2008) More on the theoretical morphospace and its relation to freshwater gomphonemoid-cymbelloid diatom (Bacillariophyta) lineages. J Biol Systems 16: 119–137doi:10.1142/S0218339008002460

47. PappasJL(2011)Graphmatchingaskeletonizedtheoretical morphospace with a cladogram for gomphonemoid-

cymbelloid diatoms (Bacillariophyta). J Biol Systems 19: 47–70doi:10.1142/S0218339011003816

48. Abbott RT (1991) Seashells of the Northern Hemisphere. New York: Gallery Books.

49. Abbott RT, Dance SP (1986) Compendium of Seashells. Melbourne, Florida: American Malacologists, Inc.

50. Kaplan W (2003) Advanced Calculus, 5th edition. Reading,Massachusetts: Addison-Wesley.

51. Baxandall PR (1981) Differential Vector Calculus. London and New York: Longman.

52. Kaplan W, Lewis DJ (1971) Calculus and Linear Algebra, Vol. 2. New York: Wiley.

53. Gould SJ (1969) Ecology and functional significance of uncoiling in *Vermicularia spirata*: an essay on gastropod form. Bull Marine Sci 19: 432–445

54. Roy K, Foote M (1997) Morphological approaches to measuring biodiversity. Trends Ecol Evol 12: 277–281

55. Foote M (1997) The evolution of morphological diversity. Annu Rev Ecol Syst 28: 129–152

56. Gould SJ (1991) The disparity of the Burgess shale arthropod fauna and the limits of cladistics analysis: why we must strive to quantify morphospace. Paleobiol 17: 411–423

57. Polly D (2008) Developmental dynamics and G-matrices: can morphometric spaces be used to model phenotypic evolution? Evol Biol 35: 83–96doi:10.1007/s11692-008-9020-0

CHAPTER 11

A COMPUTATIONAL FRAMEWORK FOR 3D MECHANICAL MODELING OF PLANT MORPHOGENESIS WITH CELLULAR RESOLUTION

Fre´de´ ric Boudon[1] , Je´ ro^ me Chopard[1] , Olivier Ali [1,2] , Benjamin Gilles[3] , Olivier Hamant[2] , Arezki Boudaoud[2] , Jan Traas[2] *, Christophe Godin[1] *

[1]Virtual Plants Inria team, UMR AGAP, CIRAD, INRIA, INRA, Montpellier, France,

[2]Laboratoire de Reproduction et De´veloppement des Plantes, Universite´ de Lyon 1, ENSLyon, INRA, CNRS, Lyon, France,

[3]Laboratoire d'Informatique, de Robotique et de Microe´lectronique de Montpellier, Universite´ Montpellier 2, CNRS, Montpellier, France

ABSTRACT

The link between genetic regulation and the definition of form and size during morphogenesis remains largely an open question in both plant and animal biology. This is partially due to the complexity of the process, involving extensive molecular networks, multiple feedbacks between different scales of organization and physical forces operating at multiple levels. Here we present a conceptual and modeling framework aimed at generating an integrated understanding of morphogenesis in plants. This framework is based on the biophysical properties of plant cells, which are under high internal turgor pressure, and are prevented from bursting because of the presence of a rigid cell wall. To control cell growth, the underlying molecular networks must interfere locally with the elastic and/or plastic extensibility of this cell wall. We present a model in the form of a three dimensional (3D) virtual tissue, where growth depends on the local modulation of wall mechanical properties and turgor pressure. The model shows how forces generated by turgor-pressure can act both cell autonomously and non-cell autonomously to drive growth in different directions. We use simulations to explore lateral organ formation at the shoot apical meristem. Although different scenarios lead to similar shape changes, they are not equivalent and lead to different, testable predictions regarding the mechanical and geometrical properties of the growing lateral organs. Using flower development as an example, we further show how a limited number of gene activities can explain the complex shape changes that accompany organ outgrowth.

SUMMARY

In recent years, much research in molecular and developmental biology has been devoted to unravelling the mechanisms that govern the development of living systems. This includes the identification of key molecular networks that control shape formation and their response to hormonal regulation. However, a key challenge now is to understand how these signals, which arise at cellular scale, are physically translated into growth at organ scale, and how these shape changes feed back into molecular regulation systems. To address this question, we developed a computational framework to model the mechanics of 3D tissues during growth at cellular resolution. In our approach, gene regulation is related to tissue mechanical properties through a constitutive tensorial growth equation. Our computational system makes it possible to integrate this equation in both space and time over the growing multicellular structure in close to interactive time. We demonstrate the

Citation: Boudon F, Chopard J, Ali O, et al. A Computational Framework for 3D Mechanical Modeling of Plant Morphogenesis with Cellular Resolution. Shvartsman S, ed. PLoS Computational Biology. 2015;11(1):e1003950. doi:10.1371/journal.pcbi.1003950.

interest of such a framework to study morphogenesis by constructing a model of flower development, showing how regulation of regional identities can, by dynamically modulating the mechanical properties of cells, lead to realistic shape development.

INTRODUCTION

The control of form and size is a central issue in developmental biology. It is commonly accepted that genetic regulation is at the basis of morphogenesis. However, while molecular genetics has provided an important number of actors required for morphogenetic events, the link between these regulators and global shape control remains largely an open question in both plant and animal biology. Furthermore, the contribution of multicellularity in shape changes and growth remains poorly explored. In fact, certain species exhibit complex shapes while being composed of only one giant and multinucleated cell. This raises the question of the exact contribution of the presence of neighboring cells in the growth of a given cell, within a tissue. So far, this issue has mainly been addressed from a signaling point of view (e.g. diffusion of morphogens [1], mechanical feedbacks [2], [3]), but the role of multicellularity in the biophysics of growth remains to be formalized.

Here we consider this issue in plants. Plant cells are under high internal turgor pressure and it is only the presence of a rigid exoskeleton that prevents them from bursting. This exoskeleton, the cell wall, is composed of a dense network of cellulose microfibrils that are cross-linked to each other by a network of polysaccharides. Sachs, as early as in 1882 (reviewed in e.g. [4], [5]) discovered that cell expansion can only take place as long as the cells are under pressure, which has led to the concept of turgor-driven cell growth. It is now widely accepted that this involves the irreversible (plastic) yielding of the cell wall to this pressure, e.g. [6].

Based on this general concept of wall yielding a now widely accepted general scenario was proposed by Lockhart to describe the growth of an isolated cell [7]. This scenario can be summarized as follows. In a non-growing isolated cell, the internal pressure is counterbalanced by the tension in the cell wall. If this pressure further increases and reaches a certain threshold, the load bearing parts of the cell wall yield. Lockhart [7]proposed to model this viscoplastic process with a simple relationship between 2 key variables, the relative rate of growth of the cell volume V, and the cell turgor pressure P: if the pressure is greater than a fixed threshold P_y and the flow of water is not a limiting factor, then the cell yields and the rate of growth is proportional to the excess of turgor pressure:

$$\frac{1}{V}\frac{dV}{dt} = \Phi(P - P_y), \qquad (1)$$

where Φ denotes the extensibility of the cell, i.e. its ability to grow under a given pressure, the inverse of its viscosity. If the turgor pressure does not reach the yield threshold P_y, no growth is achieved and the cell deformation is entirely elastic (reversible). Above P_y, the cell deformation becomes plastic (irreversible). The potential decrease in pressure due to cell growth is continuously compensated by further water uptake, thus keeping the wall under continuous tension [4]. In other words, in a single cell system, growth can be entirely described in terms of the variations of the internal turgor pressure and of the mechanical properties of the cell wall.

The initial formulation of Lockhart and subsequent models did not account for cell geometry and anisotropic properties of wall material. Recently, Dumais and coworkers applied Lockhart's model of cell growth to cell walls, and extended it to account for wall anisotropic properties [8]. They introduced a model of tip-growing cells (e.g. root hairs) that combines two key processes of cell growth, namely the deposition of material on cell walls and the mechanical deformation of the cell wall due to stresses resulting from the cell's inner turgor pressure. Interestingly, the authors show that the Lockhart growth equation can be simply extended in 3 dimensions to take into account wall anisotropy. This leads to 3 equations (instead of one) that express how the rate of deformation in the 3 directions of space are affected by mechanical anisotropy in the cell walls [8]. With the help of this model, the authors could analyze the dynamics of a tip growing cell, and how the visco-elastic properties of cell walls may impact its shape in steady or non-steady regimes.

In a multicellular context, morphogenesis relies on differential growth across tissues. Each cell may feature specific values for the various parameters (turgor pressure, yielding threshold, extensibility…) used in Eq.(1). In principle, the regulation and coordination of these parameters is achieved through the action of the molecular regulatory networks that control the composition and mechanical properties of the cell wall, as described by the black arrows in Fig. 1. For example, cell wall modifying enzymes such as expansins, xyloglucan endo-tranglycosylases or pectin modifying enzymes are known to be triggered by transcription factors such as APETALA2 [9], MONOPTEROS [10] and AGAMOUS [11]. At the scale of the cell wall, actions of such enzymes have the potential to increase or decrease the viscosity and/or the rigidity of the wall. As a consequence, extensibility Φ in Eq.(1) may be modified and affect growth.

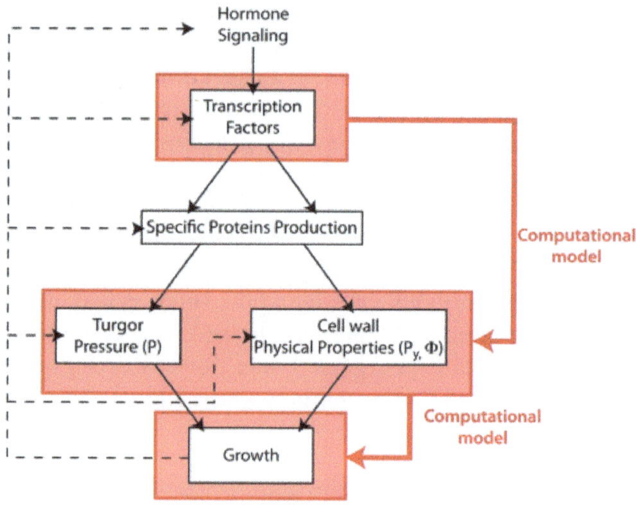

Figure 1: Schematic view of the regulation of growth in multicellular tissues.

Although the general concepts described above are widely accepted, they do not explain how genetic determinants collectively generate an organism with a particular shape. The situation is made even more complex because morphogenetic events at the multicellular level can feed back on the cellular or molecular scale. Morphogen gradients, for example, are limited by the geometry of the tissue in which they diffuse [1],[2] and mechanical stresses generated by differences in growth rate within an organ can potentially feed back on cellular growth directions and rate [3]. It is therefore not self-evident to explain how a particular gene by interfering with local cell wall properties influences the overall shape of an organ. To proceed further and to explore hypotheses linking in an intricate way gene function to morphogenesis, a computational modeling framework is required (red bold arrows in Fig. 1).

In a seminal series of papers, Coen and colleagues have proposed such a framework for tissue growth, termed the Growing Polarized Tissue (GPT) framework, that can capture overall growth rates and directions of tissues in three dimensions while taking into account mechanical interactions between different regions ([12]–[14]). With this method, the authors were able to propose hypotheses for the genetic regulation of organ formation in different species [14]. This framework was the first system able to simulate 3D organ development and shapes based on plausible genetic regulation hypotheses. However, in GPT gene functions are expressed in relatively abstract terms. This is mainly because it models events occurring at the scale of entire tissue regions. Recently, several attempts were made to develop other modeling frameworks at cellular resolution and to build mechanical models of morphogenesis for multicellular tissues. To reduce complexity, these models have been initially restricted to 2 dimensions either in the plane ([15] for algae, [16], [17] for leaves and [16], [18] for roots), or on 2D surfaces in 3D ([3] for shoot apical meristems). However, new technologies to image and segment the complete volume of multicellular tissues in 3D at wall resolution [19]and to measure cell mechanics [20], [21] now lead to new questions regarding the interactions between cells in three dimensions. Addressing these questions, requires a new generation of models, able to account for the genetic regulation of biophysical processes in 3D multicellular systems. Here, we present such a modeling framework where cell growth results from the deformation of walls that are under tension in the tissue. A tensorial formalism is used to account for the anisotropic nature of cell wall material and to model wall deformation based on a generalized Lockhart viscoplastic law. An adapted finite element method has been designed to carry out efficient numerical simulations. This computational framework is then used to analyze the development of the early flower bud and test the effect of different regulation hypotheses.

RESULTS

Prior to describing our computational framework to model multicellular tissue growth, let us analyze the physical situation of a small region - a cell or a portion of a cell wall - in a growing tissue.

Multicellular Growth Involves both Cell Autonomous and Non-Cell Autonomous Forces

Similarly to isolated cells, cells in plant tissues are growing under the action of forces that stretch their wall and make it yield. However, for single isolated cells, the only significant forces able to trigger growth are the ones generated by its turgid cytoplasm pushing on the cell wall (Fig. 2A). In a multicellular context, the situation is somehow complicated by the fact that cells are rigidly connected to each other. The deformation of one cell generates physical constraints on its neighbors and vice versa. It has been recognized by Sachs as early as in 1882 (cited by Kutschera [4]) that epidermis cells in plant tissues are experiencing external forces due to the inner turgid cell layers pushing outwards against the surface cells, and inducing tension stresses in the epidermis (Fig. 2B). In a plant tissue, the mechanical stresses undergone by one cell wall are thus not only due to their own turgid cytoplasm, that we call *cell autonomous stresses*, but also comprise the physical constraints imposed by the neighboring cells, called the *non-cell autonomous stresses*.

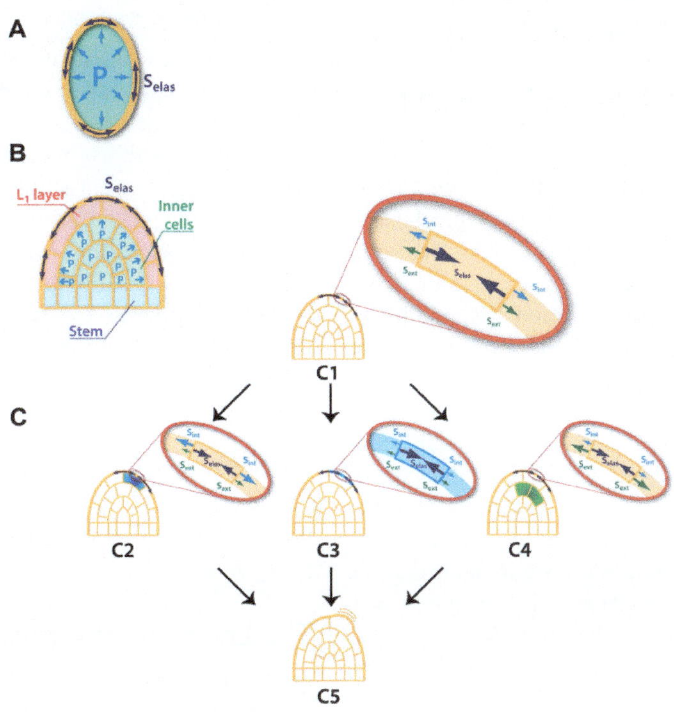

Figure 2: Origin of forces driving growth in a multicellular tissue.

Genes can regulate growth by controlling processes that modulate these stresses. Cell autonomous stresses can be modulated directly by the cell itself. This can be done in two ways: either by increasing the cell's own turgor pressure or by modulating the mechanical properties of its wall (e.g. by changing its elasticity or its yielding threshold). In contrast, non-cell autonomous stresses reflect distant mechanical interaction between cells where genes are being expressed and the ones where the consequences of this expression are observed.

Let us illustrate these different possibilities in the context of a primordium outgrowth on a growing meristematic dome, see Fig. 2C. As discussed above, any specific cell of the epidermis is suject to three types of stresses: *i*) the stress S_{int} induced by the cell's inner turgor pressure on the wall *ii*) the stress S_{ext} resulting from the action of the rest of the tissue on the cell wall and *iii*) the stress S_{elas} due to the cell wall elastic deformation in reaction to the other stresses. In growing tissue, mechanical equilibrium leads to a balance between these three stress components (Fig. 2C1):

$$S_{elas} + S_{ext} + S_{int} = 0, \qquad (2)$$

In a cell-autonomous perspective, regulation of the inner pressure of the cell would lead to a change in S_{int} (Fig. 2.C2) and modifying the mechanical properties of the wall would directly impact the elastic term S_{elas} (Fig. 2.C3). Likewise, modulating S_{ext} (e.g. by changing the turgor pressure of some neighboring cells) would correspond to a non-cell autonomous regulation of growth, (Fig. 2.4). Interestingly, new experimental evidence of the possibility of such a non-cell autonomous regulation has recently been reported by Peaucelle*et al.* [20].

The situation is complex as, if one of these stress components is affected by some regulation mechanism, all the other stresses will in turn be affected. Moreover, in real plants several of these regulations may be triggered at the same time, leading to even more complex interactions between regulation and growth. Therefore, to understand the growth of multicellular tissues, one needs to model both cell autonomous and non-cell autonomous growth. How can this be achieved?

Computational Framework to Model Local Deformations In Tissues

Both cell and non-cell autonomous growth rely on turgor-generated forces that are directly translated into mechanical stresses within the cell walls. Therefore considering a mechanical stress-based growth mechanism ensures that both types of growth are taken into consideration. Depending on their mechanical properties, the cell wall deform in response to the direction and intensity of these stresses. We assume that each small wall region of each cell in the tissue, at any time t, has a rest shape, i.e. the shape that the region would have if isolated from the rest of the tissue. Under the effect of the tissue stresses due to turgor pressure and connection to other cells, each small region is elastically deformed with respect to its rest shape. For the sake of simplicity, we assume that the thickness of the wall is kept at a constant value during growth (we do not model the details of the wall remodeling process itself). We also assume that the thickness of the walls has no major mechanical effect at this scale of analysis and therefore can be integrated in the wall's in-plane properties. Then, if the region is chosen sufficiently small, the wall deformation can be assimilated to an affine transformation and represented by a matrix $\mathbf{F}_e$ called the deformation gradient (Fig. 3*A-B*, see Modelsection for mathematical details). From $\mathbf{F}_e$, it is easy to compute the region strain $\mathbf{E}_e$:

$$\mathbf{E}_e = \frac{1}{2}(\mathbf{F}_e^T \mathbf{F}_e - \mathbf{I}), \qquad (3)$$

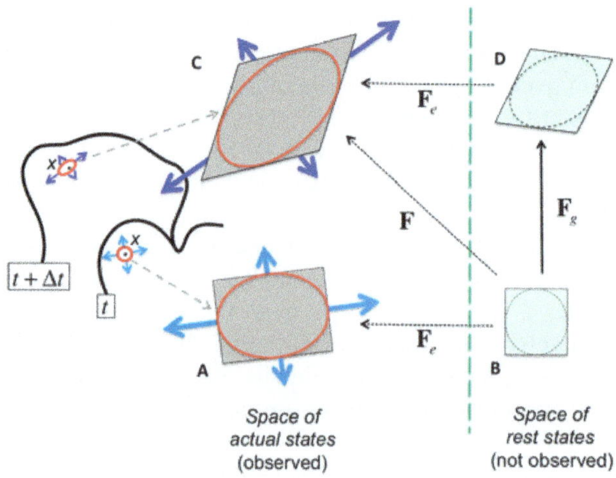

Figure 3: Formalization of plastic growth of a small region of wall.

where $\mathbf{F}^T$ denotes the transpose matrix of $\mathbf{F}$ and $\mathbf{I}$ the identity matrix. The fact that this strain is the elastic response of the material to tissue stresses is described by a constitutive law of the wall material. In the simplest case, it is a linear relation between elastic strain and stress corresponding to a generalized Hooke's law in 3 dimensions:

$$\mathbf{S}_e = \mathbf{H} : \mathbf{E}_e, \qquad (4)$$

where $\mathbf{S}_e$ is a matrix representing the stress on the small region and $\mathbf{H}$ is an order-4 tensor expressing the local elastic properties of the material. In particular, the anisotropy of the material if any is encoded in $\mathbf{H}$coefficients.

The forces that act on a region may vary throughout time, notably through either direct or indirect genetic regulation. Regions may be subject to new stress distributions (Fig. 3*C*), inducing new strains. In the spirit of Lockhart equation for cell volumes (Eq. 1, we assume that if the strains get above some threshold, the walls start to yield and the cell to remodel them. As a consequence, the reference state of the region is modified irreversibly (Fig. 3*D*). A number of studies have proposed to model this process by using a multiplicative decomposition of the overall deformation $\mathbf{F}$ [22] and then [23], [24] (Fig. 3*B-C-D*),

$$\mathbf{F} = \mathbf{F}_e . \mathbf{F}_g, \qquad (5)$$

where $\mathbf{F}_g$ denotes the irreversible modification of the rest shape of the region, called the growth tensor, and $\mathbf{F}_e$ is by a purely elastic deformation corresponding to the reversible part of the process.

The Equation of Growth

To describe growth we need a constitutive law that relates the rate of change of the matrix $\mathbf{F}_g$, called growth rate tensor, to physical processes. We assume this law to be strain-driven:

above a certain deformation threshold, the rest configuration of a region changes at a speed proportional to the strain of the region. In terms of tensors, the simplest form of such a law can be expressed as (see Model section for details):

$$\frac{d\mathbf{F}_g}{dt}.\mathbf{F}_g^{-1} = \gamma\Theta[\mathbf{E}_e - \mathbf{E}_0], \tag{6}$$

where the left-hand term defines the relative rate of variation of the reference state, γ is a constant characterizing the rate at which walls yield (extensibility) and the components of the term $\Theta[\mathbf{E}_e - \mathbf{E}_0]$correspond to non null tensor components in the directions where the elastic strain is above the threshold values encoded in $\mathbf{E}_0$. Replacing the strain components in Eq.6 by their stress counterpart using Hooke's law, leads us to the following generalized 3D *Lockhart-like* form:

$$\frac{d\mathbf{F}_g}{dt}.\mathbf{F}_g^{-1} = \gamma\Theta[\mathbf{H}^{-1}(\mathbf{S}_e(P) - \mathbf{S}_0)]. \tag{7}$$

This equation shows how growth is related to key mechanical variables: the extensibility γ controlling the rate of growth, the elastic properties of the material **H**, the turgor pressure P that appears through the stress it induces $\mathbf{S}_e$ in the tissue and $\mathbf{S}_0 = \mathbf{HE}_0$, a plastic stress yielding threshold corresponding to the strain yielding threshold $\mathbf{E}_0$. Note that the mechanical properties of the material are taken into consideration through a rigidity tensor **H** which allows the plastic deformation evolution ($\frac{d\mathbf{F}_g}{dt}.\mathbf{F}_g^{-1}$) to be non-collinear to the plastic stress ($\mathbf{S}_e(P) - \mathbf{S}_0$).

The above equation describes the plastic deformation evolution of a small region of the tissue, typically a part of a wall, during a small amount of time. To compute the deformation of the whole tissue during development, we need to integrate these local deformations over the whole tissue and throughout time, so that all these elements are assembled in a symplastic manner in the deformed tissue. This computation is made by minimizing the global mechanical energy in the tissue (see Model section). The strain and stress configuration of each region will thus be chosen so that the mechanical energy is minimal among all possible combinations of local elastic deformations that preserve the integrity of the tissue and the adjacency of cell walls. By contrast, the rest configuration of each individual small region is not necessarily compatible with that of other regions [25], i.e. there may not be physical continuity between rest configurations. The integration is carried out using a finite element method - FEM - (see Model section below).

A Comparative Analysis of the Putative Mechanisms behind Organogenesis At The Shoot Apical Meristem

We next used our modeling framework to analyze organogenesis at the shoot apical meristem (SAM). The SAM is a population of stem cells that continuously initiates new stem tissues and lateral organs, thus generating all the aerial parts of the plant. We first constructed a model of the SAM as a dome made up of polyhedra representing the 3-D cells and rigidly connected to each other (Fig. 4*A*). The faces of these polyhedra represent cell walls and are composed of 2-D elastic triangular elements whose mechanical properties are represented by tensors **H**. The stiffness of these elastic triangles is set higher in the epidermis walls than in the inner walls and may be either isotropic or anisotropic for epidermis triangles. We assume that cells are inflated with a uniform turgor pressure P_0, and that triangle mechanical properties are all initially isotropic and uniform.

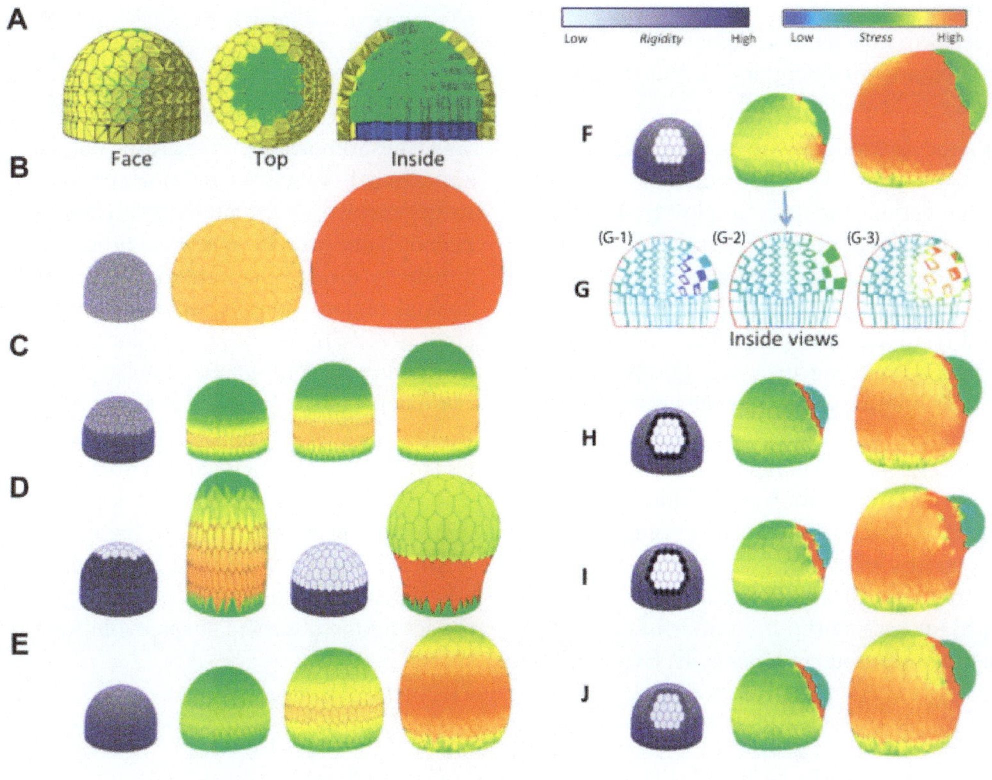

Figure 4: Growth regulation mechanisms and their impact on shape development.

In this initial configuration, the turgor pressure induces a stress that puts all the cell walls under tension. If the plastic growth threshold $\mathbf{E}_0$ is reached, the dome grows isotropically in all the directions (Fig. 4*B*). This plastic deformation is accompanied by a decrease of inner pressure in cells as their volume increases due to growth. *In vivo*, this would lead to a difference in water potential between the cells within the dome and surrounding tissues, drawing new water from below the dome inside the meristematic cells. It is generally assumed that water uptake is combined with osmolyte regulation to maintain turgor relatively constant in the cell during growth [4]. As these processes are considered fast compared with growth, we simply assume in our model that, otherwise stated, the turgor pressure is continuously kept constant at P_0.

A key feature of meristematic activity is the generation of cylindrical stems and roots. In principle there are several ways to generate axial structures, but there is overwhelming evidence that this is due to a switch from isotropic to anisotropic growth over several cell files. Indeed, when cells leave the meristematic dome, they start to generate cellulose microfibrils in highly ordered arrays, oriented along the circumference of the meristem [3], [26]. This microfibril organization creates a high circumferential rigidity that favors growth in the perpendicular axial direction. This mechanical anisotropy can be reproduced in the 3D model by imposing higher rigidity in circumferential direction using anisotropic $\mathbf{H}$ tensors (Fig. 4*C*). Interestingly, the ability of the microfibrils to create well formed axes depends on the amount of microtubule anisotropy in the dome (Fig. 4*D* dark gray). If too small, a bulge forms at the tip (Fig. 4*D*). Intermediate shapes can be obtained by using a less sharp transition between isotropic and anisotropic regions, using e.g. a gradient of anisotropy between the bottom and the top of the dome (Fig. 4*E*).

Next, using this mechanical model of the SAM, we tested different scenarios for organ emergence. As discussed above, several processes can in principle account for this phenomenon, in a cell- or non-cell-autonomous way, including changes in turgor pressure, modifications in wall stiffness and modifications in the rate of wall synthesis.

Starting from the growing dome of Fig. 4*E*, we first tested the possibility to grow an organ on the dome assuming cell autonomous regulation. For this, we lowered the outer periclinal wall rigidity in a small region close to the tip of the main dome (Fig. 4*F*). This created a bulging zone in the corresponding region of the tissue. We then tested the alternative possibility to create a primordium in a non-cell autonomous manner, by relaxing the wall rigidity up to 10-fold in a group of cells immediately below the previous surface region. No visible bulges could be obtained in this way (Fig. 4*G*–1; compare with Fig. 4*G*–2 corresponding to a transversal cut obtained from the simulation of Fig. 4*F* middle that shows a much larger bump at the same stage of development). To further explore the ability of non-cell autonomous stresses to trigger bump outgrowth locally, we also tested the possibility to obtain a bulge by increasing (by up to a 3-fold factor) the turgor pressure in the same group of inner cells. Here again, no clear bulge could be obtained (Fig. 4*G*–3). In this latter case, we could observe in the transversal cut that the tissues were actually compressed internally. We concluded from this series of simulations that bumps can more easily be generated in the context of cell autonomous regulation.

In this context however, we could observe that the growing bumps were not clearly separated from the main dome (e.g. Fig. 4*F*). We therefore constrained a ring of cells at the surface, around the bump location, to be very rigid. This resulted in a well formed bump growing on the top of the initial dome (Fig. 4*H*). A similar separation between the two organs can be obtained by stiffening the cells of the ring only in the circumferential direction. Interestingly, the cells of the ring are also supporting high stress but are left free to grow in the axial direction of the new bump (compare right images in H and I). A well marked bump can also be created by a less important decrease of the primordium zone rigidity compensated by a local increase of the cell growth rate in this region (Fig. 4*J*).

The simulations showed how simple scenarios can in principle explain important morphogenetic processes during plant development, including the formation of cylindrical stems and roots and the outgrowth of lateral organs at the SAM. Importantly, although these simulations can lead to comparable shapes, they make different predictions regarding the mechanical properties and resulting growth patterns.

A Case Study: Morphogenesis of a Flower Bud

The simulations presented above were based on abstract versions of real meristems. We therefore next applied our computational framework to perform simulations from realistic templates. Hereby, we used the floral meristem of *Arabidopsis thaliana* - which has been very well characterized - as a case study. As a reference for model construction, we used a series of confocal stacks of the same young growing floral primordium taken at 24 h intervals from early stage 1 to stage 2 (3 time-points, Fig. 5*A*-*B*-*C*). Using the Mars-Alt pipeline [19], the individual cells were identified and cell lineages were tracked in the thus segmented reconstructions (Fig. 5*D*-*E*-*F*). The confocal images and 4D reconstructions suggest that *in vivo*, the primordium first grows out from the meristem as a small radial symmetric globular structure in a direction normal to the surface of the meristem. Then the global direction of primordium growth changes progressively and the initial symmetry around the normal to the surface breaks as the the abaxial region expands more rapidly than the adaxial region (Fig. 5*G*-*H*-*I*). The sepals appear with different growth rates as the abaxial and adaxial sepals grow much faster than the two lateral ones. A recent analysis of gene expression in the flower bud showed complex spatio-temporal expression patterns with as many as 16 different domains expressing different combinations of transcription factors [27]. This raises two main questions: (i) how many different gene activities would in theory be required to produce this structure? (ii) are mechanical-related actions of these genes alone sufficient to reflect for the observed shape evolution?

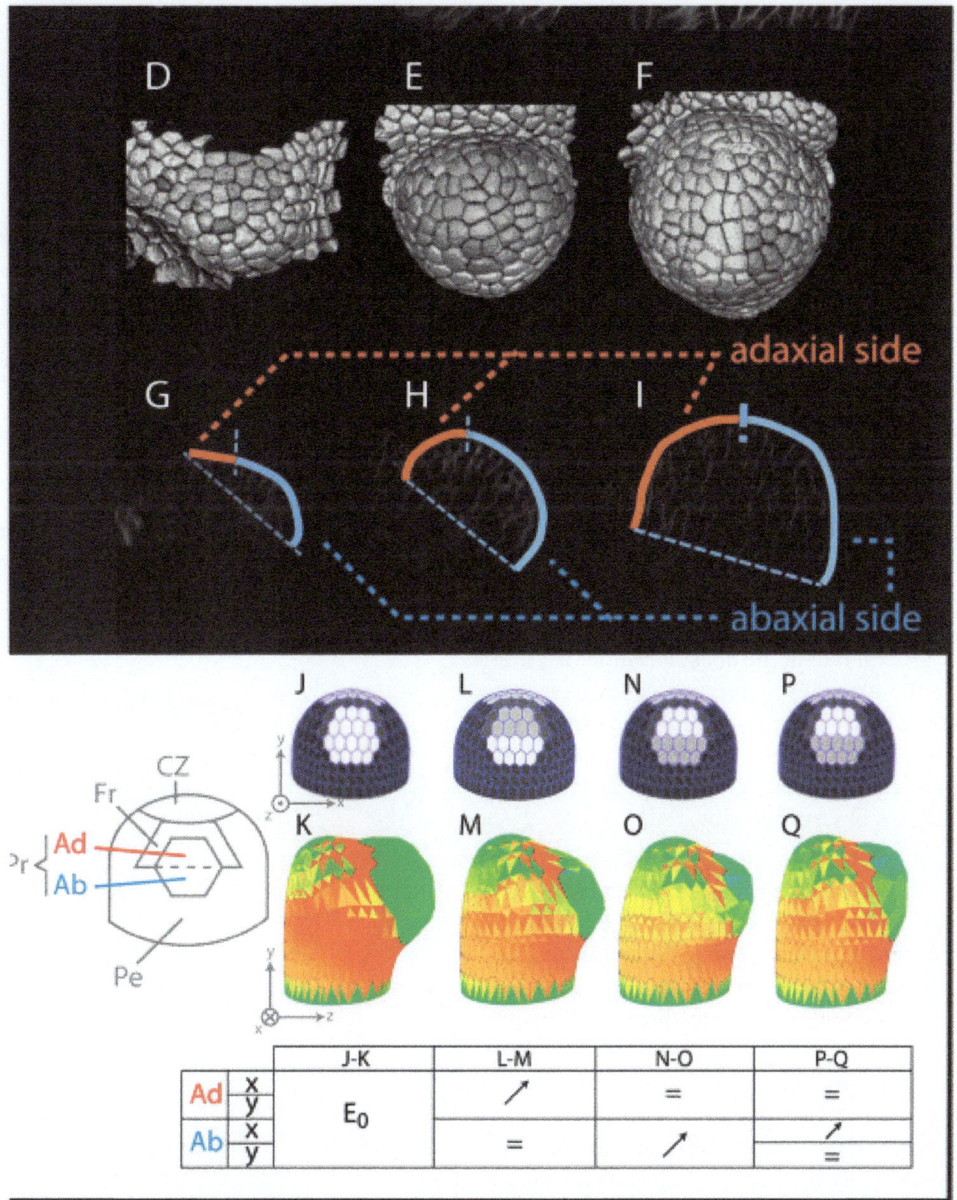

Figure 5: First stages of development of a flower bud.

To address these questions we investigated with our mechanical model how to reproduce, at first qualitatively, the developmental pattern described above. Hereby we distinguished two phases: (a) the initial outgrowth of the bud and (b) the formation of the sepal primordia. For this purpose, we made a number of assumptions based on the literature and on the first set of models described above.

First, we simulated the outgrowth of the flower bud using the following itinerary through successive scenarios:

- the first scenario, we tested the possibility to initiate organ outgrowth by a combination of local wall softening in the incipient primordium and anisotropic wall stiffening in the organ boundaries. A number of experimental evidence supports such a scenario. The young primordium is characterized by relatively high concentrations of the plant hormone auxin [28], [29]. There is evidence, that auxin influences cell wall loosening by the activation of specific enzymes. For instance, in the hypocotyl ABP1 (Auxin Binding Protein 1) controls the level of expression of glycosyl hydrolases known to modify hemicellulose within the cell wall, [30]. At the apex, atomic force microscopy measurements suggested that auxin accumulation leads to cell wall softening via a PMEI3-dependent pathway [31]. Finally, transcription factors such as AINTEGUMENTA and MONOPTEROS which are homogeneously expressed in the young outgrowing flower bud are thought to regulate the expression of expansins and xyloglucan modifying enzymes [9], [10]. Organ boundaries are also characterized by specific gene expression patterns. In particular the CUP-SHAPED COTYLEDON transcription factors are strongly expressed between the meristem and the primordium and genetic studies show that they repress growth in this region (e.g. [32], [33]). In addition, Hamant et al. [3] found that in the same region, cells are likely to have highly anisotropic wall structure. Based on these observations and on preliminary simulations shown on Fig. 4*I*, we assumed the existence of a band of anisotropic cells around the primordium's upper half. As a result, the model produced a bulge normal to the surface with a quasi symmetric shape (Fig. 5*J-K*).

- The next morphologically significant step in organogenesis happens when differential growth behavior between the adaxial and the abaxial regions emerges. At stage 2 (or P2) cells of the abaxial side of the primordium start to expand faster in the meridional direction than the ones from the adaxial region, (Fig. 5*G-I*, see also [34]). To account for that differential growth in our simulations, we assumed that mechanical properties are differentially regulated between the abaxial/adaxial regions. Although there is no strict evidence for differences in mechanical properties, both regions are characterized by specific gene expression patterns early on. For instance, transcription factors such as FIL, KANADI, YABBI or ARF3 and ARF4 are part of interlocking pathways necessary for organ polarity and proper development [35], [36], for reviews see also e.g. [27] and references therein. To account for the faster growth rate in the abaxial region, we first assumed that the abaxially expressed genes caused a lower rigidity in that region (Fig. 5*L-M*). This indeed resulted in a developmental abaxial-adaxial asymmetry. However, the organ tended to develop downwards, opposite to what is observed *in vivo*.
- To correct this behavior, we set rigidities in the primordium zone in the opposite way, i.e. with higher rigidity now in the abaxial region. The summit/adaxial region was now developing faster toward the upper direction but, contrary to observations in actual development (Fig. 5*G-I*), with little shape asymmetry (Fig. 5*N-O*). Indeed, *in vivo* observations carried out by e.g. Long et al. [34] and Fernandez et al. [19] were reporting anisotropic growth in the abaxial side of the floral primordium. The discrepancies in simulations of Fig. 5*N-O* could finally be corrected by setting a high rigidity anisotropy in the abaxial cells. (Fig. 5*P*) The dome asymmetry was then correctly oriented upward (Fig. 5*Q*), with fast development of the abaxial region upwards as expected.

Following on the simulated flower bud growth, we next simulated the outgrowth of sepal primordia (Fig. 6*A-B*). To initialize the FEM simulations we constructed a triangle-mesh of the floral meristem at stage 2 based on the segmented image. The vertices of triangles coincide with the cell vertices (Fig. 6*C*). We then reused the mechanical setting of this model to grow the four sepal primordia on the outgrowing dome obtained in phase (a). *In vivo*, this stage is characterized by auxin accumulation at the four sites of sepal initiation [37], followed by the activation of genes that promote organ outgrowth - in particular AINTEGUMENTA - first in the abaxial and adaxial sepals, then in the lateral ones (Fig. 6*B*). Simultaneously, organ boundary genes are activated at the flower meristem boundary and in between the primordia. As for the flower bud simulations presented above, we translated these gene expression domains into zones with different mechanical properties, i.e. wall loosening in the primordia and a anisotropic stiffening in the boundaries as described above. Again the rigidity in the abaxial regions was set higher than in the adaxial ones. We also assumed that the initiation of lateral sepals was slightly delayed with respect to the adaxial/abaxial pair. Based on initial geometric structures obtained from confocal microscopy, the resulting simulation was able to reproduce the developmental dynamics of a flower bud and its first lateral organs based on a spatio-temporal synchronization of the changes in mechanical properties of regions (Fig. 6*D*).

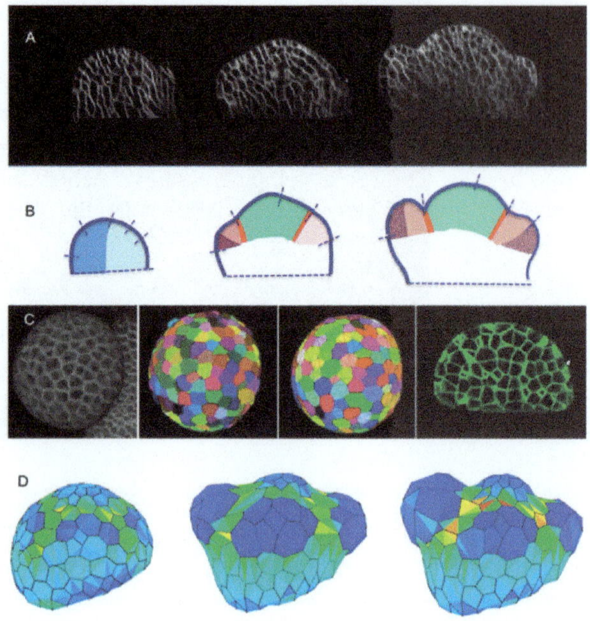

Figure 6: (A) Transverse sections of confocal images showing floral bud development between stage 1 and early stage 3.

DISCUSSION

Modeling strategies are increasingly used to help understanding the mechanisms behind shape changes and to make predictions. However, so far models have been either relatively abstract, by focusing on the instructing role of the gene network rather than on the physical constraints of cell growth, or well-defined biophysically but weakly connected to the gene network and its regulatory role. In addition, models able to simulate in a realistic way in 3 dimensions the development of organs on the basis of cellular regulation were missing.

Here we present such a conceptual and modeling framework that is able to integrate the genetic input and the mechanical determinants of the cell in a 3D multicellular context. Using this framework, we tested different scenarios relevant to shape changes and made a number of predictions for organogenesis at the shoot apical meristem. Importantly, our approach does not overlook the complexity that relates to multicellularity, but fully incorporates it by taking into account both cell autonomous and non-cell autonomous forces in 3D tissues.

In the past decade several modeling frameworks have been proposed that are aimed at providing insight at multiple levels of organization and regulation [8], [13], [16], [18]. The framework we propose here presents three significant complements to these existing models. First, we explicitly formalize growth of plant cells in tissues taking into account our current knowledge on the mechanisms controlling growth at the cellular level. The resulting equation allows us to make precise prediction on the primary function of molecular regulators in terms of four local parameters i.e. turgor pressure, wall extensibility, wall elasticity, and yield threshold. Second, taking advantage of both the availability of increased computer power and novel computing techniques we were able to design a model tissue with cellular resolution in 3D. This integrates a new data structure to handle multicellular tissues [38] used by other groups for tissue simulation (e.g. [18]) and the use of the modeling software platform SOFA for mechanical simulation in biology [39]. The platform makes it possible to express the mechanical, geometrical and simulation components in a completely modular way and optimizes the computational overhead during simulations by making use of efficient implicit integration solvers. As a result, simulation in 3D of flower development can be achieved in near interactive time. The model is able to handle several hundreds of cells, and can take into account both biochemical and physical properties of cells. Third, by fully accounting for the complexity behind multicellularity, this framework allows us to compare the contribution of cell autonomous forces and non cell autonomous forces in shape changes. Our conceptual framework notably integrates the mechanical outputs from the gene regulatory network in each individual cells and the secondary effects deriving from growth of neighboring cells. It also provides us new means to quantitatively investigate the contribution of geometrical, biochemical and mechanical interactions in growth regulation with cell wall resolution.

The simulations carried out on both synthetic and realistic templates pose a number of interesting questions that now have to be addressed experimentally. First, the plant seems to have the possibility to create similar shapes through different pathways. Growth rates in the organ boundaries, for example, can be restricted by modifying wall stiffness, wall anisotropy or combinations thereof. In parallel, organ outgrowth can be promoted by reducing wall stiffness and/or by increasing the rate of wall synthesis. Interestingly, the model may help us establish a hierarchy between developmental scenarios and organize the plausibility of developmental scenarios and to organize accordingly experimental investigations. For example, to create a realistic primordium on the meristem dome, simulations clearly indicated that the intuitive solution consisting of softening the abaxial zone is the least plausible (Fig. 5*L-M*). This forced us to explore other, less intuitive hypotheses consisting instead of loosening the adaxial zone (Fig. 5*N-Q*), and that appeared to be more realistic. Another interesting outcome concerns the number of different gene activities that are required to generate an organ primordium or a flower bud. Whereas gene expression studies have revealed a complex partitioning of the growing flower bud in different domains, the simulations carried out on realistic templates, suggest that only five domains (central zone, peripheral whorl, abaxial and adaxial domains and boundary) with specific wall modifying activities would be sufficient. These hypotheses should now be tested. This will include classical approaches such as gene expression studies and transgenic approach, but also more challenging techniques required for quantitative growth analysis. At this stage, the simulations already suggest a number of relatively straightforward experiments. Differences in wall stiffness between adaxial and abaxial domains can be measured, for example using atomic force microscopy [21]. Correlations between organ outgrowth and the expression of genes involved in wall modifications can also be made. Modifications in wall anisotropy at the abaxial side of the primordia -as suggested by the model - can even be monitored in vivo by direct observation of microtubule dynamics [3]. Metrics will also have to be developed to compare quantitatively results of mechanical simulations and the observed, actual geometry of developing organs. Based on the recent progresses in imaging protocols, cell segmentation and tracking softwares, e.g. [19],[40], it will now become possible to routinely compare the simulated development of particular cellular regions (e.g. central zone, primordium, abaxial/adaxial, frontier zone,...) with the observed ones based on a quantitative comparison of their principal direction of growth, rate of growth, and of various shape factors such as local curvature, degree of symmetry, compacity, *etc*. In turn, this opens the way to the development of inverse modeling techniques, where the estimation of some mechanical parameters that are difficult to measure, will be derived from minimizing the distance between simulated and observed developing shapes.

The current version of the model also has some limitations. In particular, it currently does not include cell division. Generic rules for cell division have been proposed in the literature, e.g. see recent works [41], [42]and references therein. However, in the meristem, the validity of these rules has not been thoroughly assessed, and in particular, the way cells divide in the inner layers of meristem is still mostly unknown. Imposing specific rules here would add an extra level of assumptions whose effect on shape development would require specific analysis. In addition, the design of a robust algorithm for cell division on real 3D meshed tissues is a challenging issue in itself and would require the development of robust 3D algorithms. For these reasons, we restricted in this paper our analysis to time-lapses that are small enough to consider that deformations and mechanical forces are not markedly modified by cell division. We used simulation duration such that the deformations of cells were kept reasonably small (cell volumes increased by less than a 3-fold factor). A second limit of the current model is related to the use of 2D finite elements with no thickness for the representation wall parts. We considered that at least in a first approach neglecting the effect of the third dimension (thickness) of these elements and the associated flexural stiffness was a reasonable approximation. A further step would then consist of using more complex 3D shell elements in the numerical method to account for situations where these non in-plane stresses cannot be neglected. Finally, in further rounds of modeling, additional hypotheses and levels of regulation can be included as well. For instance, so far we did not take into account feedback effects between mechanical stress and wall structural anisotropy during development, as reported in different recent works in cells of the L1 layer [3], [43]–[45]. Although we did not address this level of complexity here, our mechanical model provides a natural framework to model such constitutive laws as the tensor representation of the material properties is naturally adapted to the modeling of local anisotropy.

The morphogenetic model presented in this paper investigates how plant axes and branching systems emerge from basic variables at cellular level (turgor pressure, wall deformation and genetic regulation). This model provides realistic multicellular 3D shape simulations that mainly relies on the presence of a stiff epidermis that is limiting for growth.

Remarkably, this is in fact reminiscent of the way a cell wall is limiting the expansion of the protoplast. This similitude was already noticed by Kutschera: "The whole organ can [...] be regarded as a single giant cell" [4]. Using realistic 3D mechanical simulations, our computational model supports the view of multicellular axial growth underlying this statement. It shows in addition how lateral organs can be produced by locally regulating the mechanical properties of the epidermis at the tip, while relying on both cell autonomous and non-cell autonomous patterns of tension. Together with Dumais's model on single cell growth [8] also based on anisotropic growth, these models give further support for the existence of a deep self-similarity in the axial growth of plants. This self-similarity is strikingly illustrated by the similitude in the circumferential organization of cortical microtubules within the same tissue, i.e. in a single meristematic cell and in the whole meristem. It is also further supported by the existence of single cell organisms that have developed branching strategies. Caulerpa for instance, a single cell system, exhibits complex tree-like shapes comparable to fern leaves. Altogether, this suggests that plants have found ways through evolution to scale up axial growth in single cells to multicellular systems, using similar biophysical principles. Conversely, this rather supports the idea that growth in plants mainly relies on elongation. In this scenario, cell division in plants would mainly be a way to subdivide an increased volume, with little impact on growth. Such a unifying principle across scales likely reflects the existence of some common essential constraints behind growth. Identifying these core rules, and their exceptions, will be a major challenge for the future of development that only realistic mechanistic models based on the actual effectors of growth will be able to address.

MODELS

This section describes the details of our mechanical model of tissue growth. The elastic response being a pure mechanical phenomenon and the growing process a complex biochemical process, we assume that the characteristic elastic-response time ($\mathcal{T}_e$) is very small compared to the characteristic time of the growth mechanism ($\mathcal{T}_g$). Since our main focus is on the growth process, we will consider time scale Δt such as: $\mathcal{T}_e \ll \Delta t \leq \mathcal{T}_g$.

Kinematics Variables & Relations

The positions of material points of Ω are tracked throughout time. The deformation field φ relates the coordinates of a material point in the current, "*observed*" ($\mathcal{B}_t$) and reference configurations ($\mathcal{B}_0$) respectively noted $\mathbf{X}$ and $\mathbf{x}$:

$$\mathbf{x} = \varphi(\mathbf{X}, t) \tag{8}$$

This field is assumed smooth enough so that its gradient, mapping small variations of length between the material configuration and the current configuration, is defined:

$$d\mathbf{x} = \mathbf{F} \cdot d\mathbf{X}$$

$$\mathbf{F} = \nabla_{\mathbf{X}}(\varphi) \tag{9}$$

To model growth, the most common approach is to use a multiplicative decomposition of the deformation gradient [22]–[24]

$$\mathbf{F} = \mathbf{F}_e . \mathbf{F}_g, \tag{10}$$

where the global deformation gradient $\mathbf{F}$ is decomposed into a reversible part $\mathbf{F}_e$ and and irreversible part $\mathbf{F}_g$. This decomposition leads to an intermediate, "*grown*", configuration ($\mathcal{B}_g$), where material coordinates are noted $\mathbf{X}_g$. $\mathbf{F}_g$ thus maps growth-related deformations between $\mathcal{B}_0$ and $\mathcal{B}_g$ and $\mathbf{F}_e$ maps purely elastic deformations between $\mathcal{B}_g$ and $\mathcal{B}_t$, see Fig. 7. In that perspective, growth is regarded as the time evolution of the irreversible part ($\mathbf{F}_g$) of the deformation gradient field. Geometrically, the time evolution of such deformation gradient fields is estimated through their related velocity gradient fields. In the *grown*configuration $\mathcal{B}_g$, the growth-related part of the velocity gradient field reads as follows:

$$\mathbf{L}_g = \dot{\mathbf{F}}_g . \mathbf{F}_g^{-1} \tag{11}$$

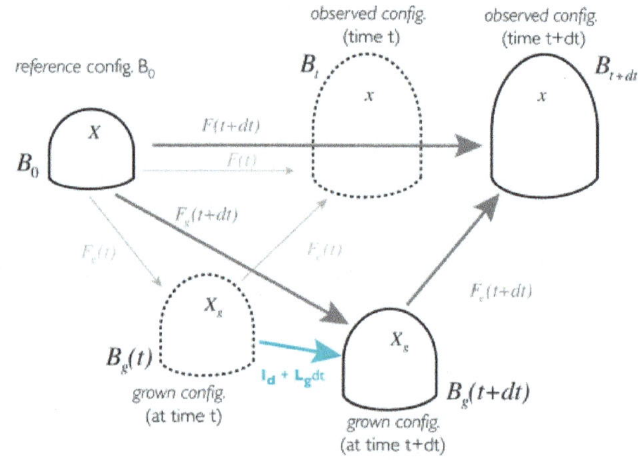

Figure 7: Schematic representation of the different configurations at different time and the deformations between them.

Conservation Equations

Mechanical Energy

We assume that the mechanical part of Ω free energy density, noted $\mathcal{W}$, is strictly elastic and only depends on $\mathbf{F}_e$, with a local minimum for $\mathbf{F}_e = \mathbf{I}$. Since it also must be rotation invariant, $\mathcal{W}$ can be expressed as a function of the Green-Lagrangian strain tensor (noted $\mathbf{E}_e$) only:

$$\mathcal{W} = \mathcal{W}(\mathbf{F}_e) = \mathcal{W}(\mathbf{E}_e) \tag{12}$$

With the following definition of $\mathbf{E}_e$:

$$\mathbf{E}_e = \frac{1}{2}(\mathbf{F}_e{}^T . \mathbf{F}_e - \mathbf{I}) \tag{13}$$

A straightforward consequence of this assumption is that all the mechanical dissipation processes depend on $\mathbf{F}_g$ only, which leads to $\dot{\mathbf{F}}_e = 0$ at mechanical equilibrium.

Forces Balance

As we assume the static equilibrium of the elastic part of the deformation gradient $\mathbf{F}$, the local formulation of the balance of linear momentum and boundary conditions in the reference 'grown' configuration reads:

$$\begin{aligned} \nabla_{\mathbf{X}_g} \cdot \mathbf{P}_e + \mathbf{b} &= 0 \quad \text{in} \quad \mathcal{B}_g \\ \varphi &= \bar{\varphi} \quad \text{on} \quad \partial\mathcal{B}_g^D, \\ \mathbf{P}_e \cdot \mathbf{N}_g &= \bar{\mathbf{T}} \quad \text{on} \quad \partial\mathcal{B}_g^N \end{aligned} \tag{14}$$

where $\mathbf{P}_e$, the first Piola-Kirchoff stress tensor, is a measure of the stress applied in the current configuration with respect to the "grown" configuration. $\mathbf{b}$ stands for any external force density field and will be neglected hereafter. $\mathbf{N}_g$ is the unit outward normal to the boundary of the undeformed body. $\bar{\mathbf{T}}$ is the prescribed traction per undeformed unit area in part of the boundary $\partial\mathcal{B}_g^N$. In our case, $\bar{\mathbf{T}} = \Delta P \cdot \mathbf{N}_g$ where ΔP stands for the turgor pressure difference between the inside and the outside of the tissue. Finally, $\bar{\varphi}$ is the prescribed deformation mapping in the rest of the boundaries $\partial\mathcal{B}_g^D$ (in our simulations, it corresponds to the fixed basis of the meristem).

Constitutive Equations

To compute the time evolution of the meristem under pressure, we need to define how the cell walls elastically deform and how they grow.

Elasticity

We assume a linear strain/stress relationship (Hooke's law): $\mathbf{S}_e = \mathbf{H} : \mathbf{E}_e$ where $\mathbf{S}_e = \mathbf{F}_e^{-1}\mathbf{P}_e$ is the second Piola-Kirchoff stress tensor and $\mathbf{H}$ is the Hooke fourth order stiffness tensor. The elastic deformation is thus characterized by a mechanical energy $\mathcal{W}$:

$$\mathcal{W} = \frac{1}{2}\mathbf{E}_e : \mathbf{S}_e = \frac{1}{2}\mathbf{E}_e : \mathbf{H} : \mathbf{E}_e \tag{15}$$

In its most general form $\mathbf{H}$ is a fourth order tensor (81 parameters) that relates stresses and strains distributed in three dimensions. For the sake of simplicity, we assume plane stress conditions (since bending forces can be neglected with respect to in-plane forces). Stresses and strains can be described as 2×2 symmetric matrices, downsizing the number of independent coefficients in $\mathbf{H}$ to 6. Using the Voigt notation, the relationship between stress and strain can be written in the following matrix form:

$$\begin{pmatrix} \sigma_{xx} \\ \sigma_{yy} \\ \sigma_{xy} \end{pmatrix} = \begin{pmatrix} E_{xx} & 0 & 0 \\ 0 & E_{yy} & 0 \\ 0 & 0 & (E_{xx}+E_{yy})/4 \end{pmatrix} \cdot \begin{pmatrix} \epsilon_{xx} \\ \epsilon_{yy} \\ 2\epsilon_{xy} \end{pmatrix} \tag{16}$$

In order to test our framework with the most simple anisotropic mechanical law possible we chose to neglect the mechanical coupling between directions (null Poisson's ratios in $\mathbf{H}$), leading to a diagonal form for $\mathbf{H}$ in the Voigt form. The non-null remaining coefficient are the two Young's moduli E_{xx} and E_{yy} and the shear modulus $G_{xy} = (E_{xx} + E_{yy})/4$. In order to reduce to the bare minimum the number of independent variables in this first set of simulations, we choose a simple expression inspired by the isotropic case.

Growth

The growth of the wall can be regarded as creep that can be modeled using the Maxwell model of viscoelasticity. In this model, reversible and irreversible phenomena are embodied, respectively, as a purely elastic spring (characterized by an effective spring constant k) and as a purely viscous dashpot (characterized by an effective viscosity μ), the two being connected in series. Adding a friction force in parallel to the viscous drag enables us to take into account the threshold phenomenon, see Fig. 8. Under loading forces, the mechanical equilibrium of such a system leads to:

$$\begin{cases} k\Delta x = \|\mathbf{f}_{load}\| \\ \mu \frac{dl_0}{dt} = k\Delta x - \|\mathbf{f}_{fric}\| \end{cases} \tag{17}$$

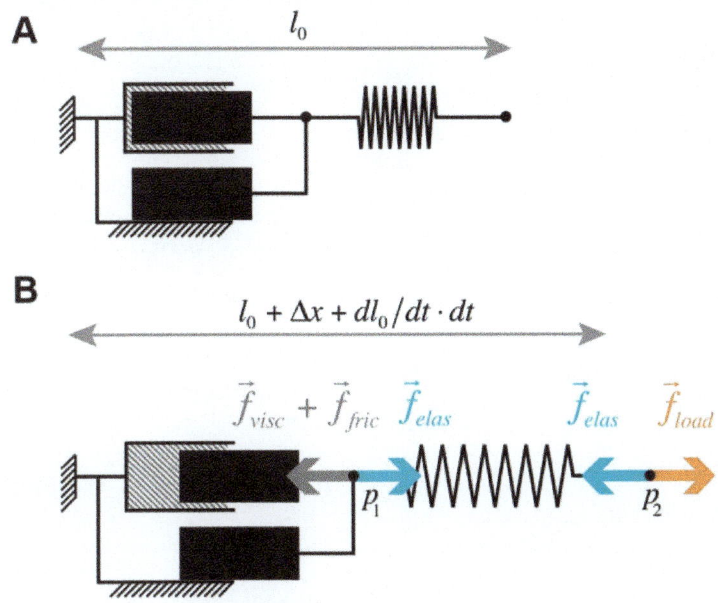

Figure 8: 1D version of a unit element of the biomechanical model.

The first line of Eq.17 refers to the mechanical equilibrium at point $P2$ on Fig. 8–*b* and links the elastic stretching of the material to its loading force. The second line refers to the mechanical equilibrium at point P1and links the creeping rate of the material to the difference between its elastic stretching and the threshold related force. Using the measure of strain $\epsilon=\Delta x/l_0$ in the system exposed at Eq. 17 leads us to the following expression:

$$\begin{cases} \epsilon=H^{-1}.P \\ \dfrac{dl_0}{dt}l_0^{-1}=\gamma(\epsilon-\epsilon_{tr}) \end{cases}, \tag{18}$$

where we assumed that the friction force is proportional to the initial rest length of the system ($\|\mathbf{f}_{fric}\|=k\epsilon_{tr}l_0$) and that the loading force can be expressed as a pressure force $\|\mathbf{f}_{load}\|=PS_0$ with S_0 the section on which the pressure P is applied. We also introduced $H=kl_0/S_0$ the effective Young's modulus of the material in the considered direction and $\gamma=k/\mu$ a coefficient characterizing the extensibility of the material in that direction. Extrapolating Eq. 18 in 3D with large deformation leads to the constitutive growth equation:

$$\begin{cases} \mathbf{E}_e=\mathbf{H}^{-1}:\mathbf{S}_e \\ \mathbf{L}_g=\dfrac{d\mathbf{F}_g}{dt}\cdot\mathbf{F}_g^{-1}=\gamma\Theta[\mathbf{E}_e-\mathbf{E}_{tr}] \end{cases}, \tag{19}$$

where $\mathbf{E}_{tr}=\epsilon_{tr}\mathbf{I}$ is the threshold matrix strain has to overcome in order to induce growth and $\Theta[\cdot]$ represents the matrix version of the primitive of the Heaviside function, defined as followed:

$$\Theta[\mathbf{A}]_{ij}=\begin{cases} 0 \;\; if \;\; A_{ij}\le 0 \\ A_{ij} \;\; if \;\; A_{ij}>0 \end{cases} \tag{20}$$

Discretization

Time Discretization

The backbone of the numerical simulation of our model is an iterative loop in which each step represents a time lap δt such as:

$$\tau_e \ll \delta t \ll \tau_g \tag{21}$$

Each time step happens as follow:

1. at the beginning of the $(n+1)^{th}$ step the system is already at mechanical equilibrium, deformed by the loading forces. If the strain tensor $\mathbf{E}_e^n$ is above the threshold value $\mathbf{E}_{tr}$ growth is initiated.
2. Once growth has been initiated, the growth-related deformation $\mathbf{F}_g$ is updated from its current value $\mathbf{F}_g^n$ to a new one $\mathbf{F}_g^{n+1}$ established by Eq. 25.
3. Once $\mathbf{F}_g$ has been updated, mechanical equilibrium is computed and the strain tensor is updated with value $\mathbf{E}_e^{n+1}$. A new step can begin.

Incremental evolution of $\mathbf{F}_g$

By discretizing time in steps of duration δt, we can express deformation $\mathbf{F}_g(t_n)$(with $t_n=n\delta t$) as the multiplication of incremental deformations $\mathbf{F}_g^i$. Moreover, deformation at time $t+\delta t$ is directly related to deformation at time t:

$$\mathbf{F}_g(t+\delta t)=\prod_{i=1}^{n+1}\mathbf{F}_g^i=\mathbf{F}_g^{n+1}\cdot\mathbf{F}_g(t) \tag{22}$$

Since $\delta t \ll \tau_g$, the instantaneous growth rate (noted thereafter $\mathbf{F}_g^i$) is supposed constant over each step. Therefore each element of the multiplicative decomposition (Eq. 22) can be expressed in terms of its corresponding instantaneous growth rate:

$$\mathbf{F}_g^i=\mathbf{I}_d+\mathbf{F}_g^i\delta t \tag{23}$$

Applying forward finite differences to $\mathbf{L}_g(t)$ leads to:

$$\mathbf{L}_g(t)=lim_{\delta t\to\tau_e}\frac{\mathbf{F}_g(t+\delta t)-\mathbf{F}_g(t)}{\delta t}\cdot\mathbf{F}_g(t)^{-1} \tag{24}$$

Combining Eqs.22),(23, 24 & 19 enables us to estimate the $(n+1)^{th}$ growth-related deformation $\mathbf{F}_g^{n+1}$knowing the strain tensor at step n:

$$\mathbf{F}_g^{n+1}=\mathbf{I}_d+\gamma(\mathbf{E}_e(n\delta t)-\mathbf{E}_{tr})\cdot\delta t \tag{25}$$

Space Discretization

We have chosen to discretize the continuous model of cell walls using first order finite elements. Nodes are placed at each wall junction, and triangles obtained by tessellation. Material points are interpolated from nodes, based on linear (barycentric) shape functions $\mathcal{N}_i(\mathbf{X}_g)$:

$$\mathbf{x}=\sum_i \mathcal{N}_i(\mathbf{X}_g)\mathbf{x}_i \tag{26}$$

where $\mathbf{X}_g$ are 2-dimensional material coordinates in the 'grown' configuration. For simplicity, we match material principal directions with mechanical anisotropy. By spatial differentiation, we obtain 3×2 elastic deformation gradient matrices, which are uniform over each triangle:

$$\mathbf{F}_e=\sum_i \mathbf{x}_i\frac{\partial\mathcal{N}_i(\mathbf{X}_g)}{\partial\mathbf{X}_g} \tag{27}$$

Nodal elastic forces in the current configuration are spatially integrated using midpoint quadrature. One evaluation of the energy density per element is sufficient since it is uniform within each triangle:

$$\mathbf{f}_i=-\left(\frac{\partial\int_S W}{\partial\mathbf{x}_i}\right)^T=-\sum_{j\in T(i)} S_j\frac{\partial\mathbf{E}_{e,j}^T}{\partial\mathbf{x}_i}:\mathbf{H}_j:\mathbf{E}_{e,j} \tag{28}$$

where $T(i)$ are the adjacent triangles of node i. $\mathbf{H}_j$, $\mathbf{E}_{e,j}$ and S_j are the Hooke tensor, strain tensor and surface of triangle j ('grown' configuration).

For inner cell walls, where no information of mechanical anisotropy is available, we assume that the representation of mechanical properties can be simplified by using one dimensional elasto-plastic elements - or springs- between cell vertices.

$$\mathbf{f}_i=-\sum_{j\in E(i)}\mathcal{L}_j\frac{\partial E_j^T}{\partial\mathbf{x}_i}.H_j.E_j=\sum_{j\in E(i)}H_j\left(1-\frac{l_j}{\mathcal{L}_j}\right)\mathbf{u}_{ij} \tag{29}$$

where $\mathbf{u}_{ij}$ is a unit vector of edge j in the direction of node

i, H_j is the spring stiffness and $\mathcal{L}_j, l_j$ the 'grown' and current length of edge j.

Turgor pressure is converted into nodal external forces by spatial integration:

$$\mathbf{f}_i^{ext} = \frac{1}{3} \sum_{j \in T(i)} \Delta P_j S_j \mathbf{N}_j \quad (30)$$

where ΔP_j is the pressure (supposed uniform) on triangle j. $\mathbf{N}_j$ is the outward unit normal (current configuration).

Time evolution

Solving the weak form of Eq. 14, turns out to minimize the total elastic. We solve this by gradient descent:

$$\delta \mathbf{x}_i = -(\mathbf{f}_i + \mathbf{f}_i^{ext})h \quad (31)$$

where h is the step size. For faster convergence and stability, we use an implicit scheme. At iteration k, we have:

$$\mathbf{x}(k+h) - \mathbf{x}(k) = (\mathbf{I} - h\mathbf{K})^{-1}\mathbf{f}(k)h \quad (32)$$

where $\mathbf{K} = \frac{\partial \mathbf{f}}{\partial \mathbf{x}}$ is the stiffness matrix. $\mathbf{x}$ and $\mathbf{f}$ are the position and force vectors of all nodes (concatenation of the $\mathbf{x}_i$ and $(\mathbf{f}_i + \mathbf{f}_i^{ext})$). This linear system is solved using the conjugate gradient algorithm.

Computational Simulation Framework

We have implemented our mechanical model in the open source software SOFA [39]. Its modularity allowed us to combine different element types (triangle and edge elements), forces (elastic forces and turgor) and positional constraints within the same model. At each step of growth, the software is used to find the static elastic equilibrium, Eq. 14 given a current configuration. The use of an implicit integration scheme makes it possible to achieve close-to interactive simulation of growth.

ACKNOWLEDGMENTS

The authors would like to thank Pradeep Das for making available the confocal images of Fig. 5.

REFERENCES

1. Jaeger J, Irons D, Monk N (2008) Regulative feedback in pattern formation: towards a general relativistic theory of positional information. Development 135: 3175–3183.
2. Shraiman BI (2005) Mechanical feedback as a possible regulator of tissue growth. Proceedings of the National Academy of Sciences of the United States of America 102: 3318–3323.
3. Hamant O, Heisler MG, Jönsson H, Krupinski P, Uyttewaal M, et al. (2008) Developmental Patterning by Mechanical Signals in Arabidopsis. Science (New York, NY) 322: 1650–1655.
4. Kutschera U (1991) Regulation of cell expansion. In: Lloyd CW, editor, The Cytoskeletal basis of plant growth and form, Academic Press, Inc. pp. 149–158.
5. Hamant O, Traas J (2010) The mechanics behind plant development. The New phytologist 185: 369–385.
6. Cosgrove DJ (1993) Wall extensibility: its nature, measurement and relationship to plant cell growth. The New phytologist 124: 1–23.
7. Lockhart JA (1965) An analysis of irreversible plant cell elongation. Journal of Theoretical Biology 8: 264–275.
8. Dumais J, Shaw SL, Steele CR, Long SR, Ray PM (2006) An anisotropic-viscoplastic model of plant cell morphogenesis by tip growth. Int J Dev Biol 50: 209–222.
9. Yant L, Mathieu J, Dinh TT, Ott F, Lanz C, et al. (2010) Orchestration of the Floral Transition and Floral Development in Arabidopsis by the Bifunctional Transcription Factor APETALA2. The Plant cell 22: 2156–2170.
10. Schlereth A, Möller B, Liu W, Kientz M, Flipse J, et al. (2010) MONOPTEROS controls embryonic root initiation by regulating a mobile transcription factor. Nature 464: 913–916.
11. Gomez-Mena C (2005) Transcriptional program controlled by the floral homeotic gene AGAMOUS during early organogenesis. Development 132: 429–438.
12. Coen E, Rolland-Lagan AG, Matthews M, Bangham JA, Prusinkiewicz P (2004) The genetics of geometry. Proceedings of the National Academy of Sciences 101: 4728–4735.
13. Green AA, Kennaway JR, Hanna AI, Bangham JA, Coen E (2010) Genetic Control of Organ Shape and Tissue Polarity. PLoS Biol 8: e1000537.
14. Kennaway R, Coen E, Green A, Bangham A (2011) Generation of diverse biological forms through combinatorial interactions between tissue polarity and growth. PLoS computational biology 7: e1002071.
15. Dupuy L, Mackenzie JP, Haseloff JP (2006) A biomechanical model for the study of plant morphogenesis: Coleocheate orbicularis, a 2D study species. In: Proceedings of the 5th Plant Biomechanics Conference, Stockholm, Sweden.
16. Dupuy L, Mackenzie J, Rudge T, Haseloff J (2008) A system for modelling cell-cell interactions during plant morphogenesis. Annals of botany 101: 1255–1265.
17. Merks RMH, Guravage M, Inz e D, Beemster GTS (2011) VirtualLeaf: An Open-Source Framework for Cell-Based Modeling of Plant Tissue Growth and Development. Plant physiology 155: 656–666.
18. Fozard JA, Lucas M, King JR, Jensen OE (2013) Vertex-element models for anisotropic growth of elongated plant organs. Frontiers in plant science 4: 233.
19. Fernandez R, Das P, Mirabet V, Moscardi E, Traas J, et al. (2010) Imaging plant growth in 4D: robust tissue reconstruction and lineaging at cell resolution. Nature methods 7: 547–553.
20. Peaucelle A, Braybrook SA, Le Guillou L, Bron E, Kuhlemeier C, et al. (2011) Pectin-Induced Changes in Cell Wall Mechanics Underlie Organ Initiation in Arabidopsis. Current biology 21: 1720–1726.
21. Milani P, Braybrook SA, Boudaoud A (2013) Shrinking the hammer: micromechanical approaches to

morphogenesis. Journal of experimental botany 64: 4651–4662.

22. Rodriguez EK, Hoger A, McCulloch AD (1994) Stress-dependent finite growth in soft elastic tissues.Journal of biomechanics 27: 455–467.
23. Goriely A, Ben Amar M (2007) On the definition and modeling of incremental, cumulative, and continuous growth laws in morphoelasticity. Biomech Model Mechanobiol 6: 289–296.
24. Menzel A, Kuhl E (2012) Frontiers in growth and remodeling. Mechanics Research Communications 42: 1–14.
25. Mandel J (1973) Equations constitutives et directeurs dans les milieux plastiques et viscoplastiques. International Journal of Solids and Structures 9: 725–740.
26. Baskin TI (2005) Anisotropic expansion of the plant cell wall. Annual review of cell and developmental biology 21: 203–222.
27. La Rota C, Chopard J, Das P, Paindavoine S, Rozier F, et al. (2011) A data-driven integrative model of sepal primordium polarity in Arabidopsis. The Plant cell 23: 4318–4333.
28. Reinhardt D, Pesce ERR, Stieger P, Mandel T, Baltensperger K, et al. (2003) Regulation of phyllotaxis by polar auxin transport. Nature 426: 255–260.
29. Barbier de Reuille P, Bohn-Courseau I, Ljung K, Morin H, Carraro N, et al. (2006) Computer simulations reveal properties of the cell-cell signaling network at the shoot apex in Arabidopsis. Proceedings of the National Academy of Sciences 103: 1627–1632.
30. Paque S, Mouille G, Grandont L, Alabadi D, Gaertner C, et al. (2014) AUXIN BINDING PROTEIN1 Links Cell Wall Remodeling, Auxin Signaling, and Cell Expansion in Arabidopsis. The Plant cell 26: 280–295.
31. Braybrook SA, Peaucelle A (2013) Mechano-Chemical Aspects of Organ Formation in Arabidopsis thaliana: The Relationship between Auxin and Pectin. PloS one 8: e57813.
32. Aida M, Ishida T, Fukaki H, Fujisawa H, Tasaka M (1997) Genes involved in organ separation in Arabidopsis: an analysis of the cup-shaped cotyledon mutant. The Plant cell 9: 841–857.
33. Aida M, Tasaka M (2006) Morphogenesis and Patterning at the Organ Boundaries in the Higher Plant Shoot Apex. Plant Molecular Biology 60: 915–928.
34. Long J, Barton MK (2000) Initiation of Axillary and Floral Meristems in Arabidopsis. Developmental biology 218: 341–353.
35. Heisler MG, Ohno C, Das P, Sieber P, Reddy GV, et al. (2005) Patterns of Auxin Transport and Gene Expression during Primordium Development Revealed by Live Imaging of the Arabidopsis Inflorescence Meristem. Current biology 15: 1899–1911.
36. Sawa S, Watanabe K, Goto K, Liu YG, Shibata D, et al. (1999) FILAMENTOUS FLOWER, a meristem and organ identity gene of Arabidopsis, encodes a protein with a zinc finger and HMG-related domains. Genes & development 13: 1079–1088.
37. van Mourik S, Kaufmann K, van Dijk ADJ, Angenent GC, Merks RMH, et al. (2012) Simulation of organ patterning on the floral meristem using a polar auxin transport model. PloS one 7: e28762.
38. Pradal C, Boudon F, Nouguier C, Chopard J, Godin C (2009) PlantGL: a Python-based geometric library for 3D plant modelling at different scales. Graphical Models 71: 1–21.
39. Faure F, Duriez C, Delingette H, Allard J, Gilles B, et al. (2012) SOFA: A Multi-Model Framework for Interactive Physical Simulation. Soft Tissue Biomechanical Modeling for Computer Assisted Surgery: 283–321.
40. Barbier de Reuille P, Robinson S, Smith RS (2014) Quantifying cell shape and gene expression in the shoot apical meristem using MorphoGraphX. Methods in molecular biology (Clifton, NJ) 1080: 121–134.
41. Sahlin P, Jönsson H (2010) A Modeling Study on How Cell Division Affects Properties of Epithelial Tissues Under Isotropic Growth. PloS one 5: e11750.
42. Besson S, Dumais J (2011) Universal rule for the symmetric division of plant cells. Proceedings of the National Academy of Sciences 108: 6294–6299.
43. Heisler MG, Hamant O, Krupinski P, Uyttewaal M, Ohno C, et al. (2010) Alignment between PIN1 polarity and microtubule orientation in the shoot apical meristem reveals a tight coupling between morphogenesis and auxin transport. PLoS Biol 8: e1000516.
44. Uyttewaal M, Burian A, Alim K, Landrein Bit, Borowska-Wykręt D, et al. (2012) Mechanical stress acts via katanin to amplify differences in growth rate between adjacent cells in Arabidopsis. Cell 149: 439–451.
45. Bozorg B, Krupinski P, Jönsson H (2014) Stress and strain provide positional and directional cues in development. PLoS computational biology 10: e1003410.
46. Wahl V, Brand L, Guo YL, Schmid M (2010) The fantastic four proteins influence shoot meristem size in arabidopsis thaliana. BMC Plant Biology 10: 285.
47. Peters WS, Hagemann W, Tomos AD (2000) What makes plants different? principles of extracellular matrix function in 'soft' plant tissues. Comparative Biochemistry and Physiology Part A: Molecular & Integrative Physiology 125: 151–167.
48. Moore T (1857) The ferns of Great Britain and Ireland. Bradbury, Henry.

CHAPTER 12 3D TREE DIMENSIONALITY ASSESSMENT USING PHOTOGRAMMETRY AND SMALL UNMANNED AERIAL VEHICLES

Demetrios Gatziolis[1], Jean F. Lienard[2], Andre Vogs[2,3], Nikolay S. Strigul[2]

[1]USDA Forest Service, Pacific Northwest Research Station, Portland, Oregon, United States of America,

[2]Department of Mathematics, Washington State University Vancouver, Vancouver, Washington, United States of America,

[3]Intel Inc., Hillsboro, Oregon, United States of America

ABSTRACT

Detailed, precise, three-dimensional (3D) representations of individual trees are a prerequisite for an accurate assessment of tree competition, growth, and morphological plasticity. Until recently, our ability to measure the dimensionality, spatial arrangement, shape of trees, and shape of tree components with precision has been constrained by technological and logistical limitations and cost. Traditional methods of forest biometrics provide only partial measurements and are labor intensive. Active remote technologies such as LiDAR operated from airborne platforms provide only partial crown reconstructions. The use of terrestrial LiDAR is laborious, has portability limitations and high cost. In this work we capitalized on recent improvements in the capabilities and availability of small unmanned aerial vehicles (UAVs), light and inexpensive cameras, and developed an affordable method for obtaining precise and comprehensive 3D models of trees and small groups of trees. The method employs slow-moving UAVs that acquire images along predefined trajectories near and around targeted trees, and computer vision-based approaches that process the images to obtain detailed tree reconstructions. After we confirmed the potential of the methodology via simulation we evaluated several UAV platforms, strategies for image acquisition, and image processing algorithms. We present an original, step-by-step workflow which utilizes open source programs and original software. We anticipate that future development and applications of our method will improve our understanding of forest self-organization emerging from the competition among trees, and will lead to a refined generation of individualtree-based forest models.

INTRODUCTION

Understanding how macroscopic patterns of forests emerge as a result of self-organization of individual plants and how ecosystems respond to environmental gradients and disturbances that occur at different spatial and temporal scales has long been reported as a largely unresolved fundamental ecological challenge[1]. The phenotypic plasticity of individual trees is regarded as the major biological determinant of self-organization, structure, and dynamics of forested ecosystems and their response to natural and anthropogenic disturbances [2,3]. Unique patterns of tree plasticity have been identified across ecological and species groups, for instance, in conifers [4,5,6] and broad-leaf trees [7,8]; and biomes, including tropical [9] and temperate ecosystems [10,11,12]. Failures to predict growth at the individual tree level with acceptable accuracy have been attributed to the heterogeneity in geomorphic and climatic phenomena affecting tree survival and growth, but

Citation: Gatziolis D, Lienard JF, Vogs A, Strigul NS. 3D Tree Dimensionality Assessment Using Photogrammetry and Small Unmanned Aerial Vehicles. Chen HYH, ed. PLoS ONE. 2015;10(9):e0137765. doi:10.1371/journal.pone.0137765.

primarily to inadequate information on the size, shape, and spatial distribution of interacting trees [3].

National Forest Inventory (NFI) systems are a major source of systematic, spatially distributed, and repeated individual tree measurements obtained during field visits of established plots. A review of NFI field protocols and data quality standards reveals that very precise measurements are prescribed for tree stem diameter at breast height, and for fixed-area field plots, distances used in determining whether a tree stem center is within the plot area. Where recorded, the relative position of tree stems within a plot and tree height is measured accurately. Some vegetation parameters such as shrub and forb percent cover, crown base height, and crown compaction ratio are assessed ocularly, and therefore should be regarded more as estimates rather than measurements. Owing to cost, complexity, and logistic constraints such as visibility, crown width and other specialized tree dimensionality measurements are obtained only during special projects.

Information on individual trees over large areas is feasible only via processing of remotely sensed data. High (submeter) resolution space- or airborne spectral imagery has been used to identify and delineate individual tree crowns[13,14,15,16], and to assess parameters of crown morphology such as height, radius, and surface curvature [17,18] using various modeling approaches. Information extracted by manual interpretation of aerial photographs has often been used as surrogate of field measurements for model development and validation [17,19]. The advent of Light Detection and Ranging (LiDAR) technology enabled 3D measurements of vegetation over forested landscapes. Operated mainly from airborne platforms, LiDAR instruments emit short pulses of light that propagate through the atmosphere as a beam of photons and are backscattered to the instrument from illuminated targets. The loci of interactions with objects or object parts along a beam's trajectory are determined with decimeter precision and reported as points georeferenced in three dimensions. The collection of points generated across all pulses is referred to as a point cloud. A typical LiDAR data set of a forested scene comprises points from the entire volume of tree crowns and ground surfaces. Models operating on metrics that describe the spatial distribution of above-ground points have been proven useful for assessing area-based forest inventory parameters such as wood volume and biomass [20,21]. With high-density LiDAR data, a single mature tree can be represented by many, up to hundreds of points, conditions conducive to a precise assessment of its dimensions, including height and crown width [22,23]. Often however, the token representation of lower canopy components and ground surfaces in LiDAR data sets caused by substantial attenuation of pulse energy in dense, multistory stands leads to less accurate estimates of tree dimensionality [24,25]. Terrestrial LiDAR systems operated from ground or near-ground locations deliver point cloud densities orders of magnitude higher than those generated by using airborne instruments, enabling detailed and precise reconstructions of individual trees [26]. Modeling of crown morphology supported by terrestrial LiDAR data has been shown effective in assessing how trees grow in response to competition between and within crowns [27]. Point clouds generated from single scanning locations always contain gaps due to partial target occlusion, either from parts of the targeted tree itself or from surrounding vegetation. As occlusion rates, gap frequency, and gap size increase with canopy height, the error levels in tree dimensionality estimates obtained by processing these point clouds also increase with height [28,29]. Ensuring that estimate precision meets established standards necessitates scanning targeted trees from multiple locations and then fusing the individual point clouds, a complication that often is logistically complex and costly.

To date, precise tree crown dimensionality and location data supportive of a rigorous modeling of individual tree growth has been inhibited by feasibility, logistics, and cost. Measuring crown characteristics by using established inventory methods is very time consuming and hardly affordable outside special projects. Existing remote sensing methods of measuring tree crowns provide only partial crown reconstructions. Airborne LiDAR data acquisitions require prolonged planning and are costly. As an example, the minimum cost for a single airborne LiDAR acquisition with common specifications in the US Pacific Northwest exceeds $20,000 irrespectively of acquisition area size [30]. Transferring to and operation of terrestrial LiDAR instruments in remote forest locations and challenging terrain is both labor intensive and time consuming. As a result, the assessment of tree growth and competition relies on numerous simplifying, albeit often unjustified, assumptions such as of trees with symmetric, vertical, perfectly geometric crowns growing on flat terrain, and illuminated by omnidirectional sunlight. These assumptions propagate through modeling efforts and ultimately reduce the validity of model predictions, thereby decreasing their utility [31,3].

Recently, unmanned aerial vehicles (UAVs) equipped with inexpensive, off-the-shelf panchromatic cameras have emerged as a flexible, economic alternative data source that supports the retrieval of tree dimensionality and location information. Flying at low altitude above the trees and with the camera oriented at a nadir view, UAVs acquire high-resolution images with a high degree of spatial overlap. In such conditions, a point on the surface of a tree crown or a small object on exposed ground is visible from many positions along the UAV trajectory and is depicted in multiple images. Automated photogrammetric systems based on computer Vision Structure from Motion (VSfM) algorithms [32] explore this redundancy to retrieve the camera location the moment an image was acquired, calculate an orthographic rendition of each original image, and ultimately produce a precise 3D point cloud that represents objects [33,34]. Application of VSfM techniques on UAV imagery has enabled accurate 3D modeling of manmade structures, bare ground features, and forest canopies [35,36,37]. Automated image processing is now supported by open-source and commercial software packages.

Image acquisitions with nadir-oriented cameras onboard UAVs, however, face the same issues as airborne imagery; the great majority of points in derived clouds are positioned near or at the very top of tree crowns. The representation of crown sides tends to be sparse and contains sizeable gaps, especially lower in the crown, a potentially serious limitation in efforts to quantify lateral crown competition for space and resources, as in the periphery of canopy openings. In this study, we extend UAV-based image acquisition configurations to include oblique and horizontal camera views and UAV trajectories around trees or tree groups at variable above-ground heights to achieve comprehensive, gap-free representations of trees. To overcome the challenges imposed by these alternative UAV/

camera configurations, we evaluated many UAV platforms and open-source VSfM software options, and developed original, supplementary programs. To determine whether comprehensive tree representations are attainable, we initially processed synthetic imagery obtained via simulation. We finally evaluated the efficacy and performance of our workflow targeting trees of different species, shapes, sizes, and structural complexity.

METHOD DEVELOPMENT AND TESTING

Image Processing

The procedure that uses a set of images exhibiting substantial spatial overlap to obtain a point cloud representing the objects present in the images contains three main steps: feature detection, bundle adjustment, and dense reconstruction. To implement this procedure, we have carefully examined a variety of software available for image processing. The workflow presented below was found by experimentation to be the most efficient for our project. We employed a sequence of computer programs, most of which are available as freeware or provide free licenses to academic institutions. The software used includes OpenCV libraries, VisualSFM, CMVS, SURE, OpenGL, and Mission Planner, with each of them accompanied by a comprehensive manual. Considering that the majority of the software listed above evolves rapidly, we intentionally refrained from duplicating here elements of associated manuals to which we refer a reader in addition to our presentation.

Feature detection is based on the identification of image regions, often called keypoints, pertaining to structural scene elements. Thanks to image overlap, these elements are present in multiple images, but because their position relative to the focal point of the camera is image-specific, they are depicted in different scale and orientation (Fig 1). Illumination differences and image resolution can impose additional feature distortions. Algorithms used in feature detection explore principles of the scale-space theory [38]. According to this theory, a high-resolution image can be perceived as a collection of scene representations, called octaves, in Gaussian scale space. The scale space can be obtained by progressively smoothing the high-resolution image, an operation analogous to a gradual reduction of its resolution. If robust against changes in scale and orientation, the characteristics of a keypoint identified on a given octave of one image can be used to identify the same keypoint on other images. The algorithms proposed for feature detection in this context include the Scale Invariant Feature Transform (SIFT) [39], the Speeded Up Robust Features (SURF) [40], and the Oriented FAST and Rotated BRIEF (ORB) [41]. We employed SIFT in our workflow as it is currently the reference approach in the field of computer vision. To identify keypoints, SIFT initially applies to each image octave an approximation of the Laplacian of Gaussian filter known as Difference of Gaussians, an efficient edge detector. Identified SIFT keypoints are circular image regions, each described by a set of parameters: the image coordinates at the center of the region, the radius of the region and an angle. The radius and angle of each keypoint serve as scale and orientation indicators respectively (Fig 1). Keypoints are further characterized by a descriptor of their neighborhood, determined from the values of pixels in the vicinity of the keypoint's center and usually encoded into a vector of 128 values. By searching for keypoints at multiple scales and positions, SIFT is invariant to image translation, rotation, and rescaling, and partially invariant to affine distortion and illumination changes. It can robustly identify scene features even in images containing substantial amounts of noise or under partial occlusion.

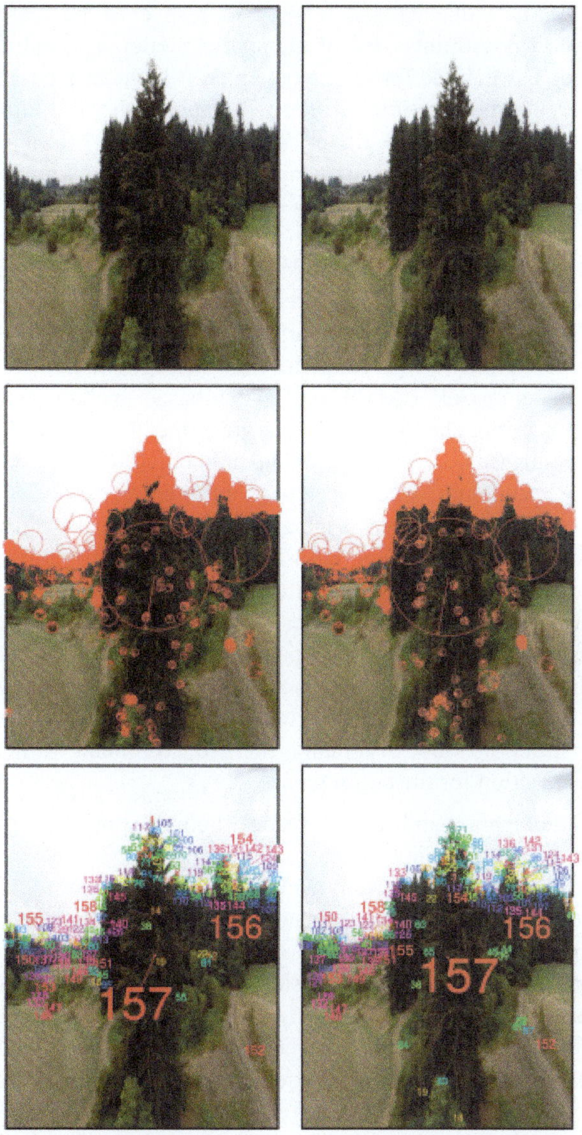

Figure 1: SIFT-based scene keypoint detection and matching on two overlapping images.

The bundle adjustment process initially compares keypoint descriptors identified across images to determine two similar images. Then, an optimization procedure is performed to infer the positions of cameras for these two images. Remaining images are added one at a time with relative positions further adjusted, until camera locations become available for all images. The optimization often uses the Levenberg-Marquardt algorithm [42,43], a general purpose non-linear optimization procedure. Heuristics and prior information, such as GPS coordinates of UAV locations at the moment an image is acquired, can be included to improve convergence speed. In the end, the spatial positions and orientations of all cameras are triangulated using the keypoints identified in the previous step. At the conclusion of bundle adjustment a so-called sparse 3D model that contains the 3D positions of all identified features becomes available. We implemented the feature detection and bundle adjustment components of our workflow in VSFM software [44,45].

In dense reconstruction, the final processing step, all image pixels, not only keypoints, along with the positions and orientations of each camera, are merged into a single high-density structure. This is achieved by matching pixels with similar value across pictures with respect to the epipolar geometry constraints [46] of the sparse model. The epipolar geometry is defined for each image pair. It includes a baseline connecting the locations of the two cameras that are known from the sparse model, the oriented image planes, the image locations where image plane and baseline intersect known as epipoles, and the epipolar lines connecting a camera location with a pixel on the image plane. By restricting searches for a pixel match along the epipolar lines, processing is greatly expedited. In our workflow, we considered CMVS [47] and SURE [48], two state-of-the-art, freely available multi-core implementations, which adopt different strategies to generating the dense model. CMVS is a patch-based method which starts from matched keypoints and generates local models of object surfaces, or patches, in the immediate neighborhood of the keypoints. These patches are then expanded until their projections on the original pictures eventually form a dense tiling. SURE's approach is based on the computation of depth maps for a set of reference images, based on the disparity between these images and other images obtained from nearby, according to the sparse model, positions. Each depth map provides a dense model of pixels equivalent to a local reconstruction from one reference viewpoint. All partial reconstructions are eventually merged to obtain a dense reconstruction for the entire scene.

The sparse and dense reconstructions obtained from a set of overlapping images are configured in the same internal coordinate system and scale. Conversion to real-world orientation and coordinate system is a prerequisite for meaningful measurements of reconstructed objects or for comparisons with ancillary spatial data. Such conversions can be performed manually on the reconstructed scene, assuming reference *in-situ*measurements of object dimensionality are available. In this study, we used an alternative, automated approach. The latitude, longitude, and elevation of camera locations recorded by a recreational-grade GPS device onboard the UAV were converted to orthographic Universal Transverse Mercator (UTM) coordinates using a GDAL reprojection function [49]. The rotation/ translation matrix linking the UTM and sparse model coordinates of the camera positions was then calculated via maximum likelihood, and applied to convert the sparse model coordinates system to UTM. All subsequent processing by CMVS and SURE were performed on the UTM version of the sparse model.

Image Calibration

All imaging systems introduce a variety of distortions onto acquired imagery. The magnitude of the distortion is usually negligible in professional systems, but it can be substantial for inexpensive, off-the-shelf cameras used in structure from motion applications [50]. Most software, including VSfM, perform internal image calibration using information on the focal length of the lens, usually stored in the header of the image, and a generic rectification process, or undistortion as it is commonly called. Departures between the actual distortion and the one anticipated by the generic rectification process reduce the spatial accuracy of reconstructed objects. Using simulated and UAV-based, nadir looking imagery featuring sparse and low vegetation on flat land, Wu [51], the author of the VSfM software, documented that scene reconstructions obtained by using the generic image calibration model present in VSfM produced a macroscopically concave ground surface, an artifact attributed to imprecise image calibration. To avoid artifacts, we first calibrated all cameras used in this study with the efficient procedure described in the OpenCV image processing library [52], and then instructed VSfM to skip the generic image calibration process. Separate calibrations were performed for each operating mode of each camera. As expected, and evident in Fig 2, calibration effects were more discernible near the periphery of the image. The convex scene horizon in the original image appears flat and horizontal after calibration and the local road pavement on the lower left part of the original image is excluded from the calibrated version.

Figure 2: Removal of lens distortion.

Simulation-Based Assessment of Image-Based Tree Reconstruction Accuracy

Upon initial consideration, the accurate and detailed reconstruction of objects characterized by complex structure and geometry, such as trees, using image-based techniques may be deemed an ill-fated effort. The main reason for pessimistic prognoses is that the aforementioned methods and algorithms used in processing the imagery anticipate planar surfaces as structural elements of the objects and well-defined edges at object surface intersections. Except for the lower part of the main stem of large trees, sizeable and homogeneous surfaces separated by crisp boundaries are absent in trees. A second reason is that trees are not opaque objects. Even in high foliage and branch density conditions, portions of scene background are clearly visible through the tree crowns. The see-through-crown phenomenon can be overlooked in nadir-oriented imagery where the forest floor is acting as tree background, but it is often rather pronounced in lateral imagery where the depth of the part of the scene situated behind the trees can be large. The term 'lateral' is used here to describe images acquired with the UAV positioned to the side of the tree and lower than the tree top. The effects of substantial differences in parallax between tree components and background depicted only pixels apart in lateral tree imagery, and high rates of component occlusion, are likely analogous to image distortion, a condition to which the SIFT algorithm is only partially invariant. Furthermore, the upper parts of tree crowns depicted in lateral imagery can have the sky as background instead of the typically darker vegetation or terrain background present in nadir-oriented imagery. Drastic changes in background brightness, for instance, from sky to vegetation and back to sky, behind a given part of a tree crown that appears across multiple overlapping lateral images, influence the red, green, and blue (RGB) values of image pixels corresponding to that crown part. The ensuing variability in pixel values often mimics effects induced by differences in diurnal solar illumination regimes. Illumination variability is another condition to which SIFT is only partially invariant.

We used simulation and synthetic images to evaluate the robustness of our standard workflow to the idiosyncrasies of lateral tree imagery described above. We relied on terrestrial LiDAR data representing a collection of free-standing trees, each scanned from multiple near-ground locations. The scanning was performed in high-density mode with the laser beams distributed in fine horizontal and vertical angular increments (0.4 mrad). Each point in the generated clouds was furnished with RGB values extracted from panchromatic imagery captured by the LiDAR instrument during the scanning. Details on the data acquisition are available in [24,53]. The RGB-colored point cloud of each tree was then visualized in an OpenGL interface [54] with perspective rendering (Fig 3A). In this virtual visualization environment, RGB-colored snapshots of each scene, henceforth referred to as synthetic images, can be obtained without limitations on image number, resolution, amount of spatial overlap, and format type. By specifying the trajectory, orientation, snapshot frequency, and field of view of the virtual camera and also the pixel dimensionality of the OpenGL interface, we can control the scale at which targeted trees, or parts of trees, are represented in the synthetic imagery. The background can be adjusted to resemble the overall scene illumination conditions effective during the acquisition of the terrestrial imagery, including illumination adjustments along azimuth and sun elevation angle gradients. Synthetic images generated by exercising combinations of these options yield very realistic approximations of imagery obtained onboard the UAVs, with the additional advantage that the dimensionality of the objects depicted in the imagery is precisely known. Point clouds generated by processing the synthetic imagery can then be compared to the original terrestrial LiDAR point cloud to evaluate the accuracy and precision of object reconstructions.

Figure 3: 3D reconstruction in simulation.

For our simulations we employed a 2500 by 2000 pixel (5 Mp) virtual camera. The camera was positioned on a circular trajectory centered on the crown of each of the trees depicted in the terrestrial LiDAR point clouds. The camera trajectory was either aligned to a horizontal plane elevated to approximately the vertical middle of the crown, or along a spiral ascent from the 15th to the 85th percentile of tree height (Fig 3A). Camera distance to the nearest part of a crown was between 10 and 15m and scene background was set to black. Between 100 and 250 synthetic images were acquired for each tree and trajectory combination, initially in BMP (bitmap) format and subsequently converted to the Joint Photographic Experts Group (JPEG) format, required by VSFM, using a maximum quality setting in ImageMagick, an open-source software suite (http://www.imagemagick.org). The synthetic imagery for each tree was processed with VSFM using standard settings, and the coordinates of the resulting point clouds generated at the sparse reconstruction stage were converted to the coordinate system of the terrestrial LiDAR data using the locations of the virtual camera known from the simulation settings. Dense reconstructions were obtained by using SURE with standard setting plus an option to ignore synthetic image regions with very low variability in pixel values, as those representing the scene background.

The original Terrestrial LiDAR and dense reconstruction point clouds for each tree were compared in voxel space [55,56]. In this setting, the bounding box of a point cloud is exhaustively partitioned into discrete, equally-sized cubical elements, called voxels. Those voxels containing one or more points are labeled 'filled', all others remain empty. By ensuring that the terrestrial and reconstruction voxel spaces have the same origin and voxel size, we were able to calculate the spatial correspondence of filled voxels between the two clouds and the rates of omission and commission, and identify parts of the voxel space where correspondence is better or worse than in other parts. The size, or resolution, of the voxels was set to 2cm, in response to the angular resolution of the terrestrial LiDAR beams at the mean distance between trees and LiDAR instrument.

UAV Platform Characteristics and Image Acquisition Procedures

After a preliminary evaluation of several commercially available UAV platforms, we focused on an APM:Copter (http://copter.ardupilot.com), a hexacopter rotorcraft (Fig 4), because of its easily modifiable architecture and open source software for flight control. We also used a commercial IRIS quadcopter developed by 3DRobotics (http://3drobotics.com). The components of the customized hexacopter and their purchasing prices are shown in Table 1. Both systems feature gyroscopes and GPS receivers. Compared to systems available in the market, our hexacopter is an inexpensive but versatile configuration whose component acquisition cost is expected to drop substantially in the future as UAV technology evolves and its popularity continues to increase.

Figure 4: Custom built UAV hexacopter used to collect imagery data in this study.

Table 1: Specifications and prices of customized UAV platform used in this study at the time of writing

Component description	March 2015 price ($)
DIJ F550 Hexacopter Frame with 6 motor controllers and brushless motors	200
3D Robotics Pixhawk flight controller. Microprocessor: 32-bit STM32F427 Cortex M4 core with FPU, 168 MHz/256KB RAM/2 MB Flash, 32 bit STM32F103 failsafe co-processor. Sensors: ST Micro L3GD20 3-axis 16-bit gyroscope, ST Micro LSM303D 3-axis14-bit accelerometer / magnetometer, Invensense MPU 6000 3-axisaccelerometer/gyroscope, MEAS MS5611 barometer.	200
3D Robotics GPS with compass	90
915 Mhz telemetry radio and transmitter to controller	30
FrSky receiver	30
Spectrum DX7 transmitter	200
Tarot T-2D brushless camera gimbal	150
GOPRO 3+ Black Edition sport camera	350
LIPO batteries	60

Both UAVs used in this study can be operated either autonomously along a predefined trajectory or manually. The manual flight control requires expertise and continuous line of sight between the system and the operator. Maintaining nearly constant planar and vertical speed and orientation of the onboard camera towards the target is challenging, even for operators with years of experience. Experimentation confirmed that imagery acquired with manual flight control exhibits variable rates of overlap between frames captured sequentially. Smaller components of the targets are sometimes depicted in too few frames or are missing completely, while others appear in an excessive number of frames. For these reasons, it was decided to rely on autonomous flights configured by prior mission planning, and reserve the manual mode only for intervention in the event of an emergency.

Characteristics of the Imaging System

We conducted extended trials with several cameras, including the sport GOPRO 3+ Black Edition (http://gopro.com/), Ilook Walkera (http://www.walkera.com/en/) and Canon PowerShot (http://www.canon.com). The evaluations involved all operating modes offered by each camera, including normal, wide, and superwide zoom settings, as well as acquiring video and then extracting individual frames with post-processing. At the conclusion of the trials, we selected the GOPRO 3+ Black Edition operated in photography mode, and normal, 5 Mp resolution. Acquired frames were stored in JPEG format to the camera's flash card. We rarely achieved event partial tree reconstruction using the alternative settings, likely because of the magnitude of distortion embedded into the imagery.

Mission Planning

The objective of the mission planning phase is to optimize the UAV trajectory, attitude, speed, and were applicable, the view angle of the camera gimbal for image acquisition. The gimbal is a hardware component which allows the orientation of the camera to be modified during the flight relative to the platform. Dynamic, trajectory-location-specific adjustments of camera orientation can be used to ensure that the target is centered on the images, especially when the UAV trajectory is not along a horizontal plane. During mission planning the image acquisition frequency is also considered. After rigorous evaluation of various UAV trajectory templates (Fig 5), we determined that the optimal reconstructions of trees are achieved when sequential images have a field-of-view overlap of approximately 70%. In this configuration, the nominal mean number of images where a part of a targeted tree would be present in is 3.4. Once determined, a trajectory template is centered on the target and scaled so that during the actual flight the mean camera-tree distance, platform speed, and image acquisition frequency will generate images exhibiting the targeted field-of-view overlap. The process is perceptually simple, but technically complex considering that all directional and attitudinal vectors of the UAV have to be converted to instructions passed to the UAV controller. Thankfully, it can be streamlined by using Mission Planner, an open-source software suite developed by Michael Osborne (http://planner.ardupilot.com). Mission Planner relies on user input and georeferenced imagery of the targeted area and tree(s), to establish the geographic (latitude and longitude) coordinates of the UAV's starting and ending position and trajectory. A small set of high-level Mission Planner commands can accomplish even complex trajectory templates. All templates shown in Fig 5 require only 5 commands (Table 2). Our typical setup uses a location positioned in the middle of an open area for both the start and end of the flight. The UAV would initially ascend vertically above its starting location to a pre-specified height, then move horizontally to the beginning of the trajectory, complete it, and finally return to the starting location. In the present development state of our system, it is the user's responsibility to ensure that the designed flight path is free of other objects, an easy to achieve requirement considering the wealth of georeferenced, high resolution, publicly available aerial photographs (Fig 6). The Mission Planer is also used to convert telemetry data of camera locations the moment images were acquired, provided by the GPS receiver stored to the onboard flash memory card, to an accessible format. As detailed in Section 2.1, these locations are later paired to those calculated during the sparse reconstruction processing phase to adjust the scale and georeference of reconstructed objects.

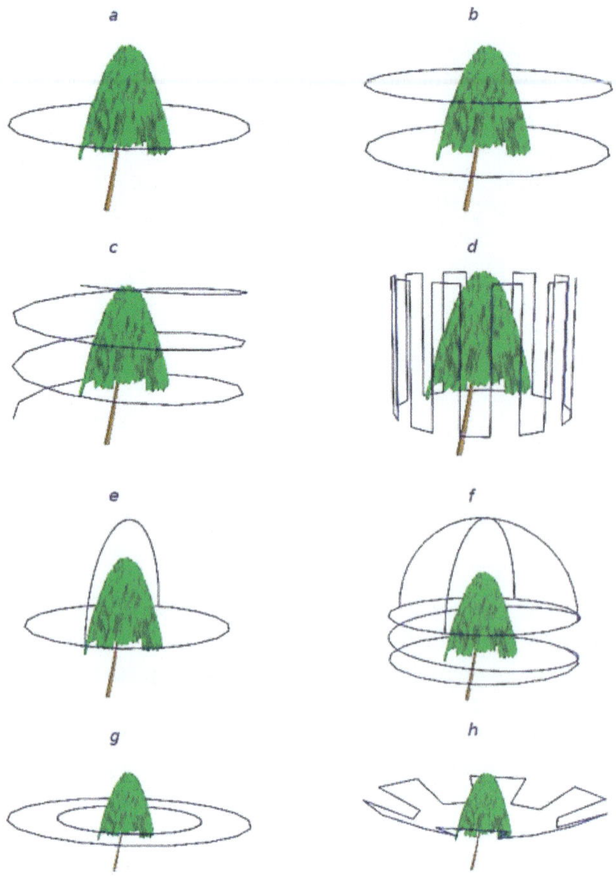

Figure 5: Different UAV trajectories tested for image acquisition.

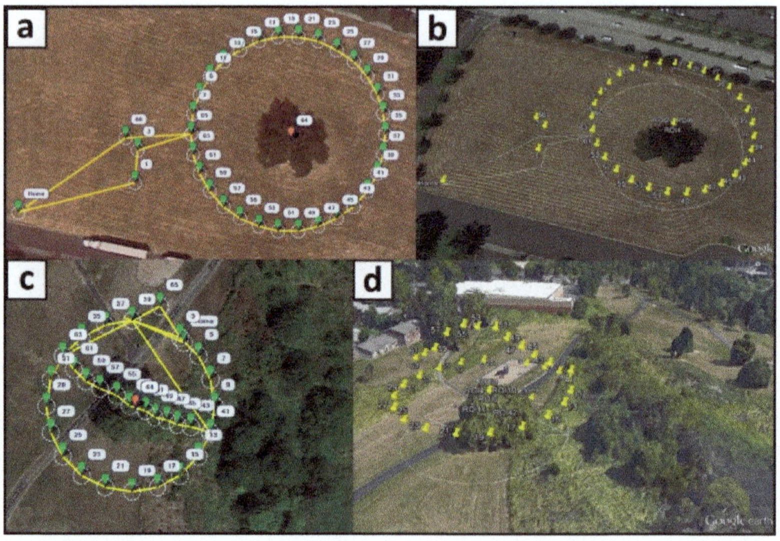

Figure 6: Visualization of designed and accomplished UAV trajectories.

Table 2: Mission Planner commands used for autonomous UAV flights

Command	Code	Description
WAYPOINT	16	Latitude, longitude (in degrees) and altitude vector (in meters) of locations visited during a flight
DO_CHANGE_SPEED	178	Speed, in meters per second. Calculated considering distance to target and image acquisition frequency, usually 2Hz. Typical speed value is 4 meters per second
DO_SET_ROI	201	Vector of UAV heading planar azimuth and gimbal angle (in degrees) that orients the camera towards relative to a specified point of interest.
RETURN_TO_LAUNCH	20	Return to launch location after flight completion
DO_SET_HOME	179	Latitude and longitude vector (in degrees) of return UAV location to use in the event of an emergency, or system anomaly

Evaluation of Tree Reconstructions

Processing of the synthetic imagery always produced complete tree reconstructions. The number of points in the reconstruction ranged between 20 and 25 percent of those present in the original terrestrial LiDAR point cloud (Fig 3B). Larger percentages could be achieved by increasing the resolution of the virtual camera, at the expense of prolonged processing time in both VSfM and SURE. Volumetric comparisons in voxel space revealed excellent agreement between LiDAR and reconstructed point clouds, with a mean of 94 percent of filled voxels collocated. Omnidirectional jittering of the voxel-rendered tree reconstructions relative to the terrestrial LiDAR equivalent always resulted in a substantial, 30 to 40 percent reduction in collocation rates, even when the jittering was limited to a single voxel. The rapid reduction in the collocation rates caused by jittering limited to one voxel suggests that the scaling and translation of the derived point cloud relative to the original terrestrial LiDAR cloud is accurate and precise. It also implies that the coordinates of the virtual camera positions deduced by VSfM during the processing of the synthetic imagery and those used in the simulation are identical up to the scale difference. Once calculated, scaling and translation of the reconstructed point cloud performed by using this relationship rendered the derived tree point cloud a thinned copy of the original terrestrial point cloud. Our simulation results suggest that the absence of planar surfaces and lack of opacity in tree crowns do not impose systemic restrictions to the surface-from-motion approach we used to obtain the 3D tree representations.

By exploring several virtual camera trajectory patterns while altering the image acquisition frequency in each of them, we were able to quantify the effects that different patterns and image field-of-view overlap percentages have on tree reconstruction accuracy (Fig 7). Even in the ideal, noise-free environment of the simulations, a minimum 30 percent image overlap was required for complete target reconstructions. For patterns involving camera locations at variable above-ground heights the minimum percentage was higher, between 35 and 40 percent. Below a mean 45 percent overlap, all simulations were susceptible to failure, pending on the image pair selected for initiating the matching process described in section 2.1. For the circular trajectory pattern, the level of volumetric correspondence between the terrestrial LiDAR and imagery-derived point clouds would increase rapidly at low field-of-view overlap percentages and then progressively decline until reaching an asymptote, usually at 90 percent volumetric correspondence or higher (Fig 7). Complete reconstructions obtained with the spiral trajectory usually required at least 35 percent image overlap. The observed volumetric correspondence to the LiDAR point cloud showed a sigmoidal increase with higher image overlap percentages until reaching an asymptote level, sometimes as high as 94 percent.

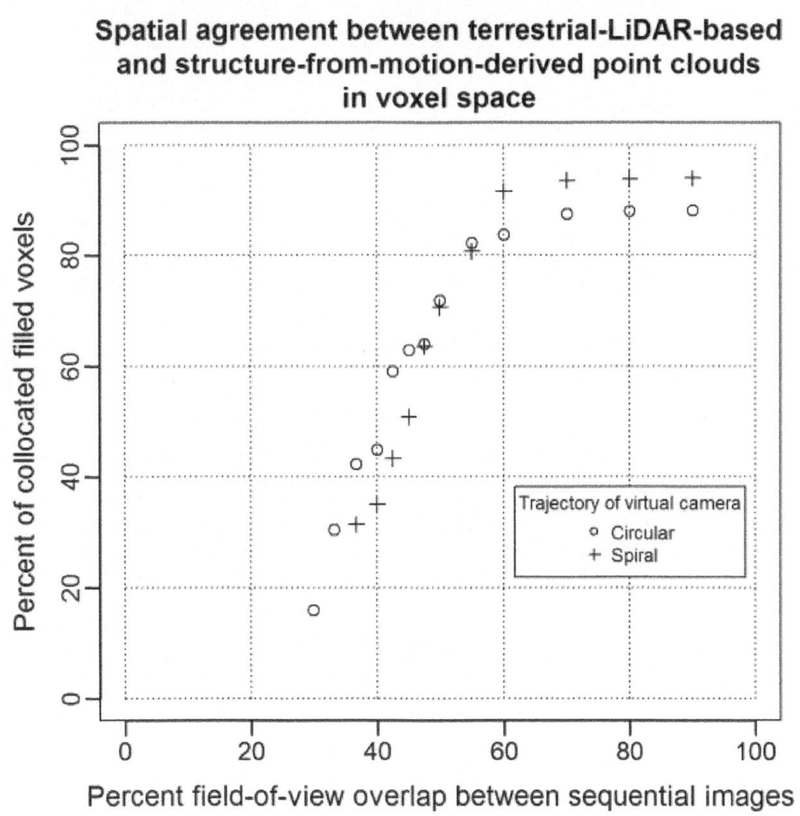

Figure 7: Accuracy and completeness of reconstruction for a *Pinus ponderosa* tree.

In a spiral acquisition trajectory yielding the same number of images of a targeted tree as a circular trajectory, the horizontal overlap percentage between two sequential images is lower. Unlike the circular trajectory, though, in the spiral there is vertical overlap with images obtained after the UAV has completed a rotation around the tree. While the overall mean overlap between the two trajectory patterns was the same in our simulations, the spiral had lower overlap percentage between any two images selected for the initiation of the matching process, and therefore more likely to fail to yield a complete reconstruction when the overall overlap image rate was low. Owing to the vertical image overlap present in spiral

UAV missions, selected parts of the tree are visible from more than one vertical viewing angles, an arrangement that reduces target occlusion rates. For tree species with dense, uniform distribution of foliage and deeply shaded crown centers, the variability in vertical view angles offered by the spiral trajectory pattern may be unimportant. For species with predominantly horizontal or angular branch arrangement and lower crown compaction rates, vertical viewing variability allows internal crown components to be represented adequately in the derived point cloud. Three out of four of the voxels accounting for the approximately 4 percent difference in reconstruction completeness between the spiral and circular UAV trajectories around a Red Pine (*Pinus ponderosa*) tree at 70 percent image overlap rates or higher (Fig 7) were located in the internal half of the crown.

Most UAV flights also produced complete tree reconstructions (Fig 8 and Fig 9).9). In the absence of detailed crown dimensionality measurements, we relied on ocular assessment of reconstruction accuracy and precision. The typical example shown on Fig 8, obtained with the spiral UAV trajectory (Fig 5C), among our most reliable for complete target reconstruction, shows that even the shaded components of the tree crown interior are represented. Many parts on the upper quarter of the crown have a light blue hue inherited from the sky background in corresponding UAV images. Although less evident, selected parts of the lower crown exhibit similar ground-influenced coloring. The coloring artifacts shown in Fig 8 appear where the image area occupied by an identified keypoint is dominated by a uniformly colored background. Sometimes these anomalies are limited to the RGB values assigned to points and can be overlooked if the main objective of the UAV mission is the retrieval of tree dimensionality. Often though they represent an overestimation of tree crown volume and must be removed (Fig 10). Accomplishing this task with manual intervention is laborious and subjective. The task can be easily automated for points pertaining to a sky background thanks to their markedly different RGB values compared to those of vegetation. Where suitable RGB value thresholds cannot be safely identified, as it is usually the case for the lower parts of trees, we found it useful to trim the depth of the part of the overall reconstructions that is derived from each image, so that only the portion nearer the camera position is retained. SURE facilitates this procedure by providing a separate dense reconstruction for each processed image organized in a common coordinate system. The complete reconstruction can be obtained by merging the trimmed parts. In the absence of precise reference data, we were unable to determine quantitatively the significance of these artifacts.

Figure 8: Orthographic horizontal view of reconstructed point cloud and UAV-based oblique perspective image.

Figure 9: Illustration of comprehensive tree reconstructions (right column) and reference UAV-based images (left column).

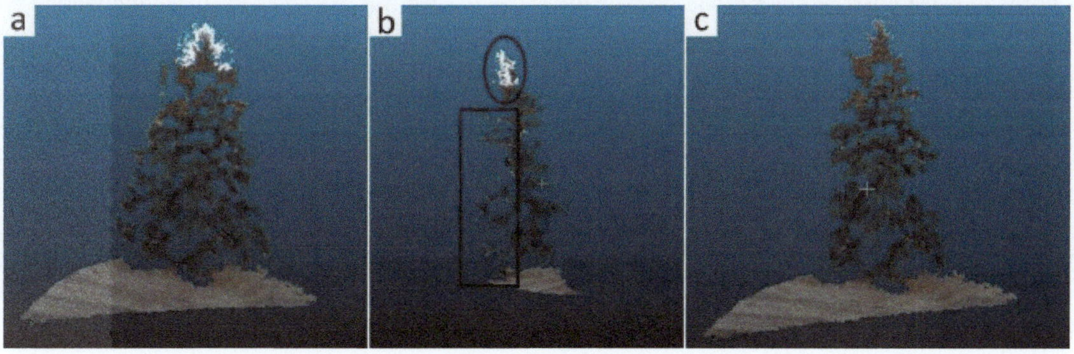

Figure 10: Demonstration of artifacts in the 3D tree reconstruction pertaining to a single UAV image.

The 'nested circle' and 'jagged saucer' trajectories (Fig 5G and 5H) produced only partial reconstructions and several disjointed models in VSfM and are, therefore, not recommended, while the altitude variability in the 'meandering' trajectory (Fig 5D) was often responsible for premature mission termination owing to rapid depletion of the UAV batteries. Partial reconstructions were the norm, rather than the exception, when for a portion of the mission the camera was positioned directly against the sun. In such conditions the shaded portion of the crown would either not be reconstructed at all, or it would be organized in separate 3D models with much lower point density and sizable gaps. In the example shown in Fig 11, the GPS recorded and process-derived positions of the camera on board the UAV show a nearly perfect correspondence for three quarters of the circular UAV trajectory. GPS recordings are half as many as the camera positions because of limitations in the recording frequency of the GPS device. Is should be noted that pending on the hardware configuration of the UAV and the number of peripheral devices connected to it, it is sometimes necessary to operate below the capacity of a particular device to either conserve energy, or to avoid overwhelming the UAV controller. Based on our experience, a close fit between recorded and derived camera positions practically guarantees that a complete target representation will be obtained during the dense reconstruction phase. The remaining part of the trajectory, where the camera is positioned against the sun, was actually derived from a separate model and shows a poor fit, resembling more of a linear transect than a circular arc. As the camera moves from partially to completely against the sun, image contrast is reduced, and the radii of identified keypoints become smaller. Radius reductions increase the uncertainty associated with keypoints orientation and descriptor. We suspect that changes in the magnitude of the mean image keypoint radius are manifested as variability in the distance between the tree and calculated camera locations, evident in the misfit part of the VSfM-derived camera trajectory shown in Fig 11.

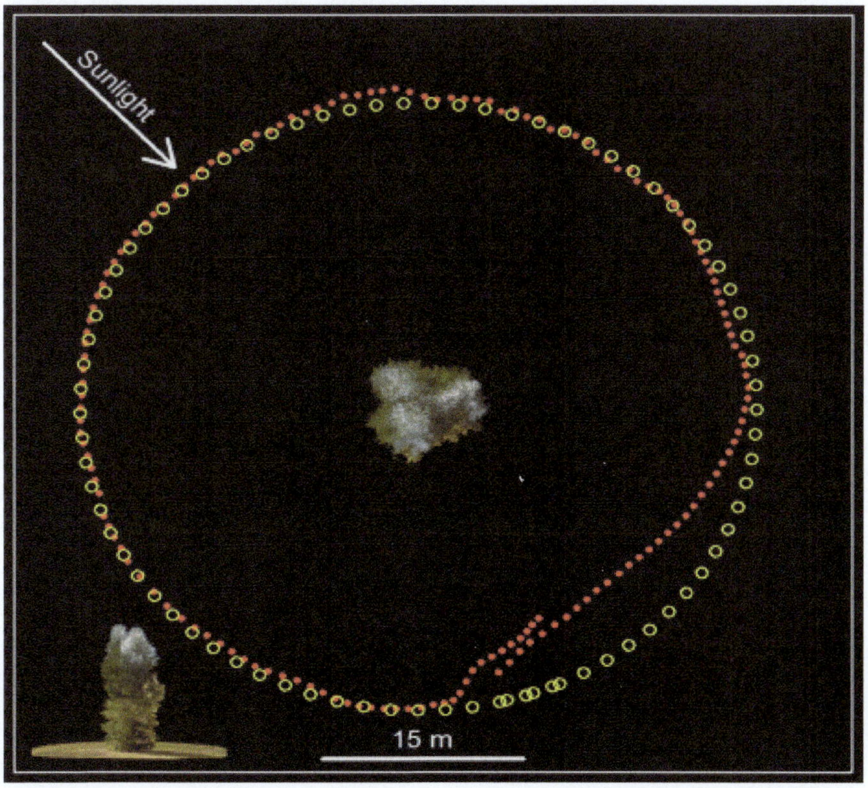

Figure 11: Comparison between real and reconstructed trajectory.

On a few occasions, we observed more than one, nearly parallel, and closely stacked layers of points representing the ground, likely an artifact of texture uniformity in those parts of the scene. The use of calibrated imagery has expedited the computations for identifying camera positions and for generating the sparse reconstructions in VSfM and has reduced the rate of partial reconstruction occurrence. However, its effect on the accuracy of the reconstruction obtained using SURE was unclear.

DISCUSSION

Our results indicate that a meticulously planned image acquisition mission, namely a judicious selection of flight trajectory, UAV speed, and image acquisition frequency, will deliver a comprehensive dense reconstruction of targeted vegetation, except perhaps in unfavorable sun illumination and wind conditions. As explained in section 2.1, our workflow relies on keypoints, most of which are identified along image discontinuities. A smooth flight trajectory around the target ensures that sequential images contain an adequate number of similar keypoints from which the camera location effective for each image capture can be calculated with adequate precision. Where the smooth change in the field of view between two sequential images is interrupted, the offending image becomes the first in a separate model. Bundle adjustments can reduce the frequency of separate model emergence but they cannot eliminate it. The often advocated practice of adding to a model image frames originally put by VSfM to a separate model without performing bundle adjustment after each frame addition may be warranted for manmade objects but is not recommended for trees because it leads to obvious reconstruction artifacts. Mission plans for flights expected to occur during bright solar illumination conditions using gimbal-equipped UAVs could be adjusted to avoid camera positioning directly against the sun. This can be accomplished by specifying a slightly downward, oblique camera orientation. The precise solar elevation angle and azimuth for any location can be obtained from the NOAA Solar Position Calculator (http://www.esrl.noaa.gov/gmd/grad/solcalc/azel.html), or can be computed as described in [57].

GPS-equipped UAV platforms not only enable preprogrammed navigation, but also, and perhaps equally importantly, can be used for a precise scaling of reconstructed tree point clouds to actual dimensions. The GPS receivers placed on the two UAVs employed in this study offer recreational grade precision, and as such, their individual position recordings may contain an absolute error of a few meters. In our trials, however, the relative error between trajectory recordings appeared to always be less than a meter, in most cases about half a meter. This is based on the observation that our UAVs, initially placed on a launch pad measuring about 60 cm on each side, would return at the completion of the mission with their landing gear partially on the launch pad. Fitting the VSfM-calculated camera locations to corresponding GPS recordings containing a relative positional error of such magnitude, would yield point cloud scaling errors of 0.5 percent or lower, a level deemed adequate for UAV imagery and structure from motion based assessment of yearly tree growth. In the absence of GPS recordings, the scaling of the point cloud would have to be performed manually using georeferenced imagery.

Except for extremes in solar illumination conditions such as sun facing camera exposures or at dusk, disparities in light distribution may actually be beneficial for structure-from-motion-based applications in natural environments because they accentuate feature edges. As it is evident in the tree portion between the red and purple colored arrows shown in Fig 8, crown parts in the penumbra are still represented, albeit with reduced point density. Image enhancements focusing on shaded or very bright parts could perhaps be used to ameliorate the direct sunlight effects or improve the reconstruction density for shaded areas.

To account for absolute GPS receiver and ancillary imagery registration errors, current UAV missions must be planned with adequate clearance from any scene objects. We were able to comply with this requirement in our trials because we mostly targeted individual trees or small groups of trees growing in open space. Extending our operations to confined areas, for instance descending into and proceeding near and along the periphery of forest openings, would require much higher navigation precision. Thankfully, obstacle avoidance has been actively researched and several solutions specific to forested environments have been proposed [58,59,60,61,62]. In particular, [62] demonstrated full flight control in forested environments using an UAV platform similar to ours. They used a low-resolution camera mounted on a quadcopter that was outsourcing via a wireless connection all computationally intensive image processing to a ground station, a standard laptop computer. Using this setup, they were able to achieve a constant speed of 1.5 meters per second while avoiding trees. The rapidly expanding onboard processing capabilities of UAVs suggest the possibility, in the near future, of coupling the 3D reconstruction methodology proposed here with autonomous flight, thereby eliminating the need for meticulous mission planning. Extensions of our approach could involve real-time modeling of terrain elevation [63] or embedded recognition of location [64,65]. The shape of reconstructed trees could be utilized in view-based 3D object retrieval [66,67].

It is often tempting to acquire images with the highest possible frequency and maximum overlap. Action cameras similar to those used in this study support high frame rates and carry ample image storage space without affecting the payload and thus compromising the UAV's flight duration or mission flexibility anyway. Large number of images though requires prolonged processing. Our simulations indicate that image field-of-view overlap higher than 70 percent, does not improve the accuracy or completeness of tree reconstructions. Visual assessments suggest that this is also true for actual UAV imagery. Mission planning designed so that target features are represented in three to four images likely maximizes the information content present in an acquisition and it is therefore recommended as an initial mission configuration.

CONCLUSION

Rapid developments in UAV technology and enhancements in structure from motion software have enabled detailed representation of manmade objects. In this paper, we describe how this technology can inexpensively be extended to representations of natural objects, such as trees or groups of trees. After extensive experimentation that involved several UAV platforms, cameras, mission planning alternatives, processing software, and numerous procedural modifications and adjustments, our workflow has been proven capable of handling most conditions encountered in practice to deliver detailed reconstruction of trees. In addition to robust performance, our imaging system can be employed rapidly in support of time-sensitive monitoring operations as, for instance, the assessment of forest fire damage or progress of forest recovery from disturbance. It is also well suited to providing tree dimensionality data through time, a prerequisite for improved models of tree growth and for an accurate assessment of tree competition and morphological plasticity.

REFERENCES

1. Levin SA (1999) Fragile dominion: complexity and the commons. Perseus Publishing, Cambridge, MA.
2. Strigul NS, Pristinski D, Purves D, Dushoff J, Pacala SW (2008) Scaling from trees to forests: Tractable macroscopic equations for forest dynamics. Ecol Monogr 78:523–545.
3. Strigul NS (2012) Individual-Based Models and Scaling Methods for Ecological Forestry: Implications of Tree Phenotypic Plasticity, Sustainable Forest Management, Dr. Julio J. Diez (Ed.), InTech, Croatia, pages 359–384, http://dx.doi.org/10.45772/29590.
4. Loehle C (1986) Phototropism of whole trees: Effects of habitat and growth form. *Am Midl* Nat 116: 190–196.
5. Umeki K (1995) A comparison of crown asymmetry between Picea abies and Betula maximowicziana.Can J For Res 25:1876–1880.
6. Stoll P, Schmid B (1998) Plant foraging and dynamic competition between branches of Pinus sylvestris in contrasting light environments. J Ecol 86:934–945.
7. Woods FW, Shanks RE (1959) Natural replacement of chestnut by other species in the Great Smoky Mountains National Park. Ecology 40:349–361.
8. Brisson J (2001) Neighborhood competition and crown asymmetry in Acer saccharum. Can J For Res31:2151–2159.
9. Young TP, Hubbell SP (1991) Crown asymmetry, tree falls, and repeat disturbance in a broad-leaved forest. Ecology 72:1464–1471.
10. Gysel LW (1951) Borders and openings of beech-maple woodlands in southern Michigan. J For 49:13–19.
11. Frelich LE Martin GL (1988) Effects of crown expansion into gaps on evaluation of disturbance intensity in northern hardwood forests. For Sci 34: 530–536.
12. Webster CR, Lorimer CG (2005) Minimum opening sizes for canopy recruitment of midtolerant tree species: A retrospective approach. Ecol Appl 15:1245–1262.
13. Wulder M, Niemann KO, Goodenough DG (2000) Local maximum filtering for the extraction of tree locations and basal area from high spatial resolution imagery. Rem Sens Env 73:103–114.
14. Leckie DG, Gougeon FA, Tinis S, Nelson T, Burnett CN, Paradine D (2005) Automated tree recognition in old growth conifer stands with high resolution digital imagery. Rem Sens Env 94(3):311–326.
15. Hirschmugl M, Ofner M, Raggam J, Schardt M (2007) Single tree detection in very high resolution remote sensing data. Rem Sens Env 110(4): 533–544.
16. Skurikhin AN, Garrity SR, McDowell NG, Cai DM (2013) Automated tree crown detection and size estimation using multi-scale analysis of high-resolution satellite imagery. Rem Sens Lett 4(5):465–474, doi:10.1080/2150704X.2012.749361
17. Gong P, Sheng Y, Biging G (2002) 3D model-based tree measurement from high-resolution aerial imagery. Photogrammetric Engineering & Remote Sensing, 68(11): 1203–1212.
18. Song C (2007) Estimating tree crown size with spatial information of high resolution optical remotely sensed imagery. Int J Rem Sens 28(15): 3305–3322, doi: 10.1080/01431160600993413
19. Coulston JW, Moisen GG, Wilson BT, Finco MV, Cohen WB, Brewer CK (2012) Modeling percent tree canopy cover: a pilot study. Photogram Eng Rem Sens 78(7): 715–727.
20. Zhao K, Popescu SC, Nelson RF (2009) Lidar remote sensing of forest biomass: A scale-invariant estimation approach using airborne lasers. Rem Sens Env 113(1):182–196.
21. Sheridan RD, Popescu SC, Gatziolis D, Morgan CLS, Ku N- W (2015) Modeling forest aboveground biomass and volume using airborne LiDAR metrics and Forest Inventory and Analysis data in the Pacific Northwest. Rem Sens 7:229–255.
22. Popescu SC, Wynne RH, Nelson RF (2003) Measuring individual tree crown diameter with lidar and assessing its influence on estimating forest volume and biomass Can J Rem Sens 29(5): 564–577.
23. Andersen H- E, Reutebuch SE, McGaughey RJ (2006) A rigorous assessment of tree height measurements obtained using airborne lidar and conventional field methods. Can J Rem Sens 32(5): 355–366.
24. Gatziolis D, Fried JS, Monleon V (2010) Challenges to estimating tree-height via LiDAR in closed-canopy forests: a parable from western Oregon. For Sci 56(2):139–155.
25. Korpela I, Hovi A, Morsdorf F (2012) Understory trees in airborne LiDAR data—Selective mapping due to transmission losses and echo-triggering mechanisms. Rem Sens Env 119: 92–104.
26. Côté J-F, Widlowski J-L, Fournier RA, Verstraete MM (2009) The structural and radiative consistency of three-dimensional tree reconstructions from terrestrial lidar. Rem Sens Env 113(5): 1067–1081.
27. Metz J, Seidel D, Schall P, Scheffer D, Schulze E-D, Ammera C (2013) Crown modeling by terrestrial laser scanning as an approach to assess the effect of aboveground intra- and interspecific competition on tree growth. Forest Ecol Manag 310: 275–288.
28. Henning JG, Radtke PJ (2006) Detailed stem measurements of standing trees from ground-based scanning Lidar. For Sci 52(1): 67–80.

29. Maas H-G, Bienert A, Scheller S, Keane E (2008) Automatic forest inventory parameter determination from terrestrial laser scanner data. Int J Rem Sens 29(5): 1579–1593, doi: 10.1080/01431160701736406

30. Erdody TL, Moskal LM (2010) Fusion of LiDAR and imagery for estimating forest canopy fuels. Rem Sens Env 114: 725–737.

31. Munro DD (1974) Forest Growth Models: A Prognosis, in Fries J. (ed.), Growth Models for Tree and Stand Simulation, Dept. of Forest Yield Research, Royal College of Forestry, Stockholm, Res. Notes Vol.30.

32. Snavely N, Seitz S, Szeliski R (2008) Modeling the world from internet photo collections. Int J Comp Vis 80: 189–210.

33. Dandois JP, Ellis EC (2013) High spatial resolution three-dimensional mapping of vegetation spectral dynamics using computer vision. Rem Sens Env 136: 259–276.

34. Rosnell T, Honkavaara E (2012) Point cloud generation from aerial image data acquired by a quadrocopter type micro unmanned aerial vehicle and a digital still camera. Sensors, 12: 453–480.[PMC free article]

35. de Matías J, Sanjosé JJ de, López-Nicolás G, Sagüés C, Guerrero JJ (2009) Photogrammetric methodology for the production of geomorphologic maps: Application to the Veleta Rock Glacier (Sierra Nevada, Granada, Spain). Rem Sens 1: 829–841.

36. Dandois JP, Ellis EC (2010) Remote sensing of vegetation structure using computer vision. Rem Sens 2: 1157–1176.

37. Dey D, Mummert L, Sukthankar R (2012) Classification of plant structures from uncalibrated image sequences. 2012 IEEE Workshop on Applications of Computer Vision (WACV) pp. 329–336.

38. Lindeberg T (1998) Feature detection with automatic scale selection. Int J Comp Vis 30(2): 79–116.

39. Lowe DG (2004) Distinctive image features from scale-invariant keypoints,» Int J Comp Vis, 60(2): 91–110.

40. Bay H, Ess A, Tuytelaars T, Gool LV (2008) SURF: Speeded Up Robust Features. Comp Vis Image Understanding 110(3): 346–359.

41. Rublee E, Rabaud V, Konolige K, Bradski G (2011) ORB: an efficient alternative to SIFT or SURF. In*Computer Vision (ICCV), 2011 IEEE International Conference on Computer Vision (ICCV)*, pp. 2564–2571, doi: 10.1109/ICCV.2011.6126544

42. Levenberg K (1944) A method for the solution of certain non-linear problems in least squares. Quart Appl Math 2: 164–168.

43. Marquardt D (1963) An algorithm for least-squares estimation of nonlinear parameters. SIAM J Appl Math 11(2): 431–441 doi: 10.1137/0111030

44. Wu C (2013) Towards linear-time incremental structure from motion. In *Proc 2013 IEEE International Conference on 3D Vision-3DV*, pp 127–134.

45. Wu C, Agarwal S, Curless B, Seitz SM (2011) Multicore bundle adjustment. In *Proc 2011 IEEE Conference on Computer Vision and Pattern Recognition (CVPR)*, pp. 3057–3064.

46. Zhang Z, Deriche R, Faugeras O, Luong Q- T (1995) A robust technique for matching two uncalibrated images through the recovery of the unknown epipolar geometry. Artificial Intelligence 78(1–2):87–119 doi:10.1016/0004-3702(95)00022-4

47. Furukawa Y, Ponce J (2010) Accurate, dense, and robust multiview stereopsis. IEEE Transcactions on Pattern Analysis and Machine Intelligence 32(8), 1362–1376, doi: 10.1109/TPAMI.2009.161

48. Rothermel M, Wenzel K, Fritsch D, Haala N (2012) SURE: Photogrammetric Surface Reconstruction from Imagery. In *Proc LC3D Workshop*, Berlin Germany, December 2012.

49. GDAL (2015) GDAL—Geospatial Data Abstraction Library: Version 1.11.2, Open Source Geospatial Foundation, http://gdal.osgeo.org.

50. Balletti C, Guerra F, Tsioukas V, Vernier P (2014) Calibration of Action Cameras for Photogrammetric Purposes. Sensors, 14(9): 17471–17490; doi: 10.3390/s140917471 [PMC free article]

51. Wu C (2014) Critical configurations for radial distortion self-calibration. In *Proc IEEE Conference on Computer Vision and Pattern Recognition, New York.*

52. Bradski G (2000) The OpenCV Library. Dr. Dobb's Journal of Software Tools. Available:http://docs.opencv.org/doc/tutorials/calib3d/camera_calibration/camera_calibration.html. Accessed 23 March 2015.

53. Gatziolis D, Popescu SC, Sheridan RD, Ku N-W (2010) Evaluation of terrestrial LiDAR technology for the development of local tree volume equations. Proc. SilviLaser 2010—The 10th International Conference on LiDAR Applications for Assessing Forest Ecosystems, Koch, B., Kändler, G., & Teguem, C. (eds), Freiburg, Germany, 14–17 September, p. 197–205.

54. Shreiner D (2009) OpenGL programming guide: The official guide to learning OpenGL, versions 3.0 and 3.1 (7th ed.). Addison-Wesley Professional.

55. Popescu SC, Zhao K (2008) A voxel-based lidar method for estimating crown base height for deciduous and pine trees. Rem Sens Env 112(3): 767–781.

56. Gatziolis D (2012) Dynamic, LiDAR-based assessment of lighting conditions in Pacific Northwest forests. In *Proceedings of Silvilaser 2012—The 12th International Conference 'First return'*, p. 16–19.

57. Reda I, Andreas A (2008) Solar Position Algorithm for Solar Radiation Applications. Technical Report NREL/TP-560-34302, National Renewable Energy Laboratory, Golden, Colorado, 56 pp.

58. Frew EW, Langelaan J, Joo S (2006) Adaptive receding horizon control for vision-based navigation of small unmanned aircraft. In *Proc IEEE American Control Conference*, 6 pp.

59. Karaman S, Frazzoli E (2012) High-speed flight in an ergodic forest. In *Proc Robotics and Automation (ICRA), 2012 IEEE International Conference*, pp. 2899–2906.

60. Mori T, Scherer S (2013) First results in detecting and avoiding frontal obstacles from a monocular camera for micro unmanned aerial vehicles. In *Proc Robotics and Automation (ICRA), 2013 IEEE International Conference*, pp. 1750–1757.

61. Roberts R, Ta D-N, Straub J, Ok K, Dellaert F (2012) Saliency detection and model-based tracking: a two part vision system for small robot navigation in forested environment. In *Proc SPIE 8387, Unmanned Systems Technology XIV*; doi: 10.1117/12.919598, pp 12.

62. Ross S, Melik-Barkhudarov N, Shankar K Sh, Wendel A, Dey D, Bagnell JA, et al. (2013) Learning monocular reactive UAV control in cluttered natural environments. In *Proc Robotics and Automation (ICRA), 2013 IEEE International Conference on*, pp. 1765–1772.

63. Luo Y, Guan T, Wei B, Pan H, Yu J (2015) Fast terrain mapping from low altitude digital imagery. Neurocomputing, 156, 105–116.

64. Guan T, He Y, Gao J, Yang J, Yu J (2013) On-device mobile visual location recognition by integrating vision and inertial sensors. Multimedia, IEEE Transactions on, 15(7), 1688–1699.

65. Guan T, He Y, Duan L, Yang J, Gao J, Yu J (2014) Efficient BOF generation and compression for on-device mobile visual location recognition. MultiMedia, IEEE, 21(2), 32–41.

66. Gao Y, Wang M, Zha Z J, Shen J, Li X, Wu X (2013) Visual-textual joint relevance learning for tag-based social image search. Image Processing, IEEE Transactions on, 22(1), 363–376.

67. Gao Y, Wang M, Ji R, Wu X, Dai Q (2014) 3D object retrieval with Hausdorff distance learning.Industrial Electronics, IEEE Transactions on, 61(4), 2088–2098.

CHAPTER 13

ANALYSIS OF LIPID EXPERIMENTS (ALEX): A SOFTWARE FRAMEWORK FOR ANALYSIS OF HIGH-RESOLUTION SHOTGUN LIPIDOMICS DATA

Peter Husen[1], Kirill Tarasov[2], Maciej Katafiasz[1], Elena Sokol[1], Johannes Vogt[3], Jan Baumgart[3], Robert Nitsch[3], Kim Ekroos[2], Christer S. Ejsing[1]

[1]Department of Biochemistry and Molecular Biology, University of Southern Denmark, Odense, Denmark,

[2]Zora Biosciences Oy, Espoo, Finland,

[3]Institute for Microscopic Anatomy and Neurobiology, Johannes Gutenberg University Mainz, Mainz, Germany

ABSTRACT

Global lipidomics analysis across large sample sizes produces high-content datasets that require dedicated software tools supporting lipid identification and quantification, efficient data management and lipidome visualization. Here we present a novel software-based platform for streamlined data processing, management and visualization of shotgun lipidomics data acquired using high-resolution Orbitrap mass spectrometry. The platform features the ALEX framework designed for automated identification and export of lipid species intensity directly from proprietary mass spectral data files, and an auxiliary workflow using database exploration tools for integration of sample information, computation of lipid abundance and lipidome visualization. A key feature of the platform is the organization of lipidomics data in "database table format" which provides the user with an unsurpassed flexibility for rapid lipidome navigation using selected features within the dataset. To demonstrate the efficacy of the platform, we present a comparative neurolipidomics study of cerebellum, hippocampus and somatosensory barrel cortex (S1BF) from wildtype and knockout mice devoid of the putative lipid phosphate phosphatase PRG-1 (plasticity related gene-1). The presented framework is generic, extendable to processing and integration of other lipidomic data structures, can be interfaced with post-processing protocols supporting statistical testing and multivariate analysis, and can serve as an avenue for disseminating lipidomics data within the scientific community. The ALEX software is available at www.msLipidomics.info.

INTRODUCTION

The lipidome of eukaryotic cells comprises hundreds to thousands of molecular lipid species that constitute and functionalize biomembranes, store metabolic energy in lipid droplets and function as signaling molecules that control cell and organism physiology [1–3]. A key tenet of contemporary mass spectrometry-based lipidomics methodology revolves around the identification and quantification of lipid species on a lipidome-wide scale [4–8]. As such, shotgun lipidomics has emerged as a powerful tool for global lipidome analysis that complements mechanistic studies of lipid metabolism, lipid homeostasis and membrane biology [9–13]. The efficacy of shotgun lipidomics stems from its relative technical simplicity where hundreds of lipid species in sample extracts can be quantitatively monitored at high throughput using direct infusion nanoelectrospray ionization combined with high-resolution Fourier transform mass spectrometry (FT MS) or/

Citation: Husen P, Tarasov K, Katafiasz M, et al. Analysis of Lipid Experiments (ALEX): A Software Framework for Analysis of High-Resolution Shotgun Lipidomics Data. Martens L, ed. PLoS ONE. 2013;8(11):e79736. doi:10.1371/journal.pone.0079736.

and tandem mass spectrometry (MS/MS) [2,14]. Notably, lipidomics analysis on a global scale generates large amounts of (spectral) data that requires software routines for automated lipid identification and quantification, and additional data management for subsequent lipidome visualization and bioinformatics analysis.

Extensive lipidome characterization by shotgun lipidomics can be achieved by executing a systematic program of mass spectrometric analyses of sample extracts in positive and negative ion mode, and by incorporating chemical derivatization procedures to specifically monitor poorly ionizing lipid molecules such as cholesterol [4–7]. Executing such an analytical program generates several mass spectral data files per sample that must be queried for lipid identification, and combined into a single dataset for lipidome quantification and visualization. Numerous software tools have been developed for the identification of lipids: LipidQA[15], LIMSA[16], FAAT[17], lipID[18], LipidSearch[19], LipidView[20], LipidInspector[21] and LipidXplorer[22]. These tools cover a broad range of applications spanning dedicated lipid identification for only certain instrumentations and specific mass analysis routines (MS and MS/MS) to cross-platform software featuring user-specified commands querying spectral data in the open-source .mzXML format. Identified lipid species are typically annotated by a shorthand nomenclature corresponding to the information detail of the mass spectrometric analysis [23,24]. The detection of lipid species by FT MS analysis or by MS/MS analysis for lipid class-specific fragment ions (e.g. *m/z* 184.0733 for phosphatidylcholine (PC) species) supports only "sum composition" annotation (e.g. PC 34:1). In comparison, annotation by the more detailed "molecular composition" (e.g. PC 16:0-18:1) requires MS/MS analysis and detection of structure-specific fragment ions [25]. To support the cataloging of lipid species, the LIPID MAPS Consortium recently developed the "Comprehensive Classification System for Lipids" which outlines an informatics framework for lipidomics [26,27]. Using a classification system enables the design of lipid databases where each lipid species is listed together with a range of accessory lipid features such as lipid category (e.g. glycerophospholipid, sphingolipid, glycerolipid, sterol lipid), lipid class, structural attributes (e.g. number of double bonds, fatty acid chain length), chemical formula, mono-isotopic mass and isotope information. These accessory lipid features can be incorporated into lipidomic data processing routines using database-orientated exploration tools to support computations and visualization of distinct lipidome hallmarks. Notably, none of the currently available software tools comprise streamlined processing routines that integrate lipid intensity data, the accessory lipid features and a full catalog of sample information.

Here we present a platform for processing, management and visualization of high-content shotgun lipidomics datasets acquired using high-resolution Orbitrap mass spectrometry. The platform features a novel software framework termed ALEX (Analysis of Lipid Experiments) that supports automated identification and export of lipid species intensity directly from proprietary mass spectral data files and the integration of accessory lipid features and sample information into a single output structured in "database table format". This design supports swift data processing and lipidome visualization across large sample sizes using an auxiliary workflow powered by the database exploration tools: Orange [28] and Tableau Software. To demonstrate the efficacy of the platform, we present a comparative neurolipidomics analysis of cerebellum, hippocampus and S1BF from wild-type and knockout mice devoid of the PRG-1 gene encoding a putative lipid phosphate phosphatase [29].

MATERIALS AND METHODS

Chemicals and Lipid Standards

Chemicals, solvents and synthetic lipid standards were purchased from Sigma-Aldrich, Rathburn Chemicals, Avanti Polar Lipids and Larodan Fine Chemicals AB.

Mouse Brain Tissue Sampling

Animal experiments were conducted in strict accordance with German law (in congruence with 86/609/EEC) for the use of laboratory animals and approved by the local animal welfare committee at the Johannes Gutenberg University Mainz. Male C57Bl/6 wild-type and PRG-1 knockout mice [29] were euthanized by an intraperitoneal injection of ketamine at an overdose. Subsequently, the mice were perfused intracardially with 4°C 155 mM ammonium acetate, and the cerebellum, hippocampus and S1BF were dissected. The tissues were immediately frozen on dry ice and stored at -80°C until further processing.

Lipid Extraction

Brain tissues were homogenized in 155 mM ammonium acetate and analyzed for total protein concentration using BCA Protein Assay Kit (Thermo Scientific). Aliquots of tissue homogenates corresponding to 10 μg of total protein were subjected to lipid extraction executed at 4°C as previously described [7]. Briefly, the tissue homogenates were spiked with 10 μl of internal mixture (providing a total spike of 54 pmol CE 19:0, 35 pmol TAG 17:1/17:1/17:1, 35 pmol DAG 19:0/19:0, 26 pmol LPA O-16:0, 35 pmol PA 17:0/14:1, 25 pmol LPS 17:1, 13 pmol PS 17:0/20:4, 50 pmol PE O-20:0/O-20:0, 30 pmol LPC O-17:0, 137 pmol PC 18:3/18:3, 35 pmol PI 17:0/20:4, 30 pmol PG 17:0/17:0, 55 pmol Cer 18:1;2/17:0;0, 69 pmol SM 18:1;2/17:0;0, 49 pmol HexCer 18:1;2/12:0;0, 28 pmol SHexCer 18:1;2/12:0;0) and diluted to 200 μl using 155 mM ammonium acetate. Samples were subsequently added 990 μl chloroform/methanol (10:1, v/v) and vigorously mixed for 2 h. The lower organic phase was collected (10:1-phase lipid extract). The remaining aqueous phase was re-extracted with 990 μl of chloroform/methanol (2:1, v/v) for 1 h and the lower organic phase was collected (2:1-phase lipid extract). The collected lower organic phases were vacuum evaporated.

Shotgun Lipidomics Analysis

Lipid extracts were dissolved in 60 μl of chloroform/methanol (1:2, v/v) and subjected to mass spectrometric analysis using an LTQ Orbitrap XL instrument (Thermo Fisher Scientific) equipped with a TriVersa NanoMate (Advion Biosciences) as previously described [4,7]. The 10:1-phase lipid extracts were analyzed by positive ion mode multiplexed FT MS analysis with scan ranges *m/z* 280-580 (monitoring lysophosphatidylcholine (LPC) and lysophosphatidylethanolamine (LPE) species) and *m/z* 500-1200 (monitoring sphingomyelin (SM), ceramide (Cer), diacylglycerol (DAG), PC, ether-linked PC (PC O-), phosphatidylethanolamine (PE), ether-linked phosphatidylethanolamine (PE O-) and triacylglycerol (TAG) species). The 2:1-phase lipid extracts were analyzed by negative ion mode multiplexed FT MS analysis with scan ranges *m/z* 370-660 (monitoring lysophosphatidic acid (LPA), lysophosphatidylserine (LPS) and lysophosphatidylinositol (LPI) species) and *m/z* 550-1700 (monitoring phosphatidic acid (PA), phosphatidylserine (PS), phosphatidylinositol (PI), phosphatidylglycerol (PG) and sulfatide (SHexCer) species). All FT MS spectra were acquired in profile mode using a target mass resolution of 100,000 (fwhm), activation of isolation waveforms, automatic gain control at 1e6, max injection time at 250 ms and acquisition of 2 µscans.

Annotation of Lipid Species

Glycerophospholipid and glycerolipid species were annotated using sum composition: <lipid class><total number of C in the fatty acid moieties>:<total number of double bonds in the fatty acid moieties> (e.g. PI 34:1). Sphingolipid species were annotated as <lipid class><total number of C in the long-chain base and fatty acid moiety>:< total number of double bonds in the long-chain base and fatty acid moiety>;< total number of OH groups in the long-chain base and fatty acid moiety> (e.g. SM 36:1;2) [5,7].

ALEX Software

The individual parts of the ALEX software were programmed using several programming languages, libraries and software frameworks. The ALEX lipid database is implemented using the library based SQLite database engine. The ALEX target list generator is written in C++ and uses the Qt framework for its user interface. The ALEX converter, ALEX extractor and ALEX unifier are written in Python 2.7 and make use of the Python packages PySide, NumPy and SciPy. Furthermore, the ALEX converter uses the package comtypes to interface with the MSFileReader library version 2.2 (Thermo Scientific), which must be installed. Finally, the standalone ALEX lipid calculator is written in common lisp and uses the GTK+ framework for its user interface, while the online version is written in PHP. The ALEX software is available at http://mslipidomics.info | software along with installation instructions. A sample dataset is also available for testing local installations of the software.

Data Processing and Visualization

Computation of molar abundance (fmol) of lipid species [20] was performed using open source software Orange 2.6 (www.orange.biolab.si) [28]. The Orange workflow is provided as part of the sample dataset available at http://mslipidomics.info | software. In addition, the Orange output with lipidomics data is available as Data S1. Visualization and calculation of mol% values were performed using commercially available Tableau Software (www.tableausoftware.com). Lipidomic data in Tableau file format is available as Data S2 and can be navigated using the freeware Tableau Reader (http://www.tableausoftware.com/products/reader).

RESULTS AND DISCUSSION

Input: High-Resolution Shotgun Lipidomics Data

Shotgun lipidomics platforms based on high-resolution Orbitrap mass spectrometry and automated nanoelectrospray ionization support high throughput analysis with high sensitivity, specificity and extensive lipidome coverage [4,5,7,30]. The extensive lipidome coverage is generated by combined analyses of sample extracts in negative and positive ion mode, and by implementing chemical derivatization procedures to monitor low abundant or poorly ionizing lipid species [31–33]. In order to maximize the sensitivity of Orbitrap mass analysis, we typically record two multiplexed FT MS scans covering a low *m/z* range and high*m/z* range (e.g. +FT MS *m/z* 280-580 and +FT MS *m/z* 500-1200). Each FT MS scan is recorded in profile mode with a target mass resolution of 100,000. The rationale for this multiplexed FT MS analysis is that the instrument injects a user-specified quantum of ions into the Orbitrap for mass analysis (defined by the automated gain control). Hence, multiplexing two or more scan ranges allows separate quanta of ions within specific *m/z* ranges to be analyzed sequentially in the Orbitrap and yields better ion statistics as compared to injecting all ions at the same time when monitoring a wider *m/z* range (e.g. +FT MS *m/z* 200-1200) [34]. We note that the boundaries of the scan ranges should be chosen to cover specific lipid classes and respective internal standards. For example, the scan range +FT MS *m/z* 280-580 is used for monitoring LPC and LPE species whereas +FT MS *m/z* 500-1200 analysis is used for monitoring Cer, SM, HexCer, PC, PC O-, PE, PE O-, DAG and TAG species. The total time of analysis is typically set to 3 min. Within this time we record 25 low *m/z* and 25 high *m/z* range FT MS spectra. Likewise, negative ion mode analysis is also executed using multiplexed FT MS acquisitions to monitor negatively charged glycerophospholipid and SHexCer species (see Material and Methods). Consequently, this lipidomics approach generates four distinct mass spectral datasets per sample (two per polarity) that need to be queried for lipid identification and export of lipid species intensity. We note that this lipidomics approach is designed for high throughput-oriented studies and supports annotation of lipid species by sum composition nomenclature (e.g. PC 34:1). Characterization of molecular lipid species (e.g. PC 16:1-18:0) requires implementation of time-consuming MS/MS analysis and lipid identification by dedicated software such as LipidXplorer [22].

Design of the ALEX Software Framework

The ALEX software framework was designed for processing of shotgun lipidomics datasets obtained by multiplexed high-resolution FT MS. The rationales for the design were: i) that ALEX should support lipidomic studies with large sample sets for which a multitude of multiplexed FT MS acquisitions have been acquired; and ii) that the output format of ALEX should be compatible with an auxiliary workflow that supports robust data processing including computation of molar abundances of lipid species across numerous lipid classes, integration of sample information, implementation of data quality control procedures and rapid lipidome visualization. To this end, we designed the ALEX software framework to utilize distinct modules that identify lipid species from proprietary .RAW spectral file format, incorporate accessory lipid features stored in a lipid database and output lipidomic data in "database table format" (Figure 1, Data S1). In this format, the lipidomic data is stored in tabulator or comma separated text files structured as database tables with a separate row for each data point. Each row separately contains fields (also termed attributes) reporting for example the originating sample (.RAW file name), the lipid species, adduct information, intensity, peak area, *m/z* values and accessory lipid features derived from the lipid database. This format also easily supports the management of multiple attributes for each data point (e.g. both peak intensity and area) and can be appended with calculated fields such as lipid abundance (e.g. both pmol and fmol, if needed), and sample information and internal standard information supplied by additional text files in database table format. Importantly, this relational database format provides a robust way to manage large datasets programmatically and avoids the need for error-prone manual alignment of data with lists of accessory features and quantification information. We note that the output of most contemporary lipidomics software tools utilizes a "spreadsheet" format where samples/injections are arranged in the columns and lipid species in the rows of a table of with either intensities or peak areas. This "spreadsheet" format is adequate for processing and visualizing simple lipidomic datasets using tools such as Microsoft Excel, but becomes exceedingly cumbersome if sample sets include more than 10 samples and monitor more than 200 lipid species across different FT MS scan ranges.

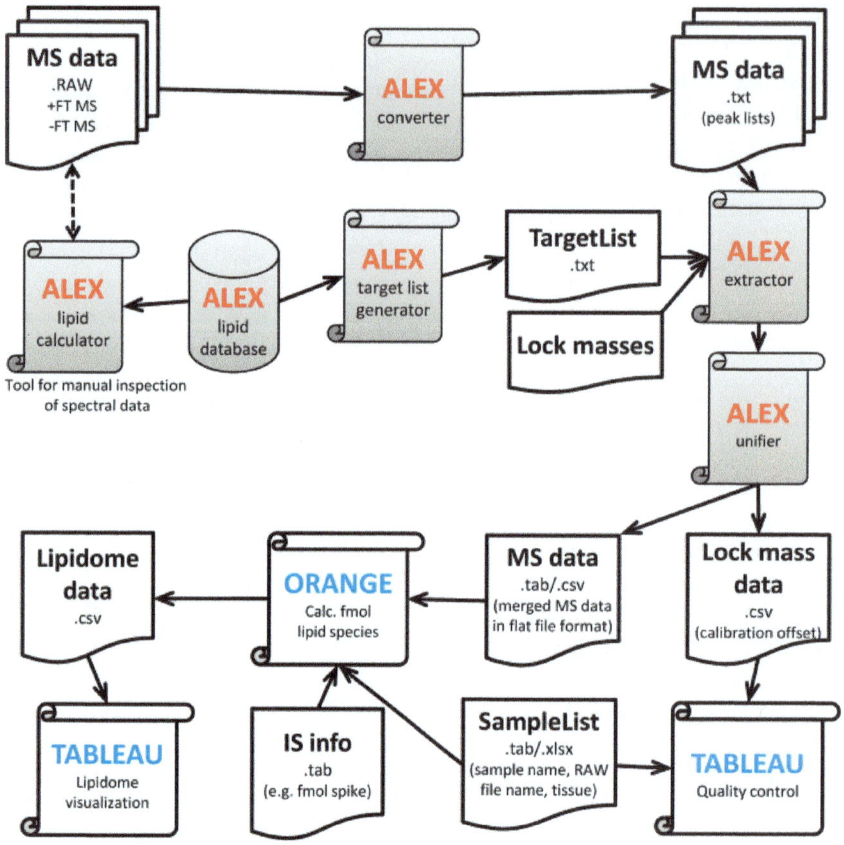

Figure 1: Overview of the ALEX software framework and auxiliary workflow.

The ALEX software framework consists of six core modules (Figure 1): i) the ALEX lipid database featuring a comprehensive range of lipid species and accessory lipid features used for lipid identification, data processing and management; ii) the ALEX lipid calculator which supports manual interpretation of mass spectra; iii) the ALEX converter which converts mass spectrometric data in proprietary .RAW format into averaged spectral peak lists in text file format; iv) the ALEX target list generator which queries the ALEX lipid database to compile target lists used for lipid species identification; v) the ALEX extractor which uses the target lists to identify lipid species and export corresponding intensities from spectral peak lists; and vi) the ALEX unifier that merges the multiple ALEX extractor outputs containing lipid species data from different FT MS scan ranges into a single data file. Further information about the function of the ALEX modules and the auxiliary workflow is outlined in the subsequent sections.

ALEX Lipid Database

To support identification of lipid species we constructed the ALEX lipid database. Currently, the database covers more than 20,000 lipid species from more than 85 lipid classes. Each lipid species in the database is annotated by sum composition and contains a range of accessory lipid features denoting its chemical formula, mono-isotopic mass, adduction in positive and negative ion mode, lipid category, lipid class, and the total number of C atoms, double bonds and hydroxyl group in fatty acid and long chain base moieties (termed C index, db index and OH index, respectively). The ALEX lipid database is available as a collection of text files and as a binary SQLite database queried by the ALEX lipid calculator and the ALEX target list generator, respectively (outlined in the next sections). The text version also serves as the source for the binary version (compiled by a supporting Python program) and can be edited by the user. To support the editing of the database, a template Microsoft Excel file can be used to assist the enumeration and calculation of the chemical formula and mass for each species, such that catalogs of entire lipid classes can readily be added. We note that the accessory lipid features serve as important attributes in the final ALEX output that facilitate data processing and lipidome visualization.

ALEX Lipid Calculator

The ALEX lipid calculator was designed to complement manual inspection of FT MS spectra when using proprietary Xcalibur software (Fig. 2). The calculator is available as executable program and as an online application at www.mslipidomics.info/lipid-calc. Both versions of the ALEX lipid calculator support querying *m/z* values of specific lipid species and searching the lipid database for candidate lipid species that match a measured *m/z* value within a user-specified tolerance window. Since the accuracy of lipid identification can be hampered by drifts of the FT MS calibration we also implemented an option to specify an *m/z* offset while querying the lipid database. To accurately specify the calibration offset requires that the user knows the identity of at least one well-characterized "lock mass" ion that can be used for estimating the*m/z* offset (Figure 2). An option to minimize potential problems with FT MS calibration drifts is to apply online lock mass calibration during sample acquisition [35]. However, the online lock mass calibration eliminates the lock mass ion(s) from the recorded FT MS spectra using waveforms that might concomitantly eliminate lipid ions having similar *m/z*. An alternative strategy is to apply an offline lock mass adjustment as implemented in the ALEX extractor (outlined below).

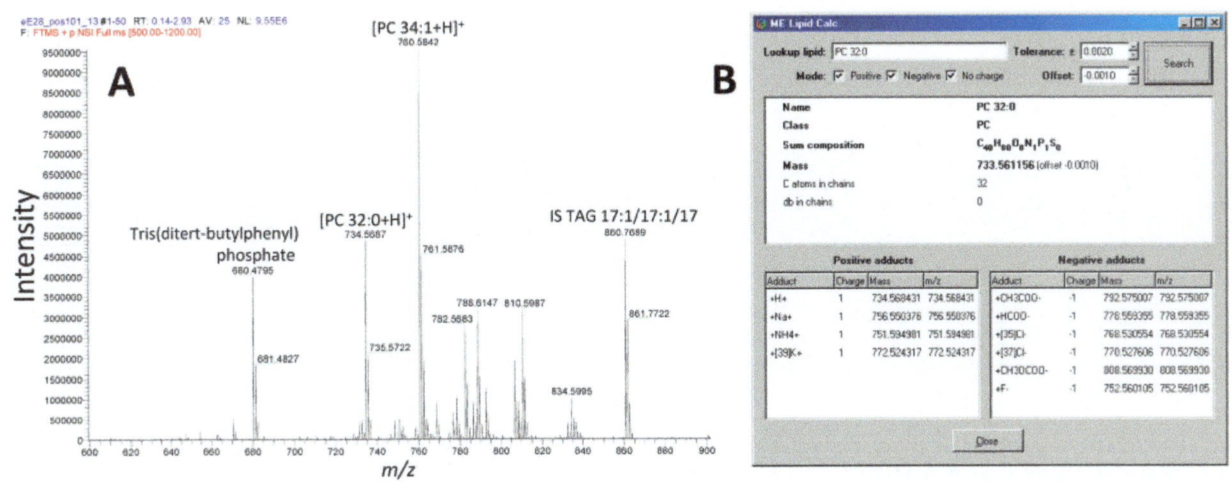

Figure 2: The ALEX lipid calculator.

ALEX Converter

A prerequisite for automated lipid identification and export of intensity is that the mass spectrometric data in proprietary .RAW format are made accessible for querying. To this end, we designed the ALEX converter to interface with the proprietary dynamic-link library MSFileReader. The MSFileReader supports export of all scan information within .RAW files including single and averaged spectral peak lists in either centroid or profile mode format. The ALEX converter was designed to export individual spectral peak lists in profile mode format, to average peak lists for specific FT MS scan ranges and to save these averaged peak lists in .txt format. The ALEX converter output consists of a directory with separate folders named according to each FT MS scan range containing corresponding .txt files named according to the originating .RAW files (i.e. the ALEX converter does not merge overlapping FT MS scan ranges). Output files in these FT MS scan range-dependent folders are queried by the ALEX extractor (outlined below). The rationales for this design were i) that no available software supports export of multiplexed FT MS data in profile mode format; and ii) that a simple output text format reporting averaged peak list data allows users to easily review data as in contrast to accessing data in stored in the encrypted .mzXML format.

ALEX Target List Generator

Lipid identification by the ALEX software framework is based on matching intensity data in the exported peak lists with lipid species information derived from the ALEX lipid database. Important for the design of the identification routine was the ability to accurately identify lipid species, and furthermore to support integration of accessory lipid features and subsequent data management for a multitude of lipid species monitored by various FT MS scan ranges across large sample sets. To this end, the ALEX framework was designed to perform a targeted identification of lipid species and export intensity by querying spectral peak lists using target lists with lipid species and respective *m/z* values. Two modules execute this routine: (i) the ALEX target list generator (Figure 3A) which compiles target lists by querying the ALEX lipid database, and (ii) the ALEX extractor which uses the target lists to identify and extract lipid species intensity (outlined in the next section).

We note that a distinct target list with appropriate lipid species should be manually compiled for each FT MS scan range. In addition to listing lipid species and respective *m/z* values, the target lists also include the accessory lipid features derived from the ALEX lipid database. Importantly, these accessory lipid features are incorporated into the final output, and used for processing and visualization by the auxiliary workflow.

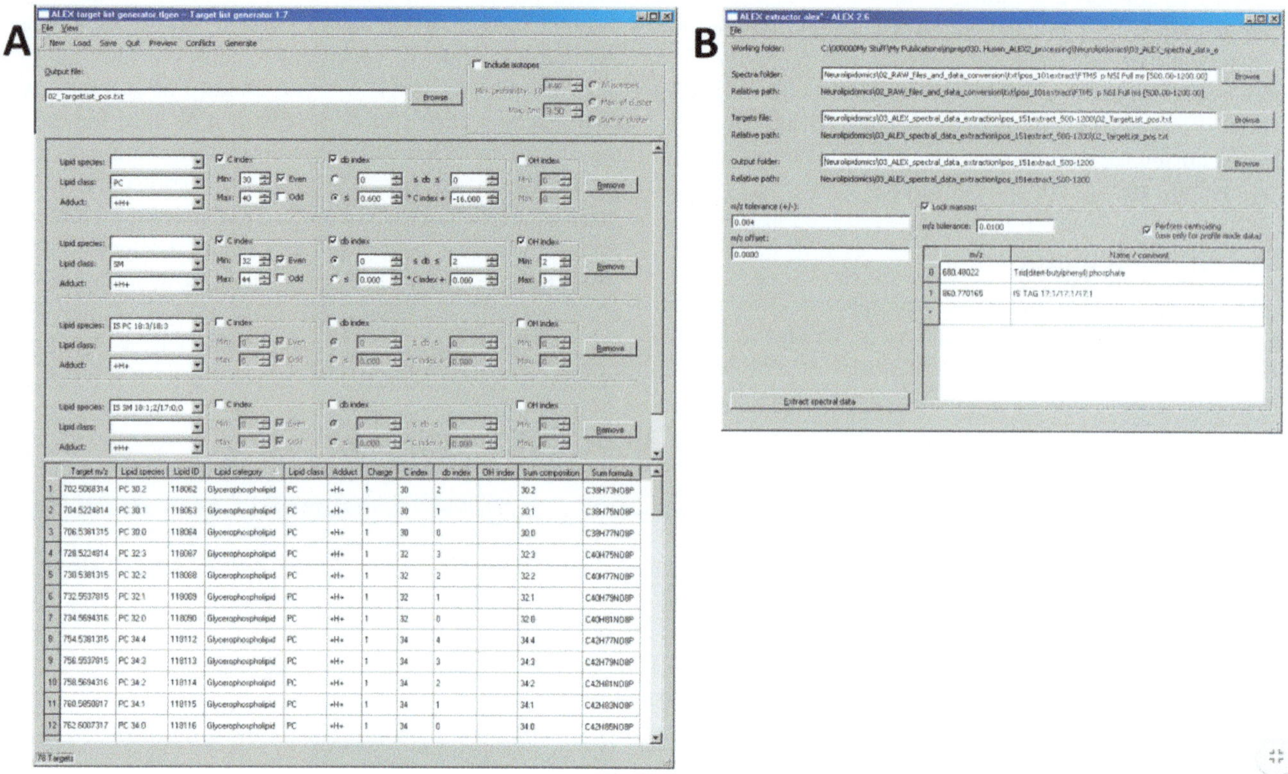

Figure 3: Screenshots of the ALEX target list generator and ALEX extractor.

ALEX Extractor and ALEX Unifier

The ALEX extractor identifies lipid species and exports intensities by querying the averaged peak lists produced by the ALEX converter (Figure 3B). As input the ALEX extractor requires the folder location containing averaged peak lists, an appropriate target list (i.e. specific for the FT MS scan range) to query the peak lists and a destination folder to deposit output text files. Notably, the ALEX extractor also requires an*m/z* tolerance window to identify lipid species. This *m/z* tolerance window is dependent on instrumental mass resolution and typically set to ±0.0020 amu when processing FT MS data acquired with a target resolution at 100,000. To export lipid species intensity the ALEX extractor selects the maximum intensity value within the specified tolerance window and reports the corresponding *m/z* bin value. As mentioned above, the FT MS calibration can drift during the sample analysis and depending on the time of analysis can yield either a constant calibration offset or a progressively changing offset (see Figure 4A). In order to monitor and adjust lipid searches for potential calibration drifts, we incorporated a novel feature in the ALEX extractor termed "lock mass adjustment". Lock mass adjustment serves to correct the *m/z* values of targeted lipid species for calibration drifts and thereby support more accurate lipid identification. This lock mass adjustment can be specified as a constant *m/z* offset and applied across all samples being processed. Alternatively, the ALEX extractor features an in-build automatic lock mass adjustment routine that calculates *m/z* calibration offsets for each sample based on selected lock mass ions (Figure 4A). The calculation of calibration offset uses a three-point quadratic interpolation for estimating the centroid *m/z* values of lock mass ions. We note that the automatic lock mass adjustment requires selection of well-characterized and ubiquitous ions as lock masses. For example, by positive ion mode FT MS analysis we always detect both the chemical background ion tris(ditert-butylphenyl) phosphate and the internal standard TAG 17:1/17:1/17:1 (Figure 2). Using these ions as lock masses enables estimation of the FT MS calibration offset for individual samples and correcting lipid searches by adjusting the *m/z* values of targeted lipid species for each sample. As exemplified in Figure 2, using these lock mass ions allows identification of endogenous lipid species with a mass error of 0.4 ppm instead of 1 ppm when ignoring the FT MS calibration offset. Hence, this automated lock mass adjustment routine serves to improve the accuracy of lipid species identification despite drifts in FT MS calibration.

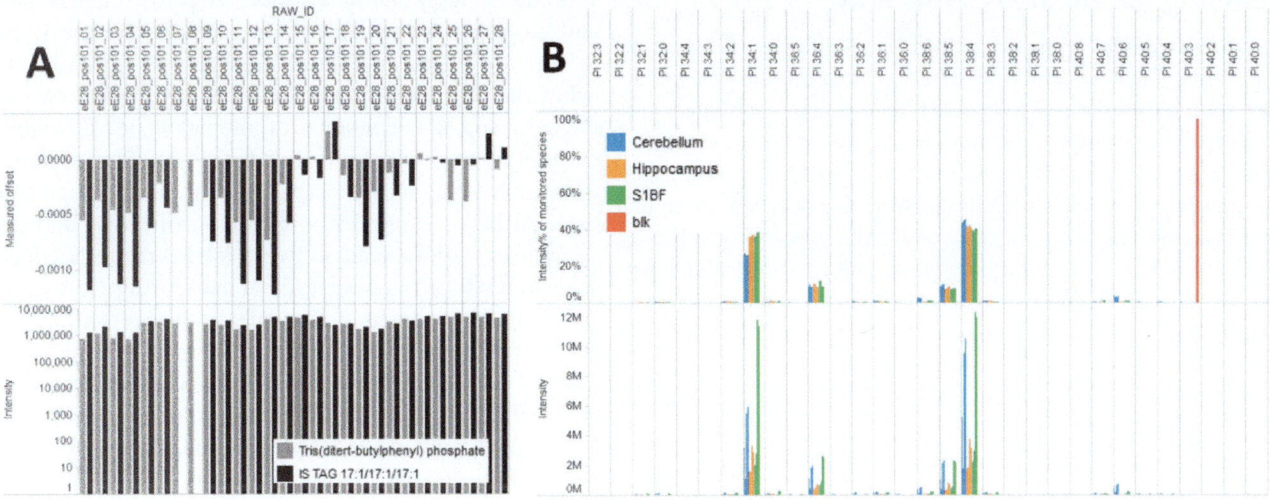

Figure 4: Quality control analysis.

The ALEX extractor outputs, for each FT MS scan range, several comma-separated value (.csv) files with lipid species intensity data and calculated lock mass adjustments for all processed samples. Notably, the lipid species intensity output is organized in database table format and includes attributes that track the originating .RAW file name, lipid species intensity and peak area, measured *m/z*, calculated *m/z*, the difference between measured and calculated *m/z*, and all the accessory lipid features included on the target list. We note that these sample attributes facilitate subsequent processing and visualization, and implementation of quality control procedures. In order to merge lipid species data from different FT MS ranges, we devised the ALEX unifier to concatenate selected .csv output files into one final .csv file which contains the union of all data and an additional column with an index, "rangeID", that tracks the FT MS scan range of the input .csv files. This output format supports further data processing including computation of lipid abundance by the auxiliary workflow using open-source Orange [28] and Tableau Software.

Application of the ALEX Software Framework

In order to demonstrate the efficacy of the ALEX software framework and describe the auxiliary workflow, we here present a neurolipidomic pilot study. Three brain tissues; cerebellum, hippocampus and S1BF from two control mice and two mice devoid of the PRG-1 gene were subjected to shotgun lipidomics analysis. Homogenates of the tissues (12 samples in total) and two blank samples were spiked with defined amounts of internal lipid standards and subjected to 2-step lipid extraction [7]. The apolar (10:1-) and polar (2:1-phase) lipid extracts were analyzed by multiplexed FT MS analysis in positive and negative ion mode, respectively. Each sample extract was analyzed twice (technical replicates). In total, this analysis produced 56 .RAW files (14 samples analyzed twice per polarity).

First, the ALEX converter was used to convert .RAW files to spectral peak lists (Figure 1). This processing produced a total of 112 peak list files (56 .RAW files, two FT MS scan ranges per polarity). The peak list files were automatically organized into 4 folders named according to polarity and scan range: +FTMS *m/z*280-580, +FTMS *m/z* 500-1200, -FTMS *m/z* 370-660, and -FTMS *m/z* 550-1700 (each folder having 28 peak list files). Next, target lists with lipid species for each of the four FT MS *m/z* ranges were generated using the ALEX target list generator (See Materials and Methods for details). These target lists were subsequently used by the ALEX extractor to identify lipid species and export corresponding intensities from the peak list files. In addition, the processing by the ALEX extractor was performed using lock mass ions for each FT MS *m/z* range in order to monitor and correct searches for FT MS calibration drifts (Figure 4A). Finally, the ALEX unifier was applied to merge the four ALEX extractor output files into single output files reporting all identified lipid species, intensities and accessory lipid features, and lock mass information and calculated FT MS calibration offsets.

As a first step in the auxiliary workflow we performed a quality control of the neurolipidomics dataset (Figure 1). To this end, we accessed the lock mass information using Tableau Software and displayed the estimated FT MS calibration offsets as function of sample injection. This quality control showed that the calibration offset was not constant across all samples (Figure 4) and thus highlighting the efficacy of the automatic lock mass adjustment. Importantly, for one sample we observed no intensity of the selected lock mass ion TAG 17:1/17:1/17:1. By manual inspection of the .RAW data, we concluded that the investigator in charge had failed to spike internal lipid standards into the particular sample. Consequently, this quality control demonstrated that the particular sample (cerebellum from a knockout mouse) could not be used for computing the molar amount of lipid species. As an additional quality control procedure we also accessed the output file with lipid species intensity data using Tableau Software, and displayed both the absolute intensity and intensity profile of monitored lipid species within lipid classes for all samples analyzed (Figure 4B). This analysis showed that a low abundant ion in the blank samples was falsely identified as PI 40:3. Due to the low intensity of this ion and its presence in the blank samples we concluded that this ion represents a chemical background ion. To minimize bias from falsely identified background ions one can implement a background subtraction during the subsequent computation of molar lipid abundance using the Orange software. Moreover, it is recommendable to perform additional tandem mass analysis to assess the identity of dubious identification.

Outlining the Auxiliary Workflow

In order to compute molar abundance (e.g. fmol) of lipid species we made use of the database exploration tool and open-source software Orange [28]. The molar abundance of lipid species is easily computed via a sequence of processing steps (depicted in Fig. 5). Step (1); the lipid species intensity data generated by the ALEX framework is specified as input (icon (a)) and merged with a second input specifying sample information (e.g. tissue type, genetic information, name of mice; icon (b)). Step (2); an intensity filter is implemented in order to remove intensities below a user-specified threshold if deemed necessary. Step (3); a third input specifying the spiked amount of internal standards is incorporated (icon (c)). Step (4); a new attribute is defined by specifying the intensities of internal standards. This attribute is used in step (6) for computing the molar abundance. Step (5); internal standards and corresponding intensities are defined for the subsequent calculation of molar lipid abundances. Step (6); an equation using the attributes lipid species intensity, internal standard intensity and spiked amount of internal standard is applied for calculating the molar lipid abundance [20]. Step (7); processed data is saved as a .csv output file (in database table format) that can be accessed for computation of mol% and lipidome visualization by Tableau Software. We here note that the Orange schema produced an output file of the neurolipidomics analysis featuring 997 targeted lipid species with accessory information across 56 sample injections producing a data matrix with a total of 1,050,504 data points (Data S1). Notably, managing and processing a dataset of this magnitude is poorly suited for Microsoft Excel and highlights the benefits of managing the data using a database exploration tools.

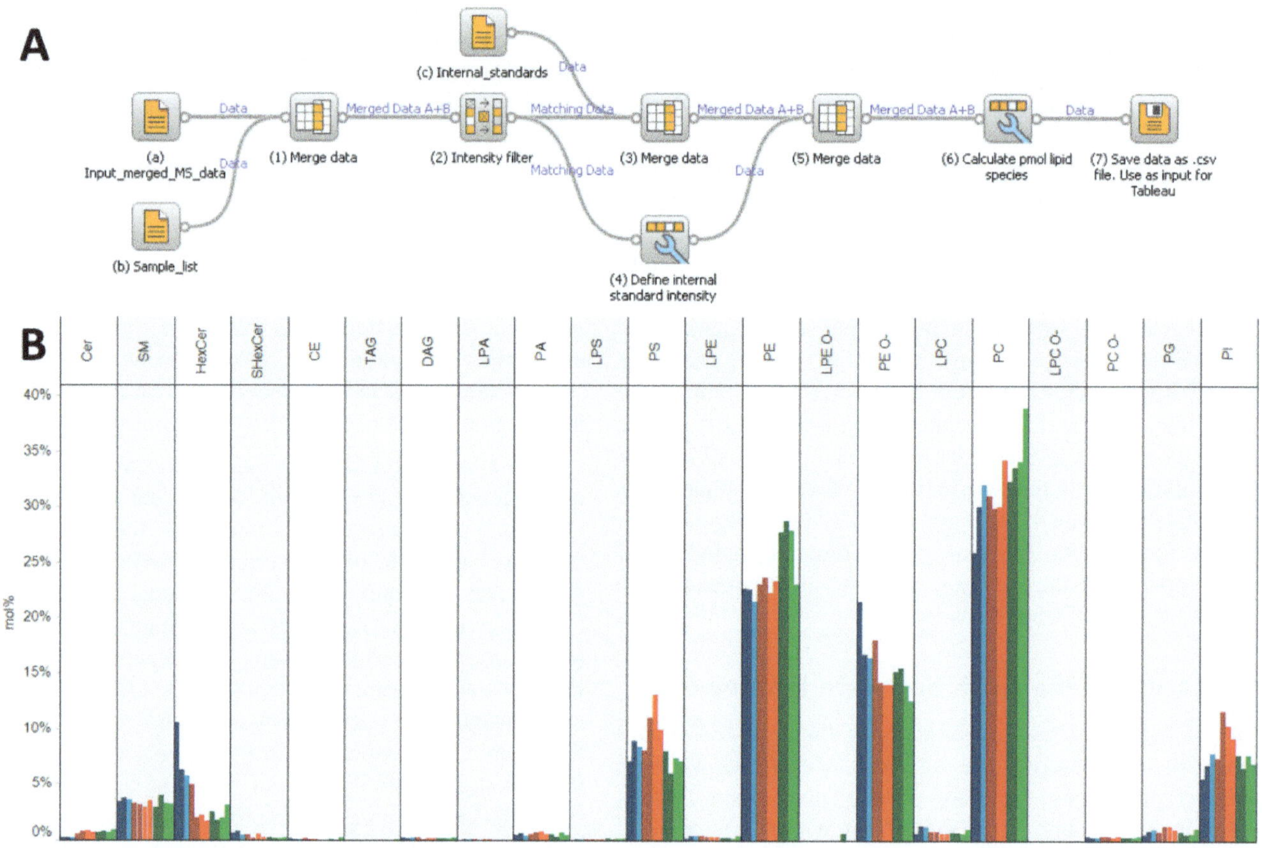

Figure 5: Outline of the auxiliary workflow used for lipidome data processing.

To support rapid and efficient visualization of large lipidome datasets, we integrated Tableau Software as part of the auxiliary workflow. Tableau Software can be dynamically linked to the Orange output files such that any potential modifications within the data processing procedure can be visualized simply by updating the Orange output file and the link to Tableau. Notably, the Tableau software includes a feature that easily allows calculation and display of "mol% of lipid species" normalized to any given set of attributes and accessory lipid features in the input file. As such, the user can rapidly display the "mol% of all monitored PE species" (Figure 6C) or "mol% of all monitored glycerophospholipid species" (Figure 6D). We note, that calculation of such data formats would require implementation of additional processing steps within Orange (and equally in Microsoft Excel) in order to calculate and output such data values. We also note that these accessory lipid features provided by the ALEX framework, and consequently the information content in database table format, facilitates lipidome visualization by Tableau software.

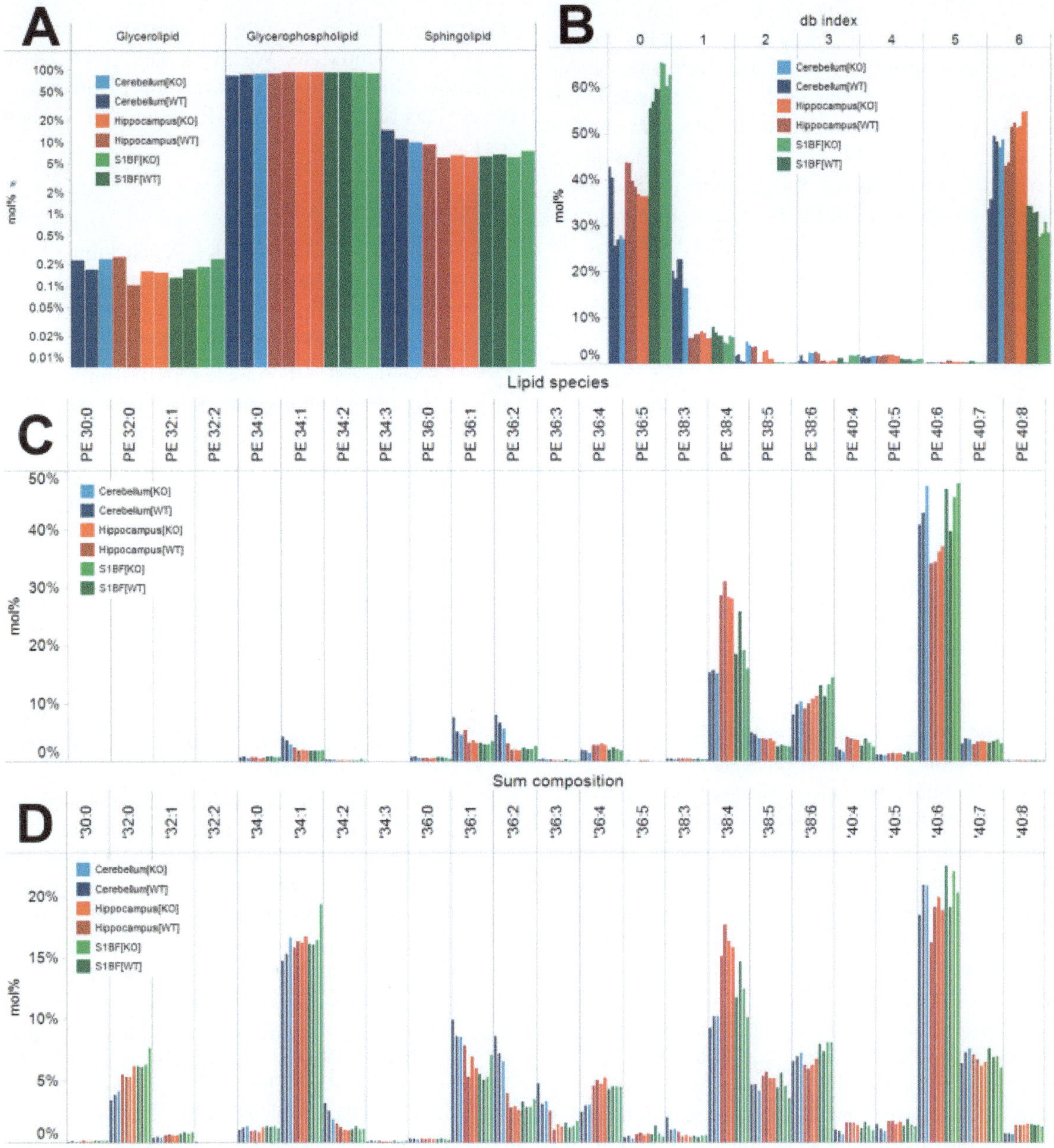

Figure 6: Lipidome visualization using different display formats.

Application of the Auxiliary Workflow

To illustrate the efficacy of the auxiliary workflow we here present the results of the neurolipidomic pilot study. First, we assessed the lipid class composition of the three mouse brain tissues from wild-type and PRG-1 knockout mice (Figure 5B). The lipid class composition of brain tissues was primarily comprised of PC, PE, PE O-, PS, PI, SM and HexCer lipids. In addition, the analysis also showed a low abundance of Cer, SHexCer, CE and TAG species (Figure 5B). This result corroborates previous reports on brain lipid composition [36–38]. Notably, comparing the lipid class composition of the brain tissues did not show any pronounced differences between the wild-type and PRG-1 knockout mice. Moreover, the analysis also did not show any major difference in lipid class composition between the three brain tissues. To further interrogate the lipidome data, we explored alternative display formats including lipid category composition (e.g. glycerophospholipids, sphingolipids and glycerolipids, Figure 6A), lipid species composition within a defined lipid class (e.g. PE species, Figure 6C), lipid species composition across a defined lipid category (e.g. all glycerophospholipids, Figure 6D) and db index within a defined lipid class (e.g. LPS, Figure 6B). Using the various modes of lipidome visualization, we observed that each of the three mouse brain tissues featured specific signatures of lipid species. Specifically, we observed that the composition of PE species was different between all three tissues; the hippocampus comprised a relatively high level of PE 38:4, the S1BF contained relatively high levels of PE 40:6, and the cerebellum contained a relatively high level of PE 40:6 and minor but systematically higher levels of PE 34:1, PE 36:1 and PE 36:2 as compared to the two other tissues. These distinct lipidome hallmarks could also be observed when assessing the collective lipid species composition across all glycerophospholipids (Figure 6D) albeit with a less pronounced differences as compared to only PE species. The specific lipid compositions of the three brain tissues were further interrogated by visualizing the distribution of double bonds within the fatty acid moieties of LPS species. This visualization revealed that the S1BF comprised relatively high levels of LPS species with 0 double bonds (primarily attributed LPS 18:0) being offset by lower levels of LPS species having 6 double bonds (attributed LPS 22:6). In comparison, the cerebellum showed a characteristic distribution of LPS species with higher levels of species with 1 double bond as compared to the two other tissues. The LPS composition of the hippocampus comprised a double bond distribution intermediate of the S1BF and the cerebellum. We note that the lipidome visualization did not reveal any pronounced differences in lipid species composition between any of the tissues from wild-type and PRG-1 knockout mice. This observation is furthermore substantiated by results of principal component analysis (data not shown).

CONCLUSIONS

Here we presented a platform for streamlined processing, computation and visualization of high-content lipidomics datasets acquired using high-resolution Orbitrap mass spectrometry. The platform features a novel routine for querying proprietary spectral data that supports identification and quantification of lipid species, execution of quality control routines and rapid visualization of lipidome data. The platform utilizes three software modules: the ALEX framework that accesses and queries mass spectral data, the visual programming tool Orange that integrates sample information and computes molar lipid abundances, and the visual analytical tool Tableau Software for lipidome visualization. A key asset of the framework is the storage of lipidome data in "database table format" that enables using a multitude of data attributes for robust data processing and visualization. To demonstrate the efficacy of the platform, we presented a comparative neurolipidomic pilot study of mouse cerebellum, hippocampus and S1BF from wild-type and PRG-1 knockout mice [29]. The analysis demonstrated a distinct lipid species signature for each of the three brain tissues, but failed to ascertain any pronounced perturbations of lipid composition in mice devoid of the PRG-1 gene. This observation can potentially be explained by the localized expression of PRG-1 in the postsynaptic density of glutamatergic synapses. Regional differences in the lipidome composition induced by PRG-1 deficiency might potentially be concealed by the complexity of the macroscopic brain tissues investigated herein. We note that the ALEX framework can easily be adapted for processing of high-resolution shotgun lipidomics data acquired by any type of instrumentation (e.g. LTQ FT, Q Exactive, Orbitrap Fusion, solariX) provided spectral peak lists are stored in .txt file format. Moreover, the Orange processing procedure and Tableau visualization can be extended to include various filters for improved lipid identification, to calculate molar abundance of lipid species per unit of sample material (e.g. pmol lipid/µg protein), to perform statistical testing or multivariate analysis, and to integrate and support processing and visualization of lipidome data acquired by MS/MS analysis. Notably, the auxiliary workflow can also be adapted to use other database-orientated exploration tools than Orange and Tableau software. Finally, we argue that storage of lipidomics data in database table format can be a future avenue for data dissemination since it enables investigators to easily access, inspect and apply such resource data.

ACKNOWLEDGEMENTS

We are grateful to Dr. Ole N. Jensen for providing access to Triversa NanoMate. We thank Dr. Albert Casanovas and Reinaldo Almeida for critical reading of the manuscript. We thank Hans Kristian Hannibal-Bach and Nikolai Schmarowski for technical assistance.

REFERENCES

1. Yetukuri L, Ekroos K, Vidal-Puig A, Oresic M (2008) Informatics and computational strategies for the study of lipids. Mol Biosyst 4: 121-127. 10.1039/b715468b PubMed: 18213405
2. Shevchenko A, Simons K (2010) Lipidomics: coming to grips with lipid diversity. Nat Rev Mol Cell Biol11: 593-598. 10.1038/nrg2850 PubMed: 20606693
3. van Meer G (2005) Cellular lipidomics. EMBO J 24: 3159-3165. 10.1038/sj.emboj.7600798 PubMed:16138081
4. Ejsing CS, Sampaio JL, Surendranath V, Duchoslav E, Ekroos K et al. (2009) Global analysis of the yeast lipidome by quantitative shotgun mass spectrometry. Proc Natl Acad Sci U S A 106: 2136-2141.10.1073/pnas.0811700106 PubMed: 19174513
5. Klose C, Surma MA, Gerl MJ, Meyenhofer F, Shevchenko A et al. (2012) Flexibility of a eukaryotic lipidome--insights from yeast lipidomics. PLOS ONE 7: e35063. 10.1371/journal.pone.0035063 PubMed:22529973
6. Carvalho M, Sampaio JL, Palm W, Brankatschk M, Eaton S et al. (2012) Effects of diet and development on the Drosophila lipidome. Mol Syst Biol 8: 600 PubMed: 22864382
7. Sampaio JL, Gerl MJ, Klose C, Ejsing CS, Beug H et al. (2011) Membrane lipidome of an epithelial cell line. Proc Natl Acad Sci U S A 108: 1903-1907. 10.1073/pnas.1019267108 PubMed: 21245337
8. Dennis EA, Deems RA, Harkewicz R, Quehenberger O, Brown HA et al. (2010) A mouse macrophage lipidome. J Biol Chem 285: 39976-39985. 10.1074/jbc.M110.182915 PubMed: 20923771
9. Breslow DK, Collins SR, Bodenmiller B, Aebersold R, Simons K et al. (2010) Orm family proteins mediate sphingolipid homeostasis. Nature 463: 1048-1053. 10.1038/nature08787 PubMed: 20182505
10. Klemm RW, Ejsing CS, Surma MA, Kaiser HJ, Gerl MJ et al. (2009) Segregation of sphingolipids and sterols during formation of secretory vesicles at the trans-Golgi network. J Cell Biol 185: 601-612.10.1083/jcb.200901145 PubMed: 19433450
11. Brügger B, Glass B, Haberkant P, Leibrecht I, Wieland FT et al. (2006) The HIV lipidome: a raft with an unusual composition. Proc Natl Acad Sci U S A 103: 2641-2646. 10.1073/pnas.0511136103 PubMed:16481622
12. Takamori S, Holt M, Stenius K, Lemke EA, Grønborg M et al. (2006) Molecular anatomy of a trafficking organelle. Cell 127: 831-846. 10.1016/j.cell.2006.10.030 PubMed: 17110340
13. Zech T, Ejsing CS, Gaus K, de Wet B, Shevchenko A et al. (2009) Accumulation of raft lipids in T-cell plasma membrane domains engaged in TCR signalling. EMBO J 28: 466-476. 10.1038/emboj.2009.6PubMed: 19177148

14. Han X, Yang K, Gross RW (2012) Multi-dimensional mass spectrometry-based shotgun lipidomics and novel strategies for lipidomic analyses. Mass Spectrom Rev 31: 134-178. 10.1002/mas.20342 PubMed:21755525

15. Song H, Hsu FF, Ladenson J, Turk J (2007) Algorithm for processing raw mass spectrometric data to identify and quantitate complex lipid molecular species in mixtures by data-dependent scanning and fragment ion database searching. J Am Soc Mass Spectrom 18: 1848-1858. 10.1016/j.jasms.2007.07.023 PubMed:17720531

16. Haimi P, Uphoff A, Hermansson M, Somerharju P (2006) Software tools for analysis of mass spectrometric lipidome data. Anal Chem 78: 8324-8331. 10.1021/ac061390w PubMed: 17165823

17. Leavell MD, Leary JA (2006) Fatty acid analysis tool (FAAT): An FT-ICR MS lipid analysis algorithm.Anal Chem 78: 5497-5503. 10.1021/ac0604179 PubMed: 16878888

18. Hübner G, Crone C, Lindner B (2009) lipID--a software tool for automated assignment of lipids in mass spectra. J Mass Spectrom 44: 1676-1683 PubMed: 19816875

19. Houjou T, Yamatani K, Imagawa M, Shimizu T, Taguchi R (2005) A shotgun tandem mass spectrometric analysis of phospholipids with normal-phase and/or reverse-phase liquid chromatography/electrospray ionization mass spectrometry. Rapid Commun Mass Spectrom 19: 654-666.10.1002/rcm.1836 PubMed: 15700236

20. Ejsing CS, Duchoslav E, Sampaio J, Simons K, Bonner R et al. (2006) Automated identification and quantification of glycerophospholipid molecular species by multiple precursor ion scanning. Anal Chem 78: 6202-6214. 10.1021/ac060545x PubMed: 16944903

21. Schwudke D, Oegema J, Burton L, Entchev E, Hannich JT et al. (2006) Lipid profiling by multiple precursor and neutral loss scanning driven by the data-dependent acquisition. Anal Chem 78: 585-595.10.1021/ac051605m PubMed: 16408944

22. Herzog R, Schwudke D, Schuhmann K, Sampaio JL, Bornstein SR et al. (2011) A novel informatics concept for high-throughput shotgun lipidomics based on the molecular fragmentation query language.Genome Biol 12: R8. 10.1186/gb-2011-12-1-r8 PubMed: 21247462

23. Ejsing CS, Husen P, Tarasov K (2012) In: Ekroos K, editor. Lipidomics: technologies and applications. Weinheim: Wiley-VCH; pp. 147-174

24. Liebisch G, Vizcaíno JA, Köfeler H, Trötzmüller M, Griffiths WJ et al. (2013) Shorthand notation for lipid structures derived from mass spectrometry. J Lipid Res 54: 1523-1530. 10.1194/jlr.M033506 PubMed:23549332

25. Ekroos K, Ejsing CS, Bahr U, Karas M, Simons K et al. (2003) Charting molecular composition of phosphatidylcholines by fatty acid scanning and ion trap MS3 fragmentation. J Lipid Res 44: 2181-2192.10.1194/jlr.D300020-JLR200 PubMed: 12923235

26. Fahy E, Subramaniam S, Brown HA, Glass CK, Merrill AH Jr. et al. (2005) A comprehensive classification system for lipids. J Lipid Res 46: 839-861. 10.1194/jlr.E400004-JLR200 PubMed: 15722563

27. Fahy E, Subramaniam S, Murphy RC, Nishijima M, Raetz CR et al. (2009) Update of the LIPID MAPS comprehensive classification system for lipids. J Lipid Res 50Suppl: S9-14 PubMed: 19098281

28. Curk T, Demsar J, Xu Q, Leban G, Petrovic U et al. (2005) Microarray data mining with visual programming. Bioinformatics 21: 396-398. 10.1093/bioinformatics/bth474 PubMed: 15308546

29. Trimbuch T, Beed P, Vogt J, Schuchmann S, Maier N et al. (2009) Synaptic PRG-1 modulates excitatory transmission via lipid phosphate-mediated signaling. Cell 138: 1222-1235.10.1016/j.cell.2009.06.050 PubMed: 19766573

30. Schwudke D, Hannich JT, Surendranath V, Grimard V, Moehring T et al. (2007) Top-down lipidomic screens by multivariate analysis of high-resolution survey mass spectra. Anal Chem 79: 4083-4093.10.1021/ac062455y PubMed: 17474710

31. Liebisch G, Binder M, Schifferer R, Langmann T, Schulz B et al. (2006) High throughput quantification of cholesterol and cholesteryl ester by electrospray ionization tandem mass spectrometry (ESI-MS/MS). Biochim Biophys Acta 1761: 121-128. 10.1016/j.bbalip.2005.12.007 PubMed: 16458590

32. Sandhoff R, Brügger B, Jeckel D, Lehmann WD, Wieland FT (1999) Determination of cholesterol at the low picomole level by nano-electrospray ionization tandem mass spectrometry. J Lipid Res 40: 126-132 PubMed: 9869658

33. Bilgin M, Markgraf DF, Duchoslav E, Knudsen J, Jensen ON et al. (2011) Quantitative profiling of PE, MMPE, DMPE, and PC lipid species by multiple precursor ion scanning: a tool for monitoring PE metabolism. Biochim Biophys Acta 1811: 1081-1089. 10.1016/j.bbalip.2011.09.018 PubMed: 22001639

34. Southam AD, Payne TG, Cooper HJ, Arvanitis TN, Viant MR (2007) Dynamic range and mass accuracy of wide-scan direct infusion nanoelectrospray fourier transform ion cyclotron resonance mass spectrometry-based metabolomics increased by the spectral stitching method. Anal Chem 79: 4595-4602.10.1021/ac062446p PubMed: 17511421

35. Schuhmann K, Almeida R, Baumert M, Herzog R, Bornstein SR et al. (2012) Shotgun lipidomics on a LTQ Orbitrap mass spectrometer by successive switching between acquisition polarity modes. J Mass Spectrom 47: 96-104. 10.1002/jms.2031 PubMed: 22282095

36. Chan RB, Oliveira TG, Cortes EP, Honig LS, Duff KE et al. (2012) Comparative lipidomic analysis of mouse and human brain with Alzheimer disease. J Biol Chem 287: 2678-2688. 10.1074/jbc.M111.274142PubMed: 22134919

37. Rappley I, Myers DS, Milne SB, Ivanova PT, Lavoie MJ et al. (2009) Lipidomic profiling in mouse brain reveals differences between ages and genders, with smaller changes associated with alpha-synuclein genotype. J Neurochem 111: 15-25. 10.1111/j.1471-4159.2009.06290.x PubMed: 19627450

38. O›Brien JS, Sampson EL (1965) Lipid composition of the normal human brain: gray matter, white matter, and myelin. J Lipid Res 6: 537-544 PubMed: 5865382

CHAPTER 14

THREE-DIMENSIONAL COMPUTER MODEL OF THE RIGHT ATRIUM INCLUDING THE SINOATRIAL AND ATRIOVENTRICULAR NODES PREDICTS CLASSICAL NODAL BEHAVIOURS

Jue Li, Shin Inada, Jurgen E. Schneider, Henggui Zhang, Halina Dobrzynski, Mark R. Boyett

Institute of Cardiovascular Sciences, University of Manchester, Core Technology Facility, Manchester, United Kingdom

ABSTRACT

The aim of the study was to develop a three-dimensional (3D) anatomically-detailed model of the rabbit right atrium containing the sinoatrial and atrioventricular nodes to study the electrophysiology of the nodes. A model was generated based on 3D images of a rabbit heart (atria and part of ventricles), obtained using high-resolution magnetic resonance imaging. Segmentation was carried out semi-manually. A 3D right atrium array model (~3.16 million elements), including eighteen objects, was constructed. For description of cellular electrophysiology, the Rogers-modified FitzHugh-Nagumo model was further modified to allow control of the major characteristics of the action potential with relatively low computational resource requirements. Model parameters were chosen to simulate the action potentials in the sinoatrial node, atrial muscle, inferior nodal extension and penetrating bundle. The block zone was simulated as passive tissue. The sinoatrial node, crista terminalis, main branch and roof bundle were considered as anisotropic. We have simulated normal and abnormal electrophysiology of the two nodes. In accordance with experimental findings: (i) during sinus rhythm, conduction occurs down the interatrial septum and into the atrioventricular node via the fast pathway (conduction down the crista terminalis and into the atrioventricular node via the slow pathway is slower); (ii) during atrial fibrillation, the sinoatrial node is protected from overdrive by its long refractory period; and (iii) during atrial fibrillation, the atrioventricular node reduces the frequency of action potentials reaching the ventricles. The model is able to simulate ventricular echo beats. In summary, a 3D anatomical model of the right atrium containing the cardiac conduction system is able to simulate a wide range of classical nodal behaviours.

INTRODUCTION

Accurate simulation of the generation and propagation of cardiac electrical activity requires detailed anatomical and electrophysiological models. A variety of heart anatomical models (human and animal) have been generated by various investigators. David et al. [1] generated a boundary-conforming mesh of the human atria comprised entirely of hexahedral elements. Bernus et al. [2] developed a human ventricular model, which reproduces geometry and fibre orientation in the right and left ventricles of the human heart. Several whole human heart models have been generated [3]–[5]. Aslanidi et al. [6] integrated a three-dimensional (3D) model of the human sinoatrial node (SAN) [7] into a 3D model of the whole atria

Citation: Li J, Inada S, Schneider JE, Zhang H, Dobrzynski H, Boyett MR (2014) Three-Dimensional Computer Model of the Right Atrium Including the Sinoatrial and Atrioventricular Nodes Predicts Classical Nodal Behaviours. PLoS ONE 9(11): e112547. doi:10.1371/journal.pone.0112547

dissected from the Visible Human dataset [8]. Apart from human heart models, a number of animal heart models have been generated. Nielsen et al. [9] developed a mathematical representation of the ventricular geometry and fibre orientation for the dog. Vetter and McCulloch [10] developed a 3D finite element model of rabbit ventricular geometry with fibre orientation. However, none of these models includes both the sinoatrial and atrioventricular (AVN) nodes. We have previously generated 3D anatomically-detailed models of the isolated SAN and isolated AVN of the rabbit[11], [12]. More recently, we have generated a 3D anatomically-detailed model of the rabbit heart, including the conduction system, using micro-CT [13], although at present the model is unsuitable for electrophysiological simulation. In this study, a 3D anatomically-detailed model of the right atrium of the rabbit heart, including the SAN and AVN and suitable for electrophysiological simulations, was generated based on magnetic resonance (MR) imaging.

There are two groups of electrophysiological models for simulation of the cardiac action potential. One comprises biophysically-detailed (ionic) models, and another comprises mathematical caricature (simplified) models. Based on patch clamp etc., a large number of biophysically-detailed models of the action potential in single cells from different regions of the heart have been developed [14]. For example, such models have been developed for: human atrial [15], [16] and ventricular [17]–[22] muscle; dog atrial [23], and ventricular [25]–[28]muscle; sheep Purkinje fibres [29], [30]; rabbit SAN [31]–[38], AVN [39], Purkinje fibres [40]and atrial [41]–[43] and ventricular [40], [44], [45] muscle; guinea-pig ventricular muscle[46]–[51]; and mouse SAN [52] and ventricular muscle [53]. Caricature models of the action potential include the cellular automaton model, coupled map lattices [54], lattices of coupled ordinary differential equations [55], FitzHugh-Nagumo models [56] and the Fenton-Karma model [57]. Action potential propagation through the heart can be simulated using caricature models as well as biophysically-detailed models. To investigate complex electrophysiological behaviour in the complex anatomical structure of the heart, the use of a set of caricature models is computationally more effective compared to the use of a set of biophysically-detailed models, as they enable testable predictions about heart electrophysiological behaviour to be made with relatively low computational resource.

The purpose of this study was to create a platform with an anatomically-detailed model of rabbit right atrium and a set of caricature action potential models to investigate nodal electrophysiology in health and disease. A 3D anatomically-detailed model of the rabbit right atrium with multiple objects, including the cardiac conduction system (SAN and AVN), was generated. This model can be used for education and research. The propagation of the action potential through the right atrium during sinus rhythm, atrial fibrillation and AVN reentry was simulated using this anatomically-detailed right atrium model, together with a set of Rogers-modified FitzHugh-Nagumo models as well as a cellular automaton model.

METHODS

Ethics Statement

New Zealand White rabbits (1.5–2.5 kg) were sacrificed humanely according to the United Kingdom Animals (Scientific Procedures) Act 1986; in addition, the investigation conformed with the Guide for the Care and Use of Laboratory Animals published by the US National Institutes of Health (NIH Publication No. 85–23, revised 1996). The rabbits were humanely sacrificed by injection of an overdose of sodium pentobarbital into the central ear vein (an approved Schedule 1 method). The heart was removed after confirmation of death of the rabbit.

Development of a 3D anatomically-detailed model of the right atrium of the rabbit including the SAN and AVN

Our aim was the reconstruction of the right atrium, and the right atrium retained its shape by retaining other structures. The atria and a part of the ventricles were dissected for imaging. 3D images were obtained using high-resolution MR imaging carried out on a 11.7 T Bruker MR system (Bruker Medical, Ettlingen, Germany) at the University of Oxford. The voxel size after reconstruction was 26.4 μm×26.4 μm×24.4 μm (anisotropic). To form a convenient platform for numerical simulations, the anisotropic images were transformed into isotropic images with a voxel size of 30 μm×30 μm×30 μm and also 60 μm×60 μm×60 μm. MATLAB (version 7; The Math Works, Inc., Matick, MA, USA) was used to analyse the data: the anisotropic images were imported into the workspace of MATLAB and transformed into isotropic 3D images (voxel size, 30 μm×30 μm×30 μm; Figure S1A) using cubic interpolation.

The superior vena cava (SVC), crista terminalis (CT) and tricuspid valve (TV) could be recognised (Figure S1A). The 3D images were rotated to be in the normal anatomical orientation (x: right-left; y: ventral-dorsal; z: caudal-cranial) for segmentation. The right atrium structure was extracted based on the global threshold; the resulting binary images are shown inFigure S1B. Isosurface rendering was used to display the overall structure of the 3D model for visualisation (Figure S1C). Figure S1D shows the 3D model after segmentation, which is discussed below.

Segmentation was carried out semi-manually by analysing the MR imaging data from three directions (along x, y and z axes). First, the best orientation for segmentation was chosen. Secondly, the object of interest was extracted from the structure every 5~10 images. Thirdly, interpolation was carried out to obtain the 3D object. Finally, the object was combined with other objects to produce a 3D multiple-object model. Eighteen objects were segmented in this study. The right atrial wall, crista terminalis, main branch (a thick muscle bundle projecting from the crista terminalis towards the atrial appendage), roof bundle (a thick muscle bundle projecting from the interatrial septum towards the atrial appendage), a small part of the right ventricle, the aorta with the aortic valve, the coronary sinus, the tricuspid valve, part of the mitral valve and the fossa ovalis were easy to recognise in the MR images and were segmented as described above.

The SAN and AVN were identified by comparing the MR images with Masson's trichrome stained and neurofilament-immunolabelled sections from the intercaval region (cut perpendicular to the crista terminalis [11]) and triangle of Koch (cut roughly perpendicular to septal leaflet of tricuspid valve [12]), as well as our previous 3D SAN and AVN models [11],[12], [58]. Neurofilament is a neuronal cytoskeletal protein that is exclusively expressed in the cardiac conduction system of the rabbit. The SAN and AVN can be recognised because they express neurofilament (brown label), whereas the surrounding atrial tissue does not (Figure 1C, F, I). To segment the SAN, MR images (Figure 1A) and Masson's trichrome stained (Figure 1C, left) and neurofilament-immunolabelled sections (Figure 1C, right) with the same features were compared. In the enlarged image within Figure 1A, it is possible to distinguish the SAN from the surrounding atrial tissue. To determine the extent of the SAN, the distance between the crista terminalis and the SAN centre and the length of the SAN centre as seen in Masson's trichrome stained and neurofilament-immunolabelled sections were measured (Figure 1C). Then the SAN was segmented accordingly. The result is shown in Figure 1B. The SAN periphery was not included in this model due to its complexity [59].

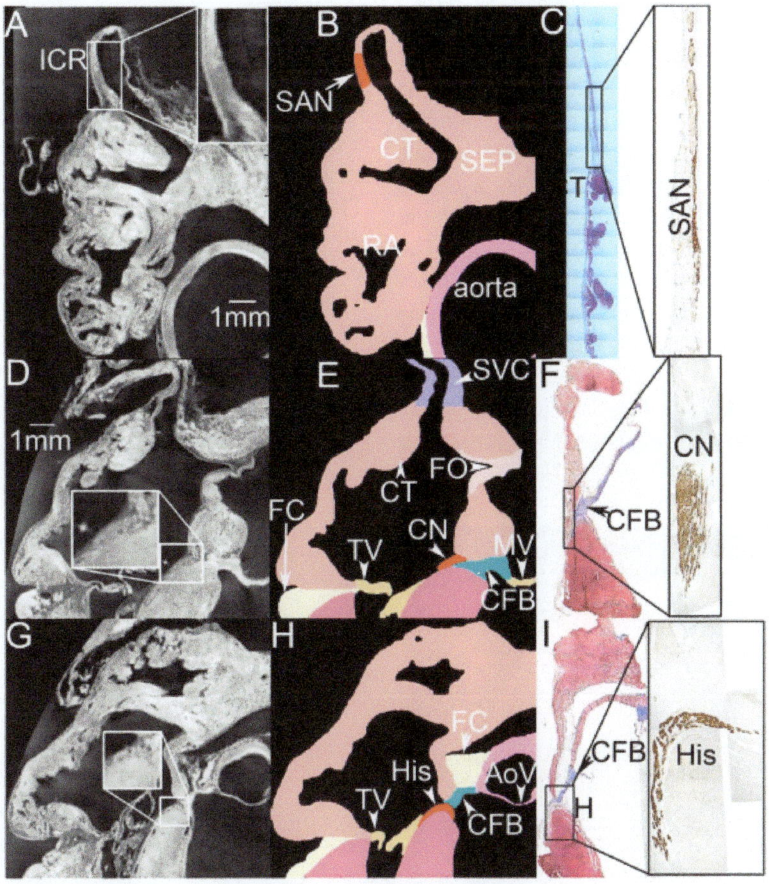

Figure 1: Identification of the SAN and AVN by comparing the MR images with Masson's trichrome stained and neurofilament-immunolabelled sections from the intercaval region and triangle of Koch region.

A, D and G, MR images including the intercaval region (A), the compact node (part of the AVN; D) and the His bundle (G). Nodal regions are enlarged (boxes). B, E and H, corresponding segmented model sections including the SAN (B) and AVN (E and H). Different segmented structures are shown in different colours. C, F and I, sections through the SAN (C) and AVN (F and I) stained with Masson's trichrome and labelled for neurofilament (inset boxes). Masson's trichrome stains myocytes red and connective tissue blue. The neurofilament-positive (brown) cells are nodal. AoV, aortic valve; CFB, central fibrous body; CN, compact AVN; CT, crista terminalis; FC, outer fatty and connective tissue; FO, fossa ovalis; His, His bundle; ICR, intercaval region; MV, mitral valve; RA, right atrium; RV, right ventricle; SEP, interatrial septum; SVC, superior vena cava; TV, tricuspid valve. http://dx.doi.org/10.1371/journal.pone.0112547.g001

Figure 1D-F shows images through the compact part of the AVN. Once again, a MR image (Figure 1D) was compared with Masson's trichrome stained (Figure 1F, left) and neurofilament-immunolabelled (Figure 1F, right) sections with the same features. The neurofilament-positive (brown) cells in Figure 1F are AVN cells. In the enlarged image within Figure 1D, the AVN can be easily distinguished from the surrounding atrial muscle. The central fibrous body, stained blue with Masson's trichrome (Figure 1F, left), can be distinguished from the surrounding atrial muscle in the MR images by comparing Figure 1D with Figure 1F. Figure 1E shows a section from the segmented multiple-object model based on Figure 1D. Figure 1G-I shows comparable images through the His bundle; an analogous method was used to segment it (Figure 1H). The location of the tendon of Todaro and the right and left sinoatrial ring bundles were determined by comparing the MR images and Masson's trichrome stained sections of a similar region. It is well known that there is a conduction block zone from the SAN towards the interatrial septum[60]. It was not possible to detect the block zone in either MR images or tissue sections. Hence, the block zone was segmented according to previously published activation maps of the rabbit SAN [61].

Finally, a 3D right atrium volume model with ~3.16 million elements and eighteen objects was constructed. The objects are the right atrial wall, the crista terminalis, the main branch, the roof bundle, the SAN, the AVN (inferior nodal extension and penetrating bundle), part of the right ventricle, the aorta with the aortic valve, the superior vena cava, the inferior vena cava, the coronary sinus, the tricuspid valve, part of the mitral valve, the fossa ovalis, the central fibrous body, the block zone, and the outer fatty and connective tissue. The isosurface rendering technique was used to visualise the 3D volume model. The tendon of Todaro and the right and left sinoatrial ring bundles were highlighted using spline lines. Figure 2 shows various views of the 3D volume model. Figure 2A shows a dorsal oblique view of the model. The outer shape of the right atrium can be seen. The atrial muscle is shown in pink. The right atrial appendage can be easily recognised. The SAN (red) is located on the dorsal side within the intercaval region (the region between the superior vena cava and inferior vena cava). The crista terminalis (purple) is located on the right side of the SAN and the block zone (grey) is located on the left side of the SAN. The coronary sinus (green) runs under the inferior vena cava (blue), approximately perpendicular to the inferior vena cava. Figure 2B shows the model with transparent atrium to reveal internal structures. The ventricles, valves and central fibrous body were removed to reveal the AVN. The main branch (purple), roof bundle (purple) and AVN (inferior nodal extension, red; penetrating bundle, orange) can be seen through the transparent right atrium. Figure 2C is an internal ventral view. The tricuspid valve (yellow) and connective tissue sheath (blue) surrounding the penetrating bundle (orange) are transparent. The tendon of Todaro (green line) and the right and left sinoatrial ring bundles (dark blue lines) can be seen. The fossa ovalis (pink) is located on the left side of the left sinoatrial ring bundle. The SAN is surrounded by the right and left sinoatrial ring bundles. The three big muscle bundles (crista terminalis, main branch and roof bundle; purple) are clearly seen. The main branch branches from the crista terminalis. The roof bundle is analogous to a roof beam connecting the interatrial septum with the right atrial free wall. Figure 2D shows an internal cranial right view. The area surrounded by a thin black line is the AVN. Part of the inferior nodal extension (red) is covered by a thin layer of atrial tissue (pink). The penetrating bundle is covered by a connective tissue sheath (transparent blue). The ventral end of the penetrating bundle lies underneath the tricuspid valve (transparent yellow). The bright yellow stars in Figure 2B, C, D mark the position of the compact AVN.

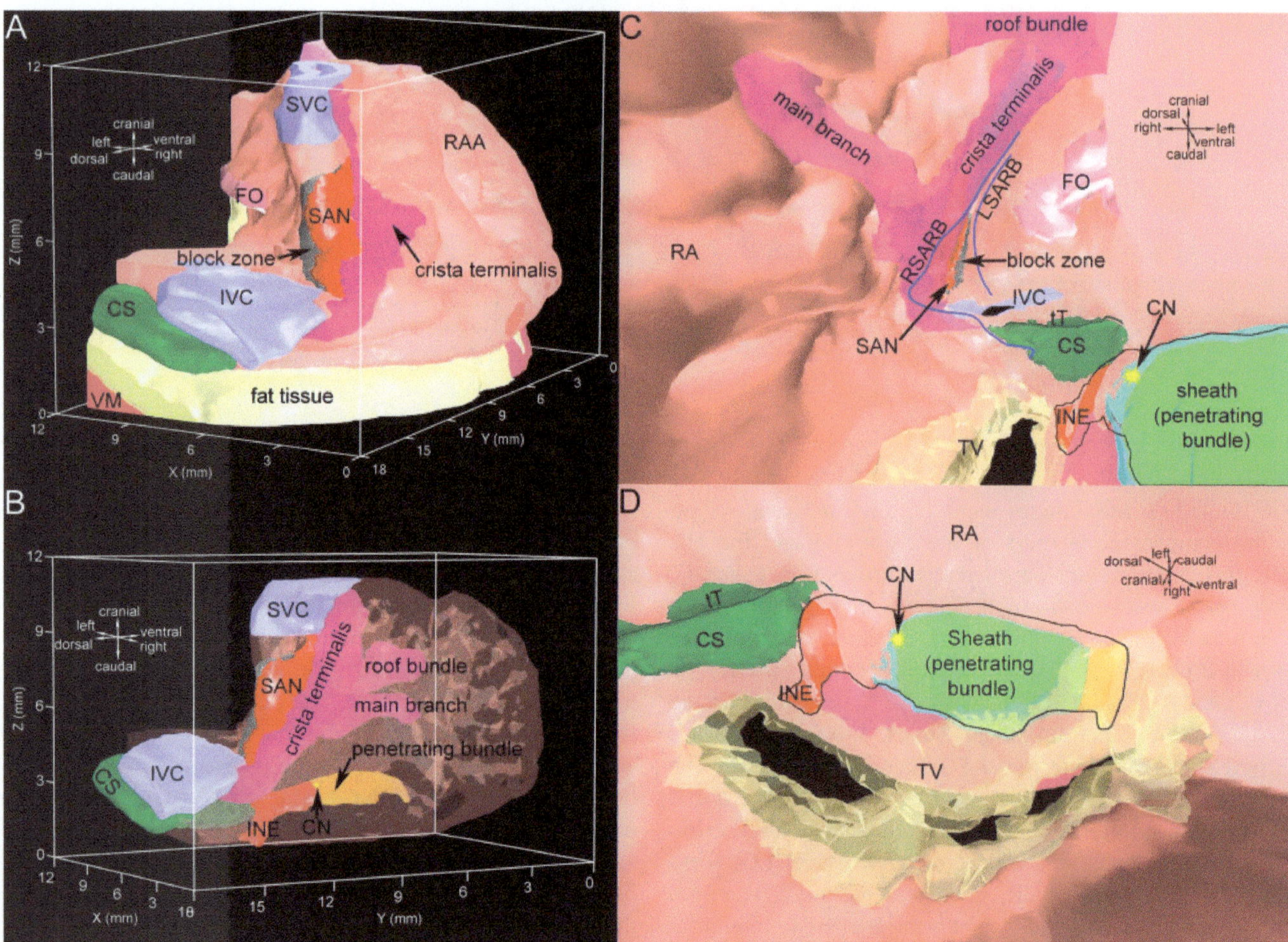

Figure 2: 3D model of the rabbit right atrium including the SAN and AVN.

A, dorsal oblique view of the model. B, model with transparent atrium to reveal internal structures. C, internal ventral view of the model. D, internal cranial right view of the model. Different segmented structures are shown in different colours. CN, compact node; CS, coronary sinus; INE, inferior nodal extension; IVC, inferior vena cava; FO, fossa ovalis; LSARB, left sinoatrial ring bundle; RAA, right atrial appendage; RSARB, right sinoatrial ring bundle; SVC, superior vena cava; tT, tendon of Todaro; TV, tricuspid valve.

http://dx.doi.org/10.1371/journal.pone.0112547.g002

Simulation of the Electrophysiological Behaviour of the Rabbit Right Atrium

Simulation Domain

For numerical simulation, a simulation domain was defined to exclude some objects which are electrically inexcitable: aorta and aortic valve, tricuspid valve, mitral valve, central fibrous body and the fatty and connective tissue. Because the model did not include Purkinje fibres, the ventricle was not included in the simulation domain. It is not possible to separate the superior and inferior vena cava from the right atrium (Figure 1A, D) and, therefore, they were treated as excitable tissue with the same properties as atrial muscle. There are cardiac myocytes in the coronary sinus area [62] and again it was treated as excitable tissue with the same properties as atrial muscle. Cardiac myocyte orientation is important for the propagation of the action potential. It was not possible to detect myocyte orientation from the MR imaging data. However, it is reasonable to assume that cardiac myocytes run longitudinally along the muscle bundles of the atria. The pectinate muscle bundles within the right atrial wall are complex and difficult to segment. Also it is not possible to infer myocyte orientation within the intercaval region and interatrial septum by segmenting muscle bundles. Hence these parts of the atrial wall were defined as isotropic. Only three major muscle bundles (crista terminalis, main branch and roof bundle) were segmented and considered as anisotropic with myocytes running longitudinally along the muscle bundles. The SAN was simulated as an anisotropic material [60]. In summary, we have six zones (~1.6 million elements) with different properties: part of the atrial wall (isotropic atrial tissue), three major muscle bundles (anisotropic atrial muscle), SAN (anisotropic), inferior nodal extension and penetrating bundle (isotropic material) and the block zone (passive atrial tissue – non excitable).

Because the right atrial model has structural complexity and consists of ~1.6 million elements, instead of biophysically-detailed models, two caricature models of the action potential were used to simulate the action potential of the right atrium.

Electrophysiological models: The original FitzHugh-Nagumo model [56] was modified to simulate the spontaneous action potential of the SAN, which shows pacemaker behaviour:

$$\frac{\partial u}{\partial t} = \nabla\cdot \boldsymbol{D}\nabla u + c(u(u-\alpha)(1-u)-v); \quad \begin{cases} c=c_{1S} & \frac{\partial u}{\partial t}\geq 0 \\ c=c_{2S} & \frac{\partial u}{\partial t}<0 \end{cases}$$

$$\frac{\partial v}{\partial t} = b(u-dv-0.1) \tag{1}$$

where u and v are the excitation variable and recovery variable, **n** is a vector normal to the boundary, ***D*** is the diffusion tensor, and α, b, c_{1S}, c_{2S}, and d are parameters that define the shape of the excitation variable u. The action potential V_m and the threshold potential V_{th} are normalised by:

$$u = \frac{V_m - V_r}{V_{os} - V_r}; \quad \alpha = \frac{V_{th} - V_r}{V_{os} - V_r} \tag{2}$$

where V_{os} is the overshoot potential and V_r is the resting potential.

The Rogers-modified FitzHugh-Nagumo model [63] is shown below:

$$\frac{\partial u}{\partial t} = \nabla\cdot \boldsymbol{D}\nabla u + c_1 u(u-\alpha)(1-u) - c_2 uv$$

$$\frac{\partial v}{\partial t} = b(u-dv)$$

with boundary condition

$$\frac{\partial u}{\partial \mathbf{n}} = 0 \tag{3}$$

The first equation of the Rogers-modified FitzHugh-Nagumo model (above) was modified further to simulate the action potential of the atrial muscle, inferior nodal extension and penetrating bundle:

$$\frac{\partial u}{\partial t} = \nabla\cdot \boldsymbol{D}\nabla u + c[u(u-\alpha)(1-u) - uv]; \quad \begin{cases} c=c_{1A} & \frac{\partial u}{\partial t}\geq 0 \\ c=c_{2A} & \frac{\partial u}{\partial t}<0 \end{cases} \tag{4}$$

In this equation, c_{1A} and c_{2A} are different from c_1 and c_2 of Equation 3.

The SAN periphery was not included in the 3D atrial model because of its complexity [11]. However, a border area of the SAN, 0.24 mm (4 elements) wide, was defined as an ideal SAN periphery using Equation 1. Its property changes gradually as follows:

$$c_{1P} = c_{1S} + disSA\frac{(c_{1A} - c_{1S})}{0.24} \tag{5}$$

where c_{1P} is c_{1S} in the SAN model (Equation 1). disSA is the minimum distance between one SAN element and atrial elements.

The block zone was modelled as a passive tissue:

$$\frac{\partial u}{\partial t} = \nabla\cdot \boldsymbol{D}\nabla u - \frac{u}{R_b} \tag{6}$$

where R_b is resistivity. R_b was set to 0.5, which is sufficient to inhibit the action potential.

To define model parameters for SAN, atrial muscle, inferior nodal extension and penetrating bundle, a 50×5×5 elements strand tissue model was used [64]. Stimulation was applied to the first three layers of elements (3×5×5). The conduction velocity was measured as the average conduction velocity calculated from the 10th element layer to the 40th element layer. Action potential duration was measured at 90% repolarization. A standard S1–S2 protocol was used to measure the refractory period. The spontaneous cycle length of the SAN (set by the modified Fitzhugh-Nagumo model) is 330 ms (corresponding to a heart rate of 182 beats/min).Table 1 lists the parameters for the SAN, atrial muscle, inferior nodal extension and penetrating bundle. Table 2 lists the conduction velocity, maximum upstroke velocity, action potential duration and refractory period of the SAN, atrial muscle, inferior nodal extension and penetrating bundle from simulations and compares them to experimentally determined values. The data show that the majority of simulation results are within the range of experimental results. Figure 3 shows action potential waveforms from simulations (left) and experiment (right). The simulated action potential waveforms are a reasonable fit to the experimental waveforms.

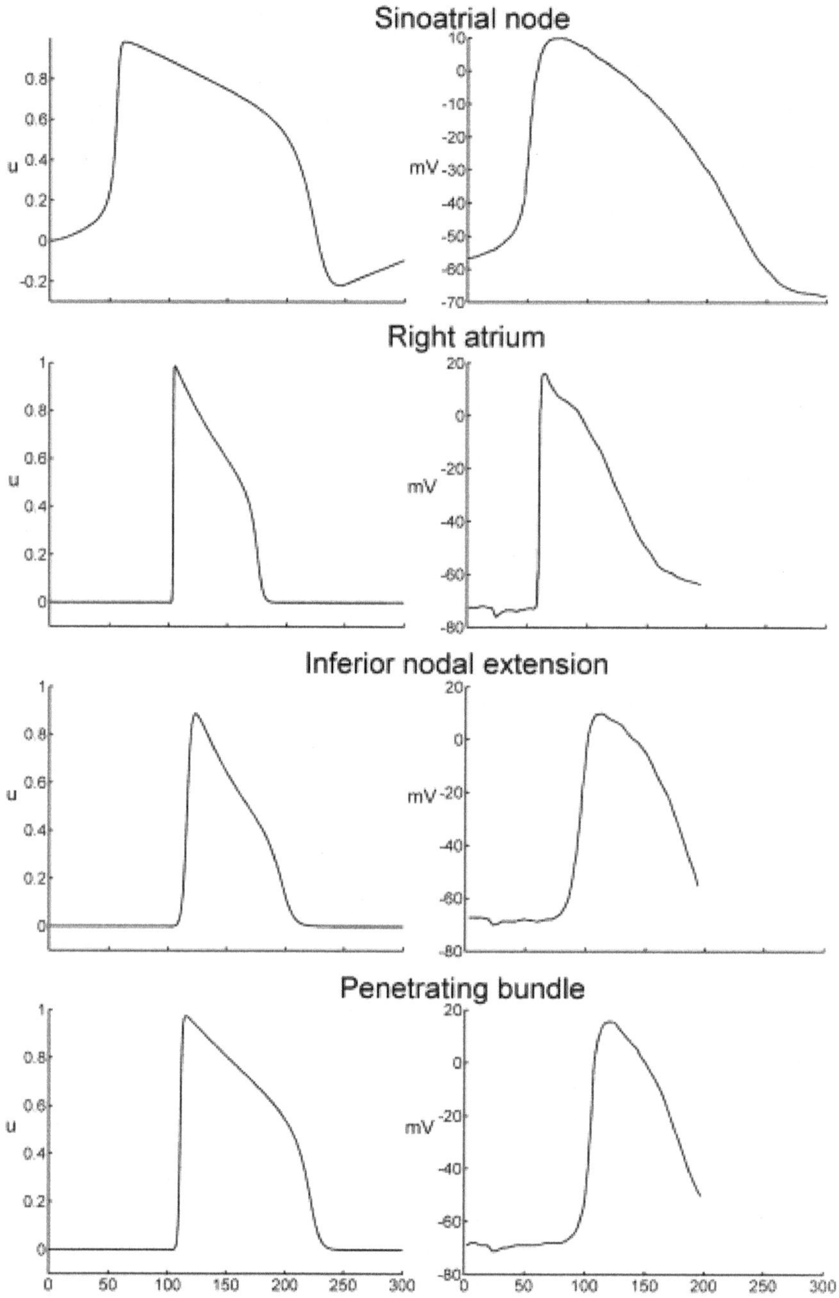

Figure 3: Comparison of action potential waveforms in the SAN, right atrium, inferior nodal extension and penetrating bundle in simulations (left) and experiment (right). Experimental recordings from Kodama et al. [95] (SAN) and de Carvalho and de Almeida [96]. http://dx.doi.org/10.1371/journal.pone.0112547.g003

Table 1: Parameters for the Fitzhugh-Nagumo model for the SAN, atrial muscle and AVN. http://dx.doi.org/10.1371/journal.pone.0112547.t001

Parameters	SAN	Atrial muscle	Inferior nodal extension	Penetrating bundle
D	2	7	2	2
c_1	1	12.7	1.45	3.05
c_2	0.22	1.84	1	1
b	0.003	0.01	0.013	0.0048
d	3.5	2.475	2.5	2

Table 2: Measurements of conduction velocity, maximum upstroke velocity of the action potential, action potential duration and refractory period of the SAN, atrial muscle and AVN from the Fitzhugh-Nagumo model and experiment. http://dx.doi.org/10.1371/journal.pone.0112547.t002

		SAN	Atrial muscle	Inferior nodal extension	Penetrating bundle
Conduction velocity (m/s)	**simulation**	0.0673	0.5333	0.0949	0.1413
	experiment	0.02–0.12 [65]	0.3–0.8 [81]	0.02–0.1 [82]	0.1–0.15
Maximum upstroke velocity (ms^{-1})	**Simulation**	0.124	1.583	0.1578	0.3755
	experiment	0.075–0.16 [83–88]	1.5±0.5 [89]	0.164 [39]	0.376 [39]
Action potential duration (ms)	**Simulation**	185	75.05	93.7	118.07
	experiment	107–151 [88,90,91]	77±5 [89]	97.2 [39]	117.5 [39]
Refractory period (ms)	**Simulation**	283	82	91	154
	experiment	166±30 [92]	68±11 [92] 81±5 [39]	91±10 [93] 91±12 [39] 100±9 [94]	–

The SAN and three major muscle bundles (crista terminalis, main branch and roof bundle) were considered as anisotropic. In the SAN, cells are reported to be oriented in different directions (forming a mesh) [11]. Experimental results suggest that the SAN is anisotropic, and the conduction velocities near the centre of SAN towards the SVC and IVC are faster than towards the crista terminalis and interatrial septum [65]–[67]. Hence it is reasonable to assume that most myocytes in the SAN run parallel to the crista terminalis. Let ***f*** denote the fibre orientation: ***f***=l***i***+m***j***+n***k***, where ***i***, ***j*** and ***k*** are unit vectors of a Cartesian coordinate system; l, mand n are direction cosines, respectively. The properties of anisotropic fibres are introduced through an anisotropic diffusion tensor $\boldsymbol{D}_f$ in the fibre coordinate system:

$$\boldsymbol{D}_f = \begin{pmatrix} D_l & 0 & 0 \\ 0 & D_t & 0 \\ 0 & 0 & D_t \end{pmatrix} \quad (7)$$

where D_l is the diffusion in the fibre direction and D_t is the diffusion in the transverse direction. Then it is transformed to the global coordinate system as

$$\boldsymbol{D} = D_t \boldsymbol{I}_3 + (D_l - D_t)\begin{bmatrix} l \\ m \\ n \end{bmatrix}\begin{bmatrix} l & m & n \end{bmatrix} \quad (8)$$

where $\boldsymbol{I}_3$ is the identity matrix. The diffusion anisotropy ratio was defined as 10 (D_l/D_t). This yields a ~3.2 conduction velocity anisotropy ratio since conduction velocity is proportional to the square root of the diffusion [68].

Cellular Automaton Model

In the cellular automaton model, each node can be in one of three states: resting, excited or refractory. In the model, each node i has a variable (u_i), which signifies node state. When the node state is resting, u_i is set to 0. Once a node is excited, the node state is changed to 1. After that the node state is gradually decreased from 1 to 0. The speed of decrease determines the refractory period. The 'action potential' in the cellular automaton model is, therefore, characterised by the time at which the node is excited and at which it returns to the resting state, and the shape of the action potential can be considered triangular (Figure 4A). In the simulation program, if a node is in the resting state, the program checks at every time step whether one of its neighbours j is in the excited state. If so, an inner variable (e_i) called the excitation counter (corresponding to ionic current from each excited neighbouring myocyte) is increased at each time step. If the excitation counter exceeds a predefined threshold ($\theta \le \sum_j e_j$), the node switches to the excited state (u_i=1). In the excited state, a node can excite its neighbours for predefined time steps (E_i) and then switches to the refractory state, where it again stays for a constant number of time steps (R_i). At the end of the refractory state, the node switches back into the resting state (u_i=0). Figure 4B shows a flow-chart of the program. Changes in conduction velocity and refractory period in different tissues (shown inTable 3) were introduced as discussed by Li et al. [12].

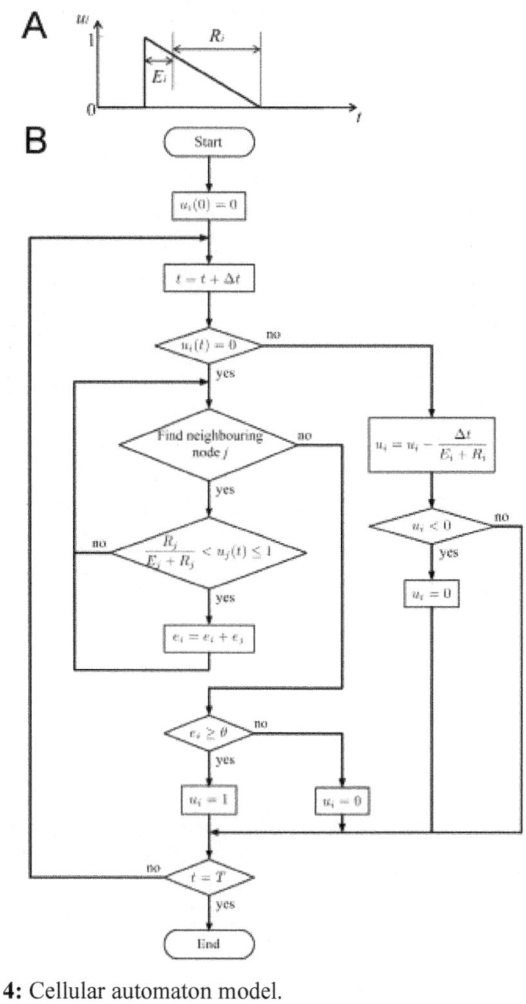

Figure 4: Cellular automaton model.

A, time course of state (u) of node i in the cellular automaton model. If node i is excited the node state u_i is changed from 0 (resting) to 1 (excited). u_i then declines back to 0. B, program flow-chart of the cellular automaton model. E_i, the excited period; R_i, the refractory period. t, time; Δt, the time increment; T, the end time of simulation; e_i, the excitation counter; θ, the threshold. http://dx.doi.org/10.1371/journal.pone.0112547.g004

Table 3: Measurements of conduction velocity and refractory period of atrial muscle and AVN from the cellular automaton model and experiment. http://dx.doi.org/10.1371/journal.pone.0112547.t003

		Atrial muscle	Inferior nodal extension	Transitional zone	Penetrating bundle
Conduction velocity (m/s)	simulation	0.27–0.55	0.05–0.114	0.154	0.104
	experiment	0.3–0.8 [81]	0.02–0.1 [82]	0.35±0.17 [70]	0.1–0.15
Refractory period (ms)	simulation	81	94	134	154
	experiment	68±11 [92] 81±5 [39]	91±10 [93] 91±12 [39] 100±9 [94]	127±9 [39] 141±15 [93]	-

All simulations were carried out on a high-performance computing cluster which has 10 quad-processors nodes.

RESULTS

Conduction of the Action Potential from the SAN to the AVN during Sinus Rhythm

The conduction of the action potential through the right atrium during sinus rhythm was simulated using the 3D anatomically-detailed right atrium model, together with a set of Rogers modified FitzHugh-Nagumo models. Figure 5 shows the activation sequence of the right atrium during sinus rhythm. The action potential was generated spontaneously in the centre of the SAN at 0 ms (Figure 5A, B). It spread preferentially in an oblique cranial direction (as a result of the orientation of nodal myocytes) and reached the crista terminalis first at ~15 ms and then spread in a radial fashion to the rest of the right atrium (Figure 5A, B). It had to propagate around the block zone to reach the interatrial septum at ~35 ms (Figure 5A, B). In the model, the action potential reached the rest of the right atrium in ~50 ms (Figure 5A, B). Propagation from the SAN to the AVN is shown in Figure 6A, B: the action potential rapidly spread down the crista terminalis (Figure 6A) and it entered the AVN at the inferior nodal extension ~40 ms after it first entered the crista terminalis (Figure 6B) and ~50 ms after its initiation in the SAN (Figure 5C). Meanwhile, the action potential also spread up the crista terminalis (Figure 6A) and down the interatrial septum and it entered the AVN via the compact node at the junction of the inferior nodal extension with the penetrating bundle again ~40 ms after it first entered the crista terminalis (Figure 6B) and ~50 ms after its initiation in the SAN (Figure 5C). This suggests that there are two principal pathways from the SAN to the AVN: one via the crista terminalis and the second via the interatrial septum. Although the action potential arrived at the AVN via both pathways at approximately the same time, the action potential arriving via the crista terminalis entered the atrioventricular conduction pathway at an upstream site, whereas the action potential arriving via the interatrial septum entered at a downstream site. Once the action potential from the crista terminalis had entered the AVN, it then began propagating along the inferior nodal extension towards the penetrating bundle (Figure 5D). Once the action potential from the interatrial septum had entered the AVN, the action potential propagated anterogradely along the penetrating bundle (Figure 5D). However, it also propagated retrogradely along the inferior nodal extension (Figure 5D) and finally collided with the action potential coming from the opposite direction (Figure 5D). Figure 5E is a summary of the activation sequence from the SAN to the AVN. The simulation suggests that the most important pathway for the action potential from the SAN to the AVN is the interatrial septum; the pathway from the SAN to the AVN via the crista terminalis is secondary. The action potential propagated anterogradely along the penetrating bundle and it reached the bundle of His at ~120 ms after its initiation in the SAN (Figure 5D).

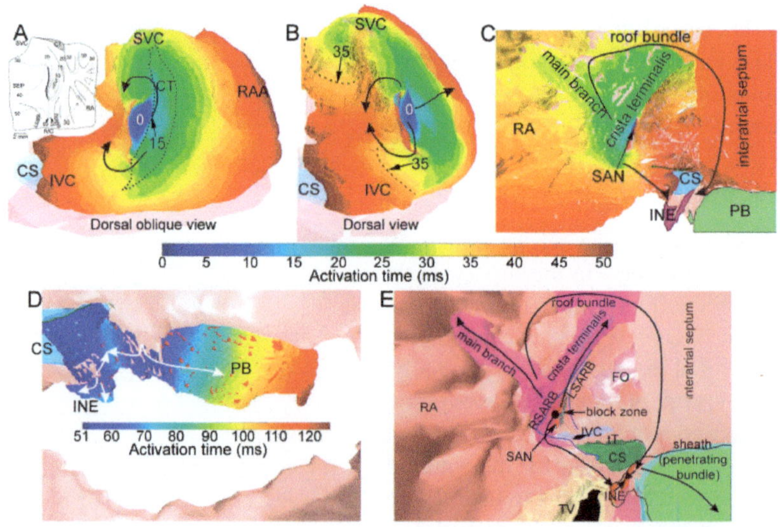

Figure 5: The sequence of right atrial activation during normal sinus rhythm in the 3D model of the rabbit right atrium.

A–E, external (A, dorsal oblique view and B, dorsal view) and internal (C–E) views of the model. In A, the crista terminalis is outlined by a dotted line. In B, the dashed line is an example isochrone at 35 ms. In A to D, the activation sequence is shown by a colour scale and the arrows show the direction of action potential conduction. In E, different segmented structures are shown in different colours and the arrows summarise action potential conduction from the SAN to the AVN. CS, coronary sinus; INE, inferior nodal extension; IVC, inferior vena cava; FO, fossa ovalis; LSARB, left sinoatrial ring bundle; PB, penetrating bundle; RA, right atrium; RAA, right atrial appendage; RSARB, right sinoatrial ring bundle; SVC, superior vena cava; TV, tricuspid valve. http://dx.doi.org/10.1371/journal.pone.0112547.g005

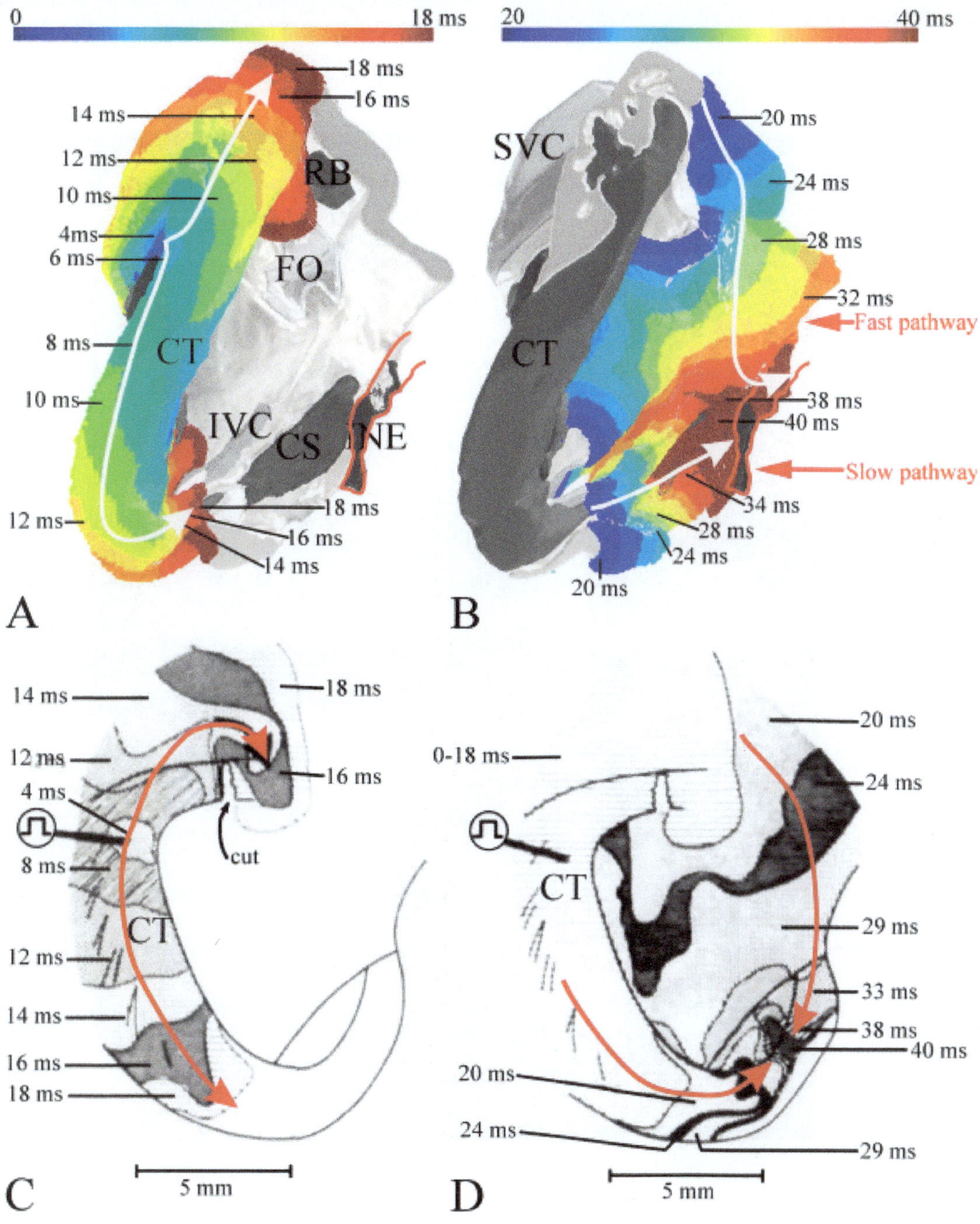

Figure 6: Action potential conduction from the SAN to the AVN.

A and B, simulation of action potential conduction from the SAN to the AVN in the model of the right atrium during sinus rhythm (A, early times; B, later times). Internal right view of the model shown. Anatomical structures are shown in grey scale. The activation sequence is shown by a colour scale and the arrows show the direction of action potential conduction. Activation times are relative to the arrival of the action potential at the crista terminalis from the SAN. C and D, equivalent experimental data for the rabbit right atrium from Spach et al. [72]. In the experiment, the crista terminalis was stimulated close to the site where the action potential is expected to arrive first from the SAN. CN, compact node; CS, coronary sinus; CT, crista terminalis; FO, fossa ovalis; INE, inferior nodal extension; IVC, inferior vena cava; PB, penetrating bundle; RB, roof bundle; SVC, superior vena cava. http://dx.doi.org/10.1371/journal.pone.0112547.g006

Nodal Activity during Atrial Fibrillation

In the model, an S1–S2 protocol was used to simulate an atrial fibrillation-like arrhythmia. TheS1 'stimulus' was a normal sinus rhythm beat initiated in the centre of SAN (Figure 7A; 39 ms). The activation time was counted from starting the simulation. The S2 stimulus was a premature planar stimulation applied to the superior vena cava (Figure 7B; 159 ms). Three reentrant circuits were observed (Movie S1, S2 and S3). The first reentrant circuit was located on the SVC and was a sustained reentrant circuit (Figure 7C; top). The second reentrant circuit was located on the IVC; it started at 495 ms and it did not stop by the end of the simulation at 1500 ms (Figure 7C; bottom). The third reentrant circuit was located on the atrial free wall and started at 780 ms and stopped at 1020 ms (Figure 7D). Figure S2 shows a set of snapshots of the action potential distribution on the epicardial surface during the atrial reentrant arrhythmia.Figure S3 shows a set of snapshots of the action potential distribution on the endocardial surface during the atrial reentrant arrhythmia.

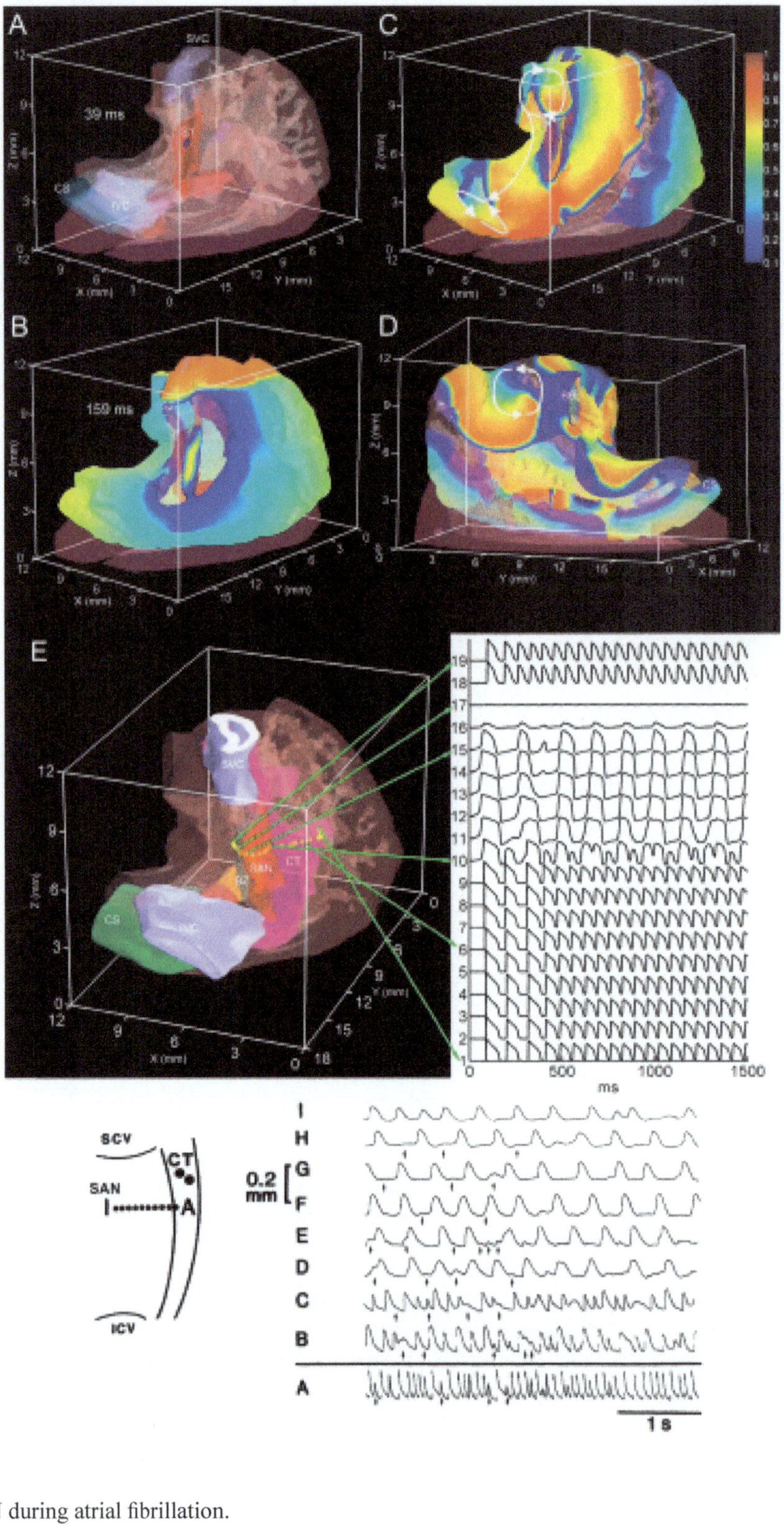

Figure 7: Behaviour of the SAN during atrial fibrillation.

Panels A–E show external views (A–C and E, dorsal oblique view; D, ventral oblique view) of the model of the right atrium with transparent atrial muscle and different segmented structures shown in different colours. A–D, snapshots of the action potential distribution on the epicardial surface on initiation of an S1 stimulus (A), on initiation of anS2 stimulus, and at two times points after the induction of atrial fibrillation (C and D). The reentry loops are highlighted by arrows. E, right atrium model showing the location of the recording sites of the action potentials shown on the right. The sites lie along a line perpendicular to the crista terminalis through the SAN. Inset, experimental data showing the behaviour of the rabbit SAN during atrial fibrillation. Left, schematic diagram of the preparation showing the location of the recording sites of the action potentials on the right. The sites lie along a line perpendicular to the crista terminalis through the SAN. From Kirchhof et al. [75]. BB, bachmann bundle; BZ, block zone; CS, coronary sinus; CT, crista terminalis; IVC (or ICV), inferior vena cava; SVC (or SCV), superior vena cava. http://dx.doi.org/10.1371/journal.pone.0112547.g007

Figure 7E shows the predicted behaviour of the SAN during atrial fibrillation. It shows action potentials recorded at different sites along a line perpendicular to the crista terminalis through the SAN: atrial muscle (e.g. site 1), SAN periphery (e.g. site 10), SAN centre (e.g. site 15), block zone (sites 16 and 17) and interatrial septum (sites 18 and 19). There is a high frequency of fibrillatory action potentials in the atrial muscle (~14 Hz; e.g. site 1), chaotic activity in the periphery of the SAN (site 10) and a slow frequency of the action potentials in the centre of the SAN (~6 Hz; e.g. site 15) as a result of alternating 2:1 and 3:1 SAN entrance block (compare action potentials in and around the periphery of the SAN at sites 9, 10 and 11). The model predicts that the SAN is protected against the high rate of fibrillatory action potentials and is not overdriven - this behaviour is due to the long refractory period of the SAN (Table 2).

Figure 8 shows the predicted behaviour of the AVN during atrial fibrillation. Figure 8A–E shows an internal view of the right atrium and atrial action potential propagation during atrial fibrillation. Action potential propagation during normal sinus rhythm (S1) is shown in Figure 8A. Figure 8B–E shows action potential propagation at different time points of the atrial fibrillation. During atrial fibrillation, the action potential entered the AVN at different times through both pathways (fast and slow). Some reentrant action potentials reached the fast pathway and the compact node from the roof bundle and the interatrial septum (Figure 8B, D), whereas others reached the fast pathway from the area of inferior vena cava and coronary sinus (Figure 8C, E). Some reentrant action potentials were blocked (dashed lines in Figure 8B, D), because of the long refractory period of the penetrating bundle (Table 2). Figure 8F shows action potentials recorded at the atrioventricular junction axis: atrial muscle (sites 1–7), inferior nodal extension (sites 8–14) and penetrating bundle (sites 15–21). It shows the high frequency (~14 Hz) fibrillatory action potentials in the atrial muscle (sites 1–7), chaotic activity in some transitional regions of the AVN (e.g. site 15) and a low frequency (~5 Hz) of action potentials in the distal penetrating bundle (site 16), as the result of ~3:1 Wenckebach block (compare recordings at sites 15 and 16). The Wenckebach block is the result of the long refractory period of the AVN (Table 2). In summary, the simulation shows that the high frequency fibrillatory activity is filtered by the AVN; this is important, because it prevents the ventricles from being paced too fast.

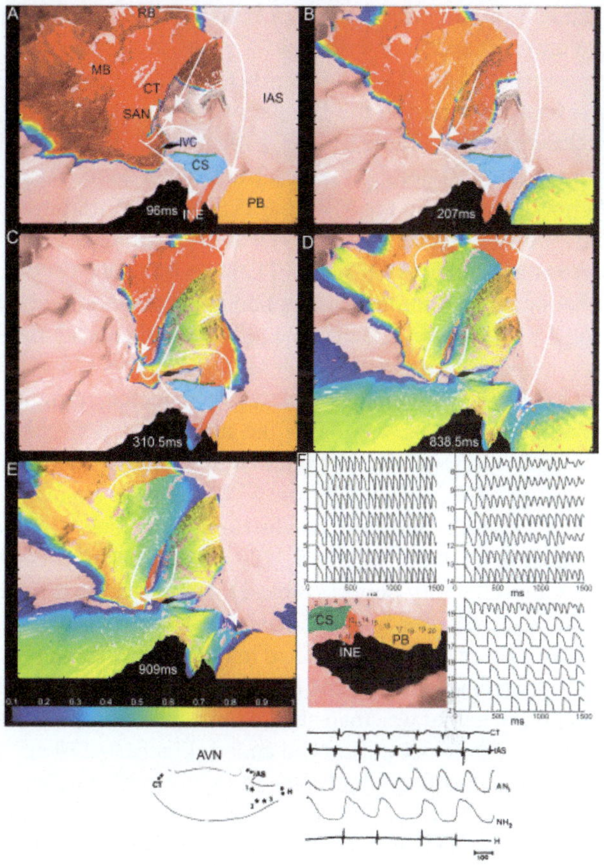

Figure 8: Behaviour of the AVN during atrial fibrillation.

Panels A–E show a ventral internal view of the model of the right atrium with different segmented structures shown in different colours. A–E, snapshots of the action potential distribution on the endocardial surface during normal sinus rhythm (A) and at different times points after the induction of atrial fibrillation (B–E). The white arrows highlight wave propagation. The white dotted line in B and D indicates a line of conduction block. F, right atrium model showing the location of the recording sites of the action potentials shown above and on the right. Inset, experimental data showing the behaviour of the rabbit AVN during atrial fibrillation. Left, schematic diagram of the preparation showing the location of the recording sites of the electrograms and action potentials on the right. From Mazgalev et al. [80]. AN, atrionodal cell; CS, coronary sinus; CT, crista terminalis; IAS, interatrial septum; INE, inferior nodal extension; IVC, inferior vena cava; H, His bundle; MB, main branch; NH, nodal-His cell; PB, penetrating bundle; RB, roof bundle. http://dx.doi.org/10.1371/journal.pone.0112547.g008

Simulation of AVN Reentry and a Ventricular Echo Beat Generated by Premature Ventricular Stimulation

A premature ventricular action potential can elicit a ventricular echo beat as a result of retrograde conduction through the AVN followed by AVN reentry and anterograde conduction through the AVN [69]. This was investigated using the 3D anatomically-detailed right atrium model together with the cellular automaton model of the action potential (less computationally intensive than the Fitzhugh-Nagumo model). AVN reentry is dependent on the fast pathway (a transitional zone contacting the compact node) and the slow pathway (inferior nodal extension) into the AVN having different refractory periods; the refractory period of the fast pathway is longer than that of the slow pathway (Table 3) [39], [70]. The model of the right atrium (Figure 2) does not have a transitional zone, because it was not detectable in the MR images. Therefore, a transitional zone (shown in green in Figure 9) was segmented based on our previous 3D rabbit AVN model [12]. To replicate ventricular activation, a S1 stimulus was applied at the penetrating bundle of the AVN (Figure 9; 20 ms time point) and then a S2stimulus was applied with a short cycle length (Figure 9; 120 ms time point). The S1 action potential was conducted retrogradely from the penetrating bundle to the atrium via both fast and slow pathways, i.e. by both the transitional zone and the inferior nodal extension (Figure 9; 70 ms time point). After applying the S2 stimulus, conduction was blocked through the transitional zone (Figure 9; 200 ms time point), because of its long refractory period (Table 3). However, the refractory period of the inferior nodal extension is shorter (Table 3) and the action potential was conducted to the atrium via the inferior nodal extension (Figure 9; 230 ms time point). From the atrium, the action potential was then able to enter the transitional zone, because its refractory period had ended (Figure 9; 270 ms time point). From the transitional zone, the action potential was then conducted anterogradely to the penetrating bundle (Figure 9; 270 ms time point); in a heart, this would result in a ventricular echo beat.

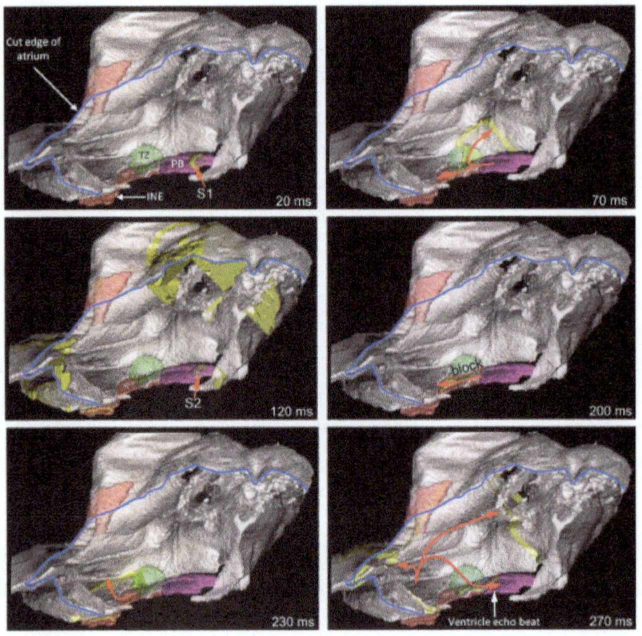

Figure 9: Simulation of a ventricular echo beat.

A right view of the model of the right atrium is shown with one face of the right atrium removed (the blue line shows the cut edge of the right atrium). Different segmented structures are shown in different colours. The panels show snapshots of activation afterS1–S2 stimulation of the penetrating bundle. The thick yellow lines show the advancing wavefront of activation. The arrows show the direction of action potential conduction. 20 ms, retrograde conduction of the action potential along the penetrating bundle followingS1 stimulation. 70 ms, retrograde conduction of the action potential along the fast (transitional zone; green) and slow (inferior nodal extension, red) pathways out of the AVN following S1 stimulation. 120 ms, retrograde conduction of the action potential along the penetrating bundle following S2 stimulation. Note the action potential followingS1 stimulation has reached the outer parts of the right atrium. 200 ms, block of conduction along the fast pathway (line of block, red dotted line) and retrograde conduction along the slow pathway following S2 stimulation. 230 ms, retrograde conduction out of the slow pathway and into the right atrium following S2 stimulation. 270 ms, retrograde conduction throughout the right atrium and reentry into the AVN – the action potential conducted anterogradely through the transitional zone to the penetrating bundle. It is this action potential that would result in a ventricular echo beat. INE, inferior nodal extension; PB, penetrating bundle; TZ, transitional zone. http://dx.doi.org/10.1371/journal.pone.0112547.g009

DISCUSSION

In this study, a 3D anatomically-detailed rabbit right atrium model including the two nodes was generated. Using this anatomical model together with mathematical caricature models of the action potential, we were able to replicate some known nodal electrophysiology: (i) the two routes of action potential conduction from the SAN into the AVN via the crista terminalis and interatrial septum; (ii) the dominance of the interatrial septum route over the crista terminalis route; (iii) little or no overdrive of the SAN during atrial fibrillation; (iv) reduction of action potential frequency by the AVN during atrial fibrillation; and (v) ventricular echo beats. The success of the simulations is a validation of the model. It also demonstrates that the phenomena can be explained on the basis of the elements incorporated in the model and in particular on anatomy, conduction velocities and refractory periods.

Comparison of Model and Experiments

Conduction of the Action Potential from the SAN to the AVN during Sinus Rhythm

In the simulation of sinus rhythm, the action potential from the SAN had to propagate around the block zone to reach the interatrial septum at ~35 ms (Figure 5A, B). A similar pattern is seen in experiments [60]. As an example, the inset in Figure 5A shows conduction from the leading pacemaker site in the rabbit SAN – the action potential spread preferentially in an oblique cranial direction, reaching the crista terminalis in 15–20 ms, and propagated around the block zone to reach the interatrial septum at 40–50 ms. In the model, the action potential reached the rest of the right atrium in ~50 ms (Figure 5A, B). There is little experimental data to compare this too. De Carvalho et al. [71] worked with a large right atrial preparation from the rabbit including both the SAN and AVN and the furthest reaches of the preparation were activated in 70–80 ms. The model suggests that there are two routes for the action potential from the SAN to the AVN: the crista terminalis and the interatrial septum (Figure 6A, B). This is consistent with experimental data from the rabbit from Spach et al. [72] shown in Figure 6C, D. In this experimental study, the crista terminalis was stimulated (at 0 ms) close to the leading pacemaker site in the SAN. As in the equivalent simulation (Figure 6A, B), the action potential rapidly spread down the crista terminalis and entered the AVN (inferior nodal extension) as well as up the crista terminalis and down the interatrial septum where it entered the AVN downstream of the inferior nodal extension. In the work of Spach et al. [72] the action potential entered the AVN at ~40 ms after stimulation of the crista terminalis (Figure 6C, D). In other work on the rabbit [71], the action potential entered the AVN at ~50 ms after its initiation in the sinus node. These timings are similar in the corresponding simulations (Figures 5C and 6A, B). Both in the human and animal models, it is well known that there are dual inputs into the AVN: the slow and fast pathways [73]. The slow pathway is the inferior nodal extension and the fast pathway is the transitional zone leading into the compact node [12]. The simulations show that the crista terminalis connects to the slow pathway, i.e. the inferior nodal extension, whereas the interatrial septum connects to the fast pathway, i.e. the region of the transitional zone and compact node (Figure 5E). The simulations suggest that the most important pathway for the action potential from the SAN to the AVN is the interatrial septum followed by the fast pathway; this is because the action potential enters the atrioventricular conduction axis at a more downstream site via this route (shortening the conduction time to the His bundle). This is consistent with what is known: both in the human and animal models, it is well known that the fast pathway is the primary pathway for atrioventricular conduction [73]. The slow pathway only operates if the fast pathway fails (for example, during premature stimulation when the fast pathway fails because of its long refractory period [39]). In the model, the action potential propagated anterogradely along the penetrating bundle and reached the bundle of His at ~120 ms (Figure 5D). This is less than an experimental value of 140 ms for the rabbit heart from De Carvalho et al. [71]. However, the PR interval is the time from atrial depolarization to the first activation of the ventricular conduction system and in the rabbit is 62 ms (Boyett, unpublished data). If the time from initiation of the action potential in the SAN to the arrival of the action potential in the atria is 15 ms (Figure 5A), then the time from the initiation of the action potential in the SAN to the first activation of the ventricular conduction system is ~80 ms. In summary, there is good agreement between model and experiment in terms of conduction of the action potential from the SAN to the AVN. The success of the simulations indicates that the principal factor determining the pattern of conduction from the SAN to the AVN is anatomy, in particular the arrangement of fast conducting atrial muscle bundles with longitudinally arranged myocytes, although the slow conduction velocity of the inferior nodal extension is another contributory factor (working against the crista terminalis-slow pathway route).

From simulations, conduction velocities in the SAN were measured as 0.028 m/s in the transverse direction and 0.089 m/s in the longitudinal direction; conduction velocities in the crista terminalis were measured as 0.183 m/s in the transverse direction and 0.417 m/s in the longitudinal direction. The conduction velocity anisotropy ratio in the SAN (~3.16) is within the range of experimental results (1.9~4.1) from rabbit SAN preparations [65]. The conduction velocity anisotropy ratio in the crista terminalis (~2.28) is similar to that measured experimentally in rabbit crista terminalis preparations (1.7~4.1 depending on age [74]).

Nodal Activity During Atrial Fibrillation

The rabbit heart has been widely used in the study of atrial fibrillation [75]–[78]. In the simulation of atrial fibrillation (Figure 7), the fibrillation frequency was 14 Hz in the atrial muscle. This is similar to that measured in the rabbit experimentally under baseline conditions (11.6 Hz[75]; 5.9–6.5 Hz [78]) and following vagal stimulation (12–13 Hz [78]). In the model, the arrhythmia was maintained by reentry located at the SVC and IVC. This is consistent with observations from patients with paroxysmal atrial fibrillation [79]. Kirchhof et al. [75] studied the SAN during atrial fibrillation in the rabbit and an example from their work is shown in the inset at the bottom of Figure 7; it shows action potentials recorded at different sites from the atrial muscle (site A), SAN periphery (e.g. site B) and SAN centre (e.g. site I). The behaviour is qualitatively similar to that predicted by the model: there is a high frequency of fibrillatory action potentials in the atrial muscle (site A), chaotic activity in the periphery of the SAN (e.g. site B) and a slow frequency of the action potentials in the centre of the SAN (e.g. site I) as a result of high degree (5:1) SAN entrance block (compare action potentials at sites A–D). Mazgalev et al.[80] studied the AVN during atrial fibrillation in the rabbit and an example from their work is shown in the inset at the bottom of Figure 8; it shows extracellular electrograms recorded from the crista terminalis (CT), interatrial septum (IAS) and His bundle (H) as well as intracellular action potentials recorded at upstream (site 1) and downstream (site 2) sites in the atrioventricular conduction axis. Again the behaviour is qualitatively similar to that predicted by the model: there is a high frequency of fibrillatory action potentials in the atrial muscle (CT and IAS), chaotic activity in a transitional region (site 1) and a slow frequency of action potentials downstream of this, including in the His bundle (corresponding to ~2:1 Wenckebach block). In summary, there is good agreement between model and experiment in terms of the behaviour of the nodes during atrial fibrillation. The success of the simulations can be attributed to the long refractory period of the nodal tissues.

AVN Reentry and Ventricular Echo Beats

A premature ventricular action potential can elicit a ventricular echo beat as a result of AVN reentry [69] and the model was able to successfully simulate this (Figure 9).

Limitations of the Study

It was not possible to obtain fibre orientation from the MR images and only three major muscle bundles (crista terminalis, main branch and roof bundle) were segmented in order to introduce the well known anisotropic property of the bundles. The SAN was also assumed to be anisotropic in order to replicate the known anisotropic conduction in the SAN, whereas the rest of the right atrium was assumed to be isotropic. Although the success of the simulations demonstrates that the level of detail in the model is sufficient to explain the various behaviours, incorporation of fibre orientation throughout the right atrium model is expected to improve the accuracy of simulations. Mathematical caricature models of the action potential were used to make computation more tractable. Once again, although the success of the simulations demonstrates that the level of detail in the models is sufficient to explain the various behaviours, use of biophysically-detailed models of the action potential is expected to improve the accuracy of simulations.

SUPPORTING INFORMATION

Figure S1. 3D segmentation. A, isotropic 3D MR images; B, converted 3D binary images; C, the 3D model before segmentation; D, the 3D model after segmentation. CT, crista terminalis; RA, right atrium; RV, right ventricle; SAN, sinoatrial node; SVC, superior vena cava; TV, tricuspid valve.

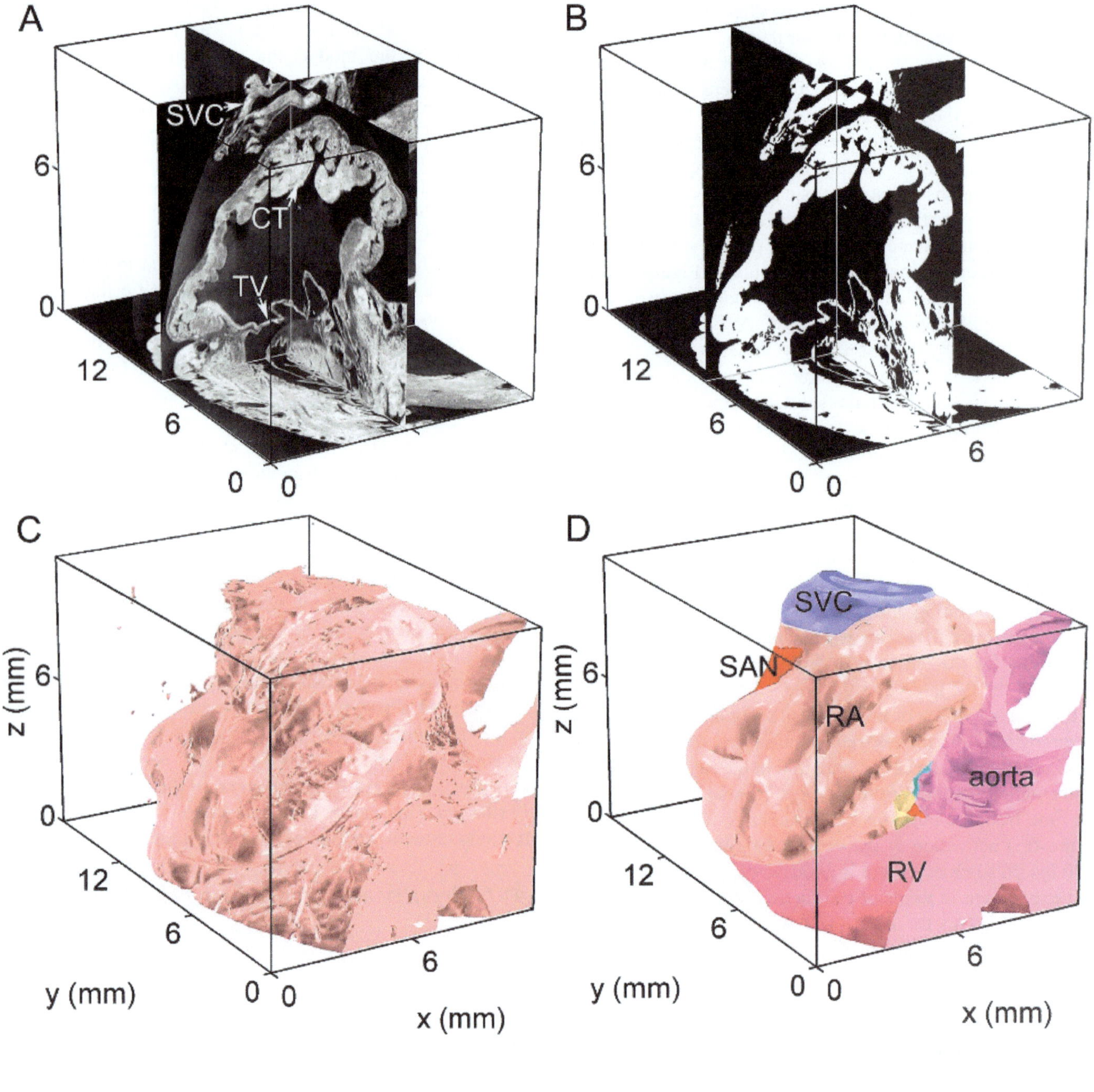

Figure S2. A set of snapshots of the action potential during the atrial reentrant arrhythmia viewed from the outside of the right atrium. The normal sinus rhythm beat (S1, 'stimulus') was initiated in the SAN at 39 ms. The action potential broke out from the SAN to the atrium at the crista terminalis (75 ms). In the opposite direction, the action potential was blocked in the block zone and propagated around the block zone to reach the interatrial septum at 90~117 ms. TheS2 stimulus at the superior vena cava was delivered at 159 ms. The first reentry wave started on the superior vena cava near the top of the SAN (175.5 ms). The action potental propagated around the SAN (202.5 ms and 211.5 ms) due to the SAN's long refractory period. The second reentry wave started at a similar position and another SAN wave was initiated in the low part of the SAN (247.5 ms). These two waves moved towards each other and finally merged (267~301.5 ms). Then the third reentry wave started, while the wave in the SAN had not faded away (367 ms). Similar as the first reentry wave, the third wave did not stimulate the SAN due to the long refractory period the SAN.

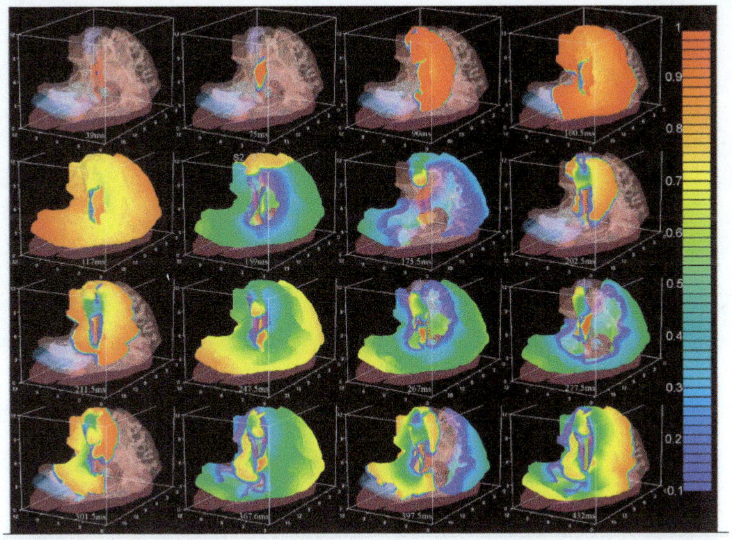

Figure S3. A set of snapshots of the action potential during the atrial reentrant arrhythmia viewed from inside of the right atrium. The action potential propagated faster along the crista terminalis and the main branch than in the rest of the atrial wall (90 ms). The action potential propagated around the block zone to reach the interatrial septum. It then reached the compact node (fast pathway) and inferior nodal extension (slow pathway) (108 ms~120 ms) at same time. The first reentry wave propagated around the SAN to the interatrial septum due to the SAN's long refractory period (202.5 ms, 207 ms). The wave front reached the compact node (fast pathway) (220 ms) first. It then reached the inferior nodal extension (slow pathway) by passing below the coronary sinus (238.5 ms). The first reentrant wave failed to propagate along the penetrating bundle (277 ms, 300 ms) due to the long refractory period of the penetrating bundle. The second reentrant wave propagated to the interatrial septum earlier than the right atrial free wall (300 ms, 315 ms). The wave front reached the penetrating bundle earlier than the inferior nodal extension and propagated successfully along the penetrating bundle (375 ms).

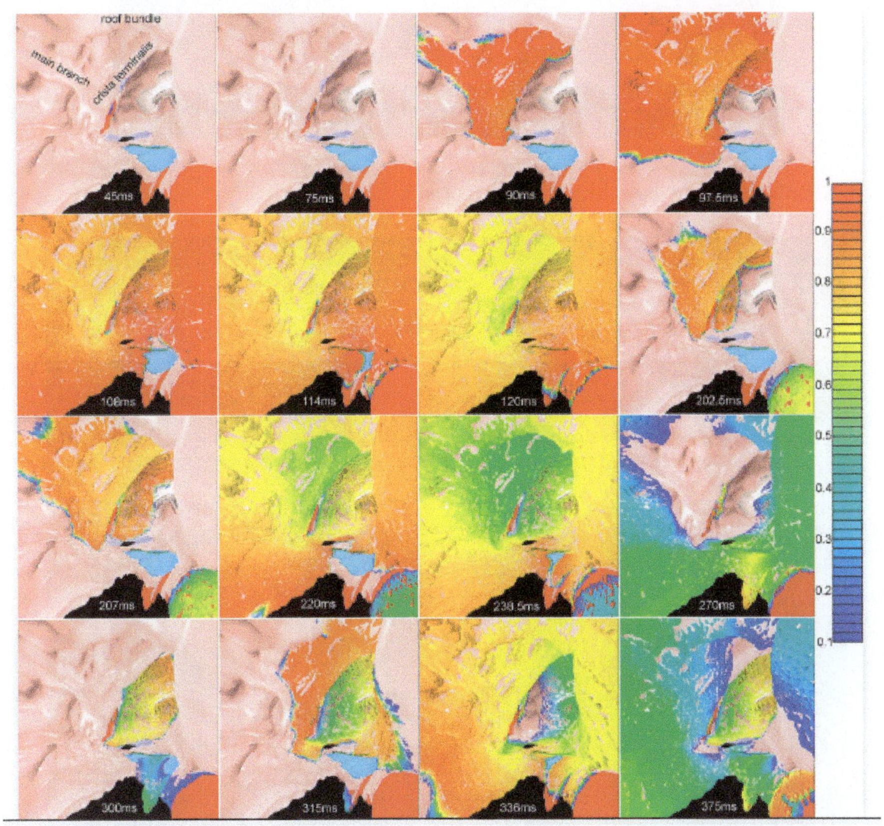

Movie S1. Dorsal right view of first reentrant circuit located on the SVC and second reentrant circuit located on the IVC.

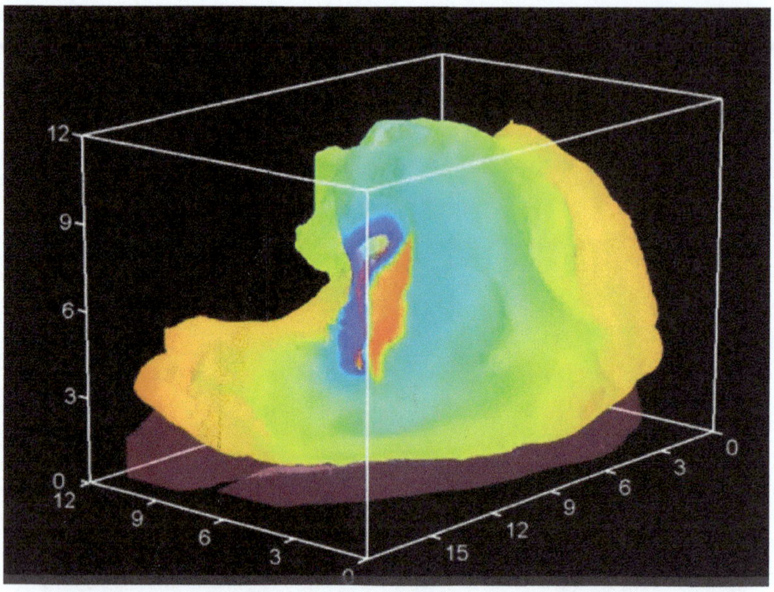

Movie S2. Dorsal left view of first reentrant circuit located on the SVC and second reentrant circuit located on the IVC.

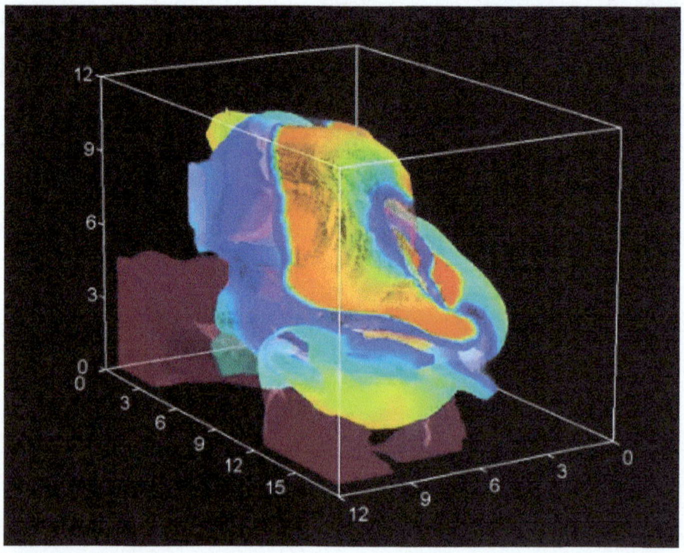

Movie S3. Left oblique view of third reentrant circuit located on the atrial free wall.

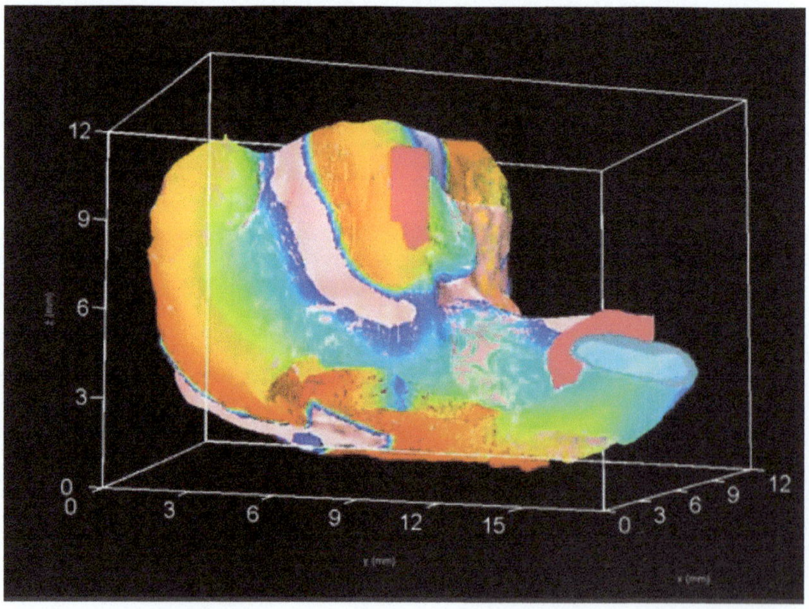

ACKNOWLEDGMENTS

We would like to acknowledge the assistance and insight of Dr. Mitsuru Yamamoto in the early part of this study.

REFERENCES

1. David MH (2000) A computer model of normal conduction in the human atria. Circ Res 87: e25–e36.
2. Bernus O, Verschelde H, Panfilov AV (2003) Reentry in an anatomical model of the human ventricles. Int J Bifurcation Chaos 13: 3693–3702.
3. Eifler WJ, Macchi E, Ritsema van Eck HJ, Horacek BM, Rautaharju PM (1981) Mechanism of generation of body surface electrocardiographic P-waves in normal, middle, and lower sinus rhythms. Circ Res 48: 168–182.
4. Lorange M, Gulrajani RM (1993) A computer heart model incorporating anisotropic propagation: I. Model construction and simulation of normal activation. J Electrocardiol 26: 245–261.
5. Sachse FB, Werner CD, Stenroos MH, Schulte RF, Zerfass P, et al.. (2000) Modeling the anatomy of the human heart using the cryosection images of the visible female dataset. Proceedings of the Third Users Conference of the National Library of Medicine's Visible Human Project.
6. Aslanidi OV, Colman MA, Stott J, Dobrzynski H, Boyett MR, et al. (2011) 3D virtual human atria: A computational platform for studying clinical atrial fibrillation. Prog Biophys Mol Biol 107: 156–168.
7. Chandler N, Aslanidi O, Buckley D, Inada S, Birchall S, et al. (2011) Computer three-dimensional anatomical reconstruction of the human sinus node and a novel paranodal area. Anat Rec 294: 970–979.
8. Seemann G, Höper C, Sachse F, Dössel O, Holden A, et al. (2006) Heterogeneous three-dimensional anatomical and electrophysiological model of human atria. Phil Trans R Soc A 364: 1465–1481.
9. Nielsen PM, LeGrice IJ, Smaill BH, Hunter PJ (1991) Mathematical model of geometry and fibrous structure of the heart. Am J Physiol Heart Circ Physiol 29: H1365–H1378.
10. Vetter FJ, McCulloch AD (1998) Three-dimensional analysis of regional cardiac function: a model of rabbit ventricular anatomy. Prog Biophys Mol Biol 69: 157–183.
11. Dobrzynski H, Li J, Tellez J, Greener ID, Nikolski VP, et al. (2005) Computer three-dimensional reconstruction of the sinoatrial node. Circulation 111: 846–854.
12. Li J, Greener ID, Inada S, Nikolski VP, Yamamoto M, et al. (2008) Computer three-dimensional reconstruction of the atrioventricular node. Circ Res 102: 975–985.
13. Stephenson RS, Boyett MR, Hart G, Nikolaidou T, Cai X, et al. (2012) Contrast enhanced micro-computed tomography resolves the 3-dimensional morphology of the cardiac conduction system in mammalian hearts. PLoS ONE 7: e35299.
14. Boyett MR, Li J, Inada S, Dobrzynski H, Schneider JE, et al. (2005) Imaging the heart: computer 3-dimensional anatomic models of the heart. J Electrocardiol 38: 113–120.
15. Courtemanche M, Ramirez RJ, Nattel S (1998) Ionic mechanisms underlying human atrial action potential properties: insights from a mathematical model. Am J Physiol Heart Circ Physiol 275: H301–H321.
16. Nygren A, Fiset C, Firek L, Clark JW, Lindblad DS, et al. (1998) Mathematical model of an adult human atrial cell: the role of K^+ currents in repolarization. Circ Res 82: 63–81.
17. Priebe L, Beuckelmann DJ (1998) Simulation study of cellular electric properties in heart failure. Circ Res 82: 1206–1223.
18. Bernus O, Wilders R, Zemlin CW, Verschelde H, Panfilov AV (2002) A computationally efficient electrophysiological model of human ventricular cells. Am J Physiol Heart Circ Physiol 282: H2296–H2308.
19. ten Tusscher KHWJ, Panfilov AV (2006) Alternans and spiral breakup in a human ventricular tissue model. Am J Physiol Heart Circ Physiol 291: H1088–H1100.
20. Iyer V, Mazhari R, Winslow RL (2004) A computational model of the human left-ventricular epicardial myocyte. Biophys J 87: 1507–1525.
21. Fink M, Noble D, Virag L, Varro A, Giles WR (2008) Contributions of HERG current to repolarization of the human ventricular action potential. Prog Biophys Mol Biol 96: 357–376.
22. Grandi E, Pasqualini FS, Bers DM (2010) A novel computational model of the human ventricular action potential and Ca transient. J Mol Cell Cardiol 48: 112–121.
23. Ramirez RJ, Nattel S, Courtemanche M (2000) Mathematical analysis of canine atrial action potentials: rate, regional factors, and electrical remodeling. Am J Physiol Heart Circ Physiol 279: H1767–H1785.
24. Kneller J, Zou R, Vigmond EJ, Wang Z, Leon LJ, et al. (2002) Cholinergic atrial fibrillation in a computer model of a two-dimensional sheet of canine atrial cells with realistic ionic properties. Circ Res 90: 73e–787.
25. Winslow RL, Rice J, Jafri S, Marbàn E, O›Rourke B (1999) Mechanisms of altered excitation-contraction coupling in canine tachycardia-induced heart failure, II: model studies. Circ Res 84: 571–586.
26. Hund TJ, Rudy Y (2004) Rate dependence and regulation of action potential and calcium transient in a canine cardiac ventricular cell model. Circulation 110: 3168–3174.
27. Benson AP, Aslanidi OV, Zhang H, Holden AV (2008) The canine virtual ventricular wall: A platform for dissecting pharmacological effects on propagation and arrhythmogenesis. Prog Biophys Mol Biol 96: 187–208.
28. Decker KF, Heijman J, Silva JR, Hund TJ, Rudy Y (2009) Properties and ionic mechanisms of action potential adaptation, restitution, and accommodation in

canine epicardium. Am J Physiol Heart Circ Physiol296: H1017–H1026.

29. McAllister RE, Noble D, Tsien RW (1975) Reconstruction of the electrical activity of cardiac Purkinje fibres. J Physiol 251: 1–59.
30. DiFrancesco D, Noble D (1985) A model of cardiac electrical activity incorporating ionic pumps and concentration changes. Philos Trans R Soc Lond B Biol Sci 307: 353–398.
31. Yanagihara K, Noma A, Irisawa H (1980) Reconstruction of sino-atrial node pacemaker potential based on the voltage clamp experiments. Jpn J Physiol 30: 841–857.
32. Demir SS, Clark JW, Murphey CR, Giles WR (1994) A mathematical model of a rabbit sinoatrial node cell. Am J Physiol Cell Physiol 266: C832–C852.
33. Dokos S, Celler B, Lovell N (1996) Ion currents underlying sinoatrial node pacemaker activity: a new single cell mathematical model. J Theor Biol 181: 245–272.
34. Demir SS, Clark JW, Giles WR (1999) Parasympathetic modulation of sinoatrial node pacemaker activity in rabbit heart: a unifying model. Am J Physiol Heart Circ Physiol 276: H2221–H2244.
35. Kurata Y, Hisatome I, Imanishi S, Shibamoto T (2002) Dynamical description of sinoatrial node pacemaking: improved mathematical model for primary pacemaker cell. Am J Physiol Heart Circ Physiol283: H2074–H2101.
36. Zhang H, Holden AV, Kodama I, Honjo H, Lei M, et al. (2000) Mathematical models of action potentials in the periphery and center of the rabbit sinoatrial node. Am J Physiol Heart Circ Physiol 279: H397–H421.
37. Oehmen CS, Giles WR, Demir SS (2002) Mathematical model of the rapidly activating delayed rectifier potassium current I_{Kr} in rabbit sinoatrial node. J Cardiovasc Electrophysiol 13: 1131–1140.
38. Wilders R, Jongsma HJ, van Ginneken AC (1991) Pacemaker activity of the rabbit sinoatrial node. A comparison of mathematical models. Biophys J 60: 1202–1216.
39. Inada S, Hancox JC, Zhang H, Boyett MR (2009) One-dimensional mathematical model of the atrioventricular node including atrio-nodal, nodal, and nodal-his cells. Biophys J 97: 2117–2127.
40. Aslanidi OV, Sleiman RN, Boyett MR, Hancox JC, Zhang H (2010) Ionic mechanisms for electrical heterogeneity between rabbit Purkinje fiber and ventricular cells. Biophys J 98: 2420–2431.
41. Hilgemann DW, Noble D (1987) Excitation-contraction coupling and extracellular calcium transients in rabbit atrium: reconstruction of basic cellular mechanisms. Proc Biol Sci 230: 163–205.
42. Lindblad DS, Murphey CR, Clark JW, Giles WR (1996) A model of the action potential and underlying membrane currents in a rabbit atrial cell. Am J Physiol Heart Circ Physiol 271: H1666–H1696.
43. Aslanidi OV, Boyett MR, Dobrzynski H, Li J, Zhang H (2009) Mechanisms of transition from normal to reentrant electrical activity in a model of rabbit atrial tissue: interaction of tissue heterogeneity and anisotropy. Biophys J 96: 798–817.
44. Shannon TR, Wang F, Puglisi J, Weber C, Bers DM (2004) A mathematical treatment of integrated Ca dynamics within the ventricular myocyte. Biophys J 87: 3351–3371.
45. Puglisi JL, Bers DM (2001) LabHEART: an interactive computer model of rabbit ventricular myocyte ion channels and Ca transport. Am J Physiol Cell Physiol 281: C2049–C2060.
46. Noble D, Kohl P, Noble P, Varghese A (1998) Improved guinea-pig ventricular cell model incorporating a diadic space, I_{Kr} and I_{Ks}, and length- and tension-dependent processes. Can J Cardiol: 123–134.
47. Beeler GW, Reuter H (1977) Reconstruction of the action potential of ventricular myocardial fibres. J Physiol 268: 177–210.
48. Luo CH, Rudy Y (1994) A dynamic model of the cardiac ventricular action potential. I. Simulations of ionic currents and concentration changes. Circ Res 74: 1071–1096.
49. Zeng J, Laurita KR, Rosenbaum DS, Rudy Y (1995) Two components of the delayed rectifier K^+current in ventricular myocytes of the guinea pig type: theoretical formulation and their role in repolarization.Circ Res 77: 140–152.
50. Shaw RM, Rudy Y (1997) Electrophysiologic effects of acute myocardial ischemia: a theoretical study of altered cell excitability and action potential duration. Cardiovasc Res 35: 256–272.
51. Viswanathan PC, Shaw RM, Rudy Y (1999) Effects of I_{Kr} and I_{Ks} heterogeneity on action potential duration and its rate dependence: a simulation study. Circulation 99: 2466–2474.
52. Kharche S, Yu J, Lei M, Zhang H (2011) A mathematical model of action potentials of mouse sinoatrial node cells with molecular bases. Am J Physiol Heart Circ Physiol 301: H945–H963.
53. Bondarenko VE, Szigeti GP, Bett GCL, Kim SJ, Rasmusson RL (2004) Computer model of action potential of mouse ventricular myocytes. Am J Physiol Heart Circ Physiol 287: H1378–H1403.
54. Holden AV, Zhang H (1993) Modelling propagation and re-entry in anisotropic and smoothly heterogeneous cardiac tissue. J Chem Soc, Faraday Trans 89: 2833–2837.
55. Winslow RL, Varghese A, Noble D, Adlakha C, Hoythya A (1993) Generation and propagation of ectopic beats induced by spatially localized Na-K pump inhibition in atrial network models. Proc Biol Sci254: 55–61.
56. FitzHugh R (1961) Impulses and physiological states in theoretical models of nerve membrane. Biophys J 1: 445–466.
57. Fenton F, Karma A (1998) Vortex dynamics in three-dimensional continuous myocardium with fiber rotation: Filament instability and fibrillation. Chaos: An Interdisciplinary Journal of Nonlinear Science 8: 20–47.

58. Li J, Schneider JE, Yamamoto M, Greener ID, Dobrzynski H, et al. (2005) A detailed 3D model of the rabbit right atrium including the sinoatrial node, atrioventricular node, surrounding blood vessels and valves.Computers in Cardiology 32: 603–606.
59. Li J, Dobrzynski H, Greener ID, Nikolski VP, Yamamoto M, et al. (2004) Development of 3-D anatomically-detailed mathematical models of the sinoatrial and atrioventricular nodes. Computers in Cardiology 31: 89–92.
60. Boyett MR, Honjo H, Kodama I (2000) The sinoatrial node, a heterogeneous pacemaker structure.Cardiovasc Res 47: 658–687.
61. Bleeker WK, Mackaay AJ, Masson-Pevet M, Bouman LN, Becker AE (1980) Functional and morphological organization of the rabbit sinus node. Circ Res 46: 11–22.
62. Coakley JB, King TS (1959) Cardiac muscle relations of the coronary sinus, the oblique vein of the left atrium and the left precaval vein in mammals. J Anat 93: 30–35.
63. Rogers JM, McCulloch AD (1994) A collocation-Galerkin finite element model of cardiac action potential propagation. IEEE Trans Biomed Eng 41: 743–757.
64. Li J, Inada S, Dobrzynski H, Zhang H, Boyett MR (2009) A modified FitzHugh-Nagumo model that allows control of action potential duration and refractory period. Computers in Cardiology 36: 65–68.
65. Fedorov VV, Hucker WJ, Dobrzynski H, Rosenshtraukh LV, Efimov IR (2006) Postganglionic nerve stimulation induces temporal inhibition of excitability in rabbit sinoatrial node. Am J Physiol Heart Circ Physiol 291: H612–H623.
66. Yamamoto M, Honjo H, Niwa R, Kodama I (1998) Low-frequency extracellular potentials recorded from the sinoatrial node. Cardiovasc Res 39: 360–372.
67. Boyett MR, Honjo H, Yamamoto M, Nikmaram MR, Niwa R, et al. (1999) Downward gradient in action potential duration along conduction path in and around the sinoatrial node. Am J Physiol Heart Circ Physiol 276: H686–H698.
68. Clayton RH (2009) Influence of cardiac tissue anisotropy on re-entrant activation in computational models of ventricular fibrillation. Physica D: Nonlinear Phenomena 238: 951–961.
69. Toshida N, Hirao K, Yamamoto N, Tanaka M, Suzuki F, et al. (2001) Ventricular echo beats and retrograde atrioventricular nodal exits in the dog heart: multiplicity in their electrophysiologic and anatomic characteristics. J Cardiovasc Electrophysiol 12: 1256–1264.
70. Nikolski VP, Jones SA, Lancaster MK, Boyett MR, Efimov IR (2003) Cx43 and dual-pathway electrophysiology of the atrioventricular node and atrioventricular nodal reentry. Circ Res 92: 469–475.
71. de Carvalho AP, de Mello WC, Hoffman BF (1959) Electrophysiological evidence for specialized fiber types in rabbit atrium. Am J Physiol 196: 483–488.
72. Spach MS, Lieberman M, Scott JG, Barr RC, Johnson EA, et al. (1971) Excitation sequences of the atrial septum and the AV node in isolated hearts of the dog and rabbit. Circ Res 29: 156–172.
73. Mazgalev TN, Tchou PJ (2000) Atrial-AV nodal electrophysiology: a view from the millennium. Wiley-Blackwell.
74. Litchenberg WH, Norman LW, Holwell AK, Martin KL, Hewett KW, et al. (2000) The rate and anisotropy of impulse propagation in the postnatal terminal crest are correlated with remodeling of Cx43 gap junction pattern. Cardiovasc Res 45: 379–387.
75. Kirchhof CJ, Allessie MA (1992) Sinus node automaticity during atrial fibrillation in isolated rabbit hearts. Circulation 86: 263–271.
76. Xiao J, Zhang H, Liang D, Liu Y, Zhao H, et al. (2010) Taxol, a microtubule stabilizer, prevents atrial fibrillation in in vitro atrial fibrillation models using rabbit hearts. Med Sci Monit 16: BR353–BR360.
77. Li H, Scherlag BJ, Kem DC, Zillner C, Male S, et al. (2014) The propensity for inducing atrial fibrillation: a comparative study on old versus young rabbits. Journal of Aging Research 2014: 684918.
78. Oliveira M, da Silva MN, Geraldes V, Xavier R, Laranjo S, et al. (2011) Acute vagal modulation of electrophysiology of the atrial and pulmonary veins increases vulnerability to atrial fibrillation. Exp Physiol96: 125–133.
79. Che X, Qu B, Wu L, Hu X, Yu J, et al. (2003) The initial study on the origin and reentrant mechanism of paroxysmal atrial fibrillation arising from right atrium with non-contact mapping system. Chinese Journal of Cardiac Pacing and Electrophysiology 17: 22–26.
80. Mazgalev T, Dreifus LS, Bianchi J, Michelson EL (1982) Atrioventricular nodal conduction during atrial fibrillation in rabbit heart. Am J Physiol 243: H754–H760.
81. Fozzard HA (1991) The heart and cardiovascular system: scientific foundations. New York: Raven Press, c1991.
82. Efimov IR, Nikolski VP, Rothenberg F, Greener ID, Li J, et al. (2004) Structure-function relationship in the AV junction. Anat Rec 280A: 952–965.
83. Nakayama T, Kurachi Y, Noma A, Irisawa H (1984) Action potential and membrane currents of single pacemaker cells of the rabbit heart. Pflugers Arch 402: 248–257.
84. Denyer JC, Brown HF (1990) Rabbit sino-atrial node cells: isolation and electrophysiological properties.J Physiol 428: 405–424.
85. van Ginneken AC, Giles W (1991) Voltage clamp measurements of the hyperpolarization-activated inward current I_f in single cells from rabbit sino-atrial node. J Physiol 434: 57–83.
86. Ono K, Ito H (1995) Role of rapidly activating delayed rectifier K^+ current in sinoatrial node pacemaker activity. Am J Physiol Heart Circ Physiol 269: H453–H462.
87. Verheijck EE, van Ginneken ACG, Bourier J, Bouman LN (1995) Effects of delayed rectifier current blockade by E-4031 on impulse generation in single sinoatrial nodal myocytes of the rabbit. Circ Res 76: 607–615.

88. Wilders R, Verheijck EE, Kumar R, Goolsby WN, van Ginneken AC, et al. (1996) Model clamp and its application to synchronization of rabbit sinoatrial node cells. Am J Physiol Heart Circ Physiol 271: H2168–H2182.

89. Yamashita T, Nakajima T, Hazama H, Hamada E, Murakawa Y, et al. (1995) Regional differences in transient outward current density and inhomogeneities of repolarization in rabbit right atrium. Circulation 92: 3061–3069.

90. Wu J, Schuessler RB, Rodefeld MD, Saffitz JE, Boineau JP (2001) Morphological and membrane characteristics of spider and spindle cells isolated from rabbit sinus node. Am J Physiol Heart Circ Physiol280: H1232–H1240.

91. Lei M, Cooper PJ, Camelliti P, Kohl P (2002) Role of the 293b-sensitive, slowly activating delayed rectifier potassium current, i_{Ks}, in pacemaker activity of rabbit isolated sino-atrial node cells. Cardiovasc Res53: 68–79.

92. Kerr CR, Prystowsky EN, Browning DJ, Strauss HC (1980) Characterization of refractoriness in the sinus node of the rabbit. Circ Res 47: 742–756.

93. Reid MC, Billette J, Khalife K, Tadros R (2003) Role of compact node and posterior extension in direction-dependent changes in atrioventricular nodal function in rabbit. J Cardiovasc Electrophysiol 14: 1342–1350.

94. Lin LJ, Billette JACQ, Medkour DJAM, Reid MC, Tremblay MAUR, et al. (2001) Properties and substrate of slow pathway exposed with a compact node targeted fast pathway ablation in rabbit atrioventricular node. J Cardiovasc Electrophysiol 12: 479–486.

95. Kodama I, Nikmaram MR, Boyett MR, Suzuki R, Honjo H, et al. (1997) Regional differences in the role of the Ca^{2+} and Na^{+} currents in pacemaker activity in the sinoatrial node. Am J Physiol Heart Circ Physiol272: H2793–H2806.

96. de Carvalho AP, de Almeida DF (1960) Spread of activity through the atrioventricular node. Circ Res 8: 801–809.

CHAPTER 15

VIRTUAL HAND ILLUSION INDUCED BY VISUOMOTOR CORRELATIONS

Maria V. Sanchez-Vives[1,2], Bernhard Spanlang[3], Antonio Frisoli[4], Massimo Bergamasco[4], Mel Slater[1,5,6]

[1] Institucio´ Catalana de Recerca i Estudis Avanc¸ats, Barcelona, Spain,

[2] IDIBAPS, Barcelona, Spain,

[3] Universitat Polite`cnica de Catalunya, Barcelona, Spain,

[4] PERCRO, CEIICP, Scuola Superiore Sant'Anna, Pisa, Italy,

[5] University of Barcelona, Barcelona, Spain,

[6] University College London, London, United Kingdom

ABSTRACT

Background

Our body schema gives the subjective impression of being highly stable. However, a number of easily-evoked illusions illustrate its remarkable malleability. In the rubber-hand illusion, illusory ownership of a rubber-hand is evoked by synchronous visual and tactile stimulation on a visible rubber arm and on the hidden real arm. Ownership is concurrent with a proprioceptive illusion of displacement of the arm position towards the fake arm. We have previously shown that this illusion of ownership plus the proprioceptive displacement also occurs towards a virtual 3D projection of an arm when the appropriate synchronous visuotactile stimulation is provided. Our objective here was to explore whether these illusions (ownership and proprioceptive displacement) can be induced by only synchronous visuomotor stimulation, in the absence of tactile stimulation.

Methodology/Principal Findings

To achieve this we used a data-glove that uses sensors transmitting the positions of fingers to a virtually projected hand in the synchronous but not in the asynchronous condition. The illusion of ownership was measured by means of questionnaires. Questions related to ownership gave significantly larger values for the synchronous than for the asynchronous condition. Proprioceptive displacement provided an objective measure of the illusion and had a median value of 3.5 cm difference between the synchronous and asynchronous conditions. In addition, the correlation between the feeling of ownership of the virtual arm and the size of the drift was significant.

Conclusions/Significance

We conclude that synchrony between visual and proprioceptive information along with motor activity is able to induce an illusion of ownership over a virtual arm. This has implications regarding the brain mechanisms underlying body ownership as well as the use of virtual bodies in therapies and rehabilitation.

Citation: Sanchez-Vives MV, Spanlang B, Frisoli A, Bergamasco M, Slater M (2010) Virtual Hand Illusion Induced by Visuomotor Correlations. PLoS ONE 5(4): e10381. doi:10.1371/journal.pone.0010381

INTRODUCTION

The problem of self-recognition is concerned with how the central nervous system distinguishes what is part of the body and what is not. Although at first this might seem to be an easy problem to solve, for example, through different patterns of neural activity distinguishing between self-generated motor actions and the motor actions of others, research into mirror neurons shows that there are similar patterns of neural firing between watching an action performed by another and carrying out that action oneself. Jeannerod [1] discusses various possible contributors to self-recognition. One is the attribution of actions to the self (agency) through correlation between the intention to move and the resulting proprioceptive, and other multisensory signals and bodily responses. Another is the sense of ownership of the body caused by multisensory correlations between stimuli on the body, such as feeling a touch on a body part and at the same time seeing the visual correlate of the cause of the touch.

A demonstration that the problem of self-recognition is not straightforward is the fact that it is easy to generate illusions that involve misattribution of a rubber hand [2] or even a hand displayed in virtual reality [3] to the self. This is achieved through tactile stimulation of the hidden real hand and corresponding and synchronous visual stimulation on the visible fake hand. This rubber hand illusion involves not just subjective attribution of the rubber hand to the self, but also a mis-localization of where the stimulated hand is felt to be after a few minutes or even seconds of such synchronous visuotactile stimulation. When asked to blindly point towards the stimulated hand subjects will typically point towards the rubber or virtual hand – the distance between the real hand position and the indicated position being termed "proprioceptive drift". Additionally, when the rubber hand is threatened, there are skin conductance responses indicating arousal, as if in preparation for pain [4]. When the visual-tactile stimulation is asynchronous, then the subjective, proprioceptive and arousal responses occur to a significantly lesser extent. For a review see [5].

Misattribution of an alien hand to the self as a result of motor actions rather than visual-tactile correlation has also been demonstrated. An experiment by Nielsen [6] showed that subjects will recognize the hand of an experimenter as their own, when their own hand is hidden and carrying out a drawing task that they see also being carried out by the experimenter's hand which is in a plausible position in relation to their own body. Moreover, when the experimenter's drawing deviates from the line that the subject is supposed to draw, the subjects tend to compensate for this, and yet remain unaware of the misattribution of the experimenter's hand as their own. Variations on this experiment [7], [8] showed that subjects tend to unconsciously and automatically follow visual cues in making corrections when the observed visual path of a stylus deviated from the path caused by their own motor actions, until the discrepancy became large enough that the conscious system took over in order to correct for bias. The point is that subjects would tolerate large mis-localization errors as well as misattribution in observations of the effects of their own motor actions.

There is also some evidence that an ownership illusion, akin to the rubber hand illusion, may be generated by synchronized vasomotor actions. Dummer et al. [9] carried out an experiment using a mechanical setup that moved a rubber hand synchronously or asynchronously with the hand movements of the subject, and compared each with a passive condition and the normal synchronous visuotactile rubber hand illusion. They found that the ownership illusion occurred with the visuomotor synchrony, although this was only demonstrated subjectively with a questionnaire. A note by Raz et al. [10] reports on an experiment using a hand projected in a stereo virtual environment (Reachin Display), where a questionnaire-based study found that a subjective illusion of ownership occurred both and separately for synchronous visuomotor and visuotactile stimulation.

In this paper we extend these results by exploiting a virtual reality system, and hand tracking with a data glove, showing that the illusion of ownership of the virtually presented hand occurs on the basis of visuomotor synchrony between movements of the real hand and the virtual hand. When there is asynchrony the illusion does not occur. This is demonstrated subjectively with a questionnaire, and behaviorally with proprioceptive drift, and additionally we observe significant positive correlations between proprioceptive drift and the questionnaire responses, akin to the original findings in [2].

MATERIALS AND METHODS

Recruitment

Fourteen male participants with mean age 22.5±5.6 (S.D.) years were recruited for the experiment by advertisement on the university campus at Scuola Superiore Sant'Anna, Pisa, Italy. They were asked to read and sign an information consent form and they were paid 10€ for their participation. Participants were naïve with respect to the virtual/rubber hand illusion.

The work on body representation using virtual reality within the EU project PRESENCCIA has been approved by the ethics committee at the Hospital Clinic (Barcelona, Spain).

Virtual Reality System

The virtual reality set-up (Fig. 1A) consisted of a tracking system with a 6 degrees-of-freedom (DOF) Polhemus (http://www.polhemus.com/)Liberty head tracker (Fig. 1B) and a 2 m×2.7 m screen, where stereoscopic 3D images were back-projected by using an Infitec system (http://www.infitec.net/). The virtual environment was developed by using the XVR virtual reality platform (VRMedia http://www.vrmedia.it/).

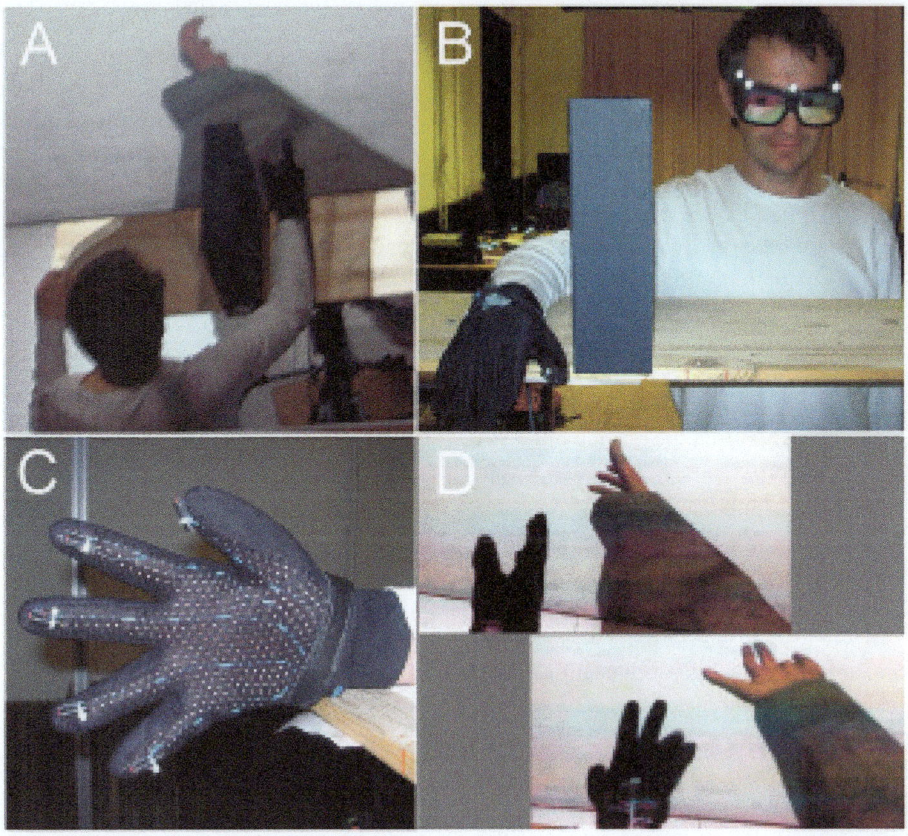

Figure 1: Experimental set up. A. The participant had his right arm resting on a tabletop. The arm was separated from view by a partition. The virtual arm was displayed on the screen in front of the participant. Its size and position was adjusted such that it looked correct from their point of view. The right hand was wearing the data glove. **B**. The participants (for the display one of the authors, BS, is represented) viewed from the front, wearing the stereo glasses and the data glove. **C**. Detail of the data glove. D. In the synchronous condition the virtual hand did follow the movements and finger position of the real hand tracked by the data glove. http://dx.doi.org/10.1371/journal.pone.0010381.g001

We used a virtual character from the AXYZ design (http://www.axyz-design.com/) character set. The virtual character was visualised and animated in XVR by using a hardware accelerated library for character animation (HALCA) [11]. To the participant only the right arm and hand of the virtual character was shown. In HALCA the body mesh was deformed with the skeleton of the virtual character by using the dual quaternion skinning method [12] in a GPU vertex shader program.

Participants wore glasses with spectral filters (Infitec) for passive stereo viewing (Fig. 1B). In the synchronous condition their right hand rotations and displacements were tracked by a second 6DOF Polhemus Liberty tracker. The rotations were mapped to the forearm and hand rotations of a virtual character. The finger movements of the participant were tracked by the dataglove (Fig. 1C) described below and were also mapped to the finger bone joints of our virtual character›s skeleton. In order to calibrate the finger movements we recorded the open and closed hand measured skeletal configurations of the finger bone angles for each subject before the experiment.

The program could log all the finger, hand and arm movements for later play back and analysis. In the asynchronous condition the arm, hand and finger movements of the virtual character were played back from a pre-recorded session.

A timer was programmed in the XVR scripting language to smoothly move for 20 s seconds the virtual character›s arm to the left by a medial rotation around the shoulder 180 s after the start of the experiment. The timer was also used to play a beep sound to signal to the participant the end of the experiment and that the subject should point to where he thought his real hand was in order to enable us to measure the proprioceptive drift.

Subjects were fitted with a data glove worn on their right hand (Fig. 1C). This tracked the movements of their fingers, which drove a 3D virtual hand. The PERCRO data glove, developed by some of the authors at PERCRO, Scuola Superiore Sant›Anna, was used for the experiment. The data glove is equipped with patented absolute goniometric sensors [13] that can measure the angular displacements of proximal (MCP) and medial phalanxes (PIP) for all fingers and abduction-adduction of the thumb. The acquired angles are acquired on-line and used to reconstruct the full hand posture that is then mapped into a virtual 3-dimensional model of the hand and displayed to the user. Latency of the system was less than 5 msec.

Experimental design and procedures

This was a repeat measures design. There was one between-group factor, which was the group to which subjects were assigned: Either they experienced the synchronous movement first followed by the asynchronous (group SA) or the other way around (group AS). There was one within groups factor – condition (synchronous or asynchronous):

Synchronous: The movements of the subject's own hand as captured by data glove determined the movements of the virtual hand (Fig. 1D).

Asynchronous: The virtual hand movements displayed were prerecorded and thus they were asynchronous with the movements of the real hand.

The experiment was carried out in a dark room, where the only light came from the screen. Volunteers stood with their right arm resting on a platform, occluded from their view by a partition (Fig. 1A). Their right hand wore the data glove (see above). The computer program generated a stereo image of a virtual arm (Fig. 1D; in the image in mono for better display). The virtual arm was positioned such that it seemed to be coming out the right shoulder, and it was also adjusted that it appeared to be the correct size. Unlike the real arm, the virtual arm was not shown resting on a shelf, but held outwards in front of the subject. The distance between the participant's real hand and the virtual hand was approximately 20 cm, the virtual hand being displaced towards the body.

Due to the head tracking, if the participant kept his body still and just moved his head as if to look at the arm from a different position, then the arm would appear to be stationary from a different point of view, as it would happen in reality. The setup was, therefore, able to powerfully induce the illusion that there was an arm pointing straight ahead, which appeared to be attached to the participant's body.

Once in the right position, subjects were told that whenever they heard a beep sound they should place one of two pieces of piece of blue-tack that they had been given and were holding in their left hand to a position under the board corresponding to where they felt the centre of their forearm to be.

The participants were then instructed to continually rotate their right hand along the prono-supination axis of their forearm and move their fingers as if they were counting, and this continued for 180 s. (The movement can seen in the video (Movie S1, S2) provided in the Supplementary Information). During this stage subjects were asked to concentrate their attention on the virtual hand in order to receive visual feedback of their movement in real time.

After this period of 180 s, the hand started drifting towards the left for 20 s, corresponding to a medial rotation of the right shoulder joint of 15 degrees with the elbow joint extended, covering a distance with the right hand of approximately 20 cm. At the end of this time there was another beep and the subject again placed a piece of blue-tack that they had been holding in their left hand under the board pointing towards where they felt position of the centre of their forearm to be.

Questionnaire

After the experience, participants filled in an 11-item questionnaire (in Italian). Most questions were adapted and translated from [2] and some new questions were added. The labels are here added for convenience for the analysis of the results (Table 1). The questionnaire contained a set of assertions and was scored according to a 7-point Likert scale, where a score of 7 was described as 'totally agree' and a score of 1 as 'totally disagree' with the assertion.

Table 1: The Post-Experiment Questionnaire. http://dx.doi.org/10.1371/journal.pone.0010381.t001

Variable Name	Assertion
Ownership	
located	I sometimes felt as if my hand was located where I saw the virtual hand to be.
own	Sometimes I felt that the virtual arm was my own arm.
Illusion of movement	
affected	I felt my own arm to be affected when I saw the virtual arm move to the left, at the end.
influencing	At some moments I felt that the movements of the virtual hand were influencing my own movements.
drifted	When the virtual arm drifted I felt that my real arm was drifting with it.
Validity	
bythemselves	The virtual hand and fingers seemed to be moving by themselves.
caused	The movements of the virtual hand and fingers were caused by my movements.
Control	
morehand	It sometimes seemed as if I might have more than one right hand or arm.
between	It sometimes seemed as if the position of the hand I was feeling came from somewhere between my own hand and the virtual hand.
resemble	The virtual hand began to resemble my own real hand, in terms of shape, skin tone, freckles or some other visual feature.
virtual	It sometimes felt as if my real hand was turning 'virtual'.

The questionnaire statements were grouped into different types: two questions that indicated ownership illusion, which were designed to be as close as possible to those of [2] given the different experimental paradigm; three that referred to the illusion of movement; the two validity statements were chosen to check that the two experimental conditions operated as designed; and there were four control questions following the style of [2].

Behavioral measure

In addition to the questionnaire, the proprioceptive drift elicited by the illusion was measured by a standard technique. Participants had been instructed to place the piece of blue-tack under the board were their forearm rested before and after the 200 s of experiment with eyes closed (see above). The position of the blue tack was immediately marked by an experimenter and then removed. The horizontal distance between both positions marked by the blue-tack corresponded to the proprioceptive drift.

RESULTS

Questionnaire Results

First we compare the results on the synchronous (Movie S1) and asynchronous (Movie S2) conditions. We use a repeat measures one-way analysis of variance with between-groups variable Group (SA or AS) and one within-groups factor Condition (asynchronous or synchronous).

Table 2 shows the means, standard deviations, and the significance level for the difference between the means of Condition. There was no evidence of interaction between Group and Condition (in other words there was no order effect).

Table 2: Mean, SD, and Significance Levels for the Difference between Means, for the Asynchronous and Synchronous Conditions (Repeated Measures ANOVA) and the correlation (r) with proprioceptive drift in relation to Figure 3A and B. http://dx.doi.org/10.1371/journal.pone.0010381.t002

	Asynch		Synch			Correlation	
Ownership	**Mean**	**SD**	**Mean**	**SD**	**P**	**r**	**P**
located	**2.8**	1.6	**4.1**	1.5	0.013	0.58	0.029
own	**2.9**	1.6	**4.9**	1.7	0.003	0.57	0.032
Illusion of movement							
affected	**3.6**	2.0	**3.6**	2.2	0.848	0.54	0.045
influencing	**4.4**	2.1	**3.6**	2.0	0.237	−0.42	0.133
drifted	**2.6**	1.6	**3.0**	1.8	0.449	0.36	0.203
Validity							
bythemselves	**5.9**	1.1	**2.1**	1.2	0.000	NA	NA
caused	**2.6**	1.5	**6.0**	1.4	0.000	NA	NA
Control							
morehand	**2.9**	1.6	**2.7**	1.8	0.709	−0.28	0.338
between	**2.5**	1.3	**3.4**	1.2	0.037	0.19	0.514
resemble	**3.4**	2.2	**4.3**	2.0	0.071	0.10	0.725
virtual	**2.9**	1.7	**5.0**	1.4	0.001	0.59	0.027

It can be seen that both the 'illusion of feeling of ownership' questions have significantly different means (higher for synchronous), however, none of the 'illusion of movement' questions have significantly different means. Amongst the control questions *between* and *virtual* were significantly different. Also there is some evidence of a difference for *resemble*. These three were also found to be significantly different between experimental and control group in a between-groups experiment using a virtual reality version of the rubber hand illusion.

The two consistency questions (*by themselves* and *caused*) were appropriately significantly different.

Note that the residual errors of all models were tested for normality using the Jarque-Bera test, and the hypothesis of normality was never rejected (the smallest significance level was 0.33).

Proprioceptive Drift

Drifts were measured before and after each experimental trial as discussed above. The measurements (cm) were the horizontal distances, i.e., along a line parallel to the direction of virtual arm movement, the average drift being 3.25 cm. Figure 2 shows the drift by each condition. It suggests that the drift is higher for the synchronous condition, although there are two outliers in that condition. Since there was no reason to suspect either of these measurements, and if we removed them we would lose the balanced experimental design, we reduce their effect by replacing all drift measurements by their ranks, and carried out the repeated measures ANOVA (this is somewhat akin to the Kruskal-Wallis non-parametric ANOVA).

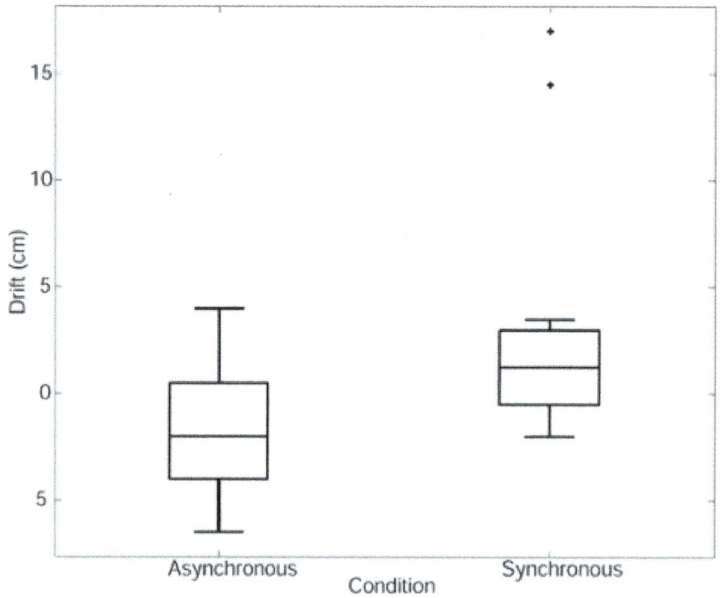

Figure 2: Standard Boxplots of the Drift for the Asynchronous and Synchronous Conditions. http://dx.doi.org/10.1371/journal.pone.0010381.g002

The ANOVA reveals that using the ranks (so that the absolute magnitudes of the outliers are not important) there is a significant difference between the two conditions as shown in Table 3, but that that there is no order effect (no difference between the groups). The Jarque-Bera test does not reject the hypothesis that the residual errors of the fit are normal (P=0.85).

Table 3: Medians and Interquartile Ranges of Drift and Significance Level of Repeated Measures ANOVA on ranks of drift, for the test between the mean asynchronous and synchronous ranks. http://dx.doi.org/10.1371/journal.pone.0010381.t003

	Asynchronous		Synchronous		
	Median	**IQR**	**Median**	**IQR**	**P**
Drift (cm)	**−2.0**	4.5	**1.25**	3.5	0.017

Drift by Questionnaire Scores

In Figure 3A and B we show scatter plots of the rank drift against two of the questionnaire scores indicating ownership, "*located*" and "*own*" respectively (Tables 1 and 2). Specifically, the plots represent the differences between the questionnaire scores for the synchronous and asynchronous conditions, plotted against the differences in the rank drift between synchronous and asynchronous. Each of these 'feeling of ownership' questions shows a positive correlation.

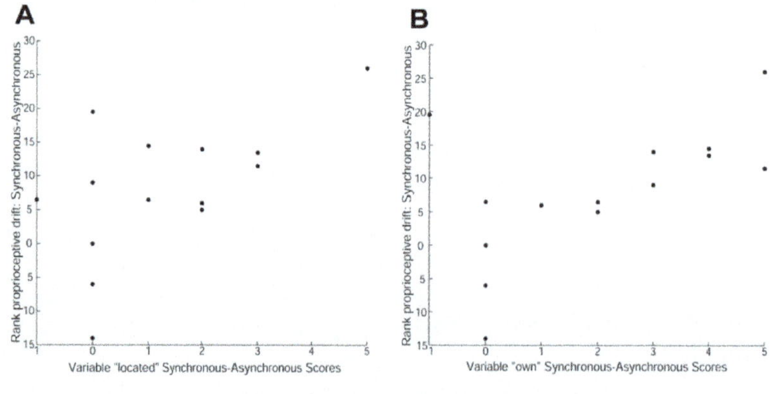

Figure 3: Scatter plots of questionnaire scores against rank drift. A. For the ownership variable "located"(Table 1). **B**. For the ownership variable "own" (Table 1). http://dx.doi.org/10.1371/journal.pone.0010381.g003

Table 2 (last 2 columns) shows that there are significant positive correlations with the two questions indicating the feeling of ownership, and also with 'virtual'.

DISCUSSION

The results we obtain with visuomotor synchrony are very similar to those of both the visuotactile synchrony based rubber hand [2] and virtual hand illusions [3]. In contrast to earlier experiments that have concentrated on movement synchrony in the rubber hand illusion [9],[10] we have also examined the behavioral proprioceptive drift measure and its correlation with the questionnaire scores. However, one difference in our experimental design compared with earlier ones was that during the period of the experiment (the first 180 s) the virtual hand was in a stable location, and then moved away from the real hand (the last 20 s). This did not lead either to a subjective illusion that the real hand was moving or to an actual movement of the real hand (as measured by the tracker on the glove). Nevertheless there was significant mis-localization of the hand when the subjects in the synchronous condition were asked to indicate their hand position at the termination of the virtual hand movement. In other words in spite of the fact that they saw the virtual hand move, did not feel their hand move, nor move it, they still blindly pointed towards the virtual hand when asked to point where they felt their hand to be.

This can be considered as a stronger result than the one obtained with the normal proprioceptive drift measure. In the latter, the rubber or virtual hand is seen to be stationary throughout the period of stimulation, so the conflict is between the stationary position of the real hand and the stationary fake hand. In these conditions it is known that there would be likely a significant misattribution of the real hand to another hand (for example, the hand of an experimenter) when either both are stationary or both are moving synchronously [14]. However, in our case the virtual hand was located approximately 20 cm away from the real hand for 180 s, and then it moved away. So to mis-localize the felt position of their hand, subjects had to negate not only the position of their real hand but also negate the fact that their real hand had not moved. Evidence for physiological changes during the RHI is given in [15] where it is shown that a drop in temperature can be observed in the hidden real hand. Although we have no physiological evidence in our experiment for this or for other physiological changes, our finding does suggest a substantial neglect of the real hand.

Makin et al. [5] presented a model for the RHI based on multisensory integration in peripersonal hand space which may be adaptable to the results presented here. In their model multisensory brain areas integrate the visual information of the fake hand with the proprioceptive information from the real hidden hand, but with the greatest weight given to the visual modality. Makin et al. mention the condition that the fake hand must be in a plausible position with respect to the body, which is satisfied in our experiment. They also suggest that the integration is weighted in favor of visual input provided that the real hand is static. In our case the real hand was not static but its movements either directly drove the movements of the virtual hand (synchronous condition) or the virtual hand made similar types of movements as the real hand but not the same movements (asynchronous condition). However, we could argue that in the synchronous condition the correlation between the visual movement and the proprioception would be enough to trigger the same recalibration of peripersonal space around the virtual hand as is the case for the RHI, so that the seen moving virtual hand triggers a unified visual-proprioceptive event centered on the virtual hand, with the real hand neglected. What is new here is that the unified visual-proprioceptive sensation is maintained even while the virtual hand moves its overall position away from the real hand (and note that the move was relatively slow - a 15 degree horizontal rotation about the shoulder lasting 20 s - and anatomically plausible). In fact it follows that if vision and proprioception become bound together, with vision dominating, then a move of the visual component should also result in a move of the proprioceptive component. Interestingly, this proprioceptive displacement was not conscious (as illustrated by the questionnaire responses) but only indicated by pointing towards the final observed hand position. It should also be noted here that it is unlikely that the proprioceptive drift occurred, for example, as a form of 'suggestion' – as a result of being induced by seeing the virtual hand move, or because the eyes of the subjects were caused to look in the direction of the moving hand with the pointing behavior following from this, since then we would have observed the same effect in the asynchronous condition. In a previous experiment [16], the proprioceptive illusion of displacement has also been induced by the realization of movement with one finger, while receiving visual feedback through the synchronously projected movement of the finger. Interestingly, the illusion of displacement occurs both with active and with passive movements of the finger, although it has different characteristics: when the movements that induce the illusion were active the perception of displacement was less than when passive, but the biases when pointing at a target were larger [16]. Here, we explored the proprioceptive displacement following active movements. However, following [16], we should have expected proportionally larger deviations had the subject been asked to point at targets. The maybe lesser degree of proprioceptive displacement that might have been expected after inducing ownership by active movements was in our case amplified by the artificial movement of the virtual hand. It should be noted that inducing ownership by visuomotor acts, or by agency, has been reported to have some different properties compared to inducing ownership by visuotactile correlations [2], [3],[17]. While visuotactile correlations of a finger may induce illusory displacements of individual fingers, visuomotor correlations of a finger movement induces an illusory drift of the arm, thus generating a more global, non-fragmented, body ownership [18].

Our final point is that virtual reality provides an excellent tool for studying body representation. The combination of stereo vision, tracking, haptic and auditory feedback, and the ability to represent even the body of the participant, generate an illusion for the participant of being and acting in an alternate virtual reality [19]. The evidence suggests that even the sense of what is their own body can be transferred to their virtual body representation. When the normal correlations between the different modalities are disrupted in a systematic way it is possible to produce the types of illusions that are important in understanding how the brain represents the body, especially when this can be combined with brain imaging.

SUPPORTING INFORMATION

Movie S1. Synchronous condition. Real hand in the data glove and virtual hand in the synchronous condition.

Movie S2. Asynchronous condition. Real hand in the data glove and virtual hand in the asynchronous condition.

ACKNOWLEDGMENTS

The authors would like to thank Adina Andreea Bizdideanu for helping with the execution of the experiments.

AUTHOR CONTRIBUTIONS

Conceived and designed the experiments: MVSV AF MB MS. Performed the experiments: MVSV BS AF MS. Analyzed the data: MVSV MS. Contributed reagents/materials/analysis tools: AF MB. Wrote the paper: MVSV MS.

REFERENCES

1. Jeannerod M (2003) The mechanism of self-recognition in humans. Behav Brain Res 142: 1–15.
2. Botvinick M, Cohen J (1998) Rubber hands 'feel' touch that eyes see;. 756 p. Feb 19.
3. Slater M, Perez-Marcos D, Ehrsson HH, Sanchez-Vives MV (2008) Towards a digital body: the virtual arm illusion. Front Hum Neurosci 2: 6.
4. Armel KC, Ramachandran VS (2003) Projecting sensations to external objects: evidence from skin conductance response. pp. 1499–1506.
5. Makin TR, Holmes NP, Ehrsson HH (2008) On the other hand: dummy hands and peripersonal space. Behav Brain Res 191: 1–10.
6. Nielsen T (1963) Volition: A new experimental approach. Scandinavian Journal of Psychology 4: 225–230.
7. Slachevsky A, Pillon B, Fourneret P, Pradat-Diehl P, Jeannerod M, et al. (2001) Preserved adjustment but impaired awareness in a sensory-motor conflict following prefrontal lesions. J Cogn Neurosci 13: 332–340.
8. Fourneret P, Jeannerod M (1998) Limited conscious monitoring of motor performance in normal subjects. Neuropsychologia 36: 1133–1140.
9. Dummer T, Picot-Annand A, Neal T, Moore C (2009) Movement and the rubber hand illusion. Perception 38: 271–280.

10. Raz L, Weiss P, Reiner M (2008) The Virtual Hand Illusion and Body Ownership. Lecture Notes in Computer Science 5024: 367.

11. Spanlang B 2009, EVENT Lab, Facultat de Psicologia, University of Barcelona: Barcelona. (2009) HALCA: A Hardware accelerated library for character animation, in PRESENCCIA Technical Report. PRESENCCIA Technical Report: EVENT Lab, Facultat de Psicologia, University of Barcelona: Barcelona.

12. Kavan L, Collins S, Žára J, O›Sullivan C (2007) Skinning with dual quaternions;. 46 p. ACM.

13. Bergamasco M, Salsedo F, Ullrich G, Villella P (2008) Goniometric sensor. Google Patents.

14. van den Bos E, Jeannerod M (2002) Sense of body and sense of action both contribute to self-recognition. Cognition 85: 177–187.

15. Moseley GL, Olthof N, Venema A, Don S, Wijers M, et al. (2008) Psychologically induced cooling of a specific body part caused by the illusory ownership of an artificial counterpart. Proc Natl Acad Sci U S A 105: 13169–13173.

16. Kammers MP, Longo MR, Tsakiris M, Dijkerman HC, Haggard P (2009) Specificity and coherence of body representations. Perception 38: 1804–1820.

17. Ehrsson HH, Holmes NP, Passingham RE (2005) Touching a rubber hand: feeling of body ownership is associated with activity in multisensory brain areas. J Neurosci 25: 10564–10573.

18. Tsakiris M, Prabhu G, Haggard P (2006) Having a body versus moving your body: How agency structures body-ownership. Conscious Cogn 15: 423–432.

19. Sanchez-Vives MV, Slater M (2005) From presence to consciousness through virtual reality. Nat Rev Neurosci 6: 332–339.

CHAPTER 16

OWNERSHIP AND AGENCY OF AN INDEPENDENT SUPERNUMERARY HAND INDUCED BY AN IMITATION BRAIN-COMPUTER INTERFACE

Luke Bashford[1,2], and Carsten Mehring[2]

[1]Department of Bioengineering, Imperial College London, London, United Kingdom

[2]Bernstein Centre Freiburg, University of Freiburg, Freiburg, Germany

University of Bologna, ITALY

ABSTRACT

To study body ownership and control, illusions that elicit these feelings in non-body objects are widely used. Classically introduced with the Rubber Hand Illusion, these illusions have been replicated more recently in virtual reality and by using brain-computer interfaces. Traditionally these illusions investigate the replacement of a body part by an artificial counterpart, however as brain-computer interface research develops it offers us the possibility to explore the case where non-body objects are controlled in addition to movements of our own limbs. Therefore we propose a new illusion designed to test the feeling of ownership and control of an independent supernumerary hand. Subjects are under the impression they control a virtual reality hand via a brain-computer interface, but in reality there is no causal connection between brain activity and virtual hand movement but correct movements are observed with 80% probability. These imitation brain-computer interface trials are interspersed with movements in both the subjects' real hands, which are in view throughout the experiment. We show that subjects develop strong feelings of ownership and control over the third hand, despite only receiving visual feedback with no causal link to the actual brain signals. Our illusion is crucially different from previously reported studies as we demonstrate independent ownership and control of the third hand without loss of ownership in the real hands.

INTRODUCTION

Illusions that create the sense of ownership for non-body objects and agency for their actions are used to investigate the neural mechanisms underpinning the above mentioned senses of the self. The classic rubber hand illusion is one common example [1]; in this illusion the experimenter brushes a rubber hand that has been placed on a table in the position of the subjects real right hand. Synchronously the experimenter brushes the subject's real hand, which is hidden from view. This leads subjects to report that they perceive the rubber hand as though it were their real hand. Illusions of ownership have also been created from more recent studies with brain-computer interface users [2] and subjects seeing virtual reality [3,4] experiencing the feeling of ownership over the artificial devices and agency of their potential or actual movements.

In many studies the feeling of ownership for the non-body object occurs as it replaces the body part which is hidden from view. It has been demonstrated however that the brain is also capable of extending the body image beyond the physical limits of the human form. Supernumerary limb illusions demonstrate the brains ability to feel ownership over an additional limb [5–7] however in such cases it appears as though the feeling

Citation: Bashford L, Mehring C (2016) Ownership and Agency of an Independent Supernumerary Hand Induced by an Imitation Brain-Computer Interface. PLoS ONE 11(6): e0156591. doi:10.1371/journal.pone.0156591

of ownership is more that of a replication of the existing limb rather than the addition of a truly independent supernumerary limb, but this distinction is not made specifically in these studies.

The feeling of an independent supernumerary limb has been presented in a medical case study [8], but recent advances in robotics and computer science suggest that artificial supernumerary limbs could be incorporated into healthy human function in the future. Supernumerary robotic limbs can be built to assist human movements [9,10] and brain-computer interfaces have been shown to work during concurrent movement execution [11]. It is therefore important to understand if the brain is capable of feeling ownership over additional and independent supernumerary limbs and agency over their actions. We therefore present here a new body ownership illusion generated by an imitation brain-computer interface, producing such feelings of ownership and agency. In our study subjects imagine moving a third virtual reality (VR) arm, while they are under the belief they control the movements of the VR arm using a brain-computer interface. We compare the strength of this illusion using both questionnaires and physiological measures: galvanic skin response (GSR) to measure ownership, hand skin temperature to measure disownership and questionnaire data to measure the feeling of ownership and agency. Furthermore we compare our illusion to the rubber hand illusions and the third rubber hand illusion. In particular we highlight the differences in terms of independence of the ownership and agency between our and the previous illusions.

MATERIALS AND METHODS

36 subjects participated in this study (16 female, 20 male, age 19–40, mean 26). All subjects performed the imitation of a BCI third hand (IBCI) and the rubber hand illusion (RHI) experiments, a subset of 17 participants additionally performed the rubber third hand illusion (RTHI) experiment and answered questions 13 and 14 for all experimental conditions. A further subset of 8 participants additionally performed a control experiment for each illusion condition. These control conditions were performed to establish a baseline measure of responses to assess the presence and strength of the illusions in the 'illusion conditions'. In the following these control conditions will, therefore, be called baseline conditions or baseline experiments. All subjects signed informed consent for participation and the experimental procedures were approved by the ethics committee at the University of Freiburg, Germany.

Imitation of BCI Third Hand—IBCI

Subjects were sat with only their hands visible on a table top and between the subjects own hands a virtual reality hand was projected, a realistically sized left hand shown facing palm up (Fig 1A top panel). Subjects were fitted with an EEG cap, skin temperature and galvanic skin response sensors.

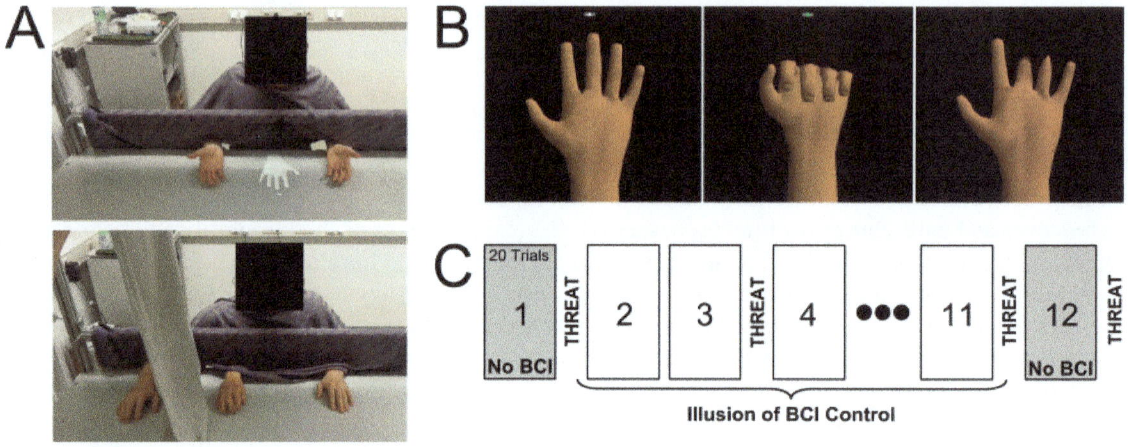

Figure 1: Experimental methods.

Subjects were shown three cues corresponding to each hand, the cue colours indicated inter-trial interval (5s), preparation (2s) and go (3s) or no-go (3s). After the go cue subjects were instructed to close their hand (clench a fist) once and return to the rest position of an open hand palm up. For the real left and right hand this was a simple go/no-go cued movement execution in the ratio (80%:20% respectively). No-go trials were used to maintain the attention of the subjects and to keep movement vs no movement trials of the real hands in the same proportion as movement vs no movement for the virtual reality third hand (also 80%:20% respectively). When cued to control the virtual third hand subjects were instructed to perform a visual imagery of the hand closing until the trial was complete, "imagine seeing the hand close". Visual imagery was chosen as it may be less likely to interfere with the movement execution component of the task, whereas perhaps a kinaesthetic imagery would. It was explained to the subjects that the EEG cap they were wearing would allow the computer to recognise and classify features of the brain signals generated specifically by the imagery, and that if classified correctly this would generate a cued movement of the virtual reality hand. The hand played a single sequence to show it closing and opening a fist from a resting open position (Fig 1B left and middle panels). In reality however the EEG was not connected and positive feedback from the 'BCI controlled' trials (i.e. movement of the virtual hand) occurred randomly with a probability of 80%.

Subjects performed 12 blocks of 20 trials per block (Fig 1C), and were offered a short break in between each block. In the first and last block subjects were told there was no control of the hand via BCI and saw only trials of cued movement of the virtual hand. Subjects were asked to always perform the imagery they should generate for the experiment as described above (at the cue and continue while the virtual hand was moving) despite knowing the BCI was not functioning at this time. In the middle 10 blocks subjects performed 200 trials containing: 40 go trials and 10 no-go each for the left and right hands and 100 trials of cued movements of the virtual middle hand under 'BCI control' with 80% positive feedback. All trial types were randomly distributed across all 200 trials.

At the end of the 1st (pre), 3rd (early), 11th (late) and 12th (post) block a threatening stimulus was shown to the subjects to probe the illusion strength measured by the galvanic skin response (GSR). The 'early' response occurred at the third rather than the second block as we wanted that the result was not due to subject's naivety, but reflected the early response after some exposure to the task. The threatening stimulus consisted of a cracking sound while the virtual middle hand changed to show broken fingers (Fig 1B right panel).

Rubber Hand Illusion—RHI

After the IBCI subjects then performed the rubber hand illusion [1]. Subjects remained seated in the same position as in the IBCI. In the rubber hand illusion experiment the right hand was shielded from view and a gender matched rubber right hand was placed on the table in front of the subject (Fig 1B bottom panel). The experimenter synchronously brushed the unseen right hand and the rubber hand following a brushing protocol consistent between subjects. Subjects were asked to comment throughout the experiment on the feeling of the illusion; firstly when they felt the illusion begin and then again if the illusion got stronger. Brushing continued for approximately 10mins apart from one case where the subject reported losing the illusion and brushing was stopped after 4mins.

Rubber Third Hand Illusion—RTHI

After the RHI we performed the rubber third hand illusion (RTHI), described by Guterstam and colleagues [7]. The rubber third hand experiment was performed exactly as the RHI (above), with the single difference that subjects were able to see both their own hands in addition to the rubber hand, i.e. the real hand was not covered from view.

Baseline Conditions

A subset of 8 subjects who performed all three illusion conditions returned for a separate session to perform baseline conditions of each illusion. To establish a baseline for the IBCI condition subjects first repeated only the first and last blocks of the IBCI as described above. Subjects made no movements, it was a static condition, but not passive as the movement imagery was still performed. Following this to establish a baseline for the RHI condition subjects repeated the RHI as described above, however in the first baseline case subjects first received 5 minutes of asynchronous brushing after which they answered the questionnaire. They then had a second baseline case with 5 minutes of synchronous brushing with the rubber hand rotated 90 degrees, viewed perpendicular to the subject's body after which they completed an additional questionnaire. Finally to establish a baseline for the RTHI condition subjects repeated the RTHI condition as described above, also with the two baseline conditions asynchronous and synchronous-rotated as described above.

Data Acquisition and Analysis

Data acquisition and analysis was treated identically for both illusion and baseline conditions. The temperature was recorded during all conditions while the GSR only during the IBCI. Measurements were made always from both hands and all recordings were made using the actiChamp (Brainproducts GmbH, Germany) at 1kHz sampling rate. Skin temperature was recorded to address the aspect of disownership in the limbs during the illusion using a physiological measure [12]. Galvanic skin response was measured using the Brainproducts GSR Module to provide a physiological measure of ownership during the illusion conditions [7,13]. The two passive electrodes were taped with skin tape to the medial phalanx of the index and middle finger, which had been covered in conductive gel for GSR. The GSR response was calculated for each subject as the difference between the maximum conductance in a 5s window after a threat and the mean conductance over 5s before the threat. Skin temperature was measured using a Greisinger GMH3210 Digital thermometer adapted for the actiChamp. The thermocouple probe was taped to the skin 2cm behind the knuckle of the middle finger. The skin temperature was calculated as the difference between the hand that was matched in the illusion condition and the other hand [12]. A baseline correction was made by subtracting the mean difference in skin temperature for a 10s window around the subjects report that they first felt the illusion. For RHI and RTHI this report was given verbally; in the IBCI condition we used the time of the early threat, at this point we had the first evidence that subjects felt the illusion. The temperature data was resampled at a frequency of 1Hz and smoothed using a first order Savitzky-Golay filter, with a window width of 20s. The temperature and galvanic skin response data for one subject was not recorded due to a battery fault in the equipment; however the questionnaire responses for this subject were retained.

Subjects had to answer several questions immediately after each experiment. The questions were similar to those typically found in ownership/control illusion literature, but also were specifically designed to explore key differences between the feelings of ownership, control and independence in the three conditions we investigated. Subjects were asked to rate their answers using the seven-step visual-analogue scale ranging from strongly agree (+++) to strongly disagree (---).

Throughout our analysis we first assessed statistical significance across groups or conditions using a one-way non-parametric ANOVA (Kruskal-Wallis test). We then performed post-hoc pairwise Wilcoxon tests to assess statistical significance between pairs of groups or conditions. All reported p-values were corrected for multiple testing where applicable using the Holm-Bonferroni method and the adjusted p-values were reported. It is important to note that this method can produce adjusted p-values larger than 1.

RESULTS

Subjects performed three experiments (Fig 1): The imitation brain computer interface (IBCI) experiment was designed to induce the illusion of a BCI controlled third hand. In this experiment healthy human subjects were fitted with an EEG cap and told that by performing visual imagery in response to a cue they could move a virtual reality hand projected on a table top in front of them. Subjects could always see both of their own hands and cues to move the virtual hand were interspersed with cues to replicate the movement with their own hands. In reality the BCI was not connected and the virtual hand moved randomly with 80% probability. We compared our illusion to the classic Rubber Hand Illusion (RHI) [1] and the Third Hand Illusion (RTHI) [7]. All conditions were additionally compared to their baseline (see materials and methods).

The illusion of ownership in IBCI

We first sought to determine whether our IBCI experiment gave subjects the feeling of ownership of a supernumerary hand. This was examined by means of questioning immediately after the experiment (Figs2and 3, Tables 1 and 2). Question (Q) 1 and 2 directly probe the feeling of ownership. For the IBCI condition significantly positive responses were recorded for Q1-2 compared to baseline (median (++), see Table 2 for p-values). In contrast for the RHI, subjects were disagreeing (median (---) for Q1 and Q2) not different from the baseline condition (Table 2). Also, subject's answers for RHI were significantly different from the IBCI condition (see Table 1 for p-values). This suggests a strong feeling of ownership of a supernumerary limb in the IBCI condition but little or no supernumerary ownership in the RHI and the baseline conditions. Between the IBCI and the RTHI we found a significant difference for Q2 with a stronger positive response for IBCI than for RTHI ((++) for IBCI, (+) for RTHI, Table 1) and no significant difference for Q1 (Table 1). Consistently, RTHI answers for Q1 and 2 were significantly more positive than during its baseline condition (Table 2). These results suggest that both our IBCI and the RTHI induce feelings of ownership of a supernumerary limb unlike the RHI, with the ownership feeling being stronger in IBCI than in RTHI.

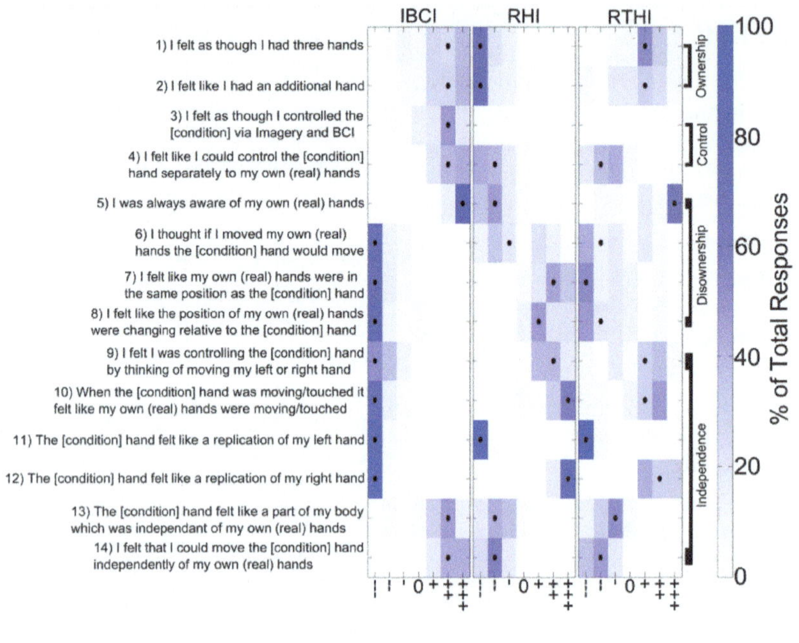

Figure 2: The questions and responses for the different experimental conditions.

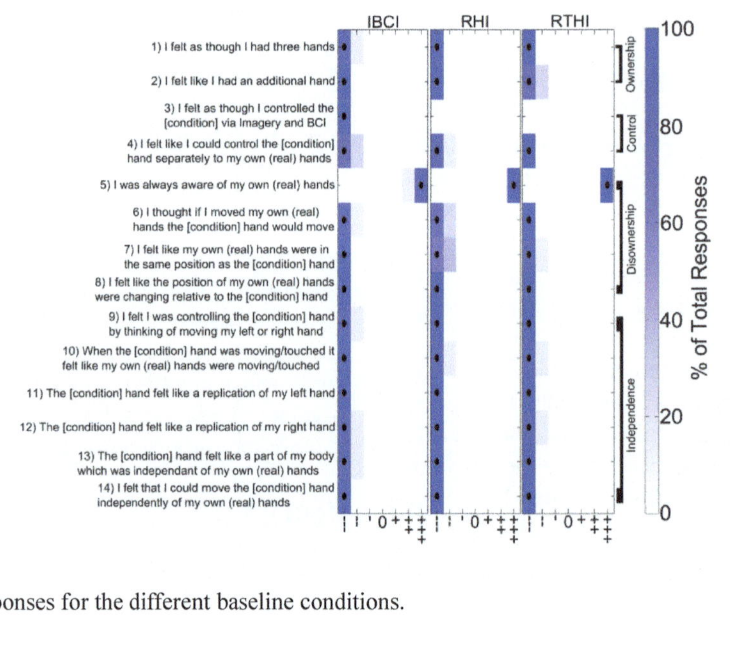

Figure 3: The questions and responses for the different baseline conditions.

Table 1: Statistical assessment of differences between conditions for each question.

Question No. and attribute addressed	Across all groups	IBCI vs RHI	IBCI vs RTHI	RHI vs RTHI
1 (Ownership)	$8.4x10^{-15}$	$4.7x10^{-7}$	0.12	$2.4x10^{-9}$
2 (Ownership)	$8.2x10^{-15}$	$3.8x10^{-7}$	$2.9\ x10^{-3}$	$7.2x10^{-8}$
3 (Control)	N/A	N/A	N/A	N/A
4 (Control)	$2.5x10^{-15}$	$2.8x10^{-7}$	$7.4x10^{-9}$	$3.0\ x10^{-2}$
5 (Disownership)	$5.4x10^{-16}$	$2.0x10^{-7}$	$3.7\ x10^{-2}$	$5.5x10^{-8}$
6 (Disownership)	$4.2x10^{-8}$	$4.3x10^{-6}$	$2.4x10^{-3}$	0.14
7 (Disownership)	$1.8x10^{-15}$	$2.3x10^{-7}$	$3.5\ x10^{-2}$	$6.4x10^{-8}$
8 (Disownership)	$3.4x10^{-15}$	$3.8x10^{-7}$	$6.0\ x10^{-3}$	$3.2x10^{-7}$
9 (Independence)	$1.2x10^{-13}$	$3.8x10^{-7}$	$2.0x10^{-7}$	$8.9\ x10^{-2}$
10 (Independence)	$3.7x10^{-16}$	$2.2x10^{-7}$	$1.0x10^{-8}$	$2.7\ x10^{-5}$
11 (Independence)	0.59	1.5	1.2	1.5
12 (Independence)	$2.7x10^{-17}$	$5.5x10^{-8}$	$3.1x10^{-11}$	$9.3\ x10^{-5}$
13 (Independence)	$3.4x10^{-7}$	$7.8x10^{-4}$	$1.2x10^{-5}$	0.38
14 (Independence)	$1.9x10^{-8}$	$5.2x10^{-4}$	$1.3x10^{-6}$	0.61

Table 2: Statistical comparison with baseline conditions.

Question No. and attribute addressed	IBCI vs IBCI baseline	RHI vs RHI baseline	RTHI vs RTHI baseline
1 (Ownership)	$2.3x10^{-2}$	0.13	$2.3x10^{-2}$
2 (Ownership)	$2.3x10^{-2}$	$6.3x10^{-2}$	$3.1x10^{-2}$
3 (Control)	$7.8x10^{-3}$	N/A	N/A
4 (Control)	$2.3x10^{-2}$	0.13	0.13
5 (Disownership)	1.0	$2.3x10^{-2}$	1.0
6 (Disownership)	1.0	$4.7x10^{-2}$	0.5
7 (Disownership)	0.50	$2.3x10^{-2}$	0.5
8 (Disownership)	0.50	$2.3x10^{-2}$	0.5
9 (Independence)	0.38	$3.1x10^{-2}$	$2.3x10^{-2}$
10 (Independence)	1.0	$2.3x10^{-2}$	$2.3x10^{-2}$
11 (Independence)	3.0	3.0	2.0
12 (Independence)	1.0	$2.3x10^{-2}$	$2.3x10^{-2}$
13 (Independence)	$2.3x10^{-2}$	$3.1x10^{-2}$	$3.1x10^{-2}$
14 (Independence)	$2.3x10^{-2}$	0.13	0.13

To provide a further physiological measure of ownership during the IBCI condition the galvanic skin response (GSR) was recorded. An increased GSR response to a threatening stimulus is an established objective measure of ownership [7,13]. Threats were made to the virtual reality third hand at four time points during the experiment (Fig 4A). Either while the subjects believed they controlled the hand (early/late) or while subjects knew there was no control (pre/post) (see Materials and Methods). A significant difference in GSR was observed across all four threats (left hand $p = 2.8x10^{-10}$, right hand $p = 7.2x10^{-10}$, Kruskal-Wallis test corrected for multiple comparisons with Holm-Bonferroni correction). There was a strong and significant difference in the GSR response to a threat during the imitation BCI control (early/late) compared to the baseline conditions where subjects saw the virtual hand move but were told they are not controlling it with the BCI (pre/post) (left hand $p = 3.4x10^{-12}$, right hand $p = 1.8x10^{-12}$, Wilcoxon signed rank test, comparing between the combined pre/post and early/late threats, $p = 0.36x10^{-11}$ after correction for multiple comparisons with Holm-Bonferroni). While there were no significant differences for either hand between the pre and post conditions there was a significant increase in the GSR response from 'early' to 'late' threats (left hand $p = 0.02$, right hand $p = 0.004$, Wilcoxon signed rank test, correction for multiple comparisons with Holm-Bonferroni). This indicates that the ownership illusion became stronger over the course of the imitation BCI control. Furthermore, there is a comparably small influence of hand on the GSR: when the data for left and right hands was compared separately at each threat there was a weakly significant difference between the hands only at the 'late' threat ($p = 0.04$, Wilcoxon sign rank test Holm Bonferroni corrected for multiple comparisons). Taken together, these results show that the strength of the ownership, as defined by the GSR, increases over time and this increase is restricted to only the period of imitation BCI control.

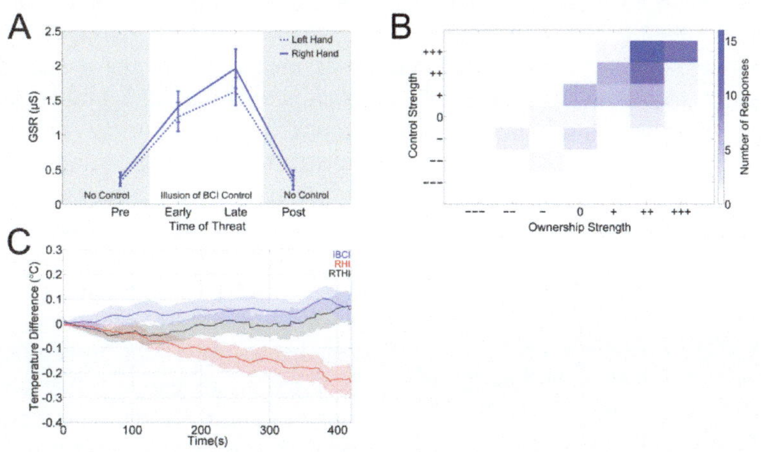

Figure 4: A) The mean across subjects GSR after a threatening stimulus during the IBCI condition. A significantly stronger response was seen during the imitation of BCI control compared to the pre and post conditions where subjects knew there was no control. Error ...

The illusion of control in IBCI

We assessed subjects feeling of control by questions Q3 & 4. Following the IBCI condition subjects answered all questions significantly positively compared to baseline conditions (median (++), Table 2), indicating a strong feeling of control. For RHI and RTHI the answers to Q4 were significantly different from the IBCI condition (Fig 2, Table 1), with subjects disagreeing to having control (median (--)), not different from baseline conditions (Table 2). This shows that there was little to no feeling of control during RHI and RTHI in contrast to IBCI. Q3 did not apply to the RHI and RTHI. In the baseline experiments subjects did not feel a sense of control in any of the conditions (Table 2, Fig 3). We recognise that the baseline experiments did not involve any overt movement. In the IBCI case subjects were performing movement imagery but were aware there is no control; we still used Q3&4 in this case to demonstrate that performing the imagery, viewing the hand and seeing it move in response to the cue did not contribute to the sense of control shown for the IBCI condition.

The correlation between the illusion of control and ownership

Q3 was asked to determine how much subjects felt like they controlled the virtual reality third hand via BCI. While the median response was significantly positive compared to baseline ((++), Table 2), the answers of individual subjects spanned both positive and negative responses. Overall most answers were positive (75th percentile response +++, 25th percentile response ++) but 7 out of 36 subjects gave answers which were negative or equal to 0. This indicates that some subjects may have realised that the BCI control was unrelated to their cognitive processes during the imitation 'BCI controlled' trials. Interestingly the strength of the illusion of ownership also varied across subjects with many subjects reporting a strong feeling of ownership while some did not report any feeling of ownership (Q1 & 2, Fig 2). We therefore hypothesized that subjects with a strong illusion of control may also have a strong feeling of ownership and subjects without the illusion of control may not report a strong feeling of ownership. To investigate this we correlated the answers to Q3 (feeling of control) with the answers to Q1 and Q2 (feeling of ownership) and found strong and significant correlations (Fig 4B, $R = 0.64$ between Q3 and Q1, $R = 0.73$ between Q3 and Q2, $p<0.001$, Spearman's rank correlation). Importantly in no case did the subjects have the feeling of owning a third or additional hand without reporting some feeling of control via the BCI, or vice versa (Fig 4B). This shows that in the IBCI condition the illusion of ownership was closely linked to the illusion of control.

Disownership of the real hands

To further explore the idea that the third hand was considered a supernumerary hand rather than the replacement of one of the real hands we investigated whether there was any disownership of the subject's real hands during the IBCI condition. Disownership refers to the loss of the awareness of ownership of one's own body parts, therefore no or little disownership of the subject's real hands would suggest that rather than replacing a limb, the illusion induces the feeling of a 'third hand'. We recorded skin temperature (see Materials and Methods) which has recently been shown to be a correlate of disownership of the real hand in the RHI [12]. We replicated the previously reported effect, as we observed a significant cooling of the matched hand between the mean of the 1st and 6th minute after the report of illusion onset in the RHI condition relative to the other hand (Fig 4C, $p = 0.02$, Wilcoxon sign rank test, Holm-Bonferroni correction for multiple comparisons). We did not see this significant cooling in either our IBCI or the RTHI ($p = 0.14$ and $p = 0.97$ respectively, Wilcoxon sign rank test, Holm-Bonferroni correction for multiple comparisons). Comparing temperatures for the 1st minute after the report of illusion onset there was no significant difference between the three conditions (for all conditions $p = 0.62$, Kruskal-Wallis test. $p = 1.1$ for IBCI vs RTHI and $p = 1.3$ RHI vs RTHI, Wilcoxon rank sum test and $p = 1.3$ IBCI vs RHI, Wilcoxon signed rank test, Holm-Bonferroni correction for multiple comparisons). When we compared the temperatures in those subjects who maintained the illusion into 6th min after the illusion onset there was a significant difference between the conditions ($p = 0.0019$, Kruskal-Wallis test, the number of subjects was N(IBCI) = 35, N(RTHI) = 11 and N(RHI) = 22); post-hoc tests showed significance between IBCI and RHI ($p = 0.002$, Wilcoxon rank sum test, Holm-Bonferroni correction for multiple comparisons), boarder line significance between RTHI and RHI ($p = 0.056$, Wilcoxon rank sum test, Holm-Bonferroni correction for multiple comparisons) and no significance between IBCI and RTHI ($p = 0.4$, Wilcoxon rank sum test, Holm-Bonferroni correction for multiple comparisons). The average duration for the illusion was 7.86min +- 19.0s in RHI and 7.33mins +- 20.8s in RTHI (mean +- SEM, 7.6s over both). During the IBCI subjects were asked not to speak, so did not report the specific onset or loss of the illusion during the experiment, this was in order to maintain the conditions of a typical BCI experiment. These results are in support of previous findings [12], suggesting that there is little to no disownership of the subject's actual hands during the IBCI and RTHI condition while there are signatures of disownership during the RHI.

In addition disownership was investigated by questions Q5-8 (Fig 2). Q5 specifically addressed the awareness subjects had of their own hands. Whilst all responses were significantly different across experiments (Table 1) the conditions IBCI and RTHI showed strongly positive results (median (+++)), not different to baseline (Table 2) indicating no or less disownership than during RHI where a negative response was measured (median (--)), which was significantly lower than during baseline (Table 2). In the baseline conditions subjects gave strongly positive results for all illusions (median (+++), Fig 3) indicating that there was no feeling of disownership during the baseline conditions. We further explored subjects' disownership of their own hands by Q6-8. These questions were again answered significantly differently for each of the three conditions IBCI, RHI and RTHI (Table 1). For the IBCI all questions had strong negative responses (median (---)), which were not different from baseline (Fig 3, Table 2), indicating full awareness of the real hands during this condition. For the RTHI responses were also negative but with broader distributions (Fig 2), though neither different from baseline (Fig 3, Table 2). For the RHI Q7-8 showed positive responses but a negative response was recorded for

Q6, in all cases the RHI response was significantly different to its baseline (Table 2). The findings from the questionnaire therefore corroborate that of the temperature data, indicating that there is no disownership of the real hands during the IBCI condition in contrast to clear disownership during the RHI. During the RTHI condition there seems to be no or little disownership, however the RTHI shows a significantly weaker and broader response range compared to the IBCI, suggesting a less clear absence of disownership than during the IBCI. Next we investigated if the supernumerary hand was also felt to be independent from the real hands during IBCI.

The independence in IBCI

To assess independence, we used several questions (Q9-Q14) which specifically addressed the feelings of independence both in terms of ownership and control. Q13 references independent ownership and showed significantly more positive answers in the IBCI than during its baseline (Table 2). This suggests a feeling of independent ownership for the IBCI condition. RHI and RTHI answers had a negative median, which was only slightly but still significantly higher than during their baselines (Fig 3, Table 2). This suggests some feeling of independent ownership also for RHI and RTHI, however, less pronounced than for IBCI where the answers were significantly more positive than in RHI or RTHI (Table 1).

Questions 11–12 addressed the extent to which the artificial hand felt like a replication i.e. a less independent limb. For the IBCI condition (where a left virtual hand was projected) the answers to Q11, probing the replication of the subjects' own left hand, were negative (median (---)) and not significantly different from baseline (Table 2), whilst for the RHI and RTHI (where a right rubber hand was used) the answers to Q12, probing the replication of the subjects' own right hand, were significantly positive compared to baseline (median (+++) and (++) respectively, Table 2). This shows that the feeling in the RHI and RTHI was a replication of the existing hand, while the IBCI condition created the feeling of a separate hand.

The feeling of independent control was assessed with Q14 and was answered significantly more positively in the IBCI condition compared to the RHI or RTHI (Table 1), with no significant difference between RHI and RTHI (Table 1). The strongly negative responses to this question for the RHI an RTHI where not different to baseline (median (--), Table 2). In contrast the positive responses during IBCI (median (++)) were significantly different to baseline (Table 2) suggesting that the feeling of being able to control the hand independently to the real hands was only felt in the IBCI condition. Q9-10 additionally probed the independence of control with positive answers indicating less independence. The positive responses reported for the RHI and RTHI condition compared to baseline (median (++) and (+) respectively, Table 2) and the negative responses for IBCI, not different from baseline (median (---), Table 2) confirmed that the control was only felt as being independent in the IBCI condition (see also Table 1 demonstrating the significance of the difference between IBCI and RHI/RTHI). This conclusion is further corroborated by Q10 where all conditions were significantly different (Table 1), with strong support for independence in the IBCI condition (median (---), not significantly different from baseline, Table 2) and no support for independence in the RHI and RTHI conditions (median (+++) and (+) respectively, significantly different from baseline, Table 2).

Baseline conditions

A subset of 8 subjects who performed all three illusion conditions were called back at a later date to perform the baseline conditions (see Materials and Methods for details). The baseline conditions were performed to establish a baseline measure of responses to assess the presence/strength of the illusions in the 'illusion conditions'. After each baseline experiment subjects were asked to answer the questionnaire again. These responses were compared to the responses given in the original experimental conditions (Table 2). Not a single subject reported an illusion in any of the three baseline conditions and the answers to the questions were very consistent across subjects (Fig 3). In the RHI and RTHI condition subjects were asked to answer questions separately for the asynchronous and the synchronous-rotated brushing conditions. Only 3 subjects answered cumulatively 5 questions differently in the rotated condition (with a maximum difference of one step on the visual analogue scale per questions in 4/5 cases, and one with a difference of 2 steps). The difference never changed the sentiment of the subjects answer, highlighting the equal ability of each baseline condition to remove any illusion of ownership and agency. For this reason the results presented in Fig 3 are for the asynchronous baseline questionnaire. Furthermore subjects were asked additional questions after the baseline conditions: 1) It felt as if my real hand were matching the texture of the [condition] hand. 2) I felt as if the illusion I was experiencing was coming from somewhere between my real hand and the [condition] hand. For these questions all subjects gave a '---' response.

DISCUSSION

We present the IBCI condition that induces the illusion of ownership in an independent supernumerary limb and the sense of agency for its actions. To distinguish this new illusion from existing body ownership illusions we compared it to the rubber hand illusion (RHI) [1]. and a variant of it, the rubber third hand illusion (RTHI) reported to give the feeling of ownership in a third hand [7].

The IBCI illusion–Ownership, Control and Independence

The IBCI illusion combines the illusion of ownership of an independent third hand and agency of its movements, with no disownership in the real hands. Our illusion demonstrates in able-bodied subjects the brains capacity to expand the body representation beyond the body's anatomy by a supernumerary limb that is not just a replication of an existing body part [5–7,14].

Questionnaire and GSR results demonstrated that subjects felt both a strong sense of ownership and control over the third hand, which was not affected by the sight or movement tasks performed with their real hands during the experiment. It has recently been suggested [12] that the skin temperature of a hand which is disembodied in the illusion cools relative to the other hand [15,16]. While this effect seems to be weak, it offers a physiological measure of disownership. Our results replicated the cooling effect in skin temperature during the RHI (Fig 4C). Importantly the cooling is not observed in our IBCI or the RTHI, which suggests no disembodiment in these conditions. We addressed the feeling of independence in the ownership and control and found a noticeable difference between the IBCI and the RTHI: subject's responses indicated strong feelings of independence for the IBCI and little to no independence in the RTHI. Furthermore there was no significant difference in the independence between the RHI, where we demonstrate disembodiment, and the RTHI. Our results, therefore, indicate that the IBCI induces the illusion of an independent supernumerary limb while the RTHI seems to feel more like a replication of the existing hand. Previously the feeling of independent ownership has only been shown in amputees [8], where three out of fourteen subjects reported that their arm shown in a mirror box felt like an additional limb, independent to their feeling of the phantom limb.

We were able to induce the illusion using an imitation BCI, showing the correct movement on 80% of the trials, with no causal connection to subject's brain activity. This is fundamentally different to previous ownership illusions, including both the RHI and RTHI, which either rely on combined visual and tactile feedback or on actual BCI control, i.e. combined motor signals and visual feedback. Subjects in the IBCI illusion received only visual feedback in response to a cued task, unrelated to their brain activity.

Relationship between control and ownership

During the IBCI condition most subjects believed they controlled the arm via the BCI, and in all these cases the illusion of control accompanied the illusion of ownership (Fig 4B). Importantly in no subjects was one felt without the other and across subjects the strength of the illusion of ownership was correlated with the strength of the illusion of control. The relationship or causality between the feelings of ownership and the sense of agency remain unclear, and some studies suggest they represent two distinct cognitive processes [17]. We have demonstrated here that the illusion of ownership and control can come from a task where the subject has no actual control. This supports the notion that success regardless of actual influence in a task [18] and observing expected congruent movements or feedback can lead to the feeling of control [19,20], but extends this notion to ownership and additionally to supernumerary limbs. In the IBCI condition, subjects received with 80% probability a positive feedback, i.e. in 80% of the trials the virtual hand moved during instructed movement imagery. It remains an open question exactly how accurate the performance must be to maintain the illusion? For example, if subjects were to receive 80% negative feedback and only 20% successful trials would they still feel the same sense of ownership and agency, or would subjects report no sense of ownership and agency as in the baseline condition where they know there is no control. The sense of agency has been shown to increase with the congruency of the expected outcome during a BCI task [21]. How the sense of ownership is affected remains to be addressed by future investigations.

In the IBCI condition the illusion of ownership as demonstrated by the GSR was present only during the perceived BCI control (Fig 4A). Even though subjects were asked to perform the exact same imagery in response to the cue in the pre and post blocks, the GSR was significantly lower compared to IBCI—the only difference being that here they were told there is no control. Hence, the illusion of ownership was independent of motor imagery but tied to the belief of control, in keeping with a recent study [21] reporting that the sense of agency in BCI is independent of motor imagery-based neural signals and dominated by visual feedback.

Relevance for brain-computer interfaces

The motivation for exploring a "third hand" illusion arises from the likely concept that BCIs could also be used to supplement existing function [22], for example by providing control of supernumerary limbs which could be used concurrently to movements of the real limbs. Our study suggests that in such BCIs subjects may feel ownership of the BCI controlled limb without disownership of the real limbs. BCI learning and control has been associated with the formation of stable cortical maps [23], and in addition several studies [24–27] suggest that brain areas and features of brain signals typically used for BCI control can be modified as a result of altering body ownership. It remains an open and interesting question as to whether a BCI controlling a supernumerary limb could induce a robust independent neural representation of the external effector.

In summary we present a new illusion that induces the feeling of independent ownership and control of a supernumerary hand. The illusion is induced by imitating the behaviour of a brain-computer interface, where subjects believe they control the movements of a virtual hand by their brain activity while in reality there is no causal link between subjects' brain activity and the movements of the hand.

Funding Statement

German Federal Ministry of Education and Research (BMBF) 01GQ0830 to Carsten Mehring.

This work was supported by the German Federal Ministry of Education and Research (BMBF) grant 01GQ0830 to BFNT Freiburg and Tübingen (http://www.bmbf.de/en/), received by CM. The article processing charge was funded by the German Research Foundation (DFG) and the University of Freiburg in the funding programme Open Access Publishing. The funders had no role in study design, data collection and analysis, decision to publish, or preparation of the manuscript.

DATA AVAILABILITY

All relevant data are within the paper and its Supporting Information files.

REFERENCES

1. Botvinick M, Cohen J. Rubber hands feels touch that eyes see. Nature. 1998;391: 756–756. doi:10.1038/35784
2. Perez-Marcos D, Slater M, Sanchez-Vives MV. Inducing a virtual hand ownership illusion through a brain computer interface. NeuroReport. 2009;20: 589–594. doi: 10.1097/WNR.0b013e32832a0a2a
3. Ma K, Hommel B. The virtual-hand illusion: effects of impact and threat on perceived ownership and affective resonance. Cognition. 2013;4: 604 doi: 10.3389/fpsyg.2013.00604

4. González-Franco M, Peck TC, Rodríguez-Fornells A, Slater M. A threat to a virtual hand elicits motor cortex activation. Exp Brain Res. 2014;232: 875–887. doi: 10.1007/s00221-013-3800-1
5. Ehrsson HH. How many arms make a pair? Perceptual illusion of having an additional limb. Perception. 2009;38: 310–312.
6. Newport R, Pearce R, Preston C. Fake hands in action: embodiment and control of supernumerary limbs.Exp Brain Res. 2010;204: 385–395. doi: 10.1007/s00221-009-2104-y
7. Guterstam A, Petkova VI, Ehrsson HH. The Illusion of Owning a Third Arm. PLoS ONE. 2011;6: e17208 doi: 10.1371/journal.pone.0017208
8. Giummarra MJ, Georgiou-Karistianis N, Nicholls MER, Bradshaw JL. The third hand: Ownership of a rubber hand in addition to the existing (phantom) hand. Cortex. 2011;47: 998–1000. doi:10.1016/j.cortex.2011.02.010
9. Parietti F, Chan K, Asada HH. Bracing the human body with supernumerary Robotic Limbs for physical assistance and load reduction. 2014 IEEE International Conference on Robotics and Automation (ICRA). 2014. pp. 141–148. doi: 10.1109/ICRA.2014.6906601
10. Abdi E, Burdet E, Bouri M, Bleuler H. Control of a Supernumerary Robotic Hand by Foot: An Experimental Study in Virtual Reality. PLoS ONE. 2015;10: e0134501 doi: 10.1371/journal.pone.0134501
11. Orsborn AL, Moorman HG, Overduin SA, Shanechi MM, Dimitrov DF, Carmena JM. Closed-Loop Decoder Adaptation Shapes Neural Plasticity for Skillful Neuroprosthetic Control. Neuron. 2014;82: 1380–1393. doi: 10.1016/j.neuron.2014.04.048
12. Moseley GL, Olthof N, Venema A, Don S, Wijers M, Gallace A, et al. Psychologically induced cooling of a specific body part caused by the illusory ownership of an artificial counterpart. Proc Natl Acad Sci. 2008;105: 13169–13173. doi: 10.1073/pnas.0803768105
13. Ehrsson HH, Rosen B, Stockselius A, Ragno C, Kohler P, Lundborg G. Upper limb amputees can be induced to experience a rubber hand as their own. Brain. 2008;131: 3443–3452. doi: 10.1093/brain/awn297
14. Folegatti A, Farnè A, Salemme R, de Vignemont F. The Rubber Hand Illusion: Two's a company, but three's a crowd. Conscious Cogn. 2012;21: 799–812. doi: 10.1016/j.concog.2012.02.008
15. Kammers MPM, Rose K, Haggard P. Feeling numb: Temperature, but not thermal pain, modulates feeling of body ownership. Neuropsychologia. 2011;49: 1316–1321. doi:10.1016/j.neuropsychologia.2011.02.039
16. Marasco PD, Kim K, Colgate JE, Peshkin MA, Kuiken TA. Robotic touch shifts perception of embodiment to a prosthesis in targeted reinnervation amputees. Brain. 2011;134: 747–758. doi:10.1093/brain/awq361
17. Kalckert A, Ehrsson HH. Moving a rubber hand that feels like your own: a dissociation of ownership and agency. Front Hum Neurosci. 2012;6: 40 doi: 10.3389/fnhum.2012.00040
18. Thompson SC, Nierman A, Schlehofer MM, Carter E, Bovin MJ, Wurzman L, et al. How Do We Judge Personal Control? Unconfounding Contingency and Reinforcement in Control Judgments. Basic Appl Soc Psychol. 2007;29: 75–84. doi: 10.1080/01973530701331189
19. Yokosaka T, Iizuka H, Yonemura T, Kondo D, Ando H, Maeda T. Alternating images of congruent and incongruent movement creates the illusion of agency. Sci Rep. 2014;4: 6201 doi: 10.1038/srep06201
20. Ninaus M, Kober SE, Witte M, Koschutnig K, Stangl M, Neuper C, et al. Neural substrates of cognitive control under the belief of getting neurofeedback training. Front Hum Neurosci. 2013;7: 914 doi:10.3389/fnhum.2013.00914
21. Evans N, Gale S, Schurger A, Blanke O. Visual Feedback Dominates the Sense of Agency for Brain-Machine Actions. PLoS ONE. 2015;10: e0130019 doi: 10.1371/journal.pone.0130019
22. Di Pino G, Maravita A, Zollo L, Guglielmelli E, Di Lazzaro V. Augmentation-related brain plasticity.Front Syst Neurosci. 2014;8 doi: 10.3389/fnsys.2014.00109
23. Ganguly K, Carmena JM. Emergence of a Stable Cortical Map for Neuroprosthetic Control. PLoS Biol. 2009;7: e1000153 doi: 10.1371/journal.pbio.1000153
24. Shokur S, O'Doherty JE, Winans JA, Bleuler H, Lebedev MA, Nicolelis MAL. Expanding the primate body schema in sensorimotor cortex by virtual touches of an avatar. Proc Natl Acad Sci. 2013;110: 15121–15126. doi: 10.1073/pnas.1308459110
25. Kanayama N, Sato A, Ohira H. Crossmodal effect with rubber hand illusion and gamma-band activity.Psychophysiology. 2007;44: 392–402. doi: 10.1111/j.1469-8986.2007.00511.x
26. Limanowski J, Lutti A, Blankenburg F. The extrastriate body area is involved in illusory limb ownership. NeuroImage. 2014;86: 514–524. doi: 10.1016/j.neuroimage.2013.10.035
27. Schaefer M, Heinze H-J, Rotte M. My third arm: Shifts in topography of the somatosensory homunculus predict feeling of an artificial supernumerary arm. Hum Brain Mapp. 2009;30: 1413–1420. doi:10.1002/hbm.20609

CHAPTER 17

DEVELOPING A NOVEL MEASURE OF BODY SATISFACTION USING VIRTUAL REALITY

Clare K. Purvis[1], Megan Jones[2], Jakki O. Bailey[3], Jeremy Bailenson[3], C. Barr Taylor[2]

[1]PGSP-Stanford Consortium, Department of Psychiatry & Behavioral Sciences, Stanford University School of Medicine, Stanford, California, United States of America,

[2]Department of Psychiatry & Behavioral Sciences, Stanford University School of Medicine, Stanford, California, United States of America,

[3]Department of Communication, Stanford University, Stanford, California, United States of America

ABSTRACT

Body image disturbance (BID), considered a key feature in eating disorders, is a pervasive issue among young women. Accurate assessment of BID is critical, but the field is currently limited to self-report assessment methods. In the present study, we build upon existing research, and explore the utility of virtual reality (VR) to elicit and detect changes in BID across various immersive virtual environments. College-aged women with elevated weight and shape concerns (n = 38) and a non-weight and shape concerned control group (n = 40) were randomly exposed to four distinct virtual environments with high or low levels of body salience and social presence (i.e., presence of virtual others). Participants interacted with avatars of thin, normal weight, and overweight body size (BMI of approximately 18, 22, and 27 respectively) in virtual social settings (i.e., beach, party). We measured state-level body satisfaction (state BD) immediately after exposure to each environment. In addition, we measured participants' minimum interpersonal distance, visual attention, and approach preference toward avatars of each size. Women with higher baseline BID reported significantly higher state BD in all settings compared to controls. Both groups reported significantly higher state BD in a beach with avatars as compared to other environments. In addition, women with elevated BID approached closer to normal weight avatars and looked longer at thin avatars compared to women in the control group. Our findings indicate that VR may serve as a novel tool for measuring state-level BID, with applications for measuring treatment outcomes. Implications for future research and clinical interventions are discussed.

INTRODUCTION

Body image disturbance (BID) refers to negative self-evaluations of one's physical appearance which lead to significant distress or impairment [1]. BID is a highly prevalent issue in the general population, particularly among young women [2,3]. BID is associated with depression and anxiety [4,5], impairment in social relationships [6], and perhaps most importantly, it is linked to the onset and maintenance of eating disorders (ED) [7–9].

While BID has been studied widely, certain important features of this phenomenon remain poorly understood. This is partly the result of existing assessment techniques, which limit the scope of BID research by focusing on BID main as a stable attitude, or trait; few measures assess the degree to which one's body image fluctuates at a given moment in time [10]. Such momentary fluctuations in BID are referred to as one's "state" body (dis)satisfaction (state BD); variability in state BD is an important predictor of the likelihood of developing ED, and an

Citation: Purvis CK, Jones M, Bailey JO, Bailenson J, Taylor CB (2015) Developing a Novel Measure of Body Satisfaction Using Virtual Reality. PLoS ONE 10(10): e0140158. doi:10.1371/journal.pone.0140158

indicator of BID severity [11]. Despite its importance, factors that influence state BD are poorly understood, partly because state BD assessments rely on retrospective accounts or lack ecological validity [12].

Recent research suggests that automatic cognitive and behavioral processes are highly involved in the state BD; however, traditional self-report measures are unable to assess automatic processes because they originate outside of conscious awareness [12]. For example, despite having a worsening effect on state BD, women with BID, compared to those without BID, look longer at thin, appearance-ideal conforming female bodies [13], and are predisposed to overestimate the prevalence of negative social feedback about their own bodies [14]. Such processing errors are believed to reinforce negative body-related self-evaluations and, as a result, may cause or maintain disordered behaviors [12,15,16]. This has not been studied directly, as tools for measuring automatic processes and implicit behaviors in BID are nascent. Researchers have developed computer-based techniques to explore visual attention [17,18], but the methods have limited ecological validity. Understanding how automatic processes in BID function in real-world contexts, and how they relate to overt behavior is of vital importance in properly conceptualizing BID and in developing effective interventions.

Virtual reality (VR) offers tools to address limitations in BID assessment. VR has gained popularity in psychological research because it provides researchers excellent environmental control, high realism, and unparalleled control over assessment timing [19]. VR is especially well suited tool to study automatic non-verbal behaviors in realistic contexts, and behavior performed in VR has been shown to closely mirror real-world behaviors [20]. Further, VR enables researchers to measure behavior automatically and fairly unobtrusively, and attitudinal measures can be completed immediately following or even during experimental manipulations. By using VR, researchers can directly measure the relationships between explicit attitudes and behaviors, typically impossible with self-report questionnaires alone [21]. These advantages listed make VR an excellent tool for exploring emerging models of BID.

VR has been applied to BID assessment and intervention in past studies, and the present research aims to build on prior work in several important ways. First, VR studies have shown that individuals with ED exhibit strong affect responses to disorder-relevant virtual environments [22], but existing paradigms emphasize food and eating, and thus are not directly relevant to BID in the absence of eating pathology. Second, the majority of studies using VR to study BID have selected mood, anxiety, and perceptual body image distortions as primary outcome variables instead of measuring body image attitudes directly [19, 20]. Little is known about the relationship between environmental cues and state BD, although there is some evidence to suggest that social environments wherein the body is highly salient–such as a virtual swimming pool–may elicit changes in state BD among non-ED women [18]. Most importantly, no existing studies have explored behaviors associated with BID. We argue that VR offers vast opportunities to expand the field's understanding of BID, particularly in terms of automatic behaviors, which may play an important role in maintaining disturbed body image.

The Present Study

The present study presents a novel paradigm for measuring BID in VR. First, we propose to validate a series of virtual environments that are relevant to BID among women without eating pathology. The environments used in existing VR exposure protocols for ED patients were developed by administering traditional measures in the context of a virtual environment. We propose to use a similar method in which we determine the extent to which one's experience in VR mirrors that expected in similar real–world situations. Observational, theoretical, and experimental studies suggest that situations in which one's body is highly salient and/or in which one engages in body-based social comparison are highly relevant to BID [11]. We intend to manipulate these factors in VR to measure their influence on self-reported state BD.

Second, we aim to investigate the extent to which automatic nonverbal social behaviors are related to BID. VR allows us to explore visual attention in a realistic setting, whereas prior studies have been conducted in highly controlled laboratory settings by measuring attention toward static images, often presented on a computer screen. Visual gaze is easily measured in VR by deriving a user's field of view as a function of her head movements (Bailenson et al., 2002). By measuring visual attention in virtual social contexts, we can determine whether attentional biases demonstrated in laboratory studies persist in situations in naturalistic settings, in which many stimuli (in addition to body-related cues) compete for one's attention.

We further aim to extend previous work in automatic processes related to BID by exploring interpersonal approach behavior, which we believe may relate to BID. Specifically, previous VR studies in a variety of disciplines have shown interpersonal distance to be a sensitive indicator of implicit attitudes and evaluations [23]. Interpersonal distance has been studied for many years as an unconscious indicator of social attitudes for many years; early studies indicated that people consistently approach more closely to positively-evaluated others, whereas they maintain greater distance from others who are members of socially derogated groups [older distance studies]. In the VR literature, interpersonal distance refers to the space a participant places between him or herself and a digital human, called an avatar. Interpretations of interpersonal distance differ depending upon the context, but it is generally regarded as an indicator of global attitudes: standing closer to another person is perceived as a positive overall appraisal, whereas standing further away indicates dislike of the target or discomfort. Further, interpersonal distance is primarily automatic. Although, interpersonal distance has never been explored as an indicator of body image attitudes we expect it is related to automatic attention biases in BID. Determining the extent to which interpersonal behavior–both distance and visual gaze–are associated with BID in realistic settings may help to identify automatic systems that underlie BID, allowing for the development of more robust interventions.

Hypothesis I: Participants with greater weight-and-shape concerns will report higher levels of state BD across all environments compared to participants with low levels of weight-and-shape concerns.

Hypothesis II: Participants will report the most state BD in a VR environment with high body salience and social presence, compared to other environmental conditions.

Hypothesis III: Participants will vary their behavior toward female avatars based on their pre-existing levels of weight-and-shape concerns. Specifically, participants with elevated weight-and-shape concerns will approach more closely and demonstrate more visual attention toward thin avatars as compared to participants without elevated weight-and-shape concerns.

METHOD

Participants

A convenience sample was recruited using the student research participant pool of the Stanford University Communication Department during the 2012–2013 Winter and Spring quarters and the 2013–2014 Fall quarter. Students received course credit for participation in the study. All students in available courses completed an online prescreen questionnaire at the beginning of the quarter to determine their eligibility. The Weight Concerns Scale (WCS) was included in the online questionnaire to assess level of body image disturbance. WCS scores greater than 47 are associated with greater BID overall, and have been used to indicate increased risk for developing an eating disorder [24,25]. Students eligible for this study were randomly invited to participate based on their WCS score. Invitations to participate in the study were extended according to departmental policies to approximately equal numbers of female students with WCS total scores above and below 47, respectively.

A total of 112 students responded to an invitation to participate in the study. Of those 112 students, 103 students provided written informed consent and completed baseline assessments. Of the 103 students who completed baseline assessments, 24 were excluded from final analysis for the following reasons: (a) failure to meet inclusion criteria (n = 3); (b) problems with the recording and/or tracking equipment in the VR system (n = 14); and (c) onset of dizziness during VR exposure (n = 7). Seventy-nine students completed all phases of the VR exposure and post-test assessments. Of these, two participants were found to have partial data on baseline assessments and were excluded from analysis. Analyses were therefore conducted on a final sample of N = 77.

At baseline, participants were screened and sorted into two groups on the basis of WCS total scores: elevated overall BID (referred to as the "weight-concerned" group; n = 39), WCS > = 47 or controls (n = 40), WCS < 47. Participants who endorsed eating-related pathology consistent with the DSM-IV-TR eating disorder criteria were not included in the study and were provided with referral information for in-person and web-based mental health services on campus.

Ethics Statement

All procedures and materials were approved by the Ethical Committee of the Institutional Review Board at Stanford University. Written consent was obtained from all participants; the individual pictured in this manuscript, who was not a study participant, has given written informed consent (as outlined in PLOS consent form) for his image to be used.

Procedures

The lead researcher and one undergraduate research assistant were present for each trial. Upon arriving at the laboratory, all participants signed a consent form. Participants completed computer-based questionnaires assessing baseline weight-and-shape concerns, eating pathology, body satisfaction, and physical appearance comparison.

The researcher provided all participants with a verbal description of the purpose of the study explaining the components of the VR experience. Participants were told that they would be participating in a study examining whether certain virtual environments influenced women's awareness of their own bodies and their behavior towards others. The researcher then assisted each participant in wearing the virtual reality headset. After completing the VR task, all participants received debriefing materials containing information about online resources for body image improvement and contact information for university counseling and psychological services.

Virtual Reality Task.

The environments for the VR assessment task were designed to elicit varying levels of state BD. We used a 2x2 design (high vs. low) to vary the degree of (a) body salience and (b) social presence, resulting in four distinct environments. These factors were chosen based on previous research on state body dissatisfaction (state BD) in nonclinical samples using both traditional methods and VR exposure [26,27]. Body salience was conceptualized as the level of suggested exposure of one's own body, and when applicable, the degree of visibility of others' bodies. A virtual beach was created as the high body salience environment, and an indoor scene reminiscent of a university building was created for low body salience conditions (Fig 1). Social presence was indicated by the presence or absence of digital humans, called avatars. In the beach environments, participants were told to imagine that they were dressed in a bathing suit, and when present, avatars were dressed in revealing bathing suits. In the indoor scenes, no specific instructions were given about the participant's appearance, and when avatars were present they were dressed in long-sleeved shirts and long pants.

Figure 1: Female avatars from beach and party environments representing thin (BMI = 18), average (BMI = 23) and overweight (BMI = 28) from left to right. http://dx.doi.org/10.1371/journal.pone.0140158.g001

In order to confirm that equipment was functioning properly, all participants were first exposed to a virtual training environment designed to mirror the physical room in which the study took place. In the training room, the participants practiced walking while wearing the equipment to gain familiarity with the experience of being immersed in the virtual environment.

After the initial training room, the participants were randomly exposed to the four virtual experimental conditions to control for order effects. All participants were exposed to an empty beach, a beach populated with avatars, an empty party scene, and a party scene populated with avatars. Populated environments contained three groups of three avatars, each composed of two female and one male avatar.

The female avatars varied according to body size between the three groups: underweight, healthy weight, and overweight body sizes were represented (the male avatar remained the same). The groups were designed to represent a BMI of approximately 18 (underweight), 23 (healthy weight), and 27.5 (overweight) respectively. The groups stood in fixed positions within virtual space, equidistant from the participant's starting position. The placement of each group relative to the participant's starting position was counterbalanced within and between participants to control for environmental and individual factors that may influence approach behavior.

In the environments containing avatars, participants viewed all three groups from their starting position and then approached ("joined") each group in the order of their preference. Participants spent 15–20 seconds observing each group. In environments without avatars, participants approached three areas of interest in the environment and observed each area for 15–20 seconds.

In the interval between exposure to each experimental environment, participants completed a measure of state BD, by reading the items on a virtual screen projected in the virtual reality headset and speaking aloud the number on the scale corresponding to their answer for each item. The research assistant was blinded to the scale items and audio-recorded participants' responses. Responses were also hand-recorded by the researcher. Immediately following exposure to the final environment and completing the final measurement of state BD, the research assistant removed the virtual reality headset and the participant completed post-test measures.

Apparatus.

Participants viewed the virtual environments using a head-mounted display (HMD), a virtual reality headset that provided three-dimensional stereoscopic views. They wore a nVisor SX111 HMD (NVIS, Reston, VA) with a resolution of 2560 x 1024 and a refresh rate of 120 frames per second (60 seconds per eye), enabling them to have naturalistic head movements while navigating the virtual worlds. An Intersense3 Cube accelerometer tracked their physical head movements (pitch, roll, and yaw) operating at 180 hertz with a 4-millisecond latency rate. In addition, participant's gross body position on x-, y-, z-axis was tracked using an optical infrared camera system (Worldviz PPT-H) operating at 180 hertz with a 20-millisecond latency rate (with a precision of 0.25 millimeters). The optical tracking system allowed participants to navigate the virtual environment by walking naturally. Fig 2 illustrates the HMD, accelerometer, and optical infrared camera system. In addition, the physical lab room generated spatialized sound using a 24-channel ambisonic auralizer sound system (i.e. the location and direction of the sound in the IVE was reflected in the same location and direction as the physical lab space). All virtual environments were generated and programmed using Worldviz's Vizard VR toolkit.

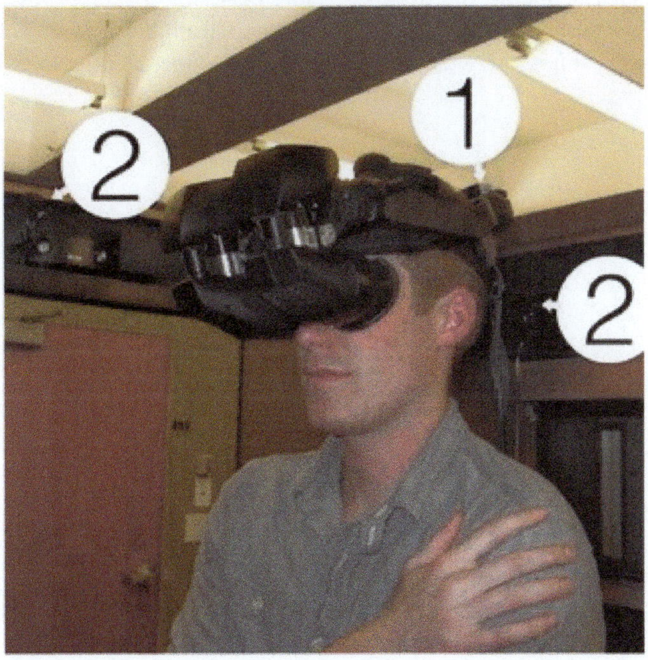

Figure 2: Subjects viewed the virtual environments using a head-mounted display (HMD). (1) An accelerometer allowed for naturalistic head movements during virtual exposure and (2) an optical infrared camera system recorded subjects' gross body position in virtual space, allowing subjects to navigate the virtual environments via natural walking patterns.

http://dx.doi.org/10.1371/journal.pone.0140158.g002

Measures: Pre-VR

Weight-and-shape concerns.

Researchers assessed weight-and-shape concerns using the five-item Weight Concerns Scale (WCS) [28]. This instrument assesses worry about weight and shape, fear of weight gain, dieting behavior, relative importance of weight, and frequency of feelings of fatness. The psychometric properties of the WCS are acceptable and the measure has been widely used in studies of eating behavior and BID [29,30].

Body mass index.

Self-reported height and weight was recorded at baseline to calculate approximate BMI [BMI = weight(kg)/height(m)^2].

Eating attitudes and behaviors.

The Eating Disorder Examination-Questionnaire (EDE-Q), a 39-item self-report version of the Eating Disorder Examination (EDE), is a well-established investigator-based interview used to diagnose and assess eating pathology [31]. The EDE-Q assesses eating disorder pathology over the past 28 days and provides a global score as well as four subscale scores (restraint, eating concerns, weight concerns, and shape concerns; [32]). The EDE-Q has shown acceptable reliability and validity as compared to the EDE interview among young adult women [33,34]. The EDE-Q was used to assess eating pathology at baseline, prior to VR exposure. Women with current EDs were not included in the present study for several reason: (a) BID is often severe among ED patients, and as such might not be expected to fluctuate according to environmental conditions in a manner comparable to non-ED women; (b) exposure to the environments could be triggering for women with ED; (c) this study aimed to explore BID among women with elevated weight-and-shape concerns but without current diagnosis of ED, as this is an important but understudied population in the BID literature.

Measures: During VR

State body dissatisfaction (State BD).

State BD was measured using the Body Parts Satisfaction Scale–Revised (BPSS-R) [35]. The BPSS-R is a 10-item self-report measure that uses a Likert-type rating scale to assess the participant's current level of satisfaction with ten body parts, resulting in a global score and two subscale scores (Satisfaction with Body scale and Satisfaction with Face scale) [35]. The measure has been shown to have acceptable psychometric properties among ethnically diverse undergraduate women [31]. The original measure was created to measure body satisfaction over the past month but it is also appropriate as an immediate measure. For the purposes of this study, only the Satisfaction with Body scale was analyzed.

Interpersonal distance from avatars.

Throughout virtual exposure, the VR tracking system collected data on participants x, y and z positions in the virtual environment at a rate of 60 frames per second, measuring participant's position relative to the virtual objects in meters. A participant's "minimum distance" was operationalized as the smallest absolute distance reached between a participant and each target avatar group. When calculating the distance between the participant and an avatar, analysis controlled for the relative body size of the avatar. First, the width of the avatar was calculated using the "Inspector" program of Vizard (WorldViz, Santa Barbara, 2012). In the analysis, the avatar was then considered to be a cylinder with a radius equal to the avatar's calculated width, and the minimum absolute interpersonal distance was considered to be the smallest distance from the center of the participant to the outside edge of this cylinder, as shown in Fig 3.

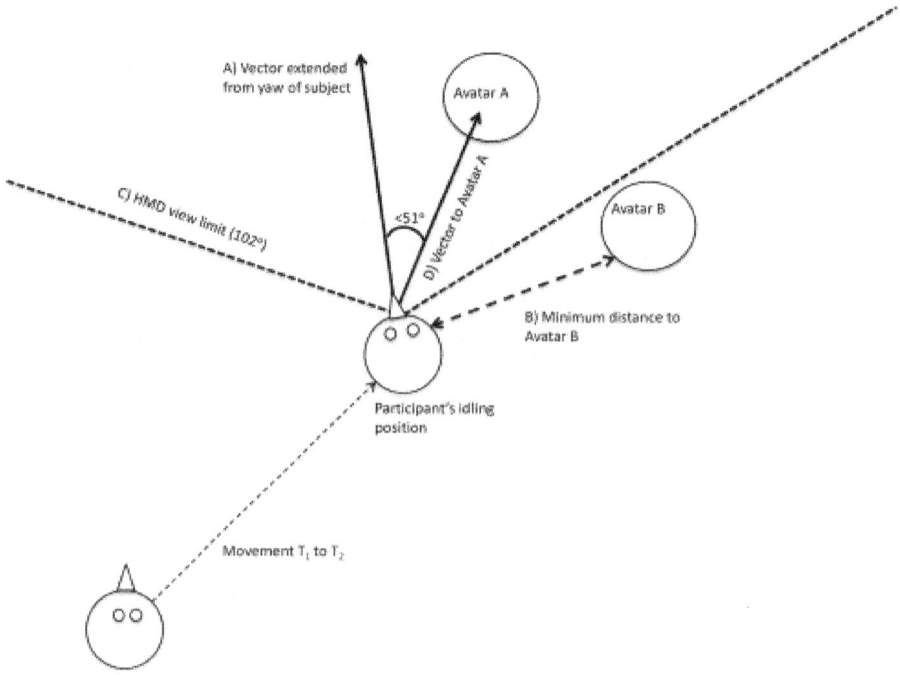

Figure 3: Aerial depiction of the method for calculating visual gaze and interpersonal distance. The participant's movement from her starting point at T_1 to her idling position at T_2 is recorded by the optical tracking system in the laboratory room. At T_2, line C indicates the participant's entire horizontal field of view of 102^0 within the HMD. Avatar A is within the participant's range of view, whereas Avatar B is not in view. A vector (line A) was extended from the center of the participant's head along the z-axis (i.e., extending out from the nose), and another vector (line D) was drawn between the participant and Avatar A, providing a measurement for angle E. The participant is considered to be looking at a given avatar if angle E is less than half of her entire field of view, or less than 51^0. Line B demonstrates the distance between the participant at T_2 and Avatar B, measured as the distance from the participant to the outside edge of a cylinder with a radius equal to the width of Avatar B, as measured in Inspector. http://dx.doi.org/10.1371/journal.pone.0140158.g003

Visual gaze toward avatars.

Visual gaze was measured as the total percentage of time within each populated virtual environment that the participant spent looking at each avatar group. Participants' head movement (yaw, rotation around the y-axis) was used to determine if they were looking at a group of avatars at a given time point (60 frames per second). Head direction, as opposed to eye movement, was used to indicate visual gaze. The technical challenges of measuring eye movement in immersive VR are significant, however head direction is very rarely decoupled from eye movement during VR exposure, and head direction is a more socially meaningful indicator of gaze than eye movement [20]. As shown in Fig 3, for each avatar in the group, a vector was drawn from the participant to the avatar. Next, the angle was calculated between the two vectors. If this angle was less than 51 degrees (half the view of the HMD worn by the participant), then the avatar was considered to be within the view of the participant. If at least one avatar was within the view of the participant, the group was considered to be in view. The participant was considered to be looking at a group only if it was the only group within her view at that frame according to this method. All time points at which avatars from more than one group were within the participant's field of view were eliminated from analysis. We therefore calculated the total time in VR during which a participant was viewing one avatar group exclusively. All sensory cues in VR were held constant except for the position of the avatar groups, which was counterbalanced within- and between-subjects. This allows us to reasonably deduce that systematic differences in visual attention toward any of the groups was indeed the result of participants looking at the avatars, and was not influenced by other sounds or visual stimuli in VR.

Approach Preference

We determined which avatar group each subject chose to approach first in each populated environment by calculating the direction of the subject's initial movement in each environment relative to the positions of each avatar group. This was also hand recorded by the researcher while observing the participant's movement (i.e., the researcher noted which group the participant chose to join first).

RESULTS

All data analyses were performed in R [36], and all mixed models were performed using lmerTest [37] in order to obtain *p*-values. Data was analyzed from participants who completed all assessments and completed all virtual exposures (N = 77). Linear mixed-effects models were used to examine the relationships between BID risk (i.e. Control, At-risk) and all relevant outcome variables (Appendices A-C). All models contained participants as random intercepts to account for within-subject correlations between repeated measures from the same person [38]. ANOVAs were used to compare models by testing the goodness of fit of the model to the data using the Maximum Likelihood method [38]. The less complex model was selected when there was no significant difference between the models in how well they fit the data. Contrasts used the average body size avatars and the empty beach scene as the comparison groups for the avatar body size and the virtual environment variables respectively. Statistical tests were two-tailed to allow for the detection of both positive and negative changes on target variables. A significance level of 0.05 was used for all analyses unless otherwise stated.

Participant Characteristics

Table 1 presents the means and standard deviations for demographic and clinical characteristics for the sample. A general linear model compared the differences in baseline WCS by risk group, controlling for BMI, age, and race/ethnicity. There was a significant effect for WCS such that the At-risk group reported significantly higher levels of weight-shape concerns at baseline with no significant differences in BMI, age, or race/ethnicity: R^2 = 0.71,$F(4, 71)$ = 42.85, p<0.001. WCS scores ranged from 5.0 to 95.0.

Table 1: Participant demographic and clinical characteristics. http://dx.doi.org/10.1371/journal.pone.0140158.t001

	Control (n = 40)	At-risk (n = 37)
Age in years, *M* (*SD*)[1]	20.07(1.47)	20.67(2.52)
Race/Ethnicity (%)		
White/European Descent	50.00	48.64
African-American	10.00	2.70
Asian-American/ Pacific Islander	12.50	18.92
Latina/Hispanic	12.50	13.51
Middle Eastern-American	2.50	0.00
Other/ Prefer not to say	12.50	16.22
BMI, *M* (*SD*)	22.19(2.45)	22.70(1.96)
Weight Concerns Scale Score, *M* (*SD*)	24.77 (12.09)	61.70 (12.57)

[1] Due to missing data for one participant, age analysis is based on N = 76.

Body Satisfaction

Linear mixed-effects models were used to examine the association between BPSS-R scores, BID risk group, and virtual environment (S1 File- Appendix). Participants were modeled as random intercepts. BID risk, virtual environment, and BPSS-R scores measured after each virtual environment acted as fixed factors. The results from showed a significant difference in mean baseline BPSS-scores, with women in the Control group reporting higher body satisfaction compared to the At-risk group, b = 1.00, t = 27.4, 12, p<0.001. After controlling for baseline body satisfaction, there was no main effect of risk group on body satisfaction, b = 0.07,t = 0.93, p = 0.35.

In examining the virtual scenes, there was a significant effect of environmental condition regardless of risk. Participants reported less body satisfaction in the populated beach scene compared to the empty beach scene (S1 File- Appendix). Contrasts were applied to the model to further examine differences between the virtual environments. There were no differences in mean BPSS-R scores in the empty environments compared to the populated environments, b= 0.02, t = 1.49, p = 0.14, and no differences in mean BPSS-R scores in the beach environments (i.e. populated and empty) compared to the party environments (i.e. populated and empty), b = -0.02, t = -1.38, p = 0.17. However, there was a significant difference In BPSS-R scores between the populated beach environment compared to all other virtual environments, b = -0.02, t = -2.15, p = 0.03. Participants reported feeling the lowest level of body satisfaction in the populated beach scene when compared to all other virtual scenes.

Mean BPSS-R Body Subscale scores for both groups across all environments are presented inFig 4.

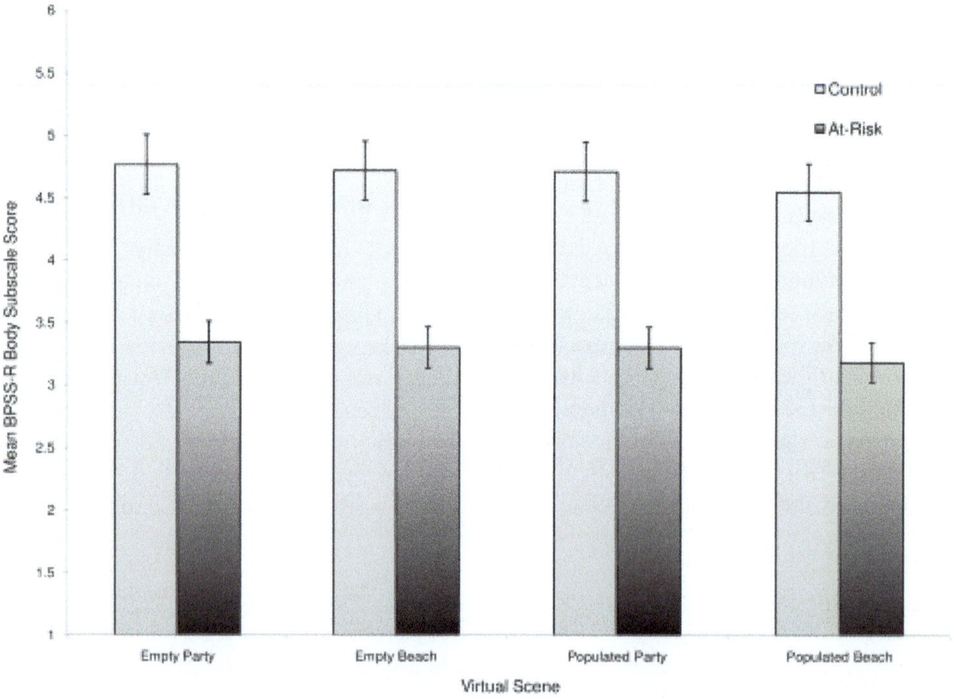

Figure 4: Mean BPSS-R Body Subscale scores for women with and without elevated weight-and-shape concerns. Error bars represent 95% confidence intervals. http://dx.doi.org/10.1371/journal.pone.0140158.g004

Interpersonal Distance

Linear mixed effects models were used to compare differences in participants' minimum interpersonal distance from the avatars of different body sizes (average, overweight, and thin) in the two populated virtual environments (beach and party; S2 File—Appendix). Table 2presents the means and confidence intervals for minimum interpersonal distance from avatars of each body size for at-risk and control group participants.

Table 2. Means and 95% confidence intervals for absolute minimum interpersonal distance from avatars of each body size for participants with elevated weight concerns versus control groups. Distances are reported in meters. http://dx.doi.org/10.1371/journal.pone.0140158.t002

	Average Weight	Thin	Overweight
Control	0.54 (0.51–0.58)	0.57 (0.53–0.60)	0.60 (0.55–0.65)
At-Risk	0.59 (0.55–0.64)	0.57 (0.52–0.61)	0.59 (0.54–0.63)

There was no significant main effects of BID risk group, b = -0.05, t = -1.35, p = 0.18, or of virtual environment, b = -0.02, t = -1.74, p = 0.08, on minimum distance. There was a significant risk group and avatar body size interaction. The Control group minimum distance to the overweight avatars compared to the average sized avatars differed significantly from the At-Risk group's comparison, b = 0.07, t = 2.29, p = 0.03. There were no differences between the average and thin sized avatar, b = 0.05, t = 1.79, p = 0.07 when compared to the At-risk group. There was no main effect of significant in minimum differences in how close participants stood to the overweight avatars compared to average sized avatars, b = -0.01, t = -0.28, p = 0.78, nor the thin avatars compared to the average sized avatars, b = -0.03, t = -1.19, p = 0.23.

Contrasts applied to the model further examine the interaction effects between BID risk group and avatar body size. The first contrast compared the interpersonal distance of the average-sized avatars against both the overweight and thin sized avatars. The second contrast compared the overweight sized avatars with the thin sized avatar. There was no main effect of avatar size comparing the overweight and thin sized avatars to the average body size group, b= -0.01, t = -0.86, p = 0.39. In addition, there was no differences in their minimum distance to the overweight avatar group compared to the thin avatar group, b = 0.01, t = 0.91, p = 0.36.

There was an interaction effect between risk group and these contrasts. The Control group demonstrated differences in minimum differences according to avatar body size compared to the At-Risk group. The Control group stood significantly closer to the average sized avatars compared to both the overweight and thin sized avatars, b = 0.02, t = 2.35, p = 0.02 in comparison to the At-risk group. The second contrast revealed that the Control group demonstrated no differences in the minimum distance between the overweight group compared to the thin sized group, $b = 0.01$, $t = 0.50$, $p = 0.62$ in comparison to the At-Risk group.

Visual Gaze

Linear mixed effects models were used to examine the effect of avatar body size and virtual scene on the percentage of time participants spent looking at the avatar groups (S3 File—Appendix). There was no difference between the beach virtual environment and the party virtual environment, $b = -0.05$, $t = -0.10$, $p = 0.91$. There was a main effect of avatar size, with the thin avatars having a greater percentage of time being viewed compared to the average, $b = 1.32$, $t = 3.04$, $p = 0.01$. There was no difference in the percentage of time between the average sized avatar and the overweight avatar, $b = -0.76$, $p = 0.32$. However there was an interaction effect of BID risk group and avatar body size. The At-risk group spent more time looking at the thin avatars compared to the average-sized avatars, whereas Control group spent more time looking at average-sized as compared to thin avatars, $b = 3.71$, $t = -3.46$, $p = 0.001$ (see Table 3). There were no differences in how the Control group looked at the average sized avatars compared to the overweight group compared to the At-Risk group, $b = -0.31$, $t = -0.29$, $p = 0.77$. The Control group overall spent more timing looking at all the avatars compared to the At-risk group. This main effect of BID risk group approached, but did not reach, the level of statistical significance, $b = 1.48$, $t = 1.96$, $p = 0.05$.

Table 3. Means and 95% confidence intervals for the total percent of time within populated environments participants spent looking at avatar dyads of each body size. http://dx.doi.org/10.1371/journal.pone.0140158.t003

	Average Weight	Thin	Overweight
Control	27.52 (26.30–28.75)	26.16 (25.17–27.15)	26.45 (25.23–27.67)
At-Risk	26.04 (25.18–26.90)	28.39 (27.26–29.51)	25.28 (24.32–26.23)

Using the average sized avatars as the comparison group, contrasts revealed additional interaction effects. The first contrast compared the average sized group against both the overweight and thin groups. The second contrasts compared the overweight avatar group with the thin sized avatar group. There was no significant difference in the amount of time that participants looked at the average sized avatars compared to both the overweight and thin avatars, $b = 0.26$, $t = 1.17$, $p = 0.24$. When comparing the overweight and thin avatars to each other, participants spent significantly higher percentage of time looking at the thin avatars compared to the overweight avatars, $b = -1.55$, $t = -4.00$, $p<0.001$. The Control group spent a significantly higher percentage looking at the average avatars compared to both the thin and overweight avatars, $b = -0.67$, $t = -2.15$, $p = 0.03$ in comparison with the At-Risk group. Finally, when only comparing the overweight avatars and the thin sized avatars, the Control group spent a significantly greater percentage of timing looking at the overweight avatars compared to the thin avatars, $b = 1.70$, $t = 3.15$, $p = 0.002$ compared to the At-Risk group.

Approach Preference

Two chi-square test of goodness-of-fit tests, were performed to determine whether the three avatar groups were equally preferred in terms of approach behavior in the two populated virtual environments. Preference for the groups was not equally distributed in the beach environment,x^2 (2, $N = 77$) = 11.342, $p = .003$, Cramer's $V = 0.27$, or in the party environment, x^2 (2, $N = 77$) = 25.66, $p < .001$, Cramer's $V = 0.41$. As shown in Fig 5, regardless of their level of weight concern, participants were less likely to initially approach overweight avatars as compared to normal weight or thin avatars in both environments.

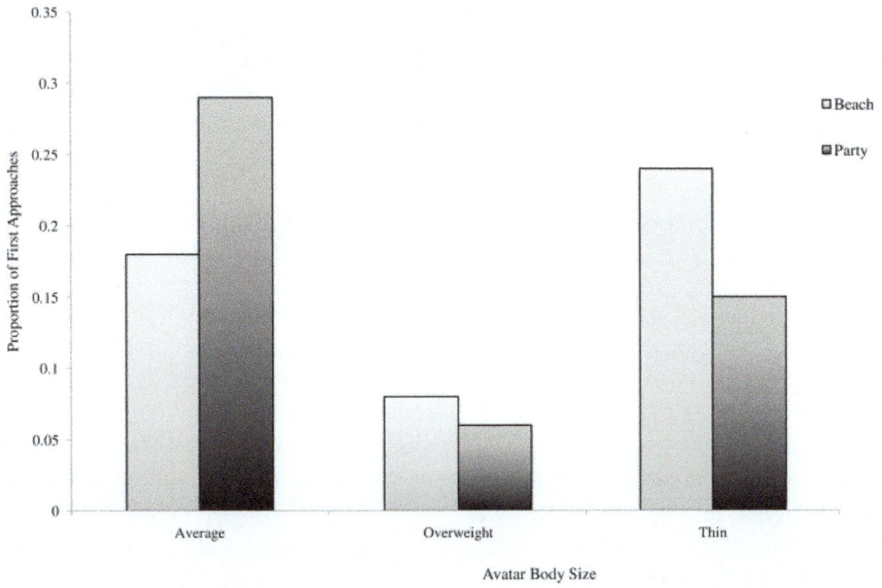

Figure 5: Proportion of avatars approached first by body size and populated environment. http://dx.doi.org/10.1371/journal.pone.0140158.g005

DISCUSSION

The present study investigated the use of VR in the study and assessment of BID. The primary aims of this study were (a) to identify environmental factors that relate to BID among women without eating pathology; and (b) to identify automatic behavioral indicators of BID. We found that all participants reported relatively stable state BD across a series of virtual environments, but greater state BD in a social environment with high body salience. With regard to behavior, we found that, compared to women without elevated weight and shape concerns, participants weight-and-shape concerns stood further from avatars of average weight and spent more time looking at thin female avatars than average or overweight female avatars. Finally, we found that all participants were reluctant to approach overweight female avatars.

Across all virtual environments, women with higher baseline weight-and-shape concerns reported lower body satisfaction than women in the control group. This was expected, and builds upon prior VR studies, which have shown that ED patients experience distress during exposure to virtual foods and eating environments [22]. We found that regardless of baseline weight-and-shape concerns, women reported slightly worse body satisfaction on the populated virtual beach. The environmental factors we manipulated in this study, body salience and social presence, had some influence on BID, although the effect was small. Interestingly, we found that state BD shifted across environments at a similar rate among women with and without pre-existing weight-and-shape concerns, suggesting that body salience in a social context may be have widely-applicable relevance for women's body image concerns. Researchers interested in VR measurement of BID in sub-clinical ED populations might consider adopting a similar paradigm.

Contrary to our hypotheses, we did not find that interpersonal distance or approach preference toward thin avatars differed depending on BID. Given that BID is characterized by thin-idealization, it is surprising that we failed to find between-group differences with regard to behavior toward a thin avatar. Previous VR research using personal space as an outcome measure in human-avatar interaction has generally concluded that smaller interpersonal distances reflect greater affiliation or sense of "intimacy" between the subject and the target avatar [39]. However, it is important to note that higher-than-average thin-idealization is extremely common among college-age women[40,41]. Thus, it is possible that both groups held similar levels of thin-idealization, consistent with cultural norms, but women with BID lack a comparable affinity for healthy or "average" looking bodies, and instead hold a narrower and more extreme definition of "thin-ideal." This could help to explain why, despite widespread cultural acceptance of the thin-ideal, not all women develop BID or struggle with disordered eating. Along these lines, we found participants with BID stood further from normal weight avatars than did participants without BID. It may be that more closely approaching normal weight avatars over other body types reflects an implicit preference for "healthy" body-cues among women in the control group. Body-dissatisfied women, on the other hand, might lack this "preference for healthy," reinforcing maladaptive social cognitions. These findings suggest that research into protective mechanisms that may function to promote healthy body image may be as elucidating as studying pathological processes.

We further found that compared to the control group women with BID spent a greater percentage of time looking at thin avatars as compared to the other avatar groups. This is consistent with existing research findings on automatic attentional biases among women with BID, which has shown that women with BID demonstrate automatic attention toward thin female bodies. It is argued that such biases in attention and processing increase women's feelings of body dissatisfaction [15,17,18]. That is, body-dissatisfied women tend to notice and attend to thin bodies more than other women do, thereby reinforcing the notion that "thin is in." In contrast, when presented with avatars with a variety of body types, women in the control group devoted the greatest attention to normal weight avatars. It may be the case that this pattern of automatic attention serves a protective function for these women. Moreover, these findings suggest that the immersive quality and affective realism of VR may provide added benefit to existing attention re-training treatment approaches for BID.

Unexpectedly, we found that participants in the present study, regardless of condition, were reluctant to approach overweight avatars. Given that overweight avatars were identical to thin and normal weight avatars on all features except weight, it is reasonable to interpret this preference as largely driven by body size. Overweight and obese individuals are frequently the targets of social stigma and, particularly among adolescents, overt social exclusion [42,43]. The detrimental impact of weight stigma on the mental and physical wellbeing of overweight and obese individuals has been well documented [44,45]. Researchers have largely relied upon the Implicit Association Test as a primary outcome in studies aimed at assessing and/or modifying weight stigma. We are aware of only one study to investigate behavioral measure of weight stigma: Bessenoff and Sherman [46] used interpersonal space in a real world waiting room paradigm to measure participants' willingness to approach an obese versus normal weight target. They found that participants with implicit anti-fat attitudes sat further away from a chair believed to be occupied by an obese woman.

The results of the present study suggest that immersive virtual environments may offer unique advantages to measure weight stigma at the behavioral level. Such methods may inform the development of more robust interventions to reduce stigmatized beliefs about overweight and obese people, which prove difficult to modify through more traditional, dissonance-based approaches [43,47]. VR has been used to study social behaviors toward other stigmatized groups; for example, researchers have measured interpersonal distance as an indicator of participants' level of stigmatization toward avatars who are perceived to be HIV-positive [48,49]. Further, one study recently investigated the use of virtual embodiment to reduce self-criticism in a non-clinical sample of young women, suggesting that VR may show promise as a tool to reduce the detrimental effects of self-stigmatization among members of marginalized groups, including people with obesity [50].

The present study may have important implications for the development of BID treatments. For example, the current set of environments could be expanded to provide an array of salient practice situations, as has been done in virtual exposure-based treatment for specific phobias and PTSD [51,52]. For instance, virtual environments could be used in exposure-based treatments; patients with high BID could practice behavioral skills in virtual environments and then apply them to real environments. The VR paradigm developed in this study was completed in less than 30 minutes, which may make it palatable to clinicians for whom time considerations are essential. There is promising evidence to suggest that implicit interventions aimed at correcting maladaptive automatic processes may be effective at improving BID [14,53,54]. The present VR task could be used to expand on existing implicit interventions by incorporating overt behavior and naturalistic settings.

Virtual spaces are increasingly common components of modern social life. Millions of people interact in virtual social spaces each day, some of which involve embodiment with an avatar such as online role-playing games [55]. As novel forms of social interaction continue to evolve, it is important for psychological research to progress in tandem. The present study may contribute to a preliminary understanding of BID as it relates to virtual social interactions. In particular, our findings suggest that as users navigate virtual social spaces, preexisting body image concerns remain salient, even when individuals have no direct feedback regarding their own virtual appearance. As this field of research develops, it will be important to consider how virtual interactions and avatar embodiment, among other factors, may influence BID within a virtual setting. Developing more nuanced understanding of BID in the context of virtual spaces, incorporating such themes as embodiment, interaction with avatars, and one's self-perspective in VR, may allow for the development of universal, disseminable virtual interventions to improve body satisfaction.

Limitations

First, our relatively homogenous sample may be a limitation of this study. Because participants were recruited only from a small number of courses within the Communications Department, the sample may not be representative of college women in general. The final sample had a high percentage of white participants, and a relatively small range in age and BMI, which may reduce the generalizability of these findings. For example, our sample contained very few women with BMI > 25, which may limit our ability to draw inferences about the influence of an individual's weight status on behavior, state BD, or their approach behaviors toward overweight avatars. In addition the present study did not include direct measurement of participants' stigmatizing beliefs or attitudes, which limits our ability to interpret participants' avoidance of overweight avatars. Future studies should recruit higher weight participants in order to better understand how personal characteristics may influence approach behavior, and more comprehensive studies are needed in order to draw strong conclusions. Finally, while men are certainly affected by BID, research on male experiences of body image is limited. Existing research indicates that BID is expressed differently in men than in women, making it difficult to extend existing models of BID to a male population. Moreover, there is no research on attention biases related to BID using male participants. In order to allow us to draw clearer conclusions based on prior work with women, we did not include male participants in the present study. Future research extending models of BID to apply to men as well as women is clearly indicated.

CONCLUSIONS

The current results suggest that VR provides effective tools for measuring BID in realistic settings. The present study provides preliminary evidence for environmental factors that are relevant for BID among non-ED women, although further research is needed to strengthen this paradigm. This is the first study to use VR to investigate implicit behavioral indicators of state BD, which may be helpful in measuring treatment outcomes or designing future behavioral exposure treatments. Emerging models increasingly indicate that automatic processes play an important role in the etiology and maintenance of BID, thus it is vital that researchers establish effective approaches by which to measure and understand such processes. The present study introduces a novel measurement of automatic behavioral processes associated with BID, lends support to growing evidence of visual attention biases in BID, and provides preliminary evidence that BID is also associated with specific patterns of automatic social behavior that may maintain the disorder.

SUPPORTING INFORMATION

S1 Dataset. Complete dataset(s) for the present study.

visualgaze.csv

bodysat.csv

distanceData.csv

bodyimage.csv

S1 **File. Appendix**. Table A. Linear Mixed-effects Models for Body Satisfaction by Virtual Scene.

			s^2	*b*	*SE*	*t*	*p*
Model: **Risk+Scene+Baseline**							
Random	Subject	Intercept	0.04				
Fixed	*Risk* (Control)			0.07	0.07	0.93	0.35
	Scene (Empty Party)			0.05	0.03	1.43	0.15
	Scene (Populated Beach)			-0.14	0.03	-4.36	<0.001
	Scene Populated Party)			-0.00	0.03	-0.01	1.00
	Baseline BPSS-R			1.00	0.04	27.40	<0.001
Model with contrasts: Risk+Scene+Baseline							
Random	Subject	Intercept	0.05				
Fixed	*Risk* (Control)			0.07	0.07	0.93	0.34
	Scene (Empty vs. Populated)			0.02	0.02	1.49	0.24
	Scene (Beach vs. Party)			-0.02	0.02	-1.38	.17
	Scene (Populated Beach vs. Other)			-0.02	0.01	-2.15	0.03
	Baseline BPSS-R			1.00	0.04	30.00	<0.001

S2 File. Appendix. Table B. Linear Mixed-Effects Models for Interpersonal Distance by Risk, Avatar Body Size, and Scene.

			s^2	*b*	*SE*	*t*	*p*
Model: Risk*Size+ Scene							
Random	*Subject*	Intercept	0.02				
Fixed	*Risk* (Control)			-0.05	0.04	-1.35	0.18
	Size (Overweight)			-0.01	0.02	-0.28	0.78
	Size (Thin)			-0.03	0.02	-1.19	0.23
	Scene (Party)			-0.02	0.01	-1.74	0.08
	Risk * *Size* (Control: Overweight)			0.07	0.03	2.29	0.02
	Risk * *Size* (Control: Thin)			0.05	0.03	1.79	0.07
Model with contrasts: Risk*Size+Scene							
Random	*Subject*	Intercept	0.02				
Fixed	*Risk* (Control)			-0.01	0.04	-1.48	0.14
	Size (Average vs. Thin/Overwt)			-0.01	0.03	-0.65	0.52
	Size (Thin vs. Overwt)			0.01	0.03	-2.13	0.03
	Scene (Party)			-0.02	0.03	-1.99	0.05
	*Risk*Size* (Control: Average vs. Thin/Overweight)			0.02	0.04	1.63	0.10
	*Risk*Size* (Control: Thin vs. Overwt)			0.01	0.04	1.86	0.06

S3 File. Appendix. Table C. Linear Mixed Models for Visual Gaze by Risk, Avatar Body Size, and Scene.

			s^2	*b*	*SE*	*t*	*p*
Model: Risk*Size + Scene							
Random	*Subject*	Intercept	0.00				
	Risk (Control)			1.48	0.76	1.96	0.05
	Size (Overweight)			-0.76	0.77	-0.99	0.32
	Size (Thin)			2.34	0.77	3.04	0.003
	Scene (Party)			-0.05	0.44	-0.11	0.91
	Risk * *Size* (Control: Overweight)			-0.31	1.07	-0.29	0.77
	Risk * *Size* (Control: Thin)			-3.71	1.07	-3.46	0.001
Model with contrasts: Risk*Size + Scene							
Subject		Intercept	0.00				

***Risk* (Control)**	0.14	0.44	0.32	0.75
***Size* (Average vs. Thin/Overwt)**	0.26	0.22	1.17.	0.24
***Size* (Thin vs. Overwt)**	-1.55	0.39	-4.00	<0.001
***Scene* (Party)**	-0.05	0.44	-0.11	0.91
***Risk* * *Size* (Control: Average vs. Thin/Overwt)**	-0.67	0.31	-2.15	0.03
***Risk* * *Size* (Control: Thin vs. Overwt)**	1.70	0.54	3.15	0.001

ACKNOWLEDGMENTS

We would like to acknowledge the Stanford University Virtual Reality Intensive Training Seminar (VRITS) students of 2012–2013 and 2013–2014, a group of exceptional undergraduate programmers who contributed to the development of this study. In particular, we wish to recognize Helen Hastings, Joanna Kim, and Melvin Low and for their individual contributions to this project.

AUTHOR CONTRIBUTIONS

Conceived and designed the experiments: CKP MJ J. Bailey J. Bailenson CBT. Performed the experiments: CKP. Analyzed the data: CKP J. Bailey CBT. Contributed reagents/materials/analysis tools: CKP MJ J. Bailey J. Bailenson CBT. Wrote the paper: CKP MJ J. Bailey J. Bailenson CBT.

REFERENCES

1. Cash TF, Smolak L (2012) Body Image: A Handbook of Science, Practice, and Prevention: Guilford Press.
2. Cash TF, Morrow JA, Hrabosky JI, Perry AA (2004) How has body image changed? A cross-sectional investigation of college women and men from 1983 to 2001. J Consult Clin Psychol 72: 1081–1089. PMID: 15612854
3. Tiggeman M (2000) Person x Situation Interaction in Body Dissatisfaction. International Journal of Eating Disorders 29: 65–70.
4. Paxton SJ, Neumark-Sztainer D, Hannan PJ, Eisenberg ME (2006) Body dissatisfaction prospectively predicts depressive mood and low self-esteem in adolescent girls and boys. Journal of Clinical Child and Adolescent Psychology 35: 539–549. PMID: 17007599
5. Stice E, Shaw HE (2002) Role of body dissatsifaction in the onset and maintenance of eating pathology: A synthesis of research findings. Journal of Psychosomatic Research 53: 985–993. PMID: 12445588
6. Cash T, Theriault J, Annis NM (2004) Body Image in an Interpersonal Context: Adult attachment, fear of intimacy, and social anxiety Journal of Social and Clinical Psychology 23: 89–103.
7. Jacobi C, Hayward C, de Zwaan M, Kraemer HC, Agras WS (2004) Coming to terms with risk factors for eating disorders: application of risk terminology and suggestions for a general taxonomy. Psychol Bull 130: 19–65. PMID: 14717649
8. Johnson F, Wardle J (2005) Dietary restraint, body dissatisfaction, and psychological distress: a prospective analysis. J Abnorm Psychol 114: 119–125. PMID: 15709818
9. Stice E, Shaw HE (2002) Role of body dissatisfaction in the onset and maintenance of eating pathology: a synthesis of research findings. J Psychosom Res 53: 985–993. PMID: 12445588
10. Menzel J, Krawczyk R, Thompson JK (2012) Assessment of body image in adolescents and adults. In: Cash TF, Smolak L, editors. Body Image. New York: Guilford.
11. Rudiger JA, Cash TF, Roehrig M, Thompson JK (2007) Day-to-day body-image states: prospective predictors of intra-individual level and variability. Body Image 4: 1–9. PMID: 18089247
12. Smeets T, Jansen A, Roefs A (2011) Bias for the (un) attractive self: On the role of attention in causing body (dis)satisfaction. Health Psychology 30.
13. Vitousek KB, Hollon SD (1990) The investigation of schematic content and processing in eating disorders. Cognitive Therapy and Research 14: 191–214.
14. Alleva J, Martijn C, Jansen A (2015) Covariation bias in women with a negative body evaluation: How is it expressed and can it be diminished?. J Behav Ther Exp Psychiatry 9: 33–39.
15. Jansen A, Smeets T, Martijn C, Nederkoorn C (2006) I see what you see: the lack of a self-serving body image bias in eating disorders. British Journal of Clinical Psychology 45: 123–135.
16. Siep N, Jansen A, Havermans RC, Roefs A (2011) Cognitions and emotions in eating disorders. Current Topics in Behavioral Neurosciences 6: 17–33. doi: 10.1007/7854_2010_82 PMID: 21243468
17. Jansen A, Nederkoorn C, Mulkens S (2005) Selective visual attention for ugly and beautiful body parts in eating disorders. Behaviour Research and Therapy 43: 183–196. PMID: 15629749
18. Roefs A, Jansen A, Moresi S, Willems P, van Grootel S, van der Borgh A (2008) Looking good: BMI, attractiveness bias and visual attention. Appetite 51: 552–555. doi: 10.1016/j.appet.2008.04.008
19. PMID: 18495295
20. Blascovich J, Loomis JM, Beall AC, Swinth KR, Hoyt CL, Bailenson JN (2002) Immersive virtual environment

technology as a methodological tool for social psychology. Psychological Inquiry 13: 103– 124.

21. Bailenson JN, Beall AC, Blascovich J (2002) Gaze and task performance in shared virtual environments. Journal of Visualization and Computer Animation 13: 313–320.
22. Bailenson JN, Aharoni EG, R., Dimov A, Blascovich J (2004) Comparing Behavioral and Self-Report Measures of Embodied Agents' Social Presence in Immersive Virtual Environments. Proceedings of the 7th Annual International Workshop on PRESENCE (Valencia, Spain).
23. Ferrer-Garcia M, Gutierrez-Maldonado J (2012) The use of virtual reality in the study, assessment, and treatment of body image in eating disorders and nonclinical samples: a review of the literature. Body Image 9: 1–11. doi: 10.1016/j.bodyim.2011.10.001 PMID: 22119329
24. Blascovich J, Mccall C (2009) Attitudes in Virtual Reality. In: Forgas JC, W.; Cooper, J., editor. Attitudes and Persuasion.
25. Taylor CB, Bryson S, Luce KH, Cunning D, Doyle AC, Abascal LB, et al. (2006) Prevention of Eating Disorders in At-Risk College-Age Women. Archives of General Psychiatry 63: 881–888. PMID: 16894064
26. Jacobi C, Fittig E, Bryson SW, Wilfley D, Kraemer HC, Taylor CB (2011) Who is really at risk? Identifying risk factors for subthreshold and full syndrome eating disorders in a high-risk sample. Psychol Med 41: 1939–1949. doi: 10.1017/S0033291710002631 PMID: 21276276
27. Haimovitz D, Lansky LM, O'Reilly P (1993) Fluctuations in body satisfaction across situations. International Journal of Eating Disorders 13: 77–84. PMID: 8477280
28. Aime A, Cotton K, Guitard T, Bouchar S (2012) Virtual Reality and Body Dissatisfaction Across the Eating Disorder's Spectrum.
29. Killen JD, Taylor CB, Hayward C, Wilson DM, Haydel KF, Robinson TN, et al. (1994) The pursuit of thinness and onset of eating disorder symptoms in a community sample of adolescent girls: A three-year prospective analysis. International Journal of Eating Disorders 16: 227–238. PMID: 7833956
30. Jacobi C, Abascal L, Taylor CB (2004) Screening for eating disorders and high-risk behavior: caution. International Journal of Eating Disorders 36: 280–295. PMID: 15478137
31. Jacobi C, Fittig E, Bryson SW, Wilfley D, Kraemer HC, Taylor CB (2011) Who is really at risk? Identifying risk factors for subthreshold and full syndrome eating disorders in a high-risk sample. Psychological Medicine 41: 1939–1949. doi: 10.1017/S0033291710002631 PMID: 21276276
32. Fairburn CG, Cooper Z (1993) The Eating Disorder Examination (12th edition). In: Fairburn CG, Wilson GT, editors. Binge Eating: Nature, Assessment, and Treatment. New York: Guilford Press.
33. Fairburn CG, Beglin SJ (1994) Assessment of eating disorders: Interview or self-report questionnaire? International Journal of Eating Disorders 16: 363–370. PMID: 7866415
34. Luce KH, Crowther JH (1999) The reliability of the eating disorder examination—self-report questionnaire version (EDE-Q). International Journal of Eating Disorders 25: 349–351. PMID: 10192002
35. Mond JM, Hay PJ, Rodgers B, Owens C (2006) Eating Disorder Examination Questionnaire (EDE-Q): norms for young adult women. Behaviour Research and Therapy 44: 53–62. PMID: 16301014
36. Petrie TA, Tripp MM, Harvey P (2002) Factorial And Construct Validity Of The Body Parts Satisfaction Scale-Revised: An Examination Of Minority And Nonminority Women. Psychology of Women Quarterly 26: 213–221.
37. Team RDC (2011) R: A Language and Environment for Statistical Computing Vienna, Austria: R Foundation for Statistical Computing.
38. Kuznetsova A, Brockhoff P, Christensen R (2014) lmerTest: tests in linear mixed effects models. R package version 2.0–20.
39. Baayen RH, Davidson DJ, Bates DM (2008) Mixed-effects modeling with crossed random effects for subjects and items. Journal of Memory and Language 59: 390–412.
40. Bailenson JN, Blascovich J, Beall AC, Loomis JM (2003) Interpersonal distance in immersive virtual environments. Pers Soc Psychol Bull 29: 819–833. PMID: 15018671
41. McCabe M, Ricciardelli L (2004) A longitudinal study of pubertal timing and extreme body change behaviors among adolescent boys and girls. Adolescence 29: 145–156.
42. McCarthy M (1990) The thin ideal, depression and eating disorders in women. Behav Res Ther 28: 205–215. PMID: 2196049
43. Crandall CS (1994) Prejudice against fat people: Ideology and self-interest. Journal of Personality and Social Psychology 66: 882–894. PMID: 8014833
44. Teachman BA, Gapinski KD, Brownell KD, Rawlins M, Jeyaram S (2003) Demonstrations of implicit anti-fat bias: The impact of providing causal information and evoking empathy. Health Psychology 22: 65–70.
45. Schvey N, Puhl RM, Brownell KD (2011) The impact of weight stigma on caloric consumption. Obesity 19: 1957–1962. doi: 10.1038/oby.2011.204 PMID: 21760636
46. Puhl RM, Heuer CA (2009) The stigma of obesity: a review and update. Obesity (Silver Spring) 17: 941–964.
47. Bessenoff G, Sherman J (2000) Automatic and Controlled Components of Prejudice Toward Fat People: Evaluation Versus Stereotype Activation. Social Cognition 18: 329–353.
48. Ciao AC, Latner JD (2011) Reducing obesity stigma: the effectiveness of cognitive dissonance and social consensus interventions. Obesity (Silver Spring) 19: 1768–1774.
49. Toppenberg HL. HIV-status acknowledgment and stigma reduction; 2014; Maastricht, The Netherlands.

50. Earnshaw VA, Chaudoir S (2009) From Conceptualizing to Measuring HIV Stigma: A Review of HIV Stigma Mechanism Measures. AIDS Behavior 13: 1160–1177. doi: 10.1007/s10461-009-9593-3 PMID: 19636699

51. Falconer CJ, Slater M, Rovira A, King JA, Gilbert P, Antley A, et al. (2014) Embodying compassion: a virtual reality paradigm for overcoming excessive self-criticism. PLoS One 9: e111933. doi: 10.1371/ journal. pone.0111933 PMID: 25389766

52. Meyerbröker K, Emmelkamp PMG (2010) Virtual reality exposure therapy in anxiety disorder: A systematic review of process-and-outcome studies. Depression and Anxiety 27: 933–944. doi: 10.1002/ da.20734 PMID: 20734361

53. Goncalves R, Pedrozo AL, Coutinho ES, Figueira I, Ventura P (2012) Efficacy of virtual reality exposure therapy in the treatment of PTSD: a systematic review. PLoS One 7: e48469. doi: 10.1371/journal. pone.0048469 PMID: 23300515

54. Martijn C, Vanderlinden M, Roefs A, Huijding J, Jansen A (2010) Increasing body satisfaction of body concerned women through evaluative conditioning using social stimuli. Health Psychology 29: 514– 520. doi: 10.1037/ a0020770 PMID: 20836606

55. Aspen V, Martijn C, Alleva J, Nagel J, Perret C, Purvis C, et al. (in press) Decreasing body dissatisfaction using a brief conditioning intervention. Behavior Research and Therapy.

56. Yee N, Bailenson JN, Urbanek M, Chang F, Merget D (2007) The unbearable likeness of being digital: The persistence of nonverbal social norms in online virtual environments. Cyberpsychology & Behavior 10: 115–121. PMID: 17305457

CHAPTER 18

DEVELOPMENT OF A THREE-DIMENSIONAL HAND MODEL USING THREE-DIMENSIONAL STEREOPHOTOGRAMMETRY: ASSESSMENT OF IMAGE REPRODUCIBILITY

Inge A. Hoevenaren,[1] J. Meulstee,[2] E. Krikken,[2,3] S. J. Bergé[2], D. J. O. Ulrich,[1] and Thomas J. J. Maal[2]

1 Department of Plastic, Reconstructive and Hand Surgery, Radboud University Nijmegen Medical Centre, Nijmegen, the Netherlands,

2 Department of Oral and Maxillofacial Surgery, Radboud University Nijmegen Medical Centre, Nijmegen, the Netherlands,

3 Technical medicine, University of Twente, Enschede, the Netherlands

ABSTRACT

Purpose

Using three-dimensional (3D) stereophotogrammetry precise images and reconstructions of the human body can be produced. Over the last few years, this technique is mainly being developed in the field of maxillofacial reconstructive surgery, creating fusion images with computed tomography (CT) data for precise planning and prediction of treatment outcome. Though, in hand surgery 3D stereophotogrammetry is not yet being used in clinical settings.

Methods

A total of 34 three-dimensional hand photographs were analyzed to investigate the reproducibility. For every individual, 3D photographs were captured at two different time points (baseline T0 and one week later T1). Using two different registration methods, the reproducibility of the methods was analyzed. Furthermore, the differences between 3D photos of men and women were compared in a distance map as a first clinical pilot testing our registration method.

Results

The absolute mean registration error for the complete hand was 1.46 mm. This reduced to an error of 0.56 mm isolating the region to the palm of the hand. When comparing hands of both sexes, it was seen that the male hand was larger (broader base and longer fingers) than the female hand.

Citation: Hoevenaren IA, Meulstee J, Krikken E, Bergé SJ, Ulrich DJO, Maal TJJ (2015) Development of a Three-Dimensional Hand Model Using Three-Dimensional Stereophotogrammetry: Assessment of Image Reproducibility. PLoS ONE 10(9): e0136710. doi:10.1371/journal.pone.0136710

Conclusions

This study shows that 3D stereophotogrammetry can produce reproducible images of the hand without harmful side effects for the patient, so proving to be a reliable method for soft tissue analysis. Its potential use in everyday practice of hand surgery needs to be further explored.

INTRODUCTION

Nowadays, three-dimensional (3D) surface imaging is being implemented in various healthcare and non-healthcare areas, such as anthropology, reconstructive surgery, craniofacial surgery and orthodontics. One of the surface imaging techniques being used more and more is 3D stereophotogrammetry [1,2]. Within the broad field of reconstructive surgery, it has already found solid applications in maxillofacial surgery [3–6] and is upcoming in breast and free flap reconstructive surgery [5,7–11]. However, 3D imaging of the hand is an unexplored field so far. The possibilities of this technique in the clinical and pre-clinical setting are numerous, for example using it for preoperative planning or postoperative evaluation of hand surgeries, and for various educational purposes.

The aim of this study was to develop a noninvasive, objective and valid method of photographing the hand using 3D stereophotogrammetry. Thereafter, making use of the standardized 3D images, we started evaluation of the reproducibility and validity of newly defined soft tissue landmarks of the hand for clinical purposes, by using our registration method on images of the hands of men and women.

SUBJECTS AND METHODS

Subjects

Seventeen adult Caucasian individuals (4 male, 13 female; mean age 39.1 years, range 26–64 years) without any preexistent hand deformities were randomly selected from our departments. All individuals gave written informed consent to participate in this study. Approval by a medical ethics review board was not obtained, since the digital photography method used was previously described by the review board as not invasive or harmful. This study was conducted in compliance with the World Medical Association Declaration of Helsinki on medical research ethics. 3D photographs of both the left and right hand were taken, resulting in a total of 34 images. The photographs were taken with the subject positioned as shown in Fig 1. The individual in this figure has given written informed consent (as outlined in PLOS consent form) to publish these case details.

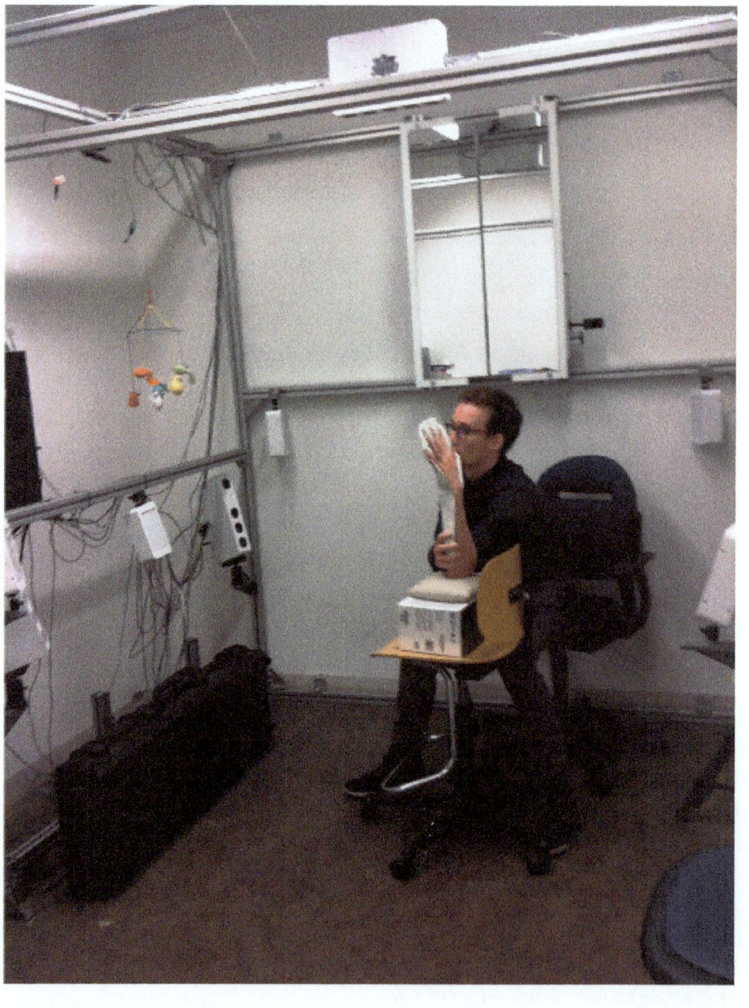

Figure 1: **3D** stereophotogrammetrical camera alignment.

Imaging methods

For every individual, 3D photographs were captured at two different time points (baseline T0 and one week later T1). The 3D photographs were captured with a 3D stereophotogrammetrical camera setup and the software program Modular System v1.0 (3dMDface System, 3dMD, Atlanta, USA). The camera setup consisted of five pods, each equipped with three digital cameras and a flash. Prior to its use, the camera was calibrated to define a 3D coordinate system for the 3D photograph, which was referred to as the original 3D coordinate system [12,13]. The 3D photographs were taken with each individual seated with the arm in a 90 degree flexion in the elbow joint and the hand kept in a pre-designed template with the fingers in a fully spreaded, upright position (Fig 1). In order to achieve exact images of both the palmar and dorsal aspect of the hand, the template was removed by an assistant, directly before the photograph was taken. The patient was instructed not to move the hand to another position and to keep it absolutely still after removal of the template.

Reproducibility of the 3D photographs

The 3D photographs of the left and right hand of the same individual acquired at different time points were registered using the Iterative Closest Point (rigid registration) algorithm [14]. After registration the difference between the two surfaces was calculated as an error or distance map indicating the amount of similarity between the registered surfaces. This distance map computes the difference (Euclidean distance) between the 3D photographs on a large number of points and is an objective measure for the reproducibility of the acquired 3D photographs. The absolute mean value as well as the 50th and 90th percentile of the registration error were computed in millimeters.

Apart from the distance map of the complete 3D model of the hand, a distance map was calculated of only the palm of the hand. Hereby it was possible to investigate whether the palm is a more rigid region compared to the individual fingers which have more variables of free movement. For the distance map of the palm of the hand, the 50th and 90th percentiles were computed.

Landmarks and registration

After investigating the reproducibility of the image acquisition method, anatomical landmarks were placed on each of the 34 photographs according to the method developed by Hoevenaren et al. [15]. These landmarks could then be used for comparing different hands and to compute an average hand model. In order to compare different hands and compute an average hand, the landmark sets of all 17 left hands and 17 right hands were registered using two different registration methods.

Method 1

The first registration method was a general Procreates registration of the complete landmark set of either the left or right hand images. At first, one of the landmark sets is randomly selected as a reference set. Towards this set, all other landmark sets are registered by matching the similar landmarks in each set. After completing this registration an average hand surface model can be computed. Thereafter, this complete process of matching similar landmarks of each set is repeated, now using the newly created average hand as reference set for the registration procedure. After complete registration, a new average hand surface model is computed.

Method 2

The second method uses an adapted version of the Iterative Closest Point algorithm in which the hand is divided into different anatomical sub regions namely the palm of the hand, the thumb, the index finger, middle finger, ring finger and the little finger. All different sets were then registered with a randomly selected reference set, resulting in a computed, average hand. Also in this method, the complete procedure was repeated with the average hand performing as the reference set (see method 1).

For both registration methods the RMS error was calculated as a measure of precision of the computed model. The reproducibility of method 1 and method 2 were evaluated using the paired t-test. Statistical analyses were performed with the IBM SPSS software program, version 20.0 for Windows (IBM Corp., Armonk, NY, USA). The level of significance for all statistical analyses was set at 5%.

Comparison of Male and Female Hands

Using the described registration methods, it was now possible to register the hands of different subjects in a clinical setting. As a pilot study, the average hand of five male and five female subjects were calculated and compared.

RESULTS

Reproducibility

The absolute mean registration error for the complete hand was 1.37 mm. The 50th and 90th percentile of the registration error were respectively 0.97 mm and 2.95 mm for the complete set of hands. Isolating the region to the inside of the hand, the absolute mean error was reduced to 0.56 mm. The 50th and 90th percentile of the registration error were respectively 0.47 mm and 1.11 mm for the isolated region of the hand palm. The registration errors indicating the reproducibility of the image acquisition method are illustrated in Fig 2.

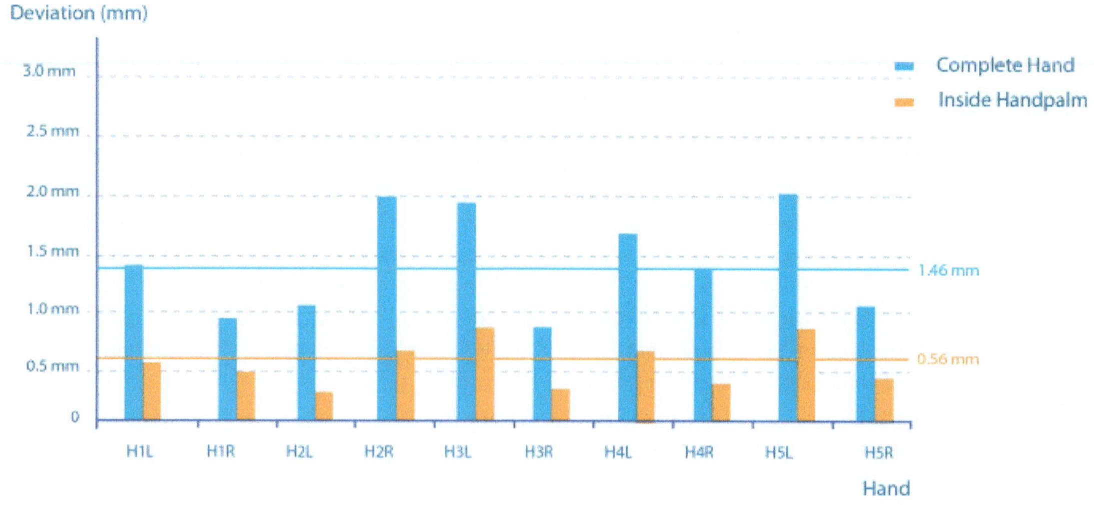

Figure 2: Registration errors.

Calculation of average hands

For the registration of method 1 a RMS error of 0.94 mm was found. For method 2, in which the fingers were registered as separate models the RMS error decreased to 0.42 mm. The variation of registered landmark sets could also be illustrated. Fig 3 illustrates the average landmark set and ellipsoids which are calculated as two times the standard deviation.

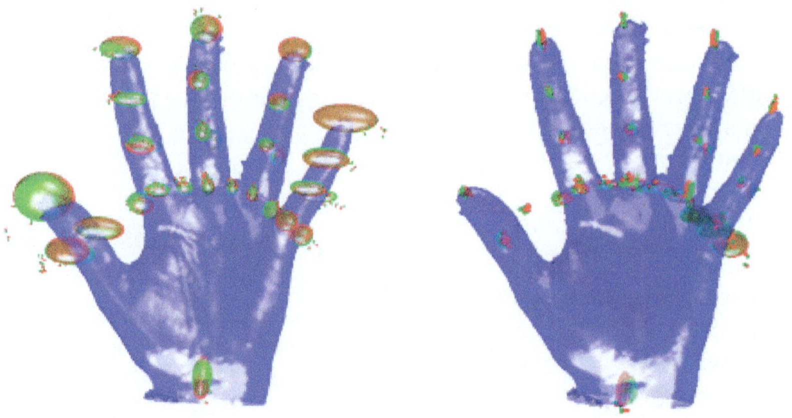

Figure 3: Landmarks and ellipsoids.

Comparison between male and female hands

The male and female hands were compared using method 2. An illustration of the difference between the male and female average hands can be found in Fig 4. In general the male hand is larger (broader base and longer fingers) than the female hand.

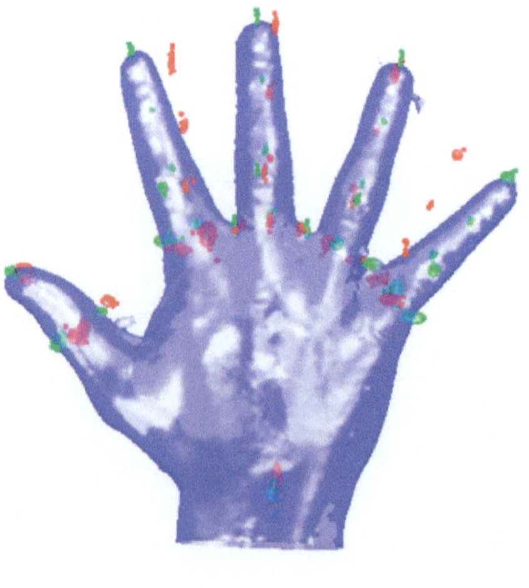

Figure 4: Comparison of male and female hands.

DISCUSSION

With the advent of digital technology, digital photography has become an increasingly important tool in reconstructive surgery, in particular in maxillofacial surgery [16]. With the introduction of systems such as the 3dMDface System (3dMD LLC, Atlanta, USA), Di3D (Dimensional Imaging, Glasgow, UK) and 3D-Sensors FaceSCAN3D (3D Shape GmbH, Erlangen, Germany), the applicability of 3D photography in daily practice has become reality. 3D stereo photogrammetry is safe, noninvasive and able to capture superior quality 'external surface' 3D photographs in less than 10 milliseconds [17]. Especially these characteristics are ideal to collect 3D data even of hands of young children or elderly patients. After processing the data, an exact digital model of the patients hand is created which can be used in a clinical setting immediately.

In the complex field of hand surgery, 3D photography is not yet being used in clinical settings as far as we know. However, the hand is a complex anatomic part of the body and its injuries are very incapacitating for the patient, making precise planning of every hand surgery procedure and predicting its outcome necessary in obtaining a satisfying postoperative result for both patient and doctor.

3D imaging

At present, the gold standard in imaging soft and bony tissues of the hand are computed tomography (CT) data or magnetic resonance imaging (MRI) scan images with their known side effects [18–20]. However, these techniques do not produce a real-life image of the outer surface of the hand. The shortcomings of CT and MRI scan images have caused a search for a minimally invasive 3D imaging technique, which could produce a realistic and exact virtual surface model of the hand. Among others, advantages of 3D stereo photogrammetry over MRI and CT are its non-invasiveness, lack of harmful radiation, immediate available images and good patient tolerance [5]. Furthermore, the usage of 3D stereo photogrammetry is highly cost effective, given the very short amount of time needed for photographing and low production costs, compared to the time consuming CT or MRI scan images. Besides the absence of potential dangerous side effects described above, there is the possibility of patient-specific modeling. 3D imaging has the potential to precisely simulate the pre-operative experience for surgery planning with reliable surface visualization leading to a more profound understanding of the anatomical structures to be reconstructed and therefore improved final surgery outcome.

The aim of this study was to develop a noninvasive, objective and valid method of photographing the hand using 3D stereo photogrammetry. Furthermore, we evaluated the reproducibility and possibility of computing an average hand and using it in a clinical pilot.

3D stereo photogrammetry

A review of literature showed there is no information on the reproducibility and validity of 3D hand photography yet. Earlier studies were performed to investigate the precision of 3D stereo photogrammetry. The majority of these studies focused on reliably measuring distances between typical anthropometric points on the 3D reconstructed images against corresponding points on live subjects or phantom models (e.g. plaster casts) as a form of validation [12,21–26]. Some other studies use more complex methods to obtain and analyze 3D shapes [27–29].

Kau et al., Maal et al. and Ma et al. investigated the reproducibility of several 3D acquisition systems. Kau et al. found an RMS error of 0.4 mm using a Minolta Vivid 900 laser scanner [30]. Maal et al. used 3D stereo photogrammetry, assessed the reproducibility and also found a value of 0.4 mm [31]. Ma et al. found a reproducibility of their structered light system of 0.2 mm [22].

Image reproducibility

In this study two different registration methods were used, method 1 matching complete hands and method 2 matching subregions of the hand. We found a significant higher registration error for method 1. An explanation for this could be found in the fact that it was difficult for the study subjects to keep their fingers completely still at time of photography. Therefore, we designed method 2, correcting for the movement of the individual fingers.

Average hand model

Using method 2 as valid registration method, an average model of both left and right hand 3D images was created. This method of landmark registration and matching has already found solid applications in maxillofacial surgery. In this study, as stated above, one of the difficulties in creating reproducible images of the hand is the movement of the individual fingers. However, we corrected this error by designing a new method of landmark based registration, matching each individual finger per image and therefore creating a more precise image.

In maxillofacial surgery 'the average face' is an instrument often used to which other facial images are being compared [32,33]. It is mainly being used in comparing preoperative images of patients and predicting treatment outcome, thereby reducing operating time. In literature, no publications exist on the average hand. Given the proven reproducibility and clinical value of calculations based on the average face, we developed a method of creating an average hand model. One of the drawbacks of this study is the small population on which this model is based. Therefore, a database of 3D hand images is being collected at the moment, created from study subjects from different age categories.

Clinical Relevance

The possibilities of the clinical and pre-clinical use of 3D photography of the hand are numerous. 3D images could be used in pre- and postoperative hand trauma surgeries, thereby quantifying the soft tissue loss and identifying the optimal reconstructive surgery type. Furthermore, during follow-up after nerve injuries of the upper extremity, soft tissue volume deficiencies can be calculated, evaluating treatment results and indicating the need for revision surgery. A normative database based on soft tissue data is not yet available, but needed. This database could be used among others in growth studies of both normal child development and pathologic growth, for example in patients with acromegaly, and in objectifying the sex-specific differences of the hand.

Using the landmark based registration method 2, we describe the first clinical pilot with this study. Though a very small population, we confirmed the reliable use of the method calculating differences in hands of both men and women. Further research with a larger study population is necessary to validate the advanced clinical use.

Besides its use in clinical practice, the developed 3D hand model can be used for various educational purposes. Computer-assisted learning has been widely established in medical anatomy learning, since it facilitates obtaining anatomic knowledge in a flexible and efficient way [34–40]. With medical imaging quickly becoming a more cost-effective source for clinical and pre-clinical training, a growing number of 3D models for different anatomic regions are being developed [40–43]. Since 3D photography produces precise images and reconstructions of the human body, these models support comprehension of spatial anatomy in diagnostic and surgical interventions [9,41,42,44,45].

Also, 3D real-life models based on 3D stereophotogrammetry could be used for exact simulation in surgical resident training, increasing the knowledge of the difficult anatomical areas of the human body and therefore improving the quality of teaching [6,39,46].

In addition, there is an application in forensic medicine, where with the aid of a database, skin reconstructions can be created out of osseous parts of the hand for identification purposes [47].

Since the hand is not a static anatomical structure and precise finger movement is one of its main functions, our future research focuses among others on combining 3D techniques with CT data. Recently, the possibility of creating fusion images with CT data is extensively being used for even more precise planning and predicting treatment outcome [9,48,49]. In conclusion, this study shows that 3D stereophotogrammetry produces precise and reproducible 3D images of the hand. This proves 3D photography to be a reliable method for soft tissue analysis. Its potential use in everyday practice of hand surgery and the concept of fusing 3D photography images with radiologic images of the interior hand structures needs to be further explored.

FUNDING STATEMENT

These authors have no support or funding to report.

DATA AVAILABILITY

All relevant data are within the paper.

REFERENCES

1. Mozzo P, Procacci C, Tacconi A, Martini PT, Andreis IAB. Original article A new volumetric CT machine for dental imaging based on the cone-beam technique : preliminary results. Eur Radiol. 1998;1564: 1558–1564.
2. Ibrahim AE, Sarhane KA, Baroud JS, Atiyeh BS. Robotics in plastic surgery, a review. Eur J Plast Surg. 2012;35: 571–578. doi: 10.1007/s00238-012-0737-8
3. Plooij JM, Swennen GRJ, Rangel FA, Maal TJJ, Schutyser FAC, Bronkhorst EM, et al. Evaluation of reproducibility and reliability of 3D soft tissue analysis using 3D stereophotogrammetry. Int J Oral Maxillofac Surg. 2009;38: 267–73. doi: 10.1016/j.ijom.2008.12.009
4. Hoehnke C, Eder M, Papadopulos NA, Zimmermann A, Brockmann G, Biemer E, et al. Minimal invasive reconstruction of posttraumatic hemi facial atrophy by 3-D computer-assisted lipofilling. J Plast Reconstr Aesthet Surg. 2007;60: 1138–44. doi: 10.1016/j.bjps.2007.01.068
5. Kovacs L, Eder M, Hollweck R, Zimmermann A, Settles M, Schneider A, et al. Comparison between breast volume measurement using 3D surface imaging and classical techniques. Breast. 2007;16: 137–45. doi: 10.1016/j.breast.2006.08.001
6. Kovacs L, Zimmermann A, Wawrzyn H, Schwenzer K, Seitz H, Tille C, et al. Computer aided surgical reconstruction after complex facial burn injuries—opportunities and limitations. Burns. 2005;31: 85–91. doi: 10.1016/j.burns.2004.07.010
7. Kovacs L, Eder M, Zimmermann A, Müller D, Schuster T, Papadopulos NA, et al. Three-dimensional evaluation of breast augmentation and the influence of anatomic and round implants on operative breast shape changes. Aesthetic Plast Surg. 2012;36: 879–87. doi: 10.1007/s00266-012-9892-3
8. Small KH, Tepper OM, Unger JG, Kumar N, Feldman DL, Choi M, et al. Re-defining pseudoptosis from a 3D perspective after short scar-medial pedicle reduction mammaplasty. J Plast Reconstr Aesthet Surg. Elsevier Ltd; 2010;63: 346–53. doi: 10.1016/j.bjps.2008.11.023
9. Gacto-Sánchez P, Sicilia-Castro D, Gómez-Cía T, Lagares A, Collell T, Suárez C, et al. Use of a three-dimensional virtual reality model for preoperative imaging in DIEP flap breast reconstruction. J Surg Res. 2010;162: 140–7. doi: 10.1016/j.jss.2009.01.025
10. Henseler H, Smith J, Bowman A, Khambay BS, Ju X, Ayoub A, et al. Objective evaluation of the latissimus dorsi flap for breast reconstruction using three-dimensional imaging. J Plast Reconstr Aesthet Surg. Elsevier Ltd; 2012;65: 1209–15. doi: 10.1016/j.bjps.2012.03.043
11. Zhang S, Pei G. Application of three-dimensional digitalized reconstruction of an anterolateral thigh flap and an arterial dorsalis pedis flap. Microsurgery. 2007;27: 553–9. doi: 10.1002/micr
12. Aldridge K, Boyadjiev SA, Capone GT, DeLeon VB, Richtsmeier JT. Precision and error of three-dimensional phenotypic measures acquired from 3dMD photogrammetric images. Am J Med Genet A. 2005;138A: 247–53. doi: 10.1002/ajmg.a.30959 [PMC free article]
13. Lane C, Harrell W. Completing the 3-dimensional picture. Am J Orthod Dentofacial Orthop. 2008;133: 612–20. doi: 10.1016/j.ajodo.2007.03.023
14. Besl PJ, McKay ND. A Method for Registration of 3-D shapes. IEEE Transactions on Pattern Analysis and Machine Intelligence; 1992. pp. 239–256.
15. Hoevenaren IA, Maal TJJ, Krikken E, Haan de A, Bergé S, Ulrich D. Development of a three-dimensional hand model using 3D stereophotogrammetry: evaluation of

landmark reproducibility. J Plast Reconstr Aesthet Surg. 2014; doi:http://dx.doi.org/10.1016/j.bjps.2014.12.025

16. Lee S. Three-dimensional photography and its application to facial plastic surgery. Arch Facial Plast Surg. 2014;6: 410–4. doi: 10.1001/archfaci.6.6.410
17. Littlefield TR, Kelly KM, Cherney JC, Beals SP, Pomatto JK. Development of a New Three-Dimensional Cranial Imaging System. J Craniofac Surg. 2004;15: 175–181.
18. Lin EC. Radiation risk from medical imaging. Mayo Clin Proc. Mayo Foundation for Medical Education and Research; 2010;85: 1142–6; quiz 1146. doi: 10.4065/mcp.2010.0260 [PMC free article]
19. Smith-Bindman R, Miglioretti DL, Johnson E, Lee C, Feigelson HS, Flynn M, et al. Use of Diagnostic Imaging Studies and Associated Radiation Exposure for Patients Enrolled in Large Integrated Health Care systems. JAMA. 2013; [PMC free article]
20. Brenner DJ, Hall EJ. Computed tomography—an increasing source of radiation exposure. N Engl J Med. 2007;357: 2277–84. doi: 10.1056/NEJMra072149
21. Ayoub A, Siebert P, Moos K, Wray D, Urquhart C, Niblett T. A vision-based three-dimensional capture system for maxillofacial assessment and surgical planning. Br J Oral Maxillofac Surg. 1998;36: 353–357.
22. Ma L, Xu T, Lin J. Validation of a three-dimensional facial scanning system based on structured light techniques. Comput Methods Programs Biomed. 2009;94: 290–8. doi: 10.1016/j.cmpb.2009.01.010
23. Kimoto K, Garrett NR. Evaluation of a 3D digital photographic imaging system of the human face. J Oral Rehabil. 2007;34: 201–5. doi: 10.1111/j.1365-2842.2006.01663.x
24. Weinberg SM, Kolar JC. Three-Dimensional Surface Imaging: Limitations and Considerations From the Anthropometric Perspective. J Craniofac Surg. 2005;16: 847–851.
25. Krimmel M, Kluba S, Dietz K, Reinert S. Assessment of precision and accuracy of digital surface photogrammetry with the dsp 400 system. Biomed Tech. 2005;50: 45–53.
26. Weinberg SM, Naidoo S, Govier DP, Martin RA, Kane AA, Marazita ML. Anthropometric precision and accuracy of digital three-dimensional photogrammetry: comparing the genex and 3dmd imaging systems with one another and with direct anthropometry. J Craniofac Surg. 2006;17: 477–483.
27. Coombes A, Moss J, Linney A, Richards R, James D. A mathematical method for the comparison of three-dimensional changes in the facial surface. Eur J Orthod. 1991;13: 95–110.
28. Johnston DJ, Millett DT, Ayoub AF, Bock M. Are facial expressions reproducible? Cleft Palate Craniofac J. 2003;40: 291–6. doi: 10.1597/1545-1569(2003)040<0291:AFER>2.0.CO;2
29. Okada E. Three-dimensional facial simulations and measurements: changes of facial contour and units associated with facial expression. J Craniofac Surg. 2001;12: 167–74. Available:http://www.ncbi.nlm.nih.gov/pubmed/11314628
30. Kau CH, Richmond S, Zhurov AI, Knox J, Chestnutt I, Hartles F, et al. Reliability of measuring facial morphology with a 3-dimensional laser scanning system. Am J Orthod Dentofacial Orthop. 2005;128: 424–30. doi: 10.1016/j.ajodo.2004.06.037
31. Maal TJJ, van Loon B, Plooij JM, Rangel F, Ettema AM, Borstlap WA, et al. Registration of 3-dimensional facial photographs for clinical use. J Oral Maxillofac Surg. Elsevier Inc.; 2010;68: 2391–401. doi: 10.1016/j.joms.2009.10.017
32. Bugaighis I, Tiddeman B, Mattick CR, Hobson R. 3D comparison of average faces in subjects with oral clefts. Eur J Orthod. 2012; 8–12. doi: 10.1093/ejo/cjs060
33. Hammond P, Hutton TJ, Allanson JE, Campbell LE, Hennekam RCM, Holden S, et al. 3D analysis of facial morphology. Am J Med Genet A. 2004;126A: 339–48. doi: 10.1002/ajmg.a.20665
34. Levine MG, Stempak J, Conyers G, Walters JA. Implementing and integrating computer-based activities into a problem-based gross anatomy curriculum. Clin Anat. 1999;12: 191–8. doi:10.1002/(SICI)1098-2353(1999)12:3<191::AID-CA8>3.0.CO;2–5
35. Jastrow H, Vollrath L. Teaching and learning gross anatomy using modern electronic media based on the visible human project. Clin Anat. 2003;16: 44–54. doi: 10.1002/ca.10062
36. Jones DG. Reassessing the importance of dissection: a critique and elaboration. Clin Anat. 1997;10: 123–7. doi: 10.1002/(SICI)1098-2353(1997)10:2<123::AID-CA9>3.0.CO;2-W
37. Adamczyk C, Holzer M, Putz R, Fischer MR. Student learning preferences and the impact of a multimedia learning tool in the dissection course at the University of Munich. Ann Anat. Elsevier; 2009;191: 339–48. doi: 10.1016/j.aanat.2009.03.003
38. Schittek M, Mattheos N, Lyon HC, Attström R. Computer assisted learning. A review. Eur J Dent Educ. 2001;5: 93–100. Available: http://www.ncbi.nlm.nih.gov/pubmed/11520331
39. Silén C, Wirell S, Kvist J, Nylander E, Smedby O. Advanced 3D visualization in student-centred medical education. Med Teach. 2008;30: e115–24. doi: 10.1080/01421590801932228
40. Reeves RE, Aschenbrenner JE, Wordinger RJ, Roque RS, Sheedlo HJ. Improved dissection efficiency in the human gross anatomy laboratory by the integration of computers and modern technology. Clin Anat. 2004;17: 337–44. doi: 10.1002/ca.10245
41. Crossingham JL, Jenkinson J, Woolridge N, Gallinger S, Tait G a, Moulton C-AE. Interpreting three-dimensional structures from two-dimensional images: a web-based interactive 3D teaching model of surgical liver anatomy. HPB (Oxford). 2009;11: 523–8. doi: 10.1111/j.1477-2574.2009.00097.x[PMC free article]

42. Pommert A, Höhne KH, Pflesser B, Richter E, Riemer M, Schiemann T, et al. Creating a high-resolution spatial/symbolic model of the inner organs based on the Visible Human. Med Image Anal. 2001;5: 221–8. Available: http://www.ncbi.nlm.nih.gov/pubmed/11524228
43. Trunk P, Mocnik J, Trobec R, Gersak B. 3D heart model for computer simulations in cardiac surgery. Comput Biol Med. 2007;37: 1398–403. doi: 10.1016/j.compbiomed.2006.11.003
44. Sierra R, Zsemlye G, Székely G, Bajka M. Generation of variable anatomical models for surgical training simulators. Med Image Anal. 2006;10: 275–85. doi: 10.1016/j.media.2005.11.003
45. Giunta RE, Eder M, Machens H-G, Müller DF, Kovacs L. Autologe Fettgewebstransplantation („Structural Fat Grafting") zur ästhetischen Verjüngung der Hand. Handchirurgie · Mikrochirurgie · Plast Chir. 2010;42: 143–147. doi: 10.1055/s-0030-1249039
46. Kowalczuk J, Meyer A, Carlson J, Psota ET, Buettner S, Pérez LC, et al. Real-time three-dimensional soft tissue reconstruction for laparoscopic surgery. Surg Endosc. 2012; doi: 10.1007/s00464-012-2355-8
47. Lefèvre P, Jan SVS, Beauthier JP, Rooze M. Hand skin reconstruction from skeletal landmarks. Int J Leg Med. 2007; 511–515. doi: 10.1007/s00414-007-0204-4
48. Gerlach NL, Meijer GJ, Maal TJJ, Mulder J, Rangel FA, Borstlap WA, et al. Reproducibility of 3 different tracing methods based on cone beam computed tomography in determining the anatomical position of the mandibular canal. J Oral Maxillofac Surg. Elsevier Inc.; 2010;68: 811–7. doi:10.1016/j.joms.2009.09.059
49. Gacto-Sánchez P, Sicilia-Castro D, Gómez-Cía T, Lagares A, Collell T, Suárez C, et al. Computed tomographic angiography with VirSSPA three-dimensional software for perforator navigation improves perioperative outcomes in DIEP flap breast reconstruction. Plast Reconstr Surg. 2010;125: 24–31. doi:10.1097/PRS.0b013e3181c4948b

CHAPTER 19

ELLIPSOID SEGMENTATION MODEL FOR ANALYZING LIGHT-ATTENUATED 3D CONFOCAL IMAGE STACKS OF FLUORESCENT MULTI-CELLULAR SPHEROIDS

Barbier M[1], Jaensch S[1], Cornelissen F[2], Vidic S[1,3,4], Gjerde K[1,3], de Hoogt R[1], Graeser R[5], Gustin E[1], Chong YT[1].

[1]Discovery Sciences, Janssen Pharmaceutical companies of Johnson & Johnson, Beerse, Belgium.

[2]Pharma R&D IT, Janssen Pharmaceutical companies of Johnson & Johnson, Beerse, Belgium.

[3]Department of Urology, Erasmus MC Rotterdam, Rotterdam, The Netherlands.

[4]Faculty of Mathematics, Natural Sciences and Information Technologies, University of Primorska, Koper, Slovenia.

[5]Translational Medicine and Clinical Pharmacology, Boehringer Ingelheim, Ingelheim am Rhein, Germany.

ABSTRACT

In oncology, two-dimensional in-vitro culture models are the standard test beds for the discovery and development of cancer treatments, but in the last decades, evidence emerged that such models have low predictive value for clinical efficacy. Therefore they are increasingly complemented by more physiologically relevant 3D models, such as spheroid micro-tumor cultures. If suitable fluorescent labels are applied, confocal 3D image stacks can characterize the structure of such volumetric cultures and, for example, cell proliferation. However, several issues hamper accurate analysis. In particular, signal attenuation within the tissue of the spheroids prevents the acquisition of a complete image for spheroids over 100 micrometers in diameter. And quantitative analysis of large 3D image data sets is challenging, creating a need for methods which can be applied to large-scale experiments and account for impeding factors. We present a robust, computationally inexpensive 2.5D method for the segmentation of spheroid cultures and for counting proliferating cells within them. The spheroids are assumed to be approximately ellipsoid in shape. They are identified from information present in the Maximum Intensity Projection (MIP) and the corresponding height view, also known as Z-buffer. It alerts the user when potential bias-introducing factors cannot be compensated for and includes a compensation for signal attenuation.

INTRODUCTION

Drug discovery and development in oncology has a success rate as low as 6% [1]. The traditional models used for oncology drug testing are monolayer cultures of tumor cells grown on glass and plastic substrates. These models are very different from solid tumor behavior in vivo. In comparison, 3D cultures are thought to more closely mimic the tumor microenvironment [2], as they allow more complex interactions between cancer and stromal cells, and expose cells to more realistic mechanic forces. Studies have demonstrated important differences in transcription profiles [3] and drug sensitivity [4–8] between 2D and 3D cell culture systems. For a recent review, see [9]. The PREDECT consortium aims to validate various 3D cell culture models by comparing their histology and protein expression with that found in patient samples [10].

Citation: Barbier M, Jaensch S, Cornelissen F, Vidic S, Gjerde K, de Hoogt R, et al. (2016) Ellipsoid Segmentation Model for Analyzing Light-Attenuated 3D Confocal Image Stacks of Fluorescent Multi-Cellular Spheroids. PLoS ONE 11(6): e0156942. doi:10.1371/journal.pone.0156942

An important category of such models are spheroid microtumors grown in a matrix, such as Matrigel®, with or without the presence of supporting stromal cells. Typically, cancer cells are seeded in the extracellular matrix in each microtiterplate well, and over time multiple spheroids of different sizes develop. These 3D multi-cellular spheroid models reflect more closely the nutrient, oxygen and drug gradients that can be found in a tumor, reproducing hypoxic, proliferative, apoptotic and necrotic regions [11, 12]. Suitable dyes or antibodies may be applied to detect these regions/processes, but an accurate quantification of the spatial distribution within the 3D spheroid, in relation to spheroid size, is challenging.

Gross qualitative differences can be detected by visual assessment of micrographs. But accurate and reproducible quantification of statistically significant differences requires automated image analysis, especially in large-scale studies. To extract all spatial information from a 3D sample, it is necessary to apply 3D image analysis, but this is a computationally expensive process. Moreover, image resolution is much poorer along the axial direction, so there is a limited information gain by treating the z axis as equivalent to the others. To circumvent this, many methods apply a hybrid form known as 2.5D analysis [13, 14]. Typically, a 2D projection, which by itself sacrifices all depth information, is combined with a method to extract and store a limited amount of depth information. In order to get meaningful information out of 2.5D image analysis, these methods then need to incorporate assumptions that permit extrapolation back to three dimensions. Orthogonal validation, using independently generated «ground truth» is essential to ensure that the analysis sufficiently represents the original data set.

Accurate extraction of quantitative features requires adequate quality images as a starting point. Acquisition of such 3D image data sets is complicated by a number of issues that are less important in, or absent from a 2D setting. These include spherical and chromatic aberrations (which, for a limited sample thickness, are usually corrected for in objective design), poor axial resolution, photobleaching and phototoxicity during long data acquisitions, and light absorption and scattering resulting in signal attenuation deeper into the sample. In this paper, we will focus on how to address the experimental errors due to light attenuation.

In many applications, the effects of light attenuation can be reduced. Tissue samples can be chemically cleared before imaging by using mounting media such as SeeDB [15], Clarity [16], ClearT [17], or Scale [18]. As the effects of light scattering are reduced at longer wavelengths, image quality is greatly improved by multiphoton microscopy, which uses infra-red excitation light that penetrates deeper into tissue [19]. Light-sheet fluorescence microscopy reduces the impact of light attenuation by illuminating and measuring a sample from multiple angles [20, 21]. However, most of these technologies are incompatible with medium and high throughput experiments on hundreds or thousands of samples. These currently need to be performed using standard high-content imagers and microtiter plates. This limits the options to spinning disk confocal microscopy, the use of far-red dyes, and the application of software techniques to correct for light attenuation after acquisition [22, 23].

Diverse tools for the analysis of fluorescent images of 3D spheroid cultures, such as the quantification of proliferation markers used here, are available. General purpose image analysis tools include both commercially available packages, such as, Imaris (Bitplane AG, Switzerland, http://bitplane.com) and Volocity (PerkinElmer, Inc., USA), and freely available open source software tools such as FIJI (ImageJ) [24] and ICY [25]. Moreover, in the context of multicellular spheroids and fluorescent spot detection, multiple tools compatible with high throughput analysis, and dedicated to specific tasks, exist: Amida (spheroid morphology) [26], AnaSP (spheroid volume) [13], MINS (nuclear segmentation) [27], CellSegm (general segmentation)[28], smart 3D-FISH (spot detection) [29], and goIFISH (spot detection) [30]. However, there is no analysis tool to date that addresses spheroid signal attenuation, and can be used in a high throughput fashion. In this paper, a 2.5D image analysis method for 3D spheroid cultures, suitable for Z-stack data acquired by confocal high-content imagers, is presented. Our method segments the entire 3D spheroid and detects the single labeled cells within. These cells are stained with a specific marker, for example a marker for cell proliferation such as EdU. It is conveniently applicable to medium throughput experiments to assess cancer cell growth under varying culture and treatment conditions. The performance of the proposed approach is compared with a simple 2D MIP analysis method, and a full 3D analysis executed using Bitplane Imaris.

RESULTS AND DISCUSSION

The proposed analysis method consists of two major parts: the segmentation of the cancer spheroids and the identification of positive cells, based on fluorescent labeling present in all cancer cells and a specific marker, respectively. Both are explained in detail in the subsequent sections. In our example data set, the cancer cells are labeled with RFP and the specific marker is EdU, a marker for proliferating cells, as detailed in the Materials & Methods section. The parameters and output features of the method are given in Tables Tables11 and and22.

Table 1: Input parameters of the algorithm.

Algorithm step	Parameter	Default value
General	**Average cell diameter:** The expectation value of the cell diameter	12 μm
2D segmentation	$max_{z\text{-}range}$: Threshold on the range-filtered height view	Calculated from the histogram of the range-filtered height view.
	r_{range}: Neighborhood radius for the range filter	3 μm (one-fourth of the cell diameter)
	min_{MIP}: threshold on the mean intensity value of the spheroids	Mean of the background intensity of the MIP
3D ellipsoid approximation	min_{radius}: The z-dimension of spheroids with radii below this threshold is put to the mean radius in the plane instead of determined from the 3D signal.	24 μm
Attenuation	**Intensity attenuation percentage** (threshold): The minimal percentage of the signal that has to be retained to be analyzable	50%
Spot detection	**Maximal spot radius:** Maximal expected spot radius, larger spots are split	8 μm (average nucleus diameter)
	Average spot radius: Scale for the spot detection	Optimal scale calculated from the LoG of the MIP

Table 2: Output features of the algorithm.

Feature	Description
Spheroid center	The coordinates of the 3D ellipsoid center of the spheroid
Ellipsoid axes	The axes and their dimensions of the fitted ellipsoid
Spheroid mean intensity	The mean intensity of the 2D spheroid mask
# spots per spheroid	The number of positive cells detected in a spheroid
Spot center	The coordinates of the center of a detected positive cell
Spot intensity	The intensity in the original image of the spot
Circularity	A Podzceck-shape factor determining the resemblance of the spheroid contour to a circle
Roundness class of the spheroid	Elliptical or non-elliptical
Attenuation class of the spheroid	Full, half, or not analyzable spheroid
Spheroid area	Area of the 2D spheroid mask
Spheroid volume	Volume of the 3D ellipsoid

Segmentation of multi-cellular spheroids

The first step is a 2D segmentation of the spheroids on maximum intensity projections (MIPs). Subsequently, ellipses are fitted to the segmented 2D spheroid-masks. Finally, these ellipses are extrapolated to 3D ellipsoids. The procedure assumes that the 3D image stacks are corrected for uneven background illumination, to eliminate any systematic bias in the height view of the background pixels, and the distance between two subsequent slices (often referred to as z-step) should be less than the average diameter of a cell.

2D segmentation

figure 1 illustrates the concepts of the MIP and corresponding height view. At each (x,y) coordinate, we obtain the highest intensity value and z-coordinate of the corresponding voxel. For the basic segmentation of the spheroids, which is illustrated in figure 2, we first select locally homogeneous regions from the height view (shown in 2(b)) [31]. Such regions correspond to object (spheroid) pixels while background pixels tend to be more random in z-coordinate. By applying a range filter, which calculates for every pixel the maximal intensity difference of the neighboring pixels within a radius r_{range}, a measure for the local inhomogeneity is obtained, as shown in figure 2(C). figure 2(E) shows the histogram of height differences in figure 2(C), indicating that the large number of pixels at the lower values belong to the spheroid regions. Applying a threshold $max_{z\text{-}range}$ on the image in figure 2(C), a value determined by the extent of the first peak in the histogram, gives rise to separation of objects from background. Finally, the resulting foreground objects are identified using connected-component labeling. Another threshold min_{MIP}, determined by the mean of the background MIP intensity, is applied on the mean intensity of each object in the MIP to ensure that the detected objects have an expected minimal fluorescence intensity, characteristic for the spheroids. Applying such a threshold prevents the detection of spurious objects.

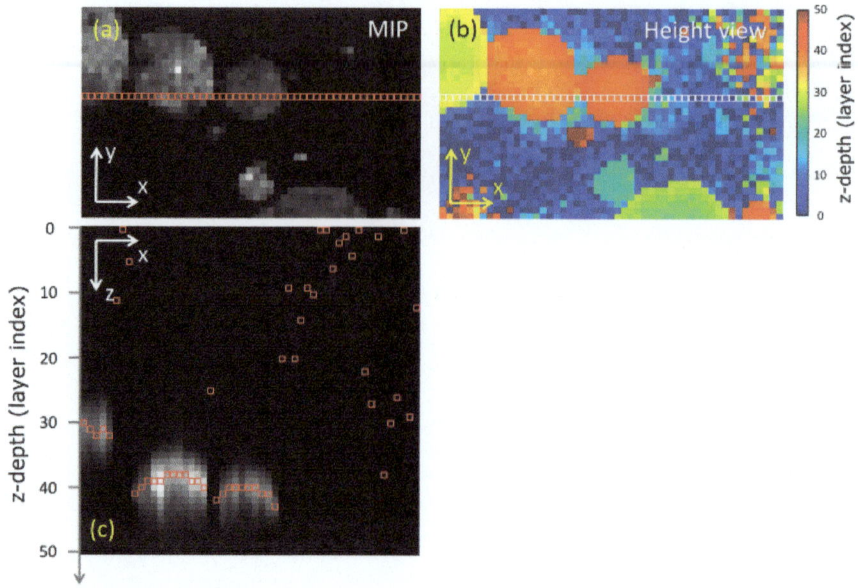

Figure 1: MIP and corresponding height view.

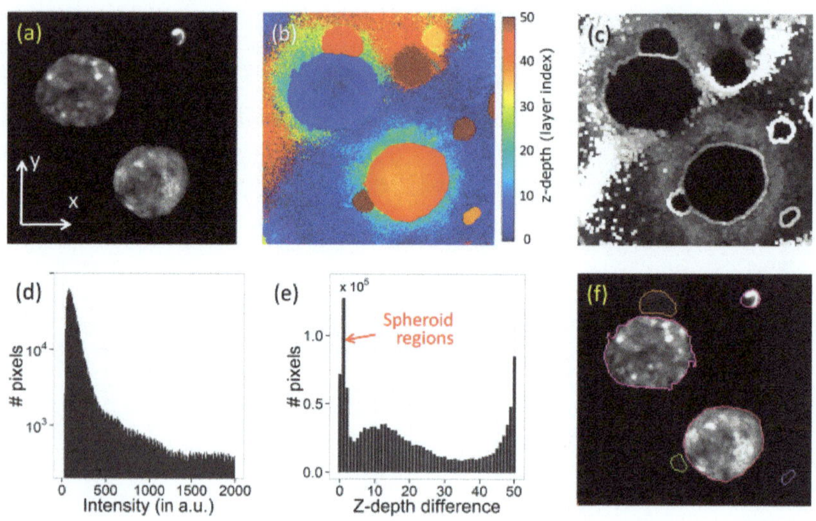

Figure 2: Segmentation of the spheroids in 2D.

Fitting of ellipses to 2D spheroid-masks

The fitting of ellipses to the spheroids segmented in 2D is done by minimizing the mean square error of the difference with the segmented 2D mask. During this step, we assume that the elliptical spheroids are verified by measuring the circularity of the 2D segmentation masks of the spheroids. Spheroid-masks with low 2D circularity do not have a 2D ellipse shape, and will not give rise to a satisfying 3D ellipsoid fit. There may be multiple reasons, beyond actual irregularity in spheroid shape, that can contribute to the production of irregular 2D masks. These include spheroids which are not fully visible in the MIP because they are obscured by other brighter spheroids, and spheroids that were incorrectly segmented during the 2D segmentation step, for example, by merging two spheroids. Since these spheroids would give rise to an incorrect 3D segmentation mask, they are filtered out based on their 2D circularity.

3D ellipsoid approximation

Next, a 3D ellipsoid is fit to each of the 2D ellipses. In employing this procedure, we assume that the spheroid has an ellipsoidal shape, which is either nearly spherical, or satisfies the less stringent condition that two of the ellipsoid axes are oriented in the xy-plane. The former case is applicable to spheroids cultivated in a homogenous matrix, and the latter case can occur for spheroid cultures seeded in a layer sandwiched between two matrices. As we assume that the 3rd ellipsoid axis is along the z-axis, the ellipsoid center position and length of the 3rd axis has to be estimated. The (x,y)-coordinates of the center of the ellipsoid in 3D are assumed to be the (x,y)-coordinates of the center of the 2D ellipse, the z-coordinate of the center and the 3rd ellipsoid axis length are derived from the intensity profile along the vertical axis. figure 3 shows the resulting fitted ellipsoids for an image stack of a 3D spheroid culture. The spheroid volume can now also be estimated.

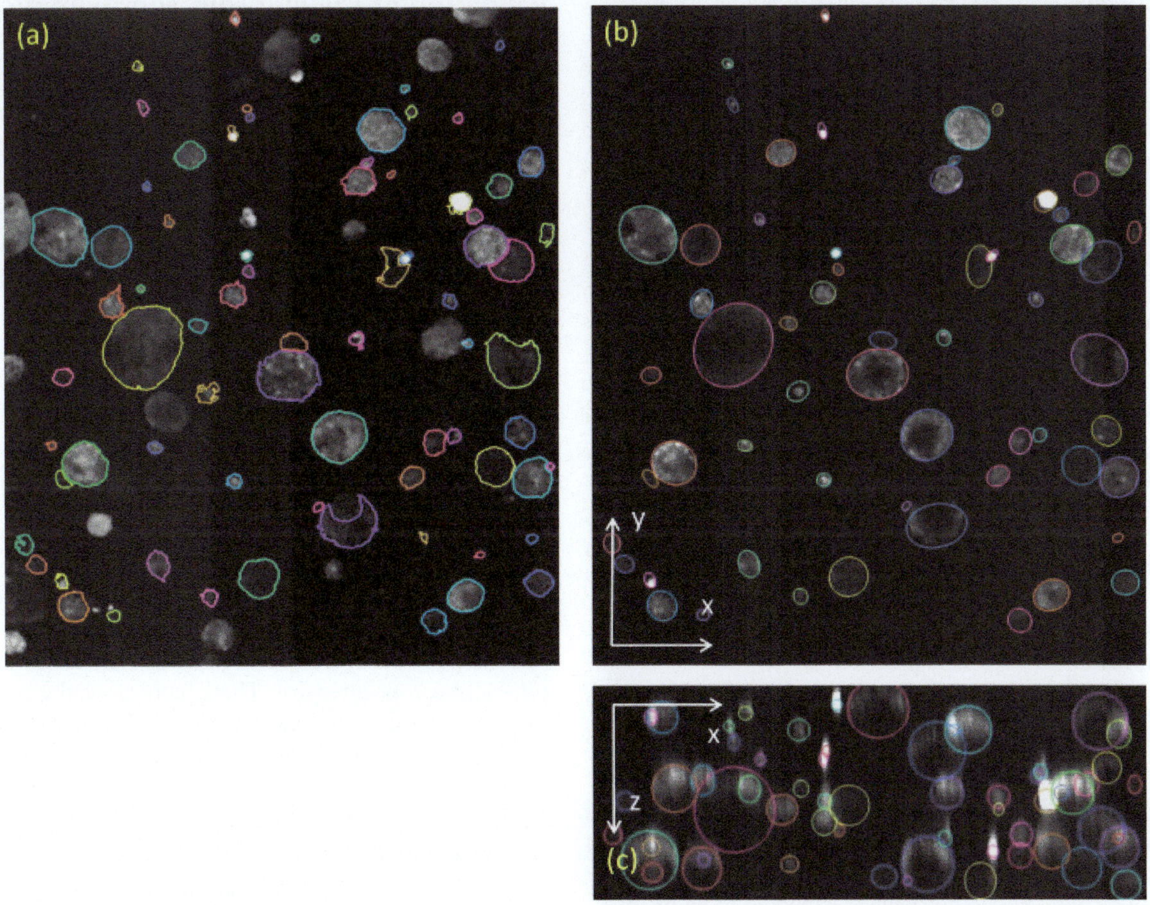

Figure 3: Fitting of 3D ellipsoids.

Determination of the analyzable ellipsoid region

There is a decreasing intensity gradient from the surface towards the center of a spheroid. Therefore, cells can be incorrectly identified. This issue is illustrated in figure 4(A), where xy- and xz-slices through the center of a spheroid show the typical intensity distribution due to signal attenuation. Note that the signal of the EdU channel is also reduced in the lower parts of the spheroid, which renders the detection of EdU positive cells impossible in that region. Notice also that signal attenuation due to deep tissue imaging is not the only reason for signal detoriation. When using for example point scanning microscopes, other factors, such as photobleaching, can also become significant.

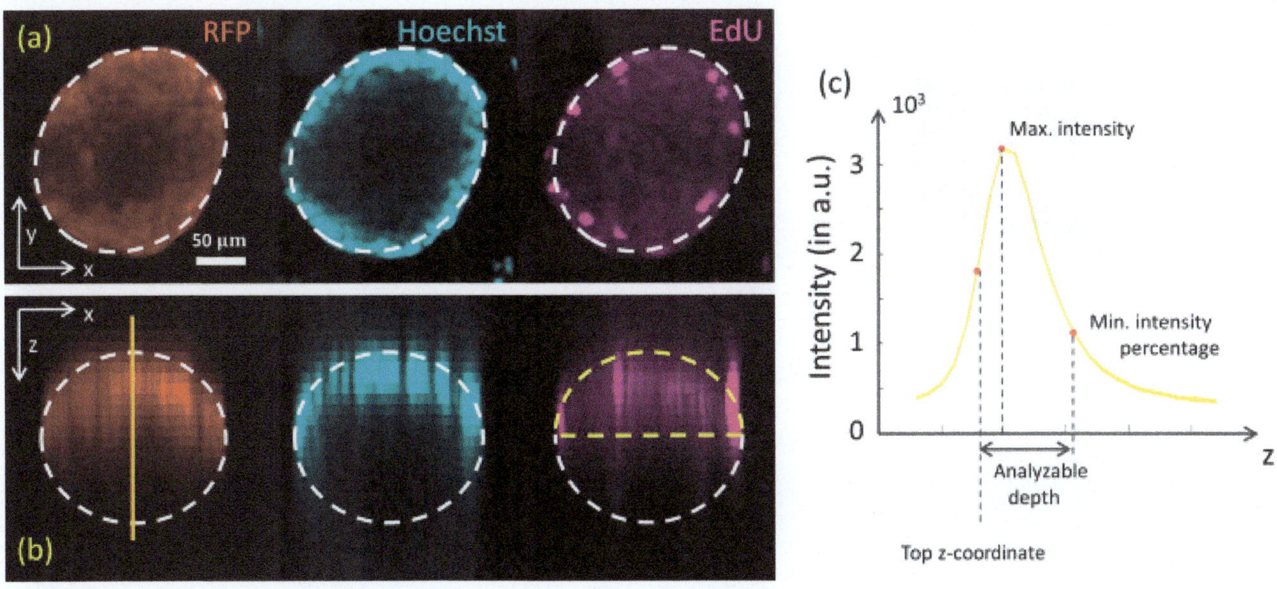

Figure 4: Light attenuation in multi-cellular spheroids.

As all cells in the spheroids carry a stable fluorescent marker (thus assumed to be constant within a single spheroid), their vertical intensity profile through the center of the spheroid can be used as a measure for the signal attenuation. Indeed, a gradual reduction of the stable marker signal (here corresponding to the RFP signal) with increasing depth in the spheroid tissue is observed in figure 4(C). In order to measure the maximum depth in the spheroid up to which positive cells can be accurately detected, we assume the attenuation of the specific fluorescent marker to be comparable to that of the stable marker. When the wavelengths of both channels do not differ significantly, this is a reasonable assumption. Given a user-defined value for the minimal percentage of intensity that has to be retained after attenuation of the signal, we can determine the depth of the analyzable region of the spheroid (see the Materials & Methods section).

The analyzable depth is subsequently compared to the spheroid vertical diameter. Three distinct cases are identified (as illustrated in S1 figure):

The entire spheroid is visible. Results are not affected by light attenuation.

More than half of the spheroid is visible. The upper hemisphere of the ellipsoid will be analyzed and the results extrapolated on the lower half, assuming similar characteristics.

Less than half of the spheroid is visible. The spheroid cannot be analyzed because positive cell counts cannot be extrapolated from the spheroid surface to the core, where the availability of oxygen and nutrients is expected to be different [11].

Since only half of a spheroid needs to be visible, our proposed analysis procedure extends the window of spheroid sizes that can be analyzed, allowing diameters to be approximately twice as large. Also, spheroids that are too large for full analysis (i.e., case three) will be identified and may be omitted from sample analysis. In the case that a significant amount of spheroids has to be omitted, the proposed analysis is not appropriate anymore and other types of analysis of the spheroid cell cultures have to be utilized.

Validation of the ellipsoid approximation

In order to test the proposed 2D segmentation step, we compared the results of the automated with a manual 2D segmentation of the data, assessing the sensitivity of the spheroid detection and the accuracy of the resulting contours as defined in detail in the Materials & Methods section. The MIPs of 8 image stacks were manually segmented as described in the Materials & Methods section, and we found an average accuracy of 0.911 ± 0.066, and a sensitivity of 0.928, within the 95% confidence interval [0.909, 0.944]. Sensitivity strongly depends on the threshold min_{MIP}, and whether a segmented contour is categorized as spheroid or as noise, especially for spheroids that have a mean intensity close to background. In S2(A) and S2(B) figure we plotted the sensitivity and the ratio of false positives (FP) and their Wilson score interval as function of the intensity threshold min_{MIP}. The sensitivity decreases together with the number of false positives with increasing threshold, as such there is a certain trade-off to consider when setting the threshold value. S2(C) and S2(D) figure illustrate the mean intensity of the spheroid as function of different categories, indicating that spheroid contours classified as false positives (FP) have generally a low intensity. The accuracy of the segmentation contours of the spheroids which are identified in the GT as "well separated spheroids" is significantly high (an exactly matching contour has an accuracy equal to 1) and shows that the non-homogeneity of the height view is a valid measure for the segmentation of this type of cell cultures.

Spheroids overlapping in the MIP are unsuitable for analysis (as illustrated in figure 5(A)). Since the circularity of overlapping spheroids is expected to be lower, identifying spheroids with a small value for circularity should allow for removal of most of these objects. To check whether there indeed is a correlation between overlaps and spheroid roundness, the circularity as determined by the 2D automatic segmentation mask was plotted against the distinct classes of the GT, i.e. well-separated spheroids (red), adjacent spheroids (cyan), and overlapping spheroids (green) in figure 5(B) where a notable decrease of circularity can be observed in overlapping spheroids (p***).

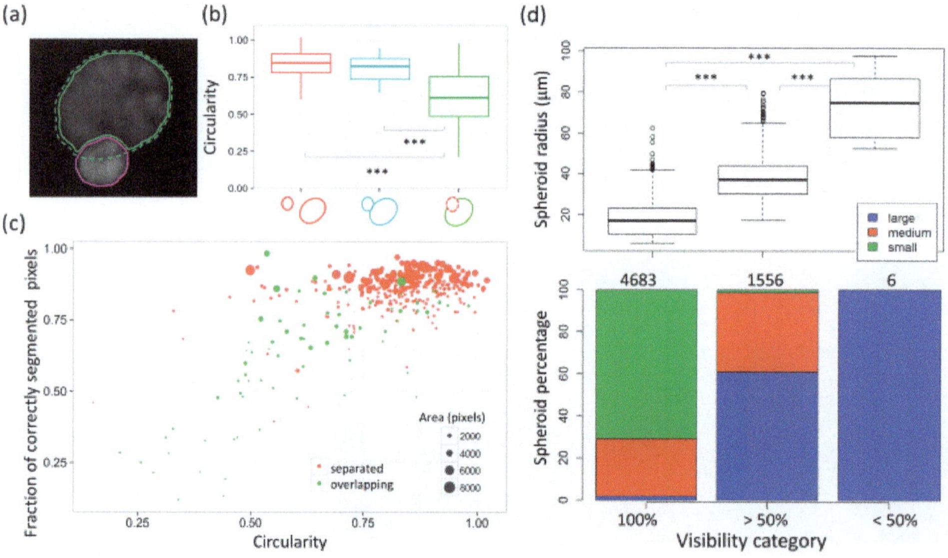

Figure 5: Validation of the ellipsoid segmentation algorithm.

To verify that a circularity threshold removes overlapping spheroids, the fraction of correctly segmented pixels of the spheroid masks was plotted as a function of their circularity for each spheroid and labeled with the manual classification of the ground truth in figure 5(C). The correlation between both is highly significant (p***), with an estimated value of $r = 0.70$ and a confidence interval [0.64, 0.75]. A MANOVA applied on the data indicates that the distributions of the two spheroid classes shown differ significantly (p***). The Figure shows further that applying a circularity threshold indeed removes most overlapping spheroids. As an undesired side-effect of this filter, some well separated but non-elliptical structures were removed. These may represent invasive spheroids rather than compact micro-tumors and are excluded from analysis. In S5 figure the origin of the spheroids which are omitted due to too low circularity is indicated. For the ground truth images, under consideration, less than 20% of spheroids were omitted, primarily due to being obscured during imaging.

Finally, in figure 5(D), the relation between spheroid size and light attenuation is illustrated, using a sample set containing control and SOC (standard of care) treated cultures, as described in the Materials & Methods section. As in the previous section, the spheroids were categorized into groups of 100%, more than 50%, or less than 50% visible spheroids. While a standard 2D MIP image analysis method would only be able to analyze spheroids in the first 100% visibility category, our approach also allows the quantification of spheroids belonging to the second category (> 50%), which in this case contains approximately 25% of the total number of spheroids, as can be seen from the spheroid count per visibility category.

Identification of cells positive for a given nuclear marker

Cells positive for a given nuclear marker are detected either via a computationally expensive 3D or a fast 2.5D spot detection algorithm based on the MIP and height view of the corresponding fluorescent channel. It is assumed that resolution is sufficient to resolve the individual nuclei, which are physically separated by the cytoplasm (as visualized in S1 Movie). In both approaches, marker positive spots are found using a Laplacian of Gaussian (LoG) filter, at the scale of the expected size of a nucleus, which reveals the coordinates of the spots in the original image as local minima. The maximal spot radius parameter is set to permit merging of multiple local minima occurring within a nucleus, while maintaining the distinction between separate nuclei. The optimal scale for the expected size of the nuclei is determined beforehand by using the 2D MIP. The total sum intensity of the LoG of the MIP is maximized by optimizing the scale parameter, between one tenth of the maximal spot radius and the value itself. A curve for the total sum intensities as function of the scale parameter is shown in S3 figure Afterwards, the spots are filtered based on their intensity, with the threshold equal to the median spheroid intensity in the MIP (resulting in a single threshold per image). As an example the results of the approach are shown in panel (b) of S3 figure where the red circles represent detected EdU positive cells. For the 2.5D spot detection algorithm, the original image stack is exchanged for its 2D MIP and the z-coordinate of a spot is obtained from the height view.

Application of the approach to a proliferation study

We applied our proposed algorithm to analyze images of proliferating cells in 3D spheroid cultures, demonstrating its performance on partially attenuated images stacks. These are images of multi-cellular spheroid prostate cancer cell cultures treated for 17 days with a cytotoxic compound (docetaxel) or a cytostatic (MDV-3100) compound. As illustrated in figure 6, tumor cells were labeled with RFP, stromal cells with GFP, and Hoechst and EdU dyes were applied to identify nuclei and proliferating cells, respectively. The spheroid volume and the number of EdU positive cells per spheroid were analyzed. The volume and number of EdU positive cells per spheroid at the endpoint of the experiment is shown for a culture of PC346-c cancer cells for both compounds in (figure 7A and 7B). Under docetaxel treatment both volume and number of EdU positive cells per spheroid decrease. The total number of EdU positive cells per image volume (four different fields were imaged in the sample well), shown in figure 7(C), indicates further that treatment with docetaxel reduces the total number of EdU positive cells in the culture.

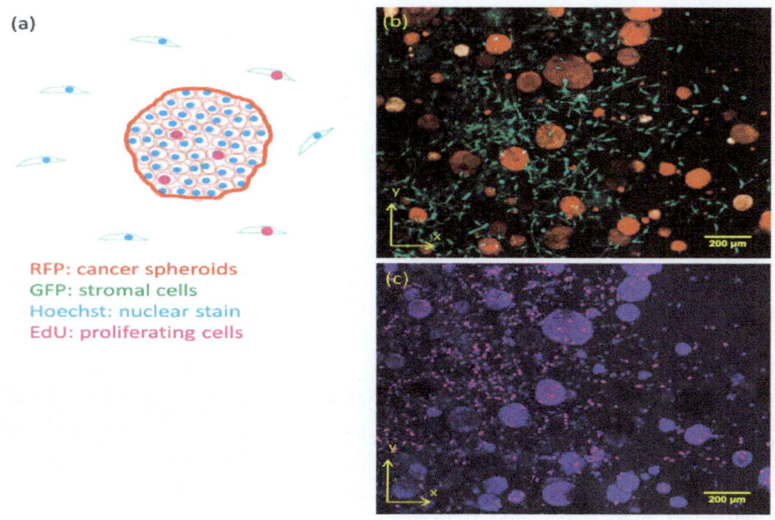

Figure 6: Proliferation analysis example data.

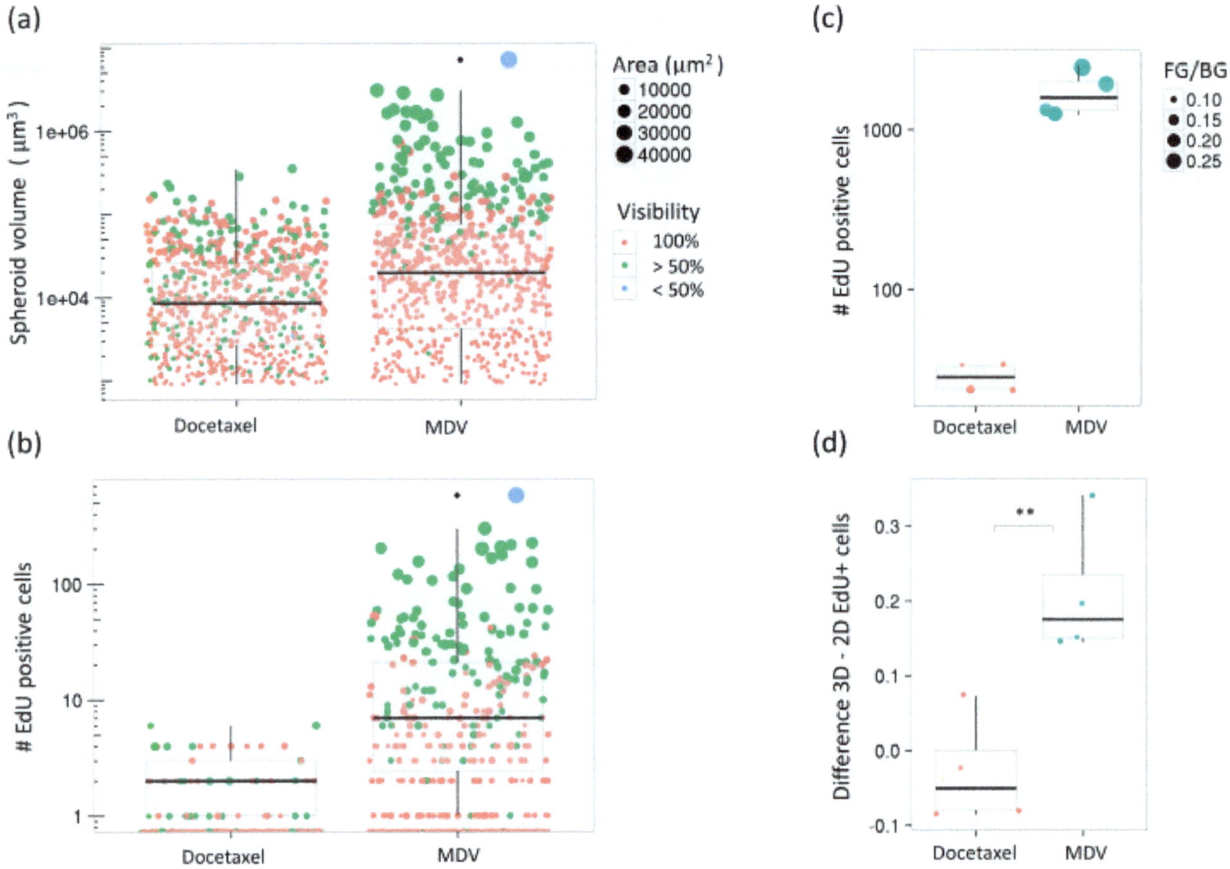

Figure 7: Comparison of the proposed analysis method with a baseline 2D MIP analysis method.

Comparison with 2D analysis results

An important difference between our approach and a standard 2D MIP analysis (as described in the Materials & Methods section) is that the number of detected EdU positive cells within a large spheroid is corrected for. The significance of this difference is shown in a comparison of the two approaches in the plots in figure 7(D). As expected, a significant nonzero difference between the two approaches is found in the MDV treated cultures representing a culture with larger spheroids. Our proposed new algorithm corrects for signals lost due to light attenuation, whereas small spheroids basically lead to the same results. As such, our approach is especially useful for spheroid cultures containing a fair amount of larger spheroids in the cultures. In particular this can lead to a significant difference when comparing two samples containing different average spheroid sizes, large vs. small, for example when imaging a culture of growing spheroids at different time points. A bias in the sample of large spheroids from a 2D MIP analysis could then lead to inaccurate conclusions.

Comparison with a full 3D analysis

We compared the proposed algorithm with a full 3D analysis executed in a widely used general purpose commercial tool, Bitplane Imaris (Bitplane AG, Switzerland,http://bitplane.com). We compared the detection of proliferating cells and the computation of the volume of matching detected spheroids. Due to differences in segmentation algorithms, the overlap of detected spheroids was 63%. In S4(A) figure the volumes from both methods are compared. The volume obtained by a full 3D approach is on average larger than in the 2.5D approach for small spheroids, but smaller for larger spheroids. In panel (b) this is illustrated. We hypothesize that due to the intensity-based segmentation in Imaris, very small spheroids tend to appear stretched along the z-direction because of poor z-resolution, increasing the total volume, while larger spheroids tend to be not fully imaged due to signal attenuation. To test this hypothesis we plotted the sphericity of the spheroids as function of their volume in panel (c). The decrease in the fitted curve indicates that the larger spheroids are not correctly segmented. To illustrate the effect on a growth analysis we plot the difference between the number of EdU positive cells obtained by our approach and Imaris. This plot in panel (d) indicates that for larger spheroids (where the visibility category is less than 100%), our approach results in more spots, corresponding to the correlation coefficient (Spearman) which was equal to 0.26.

To demonstrate the computational advantage of taking a 2.5D approach we timed both methods (on a standard workstation with an Intel® Core™ i7-3920XM CPU running at 2.9 GHz and 8 GB RAM). The image stacks used had a size of 1024 x 1256 x 91 pixels (two 16bit channels were used). In Imaris it took on average 7 minutes to perform both the spheroid segmentation and spot detection. Our 2.5D approach executed this in 3 minutes and when the simplified spot detection algorithm (based on the MIP and height view) was used this reduced to 1 minute.

MATERIALS AND METHODS

Multi-cellular spheroid culture image stacks

The cell cultures, used for testing the analysis method, contained LNCaP human prostate cancer cells (ATCC, Rockville, USA) or PC346c cells (a prostate cancer cell line obtained from the Erasmus Medical Centre, via the PREDECT consortium) and CAF-PF179T human cancer associated fibroblasts (a cell line obtained from the Weizmann Institute, via the PREDECT consortium) [32], embedded in Matrigel®(Corning #356231, Lot#3198769, growth factor reduced, phenol red free, concentration: 4 mg/ml). They were suspended in the matrix and seeded (10,000 tumor cells and 1,000 stromal cells in 60 µl/well) in 96 well plates (96 well polystyrene cell culture black microplates with a µClear® bottom; Greiner, #655090), which were pre-coated with matrix (30 µl/well). Treatment of cultures with either a cytotoxic compound (docetaxel), or a cytostatic compound (MDV-3100) started on day 6, when the cultures were in exponential phase, and the experiment was ended at day 23, when untreated cultures had reached stationary phase. Cancer and stromal cells stably expressed tRFP and eGFP, respectively. Prior to imaging, cell cultures were stained with 10 µM EdU for 2h (Click-iT EdU Alexa Fluor 647 HCS; Life Technologies) to detect proliferating cells, and Hoechst 33342 (8.1 µM, 30 min) to label all nuclei and then fixed with 4% formaldehyde.

Microscopic image stacks of multi-cellular spheroid cultures were acquired with a Yokogawa CellVoyager 7000 confocal spinning disk microscope (Wako Automation, San Diego, United States), using a UPLSAPO 10x/0.4 NA objective and a 10 µm z-separation between the slices in the stack. Excitation wavelength–filter pairs used for the tRFP, eGFP, Hoechst, and EdU channel were 561 nm—BP 600/37, 488 nm—BP 522/35, 405 nm—BP 447/45, and 640 nm—BP 676/29, respectively.

Ground truth images

Ground truth data were generated for 3D homogeneous multi-cellular LNCaP cancer spheroid cultures by manually drawing a 2D mask for each spheroid on the MIP. Moreover, each spheroid was classified to one of the cases: (1) well separated, (2) overlapping with brighter spheroids in the MIP (rendering it non-separable), (3) overlapping with less bright spheroids in the MIP (rendering it well separable), (4) merely touching other spheroids, and (5) touching the border of the 2D projection of the image. The labelling was conducted using the RoiManager plugin in FIJI (ImageJ) [24] resulting in contours for the cases (1) to (5) colored in red, green, magenta, cyan, and blue, respectively. These ground truth data were used to validate the 2D segmentation step of the algorithm by defining sensitivity and accuracy, and are provided in S2 File.

Implementation of the analysis algorithms

Analysis of the image stacks was conducted using the image analysis toolbox DIPimage (Delft University of Technology, Delft, The Netherlands) for MATLAB® (Release 2014a, The MathWorks Inc., Natick, United States). The algorithms are available as a bundle of MATLAB scripts in S1 File: Image analysis algorithms source code, containing both the code to generate the Figures, and the implementation of the method. Running the provided code requires a working MATLAB environment with installed DIPimage toolbox and the following external libraries: the Bio-formats toolbox for MATLAB, JSONlab, and ReadImageJROI. The implementation is also available on GitHub: https://github.com/mbarbie1/ellipsoids-analysis-paper.git.

Statistical analysis

Statistical analysis was executed in R (https://www.r-project.org) using the RStudio IDE (https://www.rstudio.com). The data was tested for normality using the Shapiro-Wilks test, significances were obtained using Welch's T-Test and the non-parametric Mann-Whitney-U-Test for non-normal distributions. To compare multivariate distributions, e.g. data for the spheroid size and number of proliferating cells, a MANOVA was applied. In all such cases only two groups were compared, thereby the F-statistic approximation is reduced to the Hotelling's T-square statistic. Significances were depicted as *: $p < 0.05$, **: $p < 0.01$, and ***: $p < 0.001$.

Derivation of the ellipsoid 3D fit parameters

The z-coordinate of the center and the length of the vertical axis are derived from the pixels of the original 3D image. The z-coordinate of the center of the ellipsoid is defined as the median value of the z-coordinates of the outer rim of the 2D segmentation mask. To derive the length of the vertical ellipsoid semi-axis, we consider the neighborhood which contains the 8 nearest neighbor pixels of the ellipse center (x_c, y_c) in the xy-plane, and a sufficient range (2 times the largest diameter D of the ellipse) in the z-direction:

$$(x, y, z) \in \{x, y, z: (x, y) \in [(x_c, y_c)-(1, 1), \ (x_c, y_c) + (1, 1)] \ \& \ z \in [z_c - D, \ z_c + D]\}$$

Averaging over this neighborhood yields a profile I(z) through the centroid of the 2D spheroid segmentation mask (see figure 4(C)). The z-coordinate of the top of the spheroid (which suffers the least from light attenuation) is defined from the profile curve as the z-coordinate above the spheroid center where the curvature becomes zero (the inflection point). When the 2D mask of the spheroid under consideration has an area below a specific threshold min_{radius}, the above approach is not used; instead the length of the vertical axis is approximated by the average length of the ellipse axes in the plane.

Characterization of the signal attenuation in the spheroids

The depth up to which the spheroid is analyzable is limited by the signal attenuation in the spheroid tissue. We calculate the maximal depth (z-coordinate) in the spheroid for which a user-defined percentage of the intensity is still retained after signal attenuation. To obtain this depth we assume that: (1) the fluorescent stain of the tumor cells is constant within each spheroid, (2) the scattering and absorption coefficients determining the attenuation of the signal are constant inside spheroid tissue and zero outside, and (3) intensity values of the first slice of the spheroid are free of attenuation.

The signal attenuation is expected to be largest in the (x,y)-center of the ellipsoid for each z-coordinate. Therefore, the vertical intensity profile through the center of the ellipsoid is used to characterize the attenuation of the signal. The intensity profiles $I(z)$ are determined as described in the previous section (see figure 4(C)). They typically have a steep inclination towards a single maximum $I_{max} = \max_z I(z)$ near the top of the spheroid, followed by an intensity decrease. We consider I_{max} the intensity without attenuation, and the percentage retained after attenuation is defined as $P_{att}(z) = 100\ I(z) / I_{max}$. The user-defined threshold on the minimal percentage retained P_{min} leads to a range $[z_{max}, z_{analyzable}]$ for which $P_{att}(z) > P_{min}$ is valid. The analyzable depth of the spheroid is defined as the distance between the first z-coordinate of the spheroid and $z_{analyzable}$.

Other analysis methods used for comparison

For comparative purposes, a 2D MIP analysis as well as a full 3D analysis, were utilized. Both approaches are detailed below.

Baseline 2D MIP analysis method

For the 2D analysis of proliferation in the test data, the following image analysis approach was applied: Similarly as in the proposed algorithm a 2D segmentation mask of the MIP of the RFP channel was obtained, thereafter EdU positive cells were identified by a 2D spot detection algorithm performed on the EdU channel. But unlike in the proposed analysis, all EdU positive cells within the 2D regions of the spheroid masks are considered to belong to the corresponding spheroids, and no correction for signal attenuation was attempted.

Imaris based full 3D analysis workflow

Analysis of the images was conducted using the cell extraction functionality from ImarisCell, a component of Bitplane Imaris (v8.1.2, Bitplane AG, Switzerland,http://bitplane.com), where "spheroids and proliferating cells" translate to "cells and cell vesicles". Spheroids were segmented using the surface extraction tool and morphological features of the spheroid surfaces were extracted. Proliferating cells were detected using the spot detection tool. Subsequently, the detected surfaces and spots were imported in the cell analysis module to determine the spheroid borders and proliferating cells. Finally, features relating to the location of the spheroids and number of proliferating cells were exported. As a final remark we mention that although an attenuation correction method for homogeneous signal attenuation exists (through the Imaris module ImarisXT), our spheroid cultures are largely inhomogeneous and this method cannot be applied.

Sensitivity and accuracy of the 2D segmentation mask

To quantify the deviation of an automated segmented mask from the ground truth image, we compare the 2D segmented image regions $R_{GT} = \{M_j\}$ of the ground truth image with the regions $R_A = \{m_i\}$, found by automatic segmentation. For each region M_j we search a corresponding region m_i that overlaps with M_j more than 50%. Regions M_i for which no corresponding region is found are considered as false negatives (FN), while regions m_i not belonging to a ground truth region M_j are false positives (FP). The sensitivity is given by the ratio: TP/(TP + FN), where TP stands for true positives. The error on the sensitivity is given by the confidence interval as defined by Wilson, the so-called Wilson score interval [33].

For true positives (segmented regions m_i which do belong to a ground truth region according to the above criterion) the segmentation accuracy can be quantified by assigning a penalty determined by how close the contours of the two regions are. This is done as follows: the distance of the pixels of regions $m_i \backslash M_j$ (including all pixels from m_i which are not belonging to M_j) and $M_j \backslash m_i$, from the contour of region M_j is calculated, using the distance image (in μm) of the contour mask of M_j, which weights the error contribution of misclassified pixels. A sigmoid function $f(x) = 1 / \{1 + \exp[-k(x - x_0)]\}$ with $x_0 = 1.5$ μm and $k = 2$ μm^{-1} is applied to this distance, so that very small errors in the spheroid outline contribute little to the error estimate, but large errors have a constant weight. Then the pixels are summed over and normalized by dividing by the area of the ground truth region. The resulting penalty e_i is close to zero for nearly overlapping spheroid masks, and the accuracy of spheroid mask m_i can be defined as $1 - e_i$.

SUPPORTING INFORMATION

S1 figure: Relationship between the visibility categories and the spheroid size.

From left to right spheroids of decreasing sizes are shown, as representatives of the different visibility categories: Less than half of the spheroid, more than half, or the entire spheroid is measurable. xy- and xz-slices through the middle of the spheroids are shown and overlaid with white dashed ellipsoid contours (which were fitted manually). In the xz-planes a manually drawn thin dashed contour shows the approximate boundary of the analyzable region of the spheroids. As spheroid size decreases, the Point Spread Function (PSF) of the confocal microscope leads to distortion of the spherical shape.

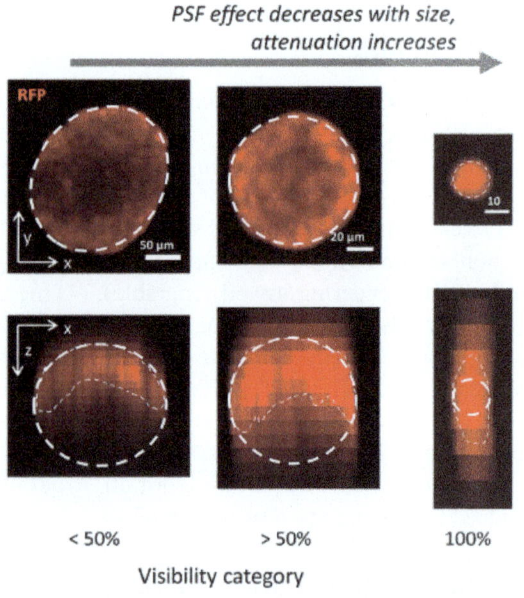

S2 figure: Segmentation sensitivity dependency on the intensity threshold min_{MIP}.

(a-b) The dependency of the sensitivity, the FP/P ratio, and their 95% confidence interval (Wilson) on the threshold set on the mean intensity of the segmented spheroids (min_{MIP}). (c)

Shows the mean intensity of each spheroid as function of their segmentation category: true positives (red), false positives (green), spheroids missed in the segmentation (blue), segmentation masks merging multiple spheroids in the GT (magenta), and spheroids which are detected as multiple spheroid masks (brown). (d) Represents the number of spheroids for each of the categories.

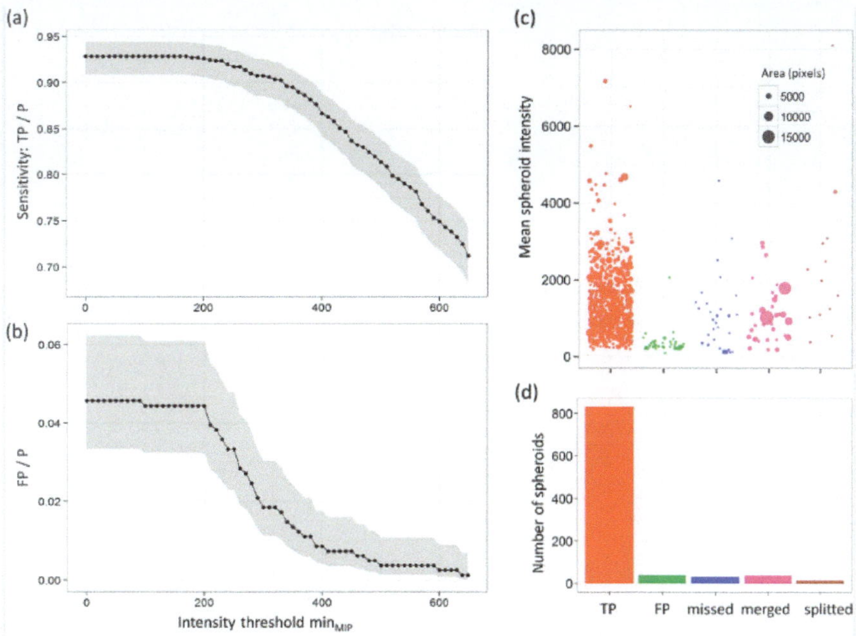

S3 figure: Spot detection in scale space.

The spot detection steps are shown using the example image stack in file Data_3, which is available from Dryad (doi:10.5061/dryad.0m9n7). (a) shows the total sum of pixel intensity of the LoG-filtered MIP image as function of the Gaussian scale parameter σ (smoothing parameter). The σ which maximizes the total sum of pixel intensity (of the negative LoG) represents the optimal scale σ for the spot detection. (b) shows a visual comparison of the results from the spot detection originating from the 3D LoG approach used in our approach (red circles), and, the 3D spot detection calculated using Bitplane Imaris (blue circles).

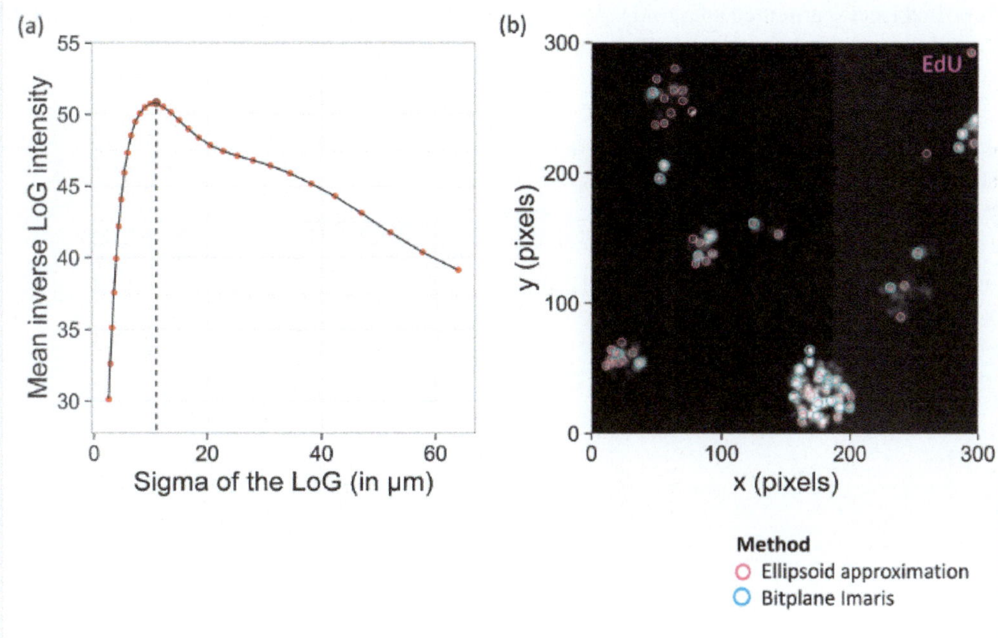

S4 figure: Comparing the proposed analysis method with a full 3D (using Bitplane Imaris) analysis approach.

(a) the obtained spheroid volume for both the full 3D method (blue dots) and our approach (red dots) is shown for corresponding spheroids. (b) the ratio of the volume obtained from Imaris over the one obtained from our approach, together with a linear fit of the data. In panel (c) the sphericity, of the spheroid surfaces obtained in Imaris, is plotted as function of the volume. Here a non-linear fit is obtained. (d) the difference of the number of proliferating cells obtained by our approach with the ones obtained in Imaris are plotted, where the colors correspond to the visibility category: 100% (red dots) and > 50% (green dots). There were no spheroids with a visibility < 50% for these samples.

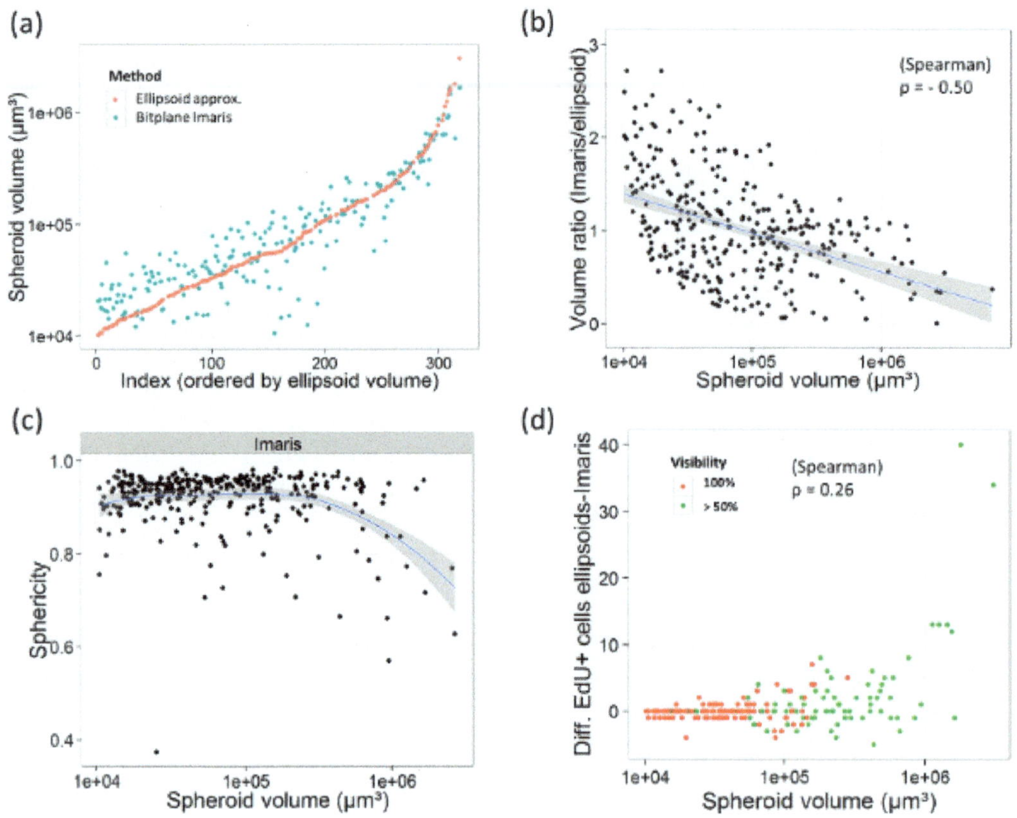

S5 figure: Spheroids omitted during ellipsoid approximation due to shape irregularity.

The percentages of spheroids that are omitted due to low circularity are categorized in spheroids with a non-spherical shape, obscured spheroids or incorrect segmented spheroids. As dataset the ground truth validation images from the file Data_4, which is available from Dryad (doi:10.5061/dryad.0m9n7), are used.

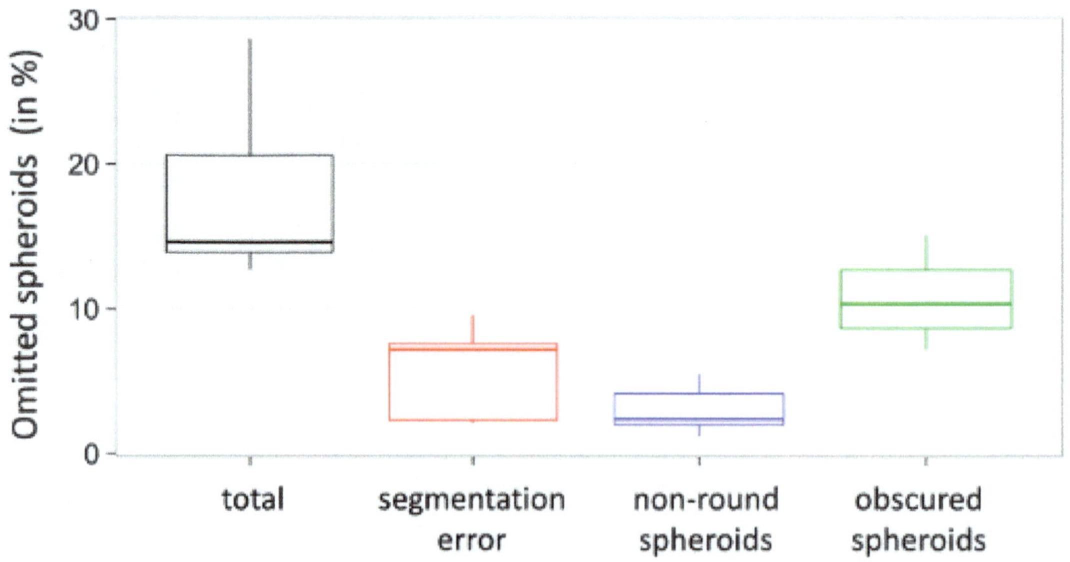

S1 File: Algorithm implementation.

This is a zip-file containing all the source code files. The code is written in MATLAB using the DIPimage toolbox. The source code is also available on GitHub: https://github.com/mbarbie1/ellipsoids-analysis-paper.git.

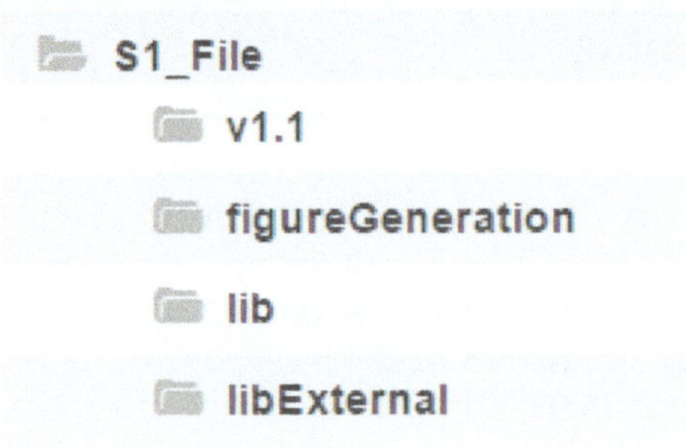

S2 File: 2D ground truth.

This is a zip-file containing manually segmented 2D ground truth labeled masks (the corresponding 3D image stacks can be found in the file Data_4 which is available from Dryad (doi:10.5061/dryad.0m9n7)). The cancer (LNCaP) spheroids are labeled in five distinct classes: (1) well separated, (2) overlapping with brighter spheroids in the MIP (rendering it non-separable), (3) overlapping with less bright spheroids in the MIP (rendering it well separable), (4) merely touching other spheroids, and (5) touching the border of the 2D projection of the image. Opening of the images in FIJI (ImageJ) with the ROI Manager allows visual inspection of the data.

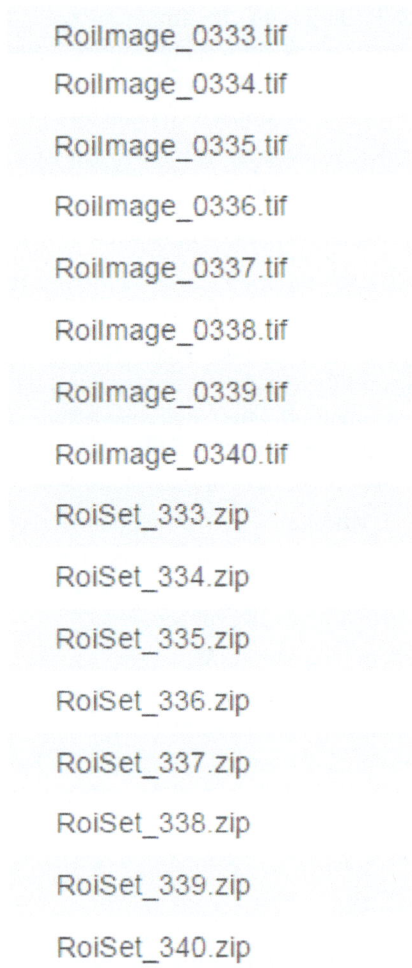

S1 Movie: Spot detection algorithm illustrated on a 3D image stack represented as a movie.

This is the 3D image stack from the file Data_6, which is available from Dryad (doi:10.5061/dryad.0m9n7), saved as a movie, in AVI format with JPEG compression. It shows the EdU channel of the 3D image stack in the file Data_3, available from Dryad (doi:10.5061/dryad.0m9n7), with the EdU positive cells annotated by small spheres. This movie can serve as an example result of the spot detection algorithm (3D version).

ACKNOWLEDGMENTS

Work Package and Platform Leaders and Representatives of the Partners in IMI PREDECT: John Hickman (Institut de Recherche Servier), Mike Burbridge (Institut de Recherche Servier), Emmy Verschuren (University of Helsinki), Olli Kallioniemi (University of Helsinki), Juha Klefström (University of Helsinki), Cathrin Brisken (Ecole Polytechnique Fédérale de Lausanne), Varda Rotter (Weizmann Institute of Science), Moshe Oren (Weizmann Institute of Science), Catarina Brito (Instituto de Biologia Experimental e Tecnológica), Jack Schalken (Radboud University Nijmegen Medical Center), Guido Jenster (Erasmus MC Rotterdam), Wystke van Weerden (Erasmus MC Rotterdam), Jaak Vilo (University of Tartu), Julia Schueler (Oncotest GMbH), Outi Monni (Biomedicum Genomics Ltd), Simon T. Barry (AstraZeneca), Sylvia Grünewald (Bayer Schering Pharma AG), Pekka Kallio (Orion Pharma), Hans-Joachim Mueller (F. Hoffmann-La Roche AG), Adam Nopora (F. Hoffmann-La Roche AG), Wolfgang Sommergruber (Boehringer Ingelheim International GmbH), Elizabeth Anderson (Boehringer Ingelheim International GmbH), Heiko van der Kuip (Robert Bosch Gesellschaft fuer medizinische Forschung mbH), Ralph Graeser (Boehringer Ingelheim International GmbH), Yolanda T. Chong (Janssen Pharmaceutical companies of Johnson & Johnson), Ronald de Hoogt (Janssen Pharmaceutical companies of Johnson & Johnson), Matthew Smalley (Cardiff University), Erwin Boghaert (AbbVie). Further, M. Barbier wants to thank the authors of the DIPimage toolbox (Delft University of Technology).

FUNDING STATEMENT

Innovative Medicines Initiative Joint Undertaking 115188.

Authors MB, SJ, FC, RH, RG, EG, and YTC were funded by Janssen Pharmaceutical Companies of Johnson & Johnson. Authors SV and KG were supported by Innovative Medicines Initiative Joint Undertaking under Grant agreement n° 115188, the resources of which are composed of financial contribution from the European Union's Seventh Framework Programme (FP7/2007–2013) and contributions in kind from the European Federation of Pharmaceutical Industries and Associations. The funders provided support in the form of salaries for the authors, but did not have any additional role in the study design, data collection and analysis, decision to publish, or preparation of the manuscript. The specific roles of these authors are articulated in the "author contributions" section.

DATA AVAILABILITY

S1 File is available as supporting information. The described files Data_3, Data_4, Data_5 and Data_6 are available from the Dryad data repository (doi:10.5061/dryad.0m9n7).

REFERENCES

1. Hay M, Thomas DW, Craighead JL, Economides C, Rosenthal J. Clinical development success rates for investigational drugs. Nat Biotechnol. 2014;32(1):40–51. doi: 10.1038/nbt.2786 . [PubMed]
2. Levinger I, Ventura Y, Vago R. Life is three dimensional-as in vitro cancer cultures should be. Adv Cancer Res. 2014;121:383–414. doi: 10.1016/B978-0-12-800249-0.00009–3 . [PubMed]
3. Wrzesinski K, Rogowska-Wrzesinska A, Kanlaya R, Borkowski K, Schwämmle V, Dai J, et al. The Cultural Divide: Exponential Growth in Classical 2D and Metabolic Equilibrium in 3D Environments. PLoS One. 2014;9(9):e106973 doi: 10.1371/journal.pone.0106973 [PMC free article] [PubMed]
4. Aljitawi OS, Li D, Xiao Y, Zhang D, Ramachandran K, Stehno-Bittel L, et al. A novel three-dimensional stromal-based model for in vitro chemotherapy sensitivity testing of leukemia cells. Leuk Lymphoma. 2014;55(2):378–91. doi: 10.3109/10428194.2013.793323 ; PubMed Central PMCID: PMC4090917. [PMC free article] [PubMed]
5. Longati P, Jia X, Eimer J, Wagman A, Witt MR, Rehnmark S, et al. 3D pancreatic carcinoma spheroids induce a matrix-rich, chemoresistant phenotype offering a better model for drug testing. BMC Cancer. 2013;13:95 doi: 10.1186/1471-2407-13-95 ; PubMed Central PMCID: PMC3617005. [PMC free article] [PubMed]
6. Perche F, Torchilin VP. Cancer cell spheroids as a model to evaluate chemotherapy protocols. Cancer Biol Ther. 2012;13(12):1205–13. doi: 10.4161/cbt.21353 ; PubMed Central PMCID: PMC3469478. [PMC free article] [PubMed]
7. Jiguet Jiglaire C, Baeza-Kallee N, Denicolai E, Barets D, Metellus P, Padovani L, et al. Ex vivo cultures of glioblastoma in three-dimensional hydrogel maintain the original tumor growth behavior and are suitable for preclinical drug and radiation sensitivity screening. Exp Cell Res. 2014;321(2):99–108. doi: 10.1016/j.yexcr.2013.12.010 . [PubMed]
8. Koshkin V, Ailles LE, Liu G, Krylov SN. Metabolic suppression of a drug resistant subpopulation in cancer spheroid cells. J Cell Biochem. 2015. doi: 10.1002/jcb.25247 . [PubMed]
9. Hickman JA, Graeser R, de Hoogt R, Vidic S, Brito C, Gutekunst M, et al. Three-dimensional models of cancer for pharmacology and cancer cell biology: capturing tumor complexity in vitro/ex vivo. Biotechnol J. 2014;9(9):1115–28. doi: 10.1002/biot.201300492 . [PubMed]
10. Hickman JA. Driving drugs to trial. Int Innov. 2013:80–2.
11. Thoma CR, Zimmermann M, Agarkova I, Kelm JM, Krek W. 3D cell culture systems modeling tumor growth determinants in cancer target discovery. Adv Drug Deliv Rev. 2014;69–70:29–41. doi: 10.1016/j.addr.2014.03.001 . [PubMed]
12. Hirschhaeuser F, Menne H, Dittfeld C, West J, Mueller-Klieser W, Kunz-Schughart LA. Multicellular tumor spheroids: an underestimated tool is catching up again. J Biotechnol. 2010;148(1):3–15. doi: 10.1016/j.jbiotec.2010.01.012 . [PubMed]
13. Piccinini F. AnaSP: a software suite for automatic image analysis of multicellular spheroids. Comput Methods Programs Biomed. 2015;119(1):43–52. doi: 10.1016/j.cmpb.2015.02.006 . [PubMed]
14. Chen W, Wong C, Vosburgh E, Levine AJ, Foran DJ, Xu EY. High-throughput image analysis of tumor spheroids: a user-friendly software application to measure the size of spheroids automatically and accurately. J Vis Exp. 2014;(89). doi: 10.3791/51639 ; PubMed Central PMCID: PMC4212916. [PMC free article] [PubMed]
15. Ke M-T, Fujimoto S, Imai T. SeeDB: a simple and morphology-preserving optical clearing agent for neuronal circuit reconstruction. Nat Neurosci. 2013;16:1154–61. doi: 10.1038/nn.3447 [PubMed]
16. Chung K, Wallace J, Kim SY, Kalyanasundaram S, Andalman AS, Davidson TJ, et al. Structural and molecular interrogation of intact biological systems. Nature. 2013;497(7449):332–7. doi: 10.1038/nature12107 ; PubMed Central PMCID: PMC4092167. [PMC free article] [PubMed]
17. Kuwajima T, Sitko AA, Bhansali P, Jurgens C, Guido W, Mason C. ClearT: a detergent- and solvent-free clearing method for neuronal and non-neuronal tissue. Development. 2013;140:1364–8. doi: 10.1242/dev.091844 [PMC free article] [PubMed]
18. Hama H, Kurokawa H, Kawano H, Ando R, Shimogori T, Noda H, et al. Scale: a chemical approach for fluorescence imaging. Nat Neurosci. 2011;14:1481–8. doi: 10.1038/nn.2928 [PubMed]
19. Helmchen F, Denk W. Deep tissue two-photon microscopy. Nat Methods. 2005;2(12):932–40. doi: 10.1038/nmeth818 . [PubMed]
20. Pampaloni F, Ansari N, Stelzer EH. High-resolution deep imaging of live cellular spheroids with light-sheet-based fluorescence microscopy. Cell Tissue Res. 2013;352(1):161–77. doi: 10.1007/s00441-013-1589-7 . [PubMed]
21. Tomer R, Khairy K, Amat F, Keller PJ. Quantitative high-speed imaging of entire developing embryos with simultaneous multiview light-sheet microscopy. Nat Methods. 2012;9(7):755–63. doi: 10.1038/nmeth.2062 . [PubMed]
22. Strasters KC. Quantitative Analysis in Confocal Image Cytometry. PhD Thesis. Delft: Delft University Press, 1994.

23. Strasters KC, van der Voort HTM, Geusebroek JM, Smeulders AWM. Fast attenuation correction in fluorescence confocal imaging: A recursive approach. Bioimaging. 1994;2:78–92.

24. Schindelin J, Arganda-Carreras I, Frise E, Kaynig V, Longair M, Pietzsch T, et al. Fiji: an open-source platform for biological-image analysis. Nat Methods. 2012;9(7):676–82. Epub 2012/06/30. doi: 10.1038/nmeth.2019 ; PubMed Central PMCID: PMCPMC3855844. [PMC free article] [PubMed]

25. de Chaumont F, Dallongeville S, Chenouard N, Herve N, Pop S, Provoost T, et al. Icy: an open bioimage informatics platform for extended reproducible research. Nat Methods. 2012;9(7):690–6. doi: 10.1038/nmeth.2075 . [PubMed]

26. Harma V, Schukov HP, Happonen A, Ahonen I, Virtanen J, Siitari H, et al. Quantification of dynamic morphological drug responses in 3D organotypic cell cultures by automated image analysis. PLoS One. 2014;9(5):e96426 doi: 10.1371/journal.pone.0096426 ; PubMed Central PMCID: PMC4014501. [PMC free article] [PubMed]

27. Lou X, Kang M, Xenopoulos P, Munoz-Descalzo S, Hadjantonakis AK. A rapid and efficient 2D/3D nuclear segmentation method for analysis of early mouse embryo and stem cell image data. Stem Cell Reports. 2014;2(3):382–97. doi: 10.1016/j.stemcr.2014.01.010 ; PubMed Central PMCID: PMC3964288. [PMC free article] [PubMed]

28. Hodneland E, Kogel T, Frei DM, Gerdes HH, Lundervold A. CellSegm—a MATLAB toolbox for high-throughput 3D cell segmentation. Source Code Biol Med. 2013;8(1):16 doi: 10.1186/1751-0473-8-16 ; PubMed Central PMCID: PMC3850890. [PMC free article] [PubMed]

29. Gue M, Messaoudi C, Sun JS, Boudier T. Smart 3D-FISH: automation of distance analysis in nuclei of interphase cells by image processing. Cytometry A. 2005;67(1):18–26. doi: 10.1002/cyto.a.20170 . [PubMed]

30. Trinh A, Rye IH, Almendro V, Helland A, Russnes HG, Markowetz F. GoIFISH: a system for the quantification of single cell heterogeneity from IFISH images. Genome Biol. 2014;15(8):442 doi: 10.1186/s13059-014-0442-y ; PubMed Central PMCID: PMCPMC4167144. [PMC free article] [PubMed]

31. Parker DL, Chapman BE, Roberts JA, Alexander AL, Tsuruda JS. Enhanced image detail using continuity in the MIP Z-buffer: Applications to magnetic resonance angiography. J Magn Reson Imaging. 2000;11(4):378–88. doi: 10.1002/(SICI)1522-2586(200004)11:4<378::AID-JMRI5>3.0.CO;2-# [PubMed]

32. Madar S, Brosh R, Buganim Y, Ezra O, Goldstein I, Solomon H, et al. Modulated expression of WFDC1 during carcinogenesis and cellular senescence. Carcinogenesis. 2009;30(1):20–7. doi: 10.1093/carcin/bgn232 ; PubMed Central PMCID: PMC2639035. [PMC free article] [PubMed]

33. Wilson EB. Probable Inference, the Law of Succession, and Statistical Inference. J Am Stat Assoc. 1927;22(158):209–12. doi: 10.2307/2276774

CHAPTER 20 ANALYSIS OF GAP GENE REGULATION IN A 3D ORGANISM-SCALE MODEL OF THE DROSOPHILA MELANOGASTER EMBRYO

James B. Hengenius[1], Michael Gribskov[1], Ann E. Rundell[2], Charless C. Fowlkes[3], David M. Umulis[4]

[1] Department of Biological Sciences, Purdue University, West Lafayette, Indiana, United States of America,

[2] Department of Biomedical Engineering, Purdue University,West Lafayette, Indiana, United States of America,

[3] Department of Computer Science, University of California Irvine, Irvine, California, United States of America,

[4] Department of Agricultural and Biological Engineering, Purdue University, West Lafayette, Indiana, United States of America

ABSTRACT

The axial bodyplan of Drosophila melanogaster is determined during a process called morphogenesis. Shortly after fertilization, maternal bicoid mRNA is translated into Bicoid (Bcd). This protein establishes a spatially graded morphogen distribution along the anterior-posterior (AP) axis of the embryo. Bcd initiates AP axis determination by triggering expression of gap genes that subsequently regulate each other's expression to form a precisely controlled spatial distribution of gene products. Reaction-diffusion models of gap gene expression on a 1D domain have previously been used to infer complex genetic regulatory network (GRN) interactions by optimizing model parameters with respect to 1D gap gene expression data. Here we construct a finite element reaction-diffusion model with a realistic 3D geometry fit to full 3D gap gene expression data. Though gap gene products exhibit dorsal-ventral asymmetries, we discover that previously inferred gap GRNs yield qualitatively correct AP distributions on the 3D domain only when DV-symmetric initial conditions are employed. Model patterning loses qualitative agreement with experimental data when we incorporate a realistic DV-asymmetric distribution of Bcd. Further, we find that geometry alone is insufficient to account for DV-asymmetries in the final gap gene distribution. Additional GRN optimization confirms that the 3D model remains sensitive to GRN parameter perturbations. Finally, we find that incorporation of 3D data in simulation and optimization does not constrain the search space or improve optimization results.

INTRODUCTION

Embryonic development in Drosophila melanogaster is initiated with the formation of spatial morphogen distributions in the early embryo. The dynamic spatial patterns of diffusive morphogens encode information which specifies organism-scale development [1], [2]. Nonuniform initial spatial distributions of maternally deposited morphogen mRNAs, coupled with diffusion, decay, and complex genetic regulatory interactions, give rise to finer patterns that subdivide the dorsal-ventral (DV) [3]–[5] and anterior-posterior (AP) axes [2], [6] into distinct developmental regions.

The gap gene system is one of the most widely studied morphogen systems in Drosophila and is involved in delineation of boundaries of gene expression within the AP body plan [2]. AP patterning events begin approximately one hour post-fertilization. This patterning foreshadows the subsequent segmentation of the embryo [1], [2], [6]–[9]. During early development, the embryo is a polynucleated syncytium; most

Citation: Hengenius JB, Gribskov M, Rundell AE, Fowlkes CC, Umulis DM (2011) Analysis of Gap Gene Regulation in a 3D Organism-Scale Model of the Drosophila melanogaster Embryo. PLoS ONE 6(11): e26797. doi:10.1371/journal.pone.0026797

nuclei are arrayed in a thin layer near the surface of the embryo. Due in part to a cytoplasmic viscosity gradient common to insect embryos [10], morphogens (here, gap gene products) are thought to diffuse freely through periplasm near the embryonic surface and less substantially through the interior. Here, they regulate transcription within the periplasmic nuclei [2]. The process is initiated by the gene products of maternally-deposited, spatially-heterogeneous bicoid (Bcd), caudal (Cad), and nanos mRNAs [2], [11],[12]. Maternally deposited RNA species regulate expression of the gap genes: Hunchback (Hb, with a maternal mRNA contribution), Giant (Gt), Tailless (Tll), Krüppel (Kr), and Knirps (Kni) (see **Fig. 1a**) [11], [13], [14]. The gap genes, in turn, regulate the pair-rule genes which in turn control segment-polarity genes and embryonic segmentation [1], [2], [6], [15].

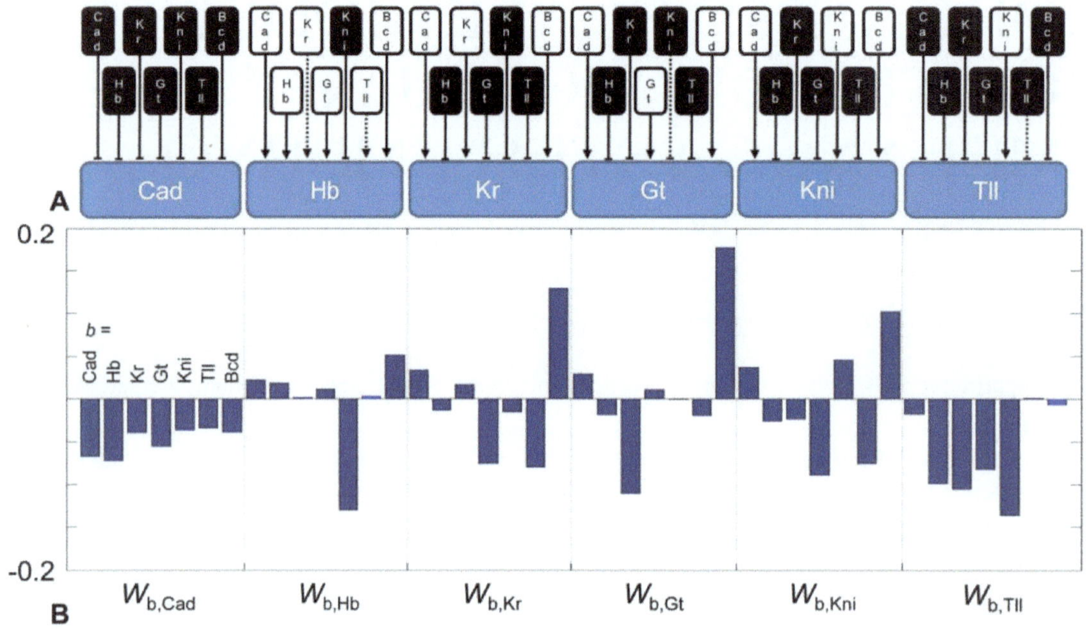

Figure 1: Gap gene genetic regulatory network. The model representation of the gap gene network. The network topology in (**A**) represents negative (black box, flat line) and positive (white box, arrowhead line) regulatory effects on each target gene (blue). Dashed lines represent near-zero regulatory inputs that may be negligible. This qualitative topology is quantified in (**B**) as a set of genetic regulatory network (GRN) weight parameters $w_{b,a}$, the influence of gene b on gene a. From left to right, each set of seven inputs represent Cad, Gt, Hb, Kni, Kr, Tll, and Bcd. Each cluster of seven interactions represents a target gene Cad, Gt, Hb, Kni, Kr, and Tll. http://dx.doi.org/10.1371/journal.pone.0026797.g001

Most inferences regarding the gap genetic regulatory network (GRN) have been drawn from mutant and gene dosage studies in which the effects on morphology, gap, pair-rule, or segment polarity genes are observed [12], [16]–[36]. While these experiments are informative, it is difficult to unambiguously derive genetic regulatory interactions from such data; phenotypic changes may arise via direct action of the perturbed gene or via downstream targets of that gene. In contrast, Reinitz, Jaeger, and others applied a reverse engineering approach using dynamic wild-type data. Computational studies have modeled gap gene patterning using 1D partial differential equation (PDE) systems or ordinary differential equation systems that include an implicit approximation to the PDE [13], [14], [37]–[40] and logical rule sets [41]. These models represent the lateral trunk region of the Drosophila embryo along the AP axis, typically omitting the anterior and posterior end regions (with the exception of [40]). GRN topology is represented by a regulatory weight matrix and gene expression is modeled by a transfer function that sums the regulatory impact of each regulatory protein on expression of the others (see **Fig. 1b**) [13]. Model-driven inferences about GRN topology (i.e., inferring whether and to what degree one morphogen regulates expression of other morphogens) have been obtained by inverse modeling: optimizing the regulatory weight matrix against experimental gap gene expression data in hopes of recovering "true" GRNs [14], [42]–[45]. Findings have been mixed. Biological systems are thought to be robust (and thus insensitive) to perturbations. Some GRN parameters are highly sensitive while considerable uncertainty is associated with others [44],[45].

Previous 1D PDE models have been used effectively to infer network topology and investigate patterning regulation [13], [14], [42], [45], [46], but there are some questions that are better investigated using a full 3D spatial patterning model. Many important 3D effects, including variable diffusive path lengths around the embryo surface and optimization against 3D data, cannot be observed in a 1D model domain. DV asymmetries in gap gene distribution and possible interactions between the gap gene system and DV patterning systems are also neglected. Further, these 3D data may serve to constrain GRN optimization and inference.

Quantitative spatiotemporal atlases of gene expression data in the Drosophila embryo have been published and provide the starting point for quantitative analysis. [47]–[49]. The atlas includes measurement of gap gene expression collected from hundreds of individual embryos and registered onto a standardized 3D mesh of nuclei coordinates using pair-rule gene expression patterns as fiduciary points (mesh coordinates available in . This composite VirtualEmbryo (VE) is a logical starting point for the development of 3D embryonic GRN models. It provides a ready-made embryonic geometry for full spatial PDE representations of the gap gene system. It also contains quantitative expression data against which we can optimize model parameters (and thus infer GRNs).

Using the VE data, we evaluate the impact of 1D model assumptions, conversion from 1D to 3D geometries, and incorporation of fully 3D protein distribution data in model simulation. Herein we reconstruct the 1D gap gene model of Jaeger et al. [13] using the finite element method (FEM) and extend it to the 3D VE geometry. The 1D model of Jaeger et al. [13], M^{1D-P} (see Table 1 for model definitions), is refit to lateral expression data from the VE. We then extend the 1D model PDEs to the full 3D embryonic geometry described by Fowlkes et al.and compare GRNs inferred from 1D and 3D models. Though 1D models focus on the lateral AP axis in 1D simulations, gap genes are not uniformly distributed along the DV axis. Coupled with the 3D geometry, DV asymmetries in initial conditions may encode positional information partially responsible for the observed AP patterning. As a preliminary exploration of asymmetric DV effects in an embryonic geometry, we evaluate the model using DV-asymmetric Bcd concentration data from thirteen embryos compiled in the VE.

Table 1: Model Variants and Corresponding Optimal Parameter Sets. http://dx.doi.org/10.1371/journal.pone.0026797.t001

Model	Geometry	Initial Conditions	Optimal GRN Parameters* (M^{Model}_{Data})
M^{1D-P}	1D domain representing partial 35%–92% AP axis	$Gt_0 = Kni_0 = Kr_0 = Tll_0 = 0$; Bcd_{SS}, Hb_0, Cad_0 values in Jaeger *et al.* [13]	*N/A* (evaluated with Jaeger *et al.*'s reported GRN, P^0)
M^{1D-F}	1D domain representing full 0%–100% AP axis	$Gt_0 = Kni_0 = Kr_0 = Tll_0 = 0$; Bcd_{SS}, Hb_0, Cad_0 values in Jaeger *et al.*	P^{1D-F}_{1D-P} (fit to Jaeger's model output); P^{1D-F}_{VE} (fit to VirtualEmbryo data)
M^{3D}_{Bcd1D}	VE 3D domain	$Gt_0 = Kni_0 = Kr_0 = Tll_0 = 0$; Bcd_{SS}, Hb_0, Cad_0 values in Jaeger *et al.* projected about AP axis (see **Fig. 3b–d**)	P^{3D}_{Bcd1D} (evaluated with GRN P^{1D-F}_{VE})
M^{3D}_{Bcd3D}	VE 3D domain	$Gt_0 = Kni_0 = Kr_0 = Tll_0 = 0$ Hb_0, Cad_0 values in Jaeger *et al.* projected about AP axis. Bcd_{SS} interpolated from VE data (**Fig. 3e**).	P^{3D}_{Bcd3D} (evaluated with GRN P^{1D-F}_{VE})
$M^{3D}_{Bcd3D-S}$	VE 3D domain	$Gt_0 = Kni_0 = Kr_0 = Tll_0 = 0$; Hb_0, Cad_0 values in Jaeger *et al.* projected about AP axis. $[Bcd]_{SS}$ interpolated from VE data and smoothed (**Fig. 3f**).	$P^{3D}_{Bcd3D-S}$

In addition to GRN sensitivities highlighted by previous 1D analyses [14], [38], [39], [44], [50], we find that the 3D model exhibits fragility with respect to the shape of maternal gradients: GRNs which were inferred by optimization of 1D models showed similar gap gene patterning when applied to 3D models with DV-symmetric Bcd. However, these GRNs gave rise to qualitatively different patterns in DV-asymmetric models. These realistic Bcd gradient models also captured some of the DV-asymmetries in gap gene patterning. The 3D models were also sensitive to small perturbations in GRNs; regulatory networks which were qualitatively similar (i.e., all network interactions maintained the same excitatory or inhibitory relationships and differed only by small changes in magnitude) led to qualitatively different gap gene patterns. Refitting of the DV-asymmetric 3D model to VE data produced a GRN which was similar to 1D GRNs but which produced an improved fit.

Another question addressed in this study is whether inclusion of 3D data improves optimization by the inclusion of additional constraints without increasing the degrees of freedom in the model. Unexpectedly, we found that the incorporation of additional 3D information in the form of a realistic DV-asymmetric Bcd worsened the error between optimized 3D models and data. This suggests the involvement of additional regulators in the formation of DV-asymmetries and indicates a direction for future modeling studies.

RESULTS

One-Dimensional Model Analysis

Before analyzing the effects of embryonic geometry and DV-asymmetric positional information, we reimplemented the 1D model of Jaeger et al. using the finite element method. In this work we denote model variants with M; superscripts represent model domains and subscripts signify initial conditions if multiple initial conditions are used. The 1D model of Jaeger et al. is called M^{1D-P} (using a 1D domain representing a partial AP length of 35%–92%; full model nomenclature available in **Table 1**). We verified M^{1D-P} against Jaeger et al. 's simulated output. Whereas the original model limited gene expression to a finite number of discrete nuclear coordinates along the 35–92% region of the embryonic AP axis, the FEM approximates a continuous solution to these equations along this domain. Discrete versus continuous model comparisons by Gursky et al. suggest that embryonic patterning is not strongly coupled to nuclear position and that continuous models are comparable to discrete models of gene expression [51]. Our results agree with this finding. FEM simulations produce model output comparable to Jaeger et al. 's discrete 1D model (**Fig. 2a**, dashed line, cf. Figure S20 in [13]).

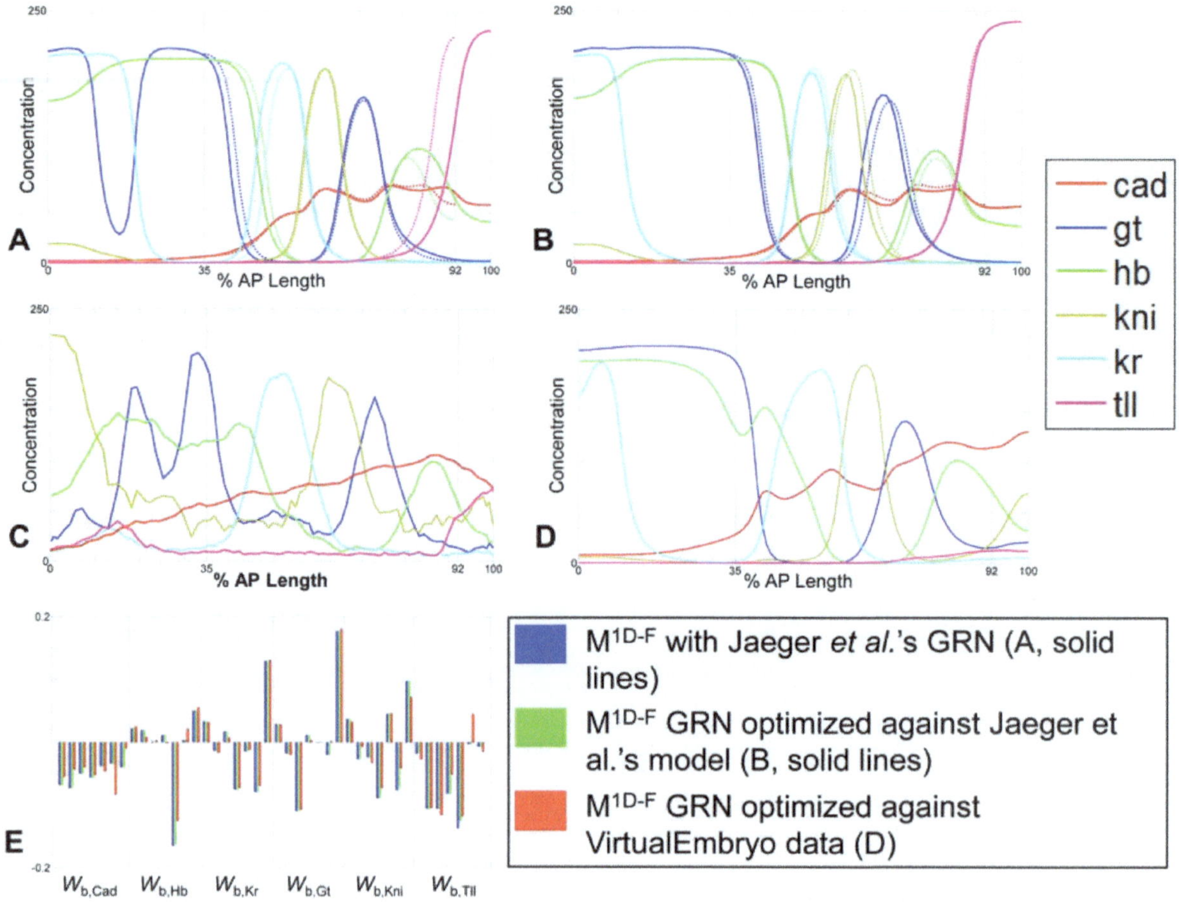

Figure 2: One-dimensional model results. Model output was simulated over a 0–100% AP length domain using the optimal GRN reported by Jaeger et al. Solid vertical lines represent the original model boundaries, not used in this simulation. (**A**) M^{1D-F} (solid lines) shows qualitative agreement with the Jaeger model M^{1D-P} (dashed lines) in the 35–92% AP range, but shows discrepancies at either end of the domain due to the movement of boundaries; all species displayed at t=70 min. (**B**) The best-fit GRN from Jaeger et al. was locally optimized to improve the agreement of the 0–100% AP length, model M^{1D-F} (solid lines), and the original Jaeger et al. original model (M^{1D-P} dashed lines); all species displayed at t=70 min. (**C**) VE protein data for Gt, Hb, Kni, Kr at t=70 min; VE mRNA data for Tll at t=70 min; protein data from Jaeger et al. for Cad at t=56 min. (**D**) Model output (M^{1D-F}) was also optimized against VE data (RMSE=13.992); Gt, Hb, Kni, Kr, Tll at t=70 min; Cad at t=56 min. Despite modest improvements in model agreement in the 35% and 92% region (**C–D**), the resulting changes in parameter values were small. (**E**) Optimized parameter magnitudes vary but signs remained the same in most cases (blue - P^0; green - P^{1D-F}_{1D-P}; red - P^{1D-F}_{VE}). http://dx.doi.org/10.1371/journal.pone.0026797.g002

Though M^{1D-P} recapitulated previous results when simulated in the region from 35–92% on the AP axis, we sought to determine whether moving the boundaries to the embryo ends perturbed gap gene patterning in the trunk region. It is unclear a priori how modification of boundary conditions might impact the model output, because the selection of boundaries at 35% and 92% in earlier work appears to coincide with either maxima or minima of gap gene distributions; at these positions, spatial derivatives are near zero and diffusive flux may be negligible. Using no-flux boundaries at 0% and 100% EL, coupled with the parameters and initial conditions specified in the original model [13], we evaluated M^{1D-P} and M^{1D-F} using the GRN parameters P^0 reported by Jaeger et al. [13]. Herein, parameter sets are denoted Pand super- and subscripts have model-specific meanings. The simulated patterns from the original 35–92% AP and the 0–100% AP domains are shown in **Figure 2a**'s dashed and solid lines, respectively.

Pronounced shifts in Tll and Kr distributions, coupled with the qualitative change in the anterior Gt distribution, demonstrate the role boundary conditions play in the in the distribution of gap gen products for a given set of parameters. Though the output of M^{1D-P} qualitatively resembles the expression data collected previously when evaluated at P^0 [13], these findings suggest that M^{1D-P}'s agreement with data arises from a combination of the inferred GRN and the domain's boundary conditions. Thus, the internal zero-flux boundary conditions used in previous models may bias GRN inference. To evaluate the impact of boundary placement on GRN inference, we performed a numerical gradient descent search of the parameter space to minimize the root mean squared error (RMSE) between M^{1D-F} and M^{1D-P} (represented by the dashed line in **Figure 2a**). The search was initialized with the previously reported optimalP^0. The result of this search, optimized GRN P^{1D-F}_{1D-P} (superscript denotes the model being optimized and subscript denotes data with which the model is optimized), is illustrated in **Figure 2b**. Here, the output of M^{1D-P} represents extant models' with internal no-flux boundaries.

Though domain boundary placement affects the banding pattern, **Figure 2b** suggests that these constraints have a limited effect on GRN inference. Optimizing the GRN parameters of M^{1D-F} to fit the original model output recovered a quantitatively similar patterning within the 35–92% AP length of the full 1D domain. Additionally, the optimized GRN P^{1D-F}_{1D-P} was qualitatively similar to P^0 (e.g., though optimized parameters underwent small changes in magnitude, all parameters maintained the same sign, **Fig. 2e**).

To facilitate a direct comparison between 1D and 3D models presented herein, we first evaluated the goodness-of-fit between the 35–92% AP (M^{1D-P}) and full AP domain (M^{1D-F}) 1D models using VE data. When possible, we use protein expression data from the VE: Gt, Hb, Kni, and Kr protein data is available across six equidistant time points spanning 50 minutes. Tll protein data is unavailable and we use Tll mRNA data as a surrogate for the protein distributions. Cad protein distributions are also unavailable in the VE; we substitute 1D Cad data from Jaeger et al. [13] that spans 45 minutes with seven time points. Because the 1D model domains represent the lateral region of the full embryo, we extracted expression data from this region of the VE (**Fig. 2c**). We performed a constrained search of GRN parameters initialized at P^{1D-F}_{1D-P} to yield an optimized GRN P^{1D-F}_{VE} (subscript VE denotes VE training data). The resulting model output and a comparison of model parameters are shown in**Figures 2d–e**.

Though M^{1D-F} was capable of recovering the output of M^{1D-P} (with parameter set P^{1D-F}_{1D-P}) and VE data (P^{1D-F}_{VE}) within the 35–92% AP axis, poor fit to VE data persisted outside of this region. The 0–35% and 92–100% AP regions exhibit qualitative disagreement with VE data in these regions consistent with the biological requirement for additional head and tail patterning genes (**Fig. 2c–d**).

Three-Dimensional Model Analysis

Beginning with the GRN optimized on the full 1D domain, we extended the model to a 3D domain using the geometry in the VE. This was performed by implementing the system of PDEs on a 2D surface "wrapped" around the VE geometry. We used this model to evaluate the effects of both model geometry and DV-asymmetric initial conditions on model output.

To assess the effects of model geometry on patterning independent of initial conditions, the model was first simulated using DV-symmetric initial conditions (M^{3D}_{Bcd1D}): Bcd, Hb, and Cad distributions at time zero were obtained from the original 1D model and projected around the surface of the embryo (**Fig. 3a–d**). Evaluated at the previously inferred optimal 1D GRN (P^{1D-F}_{VE}), model M^{3D}_{Bcd1D} yielded patterning qualitatively similar to the full length 1D model output (**Fig. 4a–g, column 2**). To confirm our derivation of the diffusion constants (seeMethods) and rule out unintentional adjustment of the diffusive length constant ($\sqrt{D_a/\lambda_a}$), we performed a continuation of diffusion constants while holding decay parameters (λ_a) values constant . While band overlap does vary with diffusion constants, they are quantitatively similar. Interestingly, symmetric Bcd models appear robust against increased diffusion while increased diffusion disrupted patterning in asymmetric Bcd models. The pattern formation timecourse for Bcd-symmetric patterning is animated in

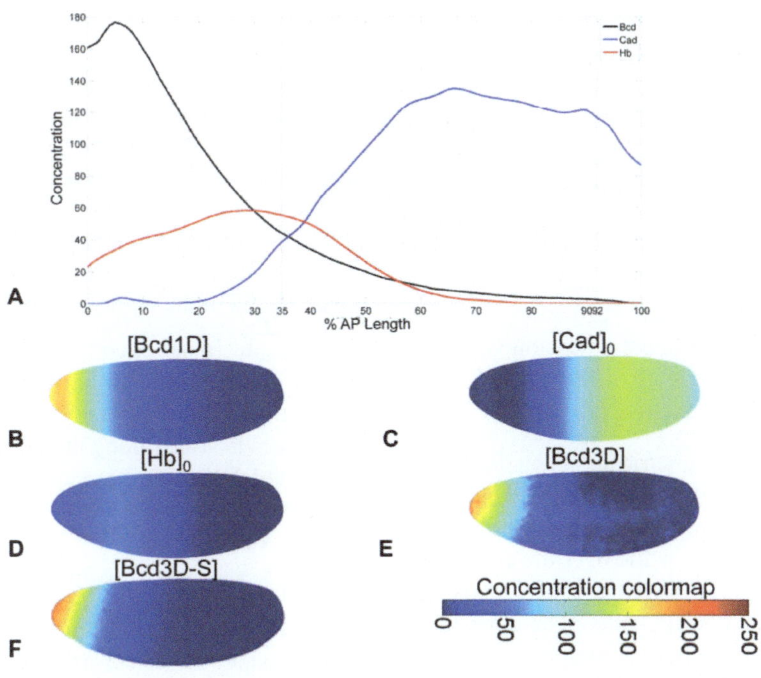

Figure 3: 1D and 3D initial conditions. Initial conditions in various models. (**A**) 1D model initial conditions, reported by Jaeger et al., and used in models M^{1D-P} and M^{1D-F}. (**B**) 1D initial conditions were mapped onto the 3D embryonic geometry (M^{3D}_{Bcd1D}). (**C**), 1D initial Cad protein distribution, (**D**) 1D initial Hb protein distribution. Subsequent models incorporated (**E**) DV-asymmetric interpolated [Bcd] distribution (M^{3D}_{Bcd3D}) or (**F**) smoothed DV-asymmetric interpolated [Bcd] distribution ($M^{3D}_{Bcd3D-S}$). http://dx.doi.org/10.1371/journal.pone.0026797.g003

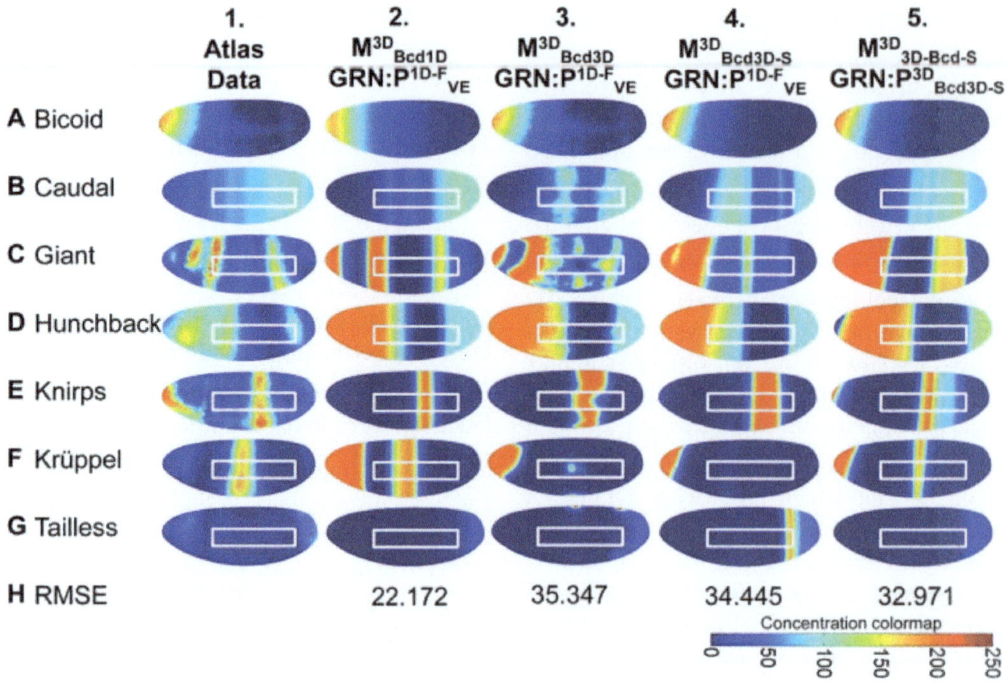

Figure 4. Three-dimensional model results. Simulation results in the 3D model. (**A–H**) Lateral view of VE geometry is shown in rows**A–G** (Gt, Hb, Kni, Kr, Tll at t=70 min, Cad at t=56 min); row **H** displays RMSE difference between model and VE data summed with all-time points. Column 1 shows scaled VE data. Column 2 displays output from M^{3D}_{Bcd1D} evaluated with GRN P^{1D-F}_{VE}. Column 3 contains output from M^{3D}_{Bcd3D} incorporating DV-asymmetric Bcd data and GRN P^{1D-F}_{VE}; Column 4 illustrates the effect of the smoothed Bcd interpolant in $M^{3D}_{Bcd3D-S}$ while considering the same GRN P^{1D-F}_{VE}. Column 5 displays output from $M^{3D}_{Bcd3D-S}$ with reoptimized parameters $P^{3D}_{Bcd3D-S}$. White boxes indicate the lateral areas where Jaeger et al. optimized their 1D model. Animations of pattern development are available for column 2 (M^{3D}_{Bcd1D},) and column 5 ($M^{3D}_{Bcd3D-S}$,) in the supplementary material. http://dx.doi.org/10.1371/journal.pone.0026797.g004

Though there are some DV-asymmetries present in the output (e.g., slight curvature of the anterior Gt stripe), 1D versus 3D domain geometry alone has only a modest impact on DV patterning of gap genes. This suggests that the pronounced DV-asymmetries present in the final distributions of the proteins at the onset of nuclear division 14 (**Fig. 4, column 1**) stem from other sources. We consider the effect of spatial information encoded in initial DV asymmetries of protein distributions. The coupling of gap gene regulation with DV-patterning systems [5], [52], [53] is another possibility.

Effect of Dorsal-Ventral Asymmetric Bcd

To evaluate the impact of DV-asymmetric inputs on the model, we modified the steady-state Bcd distribution shown in **Figure 3b** to incorporate a realistic DV gradient (**Fig. 3e**). Unlike other morphogens, the Bcd distribution is static over the entire time course of model simulation. This allowed us to create a single interpolant of VE Bcd data and use it as a model input for all 70 minutes of the simulation. The pattern formation timecourse for Bcd-asymmetric patterning is animated in

Evaluated at the optimal 1D GRN P^{1D-F}_{VE}, model M^{3D}_{Bcd1D} produces patterning that is radically different from DV-symmetric 1D (M^{1D-F}) and 3D (M^{3D}_{Bcd1D}) models (**Figs. 4a–g, column 3**). The most striking example of this is the Kr model output; whereas Kr forms a full band in vivo, M^{3D}_{Bcd3D} lacks full lateral expression of Kr and has an anomalous region of expression at the anterior end of the embryo (**Fig. 4f, column 3**). Similarly, the simulated Hb concentrations remain above observed levels (**Fig. 4d, column 3**). The posterior Hb band also shifts to the posterior end of the embryo. Gt exhibits qualitative disagreement with the VE data; whereas anterior Gt expression is observed only in a limited dorsal region of the embryo (**Fig. 4c, column 1**), the anterior of the M^{3D}_{Bcd3D} is saturated with Gt (**Fig. 4c, column 3**). Further, though the experimentally observed posterior Gt band (**Fig. 4c, column 1**) is predicted by simulation, it exhibits unusual differences in width along the DV axis. As in previous versions of the model, the best agreement between model and data was found in the lateral 35–92% AP region (**Fig. 4b–g, column 3 white boxes**).

The cell-to-cell variability in patterning found for many simulated proteins (e.g., Gt, Cad, and Kni) in M^{3D}_{Bcd3D} led us to consider the effect of noise in the VE Bcd distribution. Diffusion of Bcd may serve to smooth this variation in vivo; our use of a single static Bcd interpolant in M^{3D}_{Bcd3D} leads to an artificial persistence of the noise found in VE data (**Fig. 3e**). To test for and remove this artificial condition, we created a regularized version of the Bcd interpolant (**Fig. 3f**). This was constructed by building a simple source diffusion decay (SDD) reaction-diffusion model of Bcd alone [18]. This SDD model was fit to VE data and the steady-state solution was used as the smoothed Bcd interpolant. The model incorporating regularized Bcd, $M^{3D}_{Bcd3D-S}$, did not show significant improvement over M^{3D}_{Bcd3D} when evaluated with P^{1D-F}_{VE} (**Fig. 4 a–g, column 4**). However, it did eliminate the cell-to-cell variability present in M^{3D}_{Bcd3D}. The model's artificial sensitivity to Bcd noise was especially evident in Gt (**Fig. 4c, columns 3–4**). Two anterior and one posterior Gt bands in M^{3D}_{Bcd3D} changed width and AP position after smoothing of Bcd. This result suggests that while diffusion may serve as a buffer against transient stochastic variations in protein expression and local concentration (in agreement with stochastic simulation [54]), sustained cell-to-cell variability has the potential to disrupt patterning.

Having observed that a GRN inferred on the 1D domain (and lacking DV asymmetries) produces a qualitatively incorrect fit compared to 3D data, we attempted to optimize the GRN with Matlab's constrained search function fmincon() initialized at P^{1D-F}_{VE} (previously used to estimate P^{1D-F}_{1D-P} and P^{1D-F}_{VE}). This approach failed to reduce model error. Fomekong-Nanfack et al. demonstrated that 1D gap gene systems are amenable to optimization by evolutionary algorithms [45]. We therefore employed a genetic algorithm (GA) to more broadly survey the parameter space. Do to computational cost, we used a small population size of 20 genomes to search the GRN parameter space (42 parameters), the GA identified an optimal GRN for $M^{3D}_{Bcd3D-S}$. The resulting GRN, $P^{3D}_{Bcd3D-S}$, led to a reduction in model error and a modest qualitative improvement with respect to 3D data (**Fig. 4, column 5**). The lateral Kr band missing from the 1D-inferred GRN P^{1D-F}_{VE} (**Fig. 4f, columns 3–4**) is restored (**Fig. 4f, column 5**), though it is not as wide as the experimentally observed band. Tll no longer shows relative over-expression at the posterior end of the embryo (**Fig. 4g, column 5**). Hb continues to exhibit relative over-expression at the anterior end of the embryo, though its posterior band is shifted closer to its correct position (**Fig. 4d, column 5**). Similarly, the anterior distribution of Gt extends beyond the dorsal region observed in the VE (**Fig. 4c, column 1**). However, its posterior band is now located correctly in **Figure 4c, column 5** (though it is wider than the observed protein band). Beyond differences in concentration of individual proteins, DV-asymmetric Bcd causes a notable qualitative difference in the AP position and emergence of protein bands. Compared to the M^{3D}_{Bcd1D} (**Fig. 4, column 2**), the DV-asymmetric GRNs (**Fig. 4, columns 4–5**) exhibit DV-asymmetries in their output. For example, the dorsal terminus of the anterior Gt band is posterior to its ventral terminus; it is splayed toward the anterior. This behavior agrees with observed data in the anterior half of the embryo, but the expected DV curvature is either absent (posterior Hb **Fig. 4d, column 5**) or inverted in the posterior half of the embryo. For example, Kni, whose dorsal terminus should exhibit posterior-splaying (**Fig. 4e, column 5**), is inverted. This DV curvature corresponds in direction to the DV asymmetry of Bcd. The absence of reversed splaying in the output in the posterior portion of the model (though present in the data) suggests that the model may be lacking additional posterior determinant(s) affecting the gap gene system.

In the 3D regime, $M^{3D}_{Bcd3D-S}$ demonstrated considerable sensitivity to small changes in GRN parameter values. The model was simulated after adding normally distributed noise scaled by each parameter value, p_i, across a range of magnitudes (sample model output in **Fig. 5**). The model gives output qualitatively similar to the optimal GRN $P^{3D}_{Bcd3D-S}$ only when parameter noise is low (e.g., 0.1% p_i in **Fig. 5, column 1**). All other simulations, with noise terms of 1%p_i and higher, yielded drastically and qualitatively different outputs.

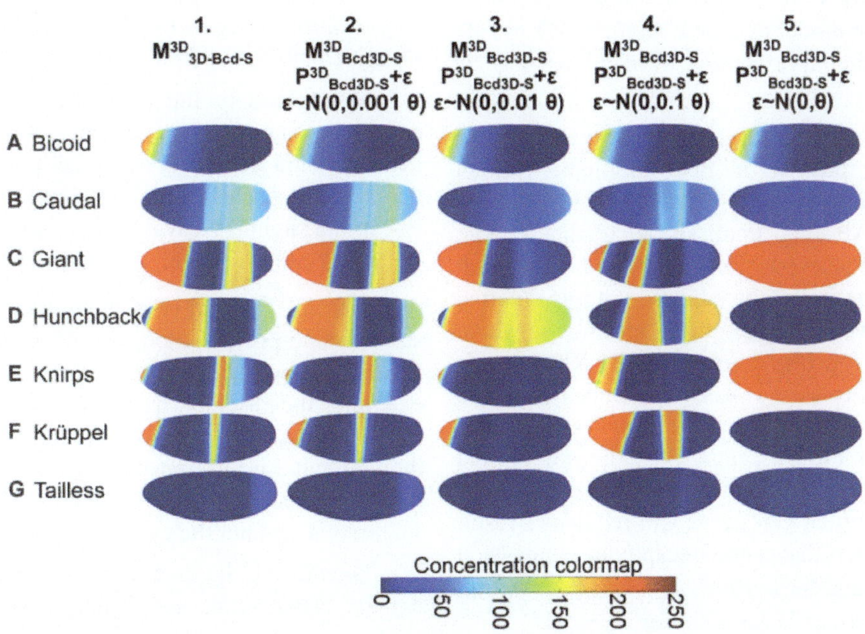

Figure 5: Model $M^{3D}_{Bcd3D-S}$ is not robust to noise in GRN parameters. Parametric noise alters model output. Lateral view of VE geometry for all genes is shown in rows **A–G** (all outputs at t=70 min). Each column displays $M^{3D}_{Bcd3D-S}$ output at t=70 min evaluated with

GRN $P^{3D}_{Bcd3D-S}$. Columns 2–5 represent randomly chosen sample output when a normally distributed noise vector ε is added to the GRN parameter set (denoted θ). ε has mean of 0 and variance that scales with θ. http://dx.doi.org/10.1371/journal.pone.0026797.g005

>In summary, the GRNs we inferred in this study are qualitatively similar: magnitudes of parameters vary by approximately 10% and parameter sign stays the same in all but a few low-magnitude parameters. A notable exception is the regulatory parameter for the Kni→Tll interaction; here the sign of the parameter (and thus the regulatory relationship) is reversed. However, we acknowledge that the treatment of Tll as a state variable under gap gene regulation is artificial and this biological relevance of this observation is questionable. Optimization leaves most regulatory parameters with the same sign and changes only the magnitudes, and those regulatory weights which change sign have small magnitudes (i.e., small regulatory effects). The use of a global search method (GA) to optimize $M^{3D}_{Bcd3D-S}$ did not recover a superior GRN that differed qualitatively from the original P^0.

DISCUSSION

The understanding of Drosophila developmental gene regulation has benefited from advances in quantitative modeling of gene regulation. However, existing PDE models of AP patterning have been limited to 1D approximations of the 3D geometry. By extending a model of gap gene regulation to a 3D embryonic geometry and adding realistic DV-asymmetry to upstream maternal Bcd, this work allows us to pose new questions about the effects of embryonic shape and DV gradients on gap gene patterning. Jaeger et al. 's 2004 model has been succeeded by more recent models of gap gene development incorporating additional regulatory inputs[37]–[39], [46], [55]–[58]. However, recent models of AP patterning retain partial domains (e.g., 35%–92% AP) with internal no-flux boundary conditions and use regulatory schema similar to**eqns 1–3** (see **Methods**) to represent GRNs. We chose the Jaeger et al. 's 2004 model as a case study in 1D vs. 3D modeling because it is the representative of many existing 1D models.

Before comparing 1D and 3D geometries, we examined the effect of boundary position in PDE solutions. Though embryos do not contain physical barriers to diffusion at 35% and 92% of the AP axis, small spatial gradients (**Fig. 2a, dashed lines**) at those positions suggested that small diffusive flux would minimize the effects of these internal boundaries. However, we found that the system was sensitive to boundary placement (cf. **Fig. 2a, solid lines**). Though this finding indicates the importance of using biologically realistic boundary conditions (i.e., no-flux boundaries at 0% and 100% AP), the simulations in **Figure 2** also illustrate our limited representation of regulation beyond the 35%–92% trunk region: Omission of terminal gap genes and regulators result in optimized parameter sets that cannot recapitulate expression patterns from 0%–35% and 92%–100% AP in M^{1D-F} (**Fig. 2a,c**). Optimization to correct the boundary artifacts (M^{1D-F} with P^{1D-F}_{1D-P}) likewise fail to improve agreement with data outside of the 35%–92% region (**Fig. 2b**). The inclusion of terminal gap genes such as Huckebein in 1D gap gene models [37] provides a basis for extension to full 100% AP 1D and 3D models, though inclusion of Huckebein in a recent 3D modeling study yielded only modest improvements in overall cost and qualitative agreement at the AP extrema [59].

Prior analyses demonstrated the sensitivity of gap gene models to GRN parameter values [14],[43], [44] and examination of boundary conditions support this finding: GRN parameter optimization corrected boundary artifacts with extremely small changes to parameter values (**Fig. 2e**). Optimization against VE data produced similar small changes in GRN parameters (**Fig. 2e**). The GRN sensitivity of 1D models M^{1D-P} and M^{1D-F} was also found in 3D models. collects all parameter values and reports the standard deviation for each parameter across 1D and 3D model optimizations. Parameter $w_{Gt,Bcd}$ exhibits the highest deviation across models with a standard deviation of 0.05, but this represents only 13% of the total parameter range ([−0.2,0.2]). These small changes in GRN parameters do more than shift protein band location as observed in **Figure 2**; they are capable of effecting qualitative patterning changes (e.g., changing the number of protein bands present on the embryo). For example, the transition from P^{1D-F}_{VE} to $P^{3D}_{Bcd3D-S}$ in model $M^{3D}_{Bcd3D-S}$ leads to the loss of a posterior Gt band and the creation of a posterior Kr band (**Fig. 4c,f, columns 4–5**). **Figure 5**shows randomly selected sample model outputs at t=70 min with increasing levels of normally distributed noised added to the GRN parameter vector. One percent noise was sufficient to induce qualitatively different banding patterns on the 3D geometry. The qualitative changes in patterning for all but the smallest levels of noise confirm the observations of parameter sensitivity in 1D and 3D models. The extreme sensitivity of model outputs to small changes in GRN parameters challenges analyses of GRN evolution positing phenotypically robust fitness landscapes [60]–[62]. Unfortunately, the computational expense of PDE models prevented an exhaustive exploration of the GRN parameter space and corresponding approximation of a fitness landscape. The fragility of the gap gene system to GRN perturbations bears further study, especially in its contrast to prevailing thoughts that evolution occurs on networks with highly-connected neutral (selectively equivalent) genotypes.

In addition to the parameter sensitivity and boundary conditions, our work also demonstrate the use of accurate 3D geometry and its effects on model predictions. We found that geometry alone has a limited effect on gap gene patterning: Excepting slight DV-asymmetry brought about by the curvature of the 3D embryo, 1D output from M^{1D-F} (**Fig. 2d**) and 3D output from M^{3D}_{Bcd1D} (**Fig. 4, column 2**) display qualitatively similar band position along the AP axis. The path length from anterior to posterior extrema differs with DV position: For example, the distance from anterior to posterior extrema is shorter along the dorsal surface than the ventral surface. We thought that this difference in diffusion distance might account for the anterior splaying displayed in VE data (**Fig. 4, column 1**), but this was not the case.

Though the 3D embryonic geometry was insufficient to explain DV-asymmetries in gap gene data, it allowed us to explore the effect of DV-asymmetric protein distributions on patterning. Notably, the 1D Bcd distribution of M^{3D}_{Bcd1D} (**Fig. 3b**) differed from the typical dorsal-anterior distribution [63], [64] also found in the VE (**Fig. 3e**). Experimental noise in this data led to aberrant patterning in most gap genes in M^{3D}_{Bcd3D} **(Fig. 4, column 3)**, but a regularized version of the distribution **(Fig. 3f)** produced cleaner (though qualitatively incorrect) band appearance and position in $M^{3D}_{Bcd3D-S}$ **(Fig. 4, column 4)**. It also produced anterior-splaying in the anterior bands of Gt, Hb, Kni, and Kr. As previously noted, optimization of the sensitive GRN parameters improved qualitative agreement in model patterning with only small changes to parameter values .

When considering 3D models and the data associated with them, we endeavored to identify any constraints on model optimization. This model has many degrees of freedom and additional information encoded in the DV asymmetries of gap genes might better guide parameter searches toward accurate GRNs. However, we observed no improvement in RMSE values and failed to find any novel GRNs for DV-asymmetric models.

Though our ensemble of models has led to interesting findings, we acknowledge model limitations. Recent modeling studies recognize that Cad and Tll patterning cannot be completely accounted for by gap genes in existing models; maternal mRNA complicates Cad expression and Tll is under the regulation of additional proteins [38]. Instead, newer models use data interpolants to represent these proteins [38]. The absence of these interpolants in our models may contribute to the unrealistic sensitivity of the 3D model parameters and DV-information. 3D interpolating functions incorporating VE data for Cad and Tll are under development; we will use these to explore the behavior of more recent 1D models on the 3D embryonic geometry.

The primary focus of this work is the comparison of 1D and 3D model geometries. **Figures 2d**and **4, column 2** reveal that differences in model geometry can be accommodated by relatively minor adjustments to GRN parameters. The 3D implementation (M^{3D}_{Bcd1D}) exhibits minor DV-asymmetries but otherwise mirrors M^{1D-F}. However, consideration of AP patterning in three dimensions allows us to address the experimentally observed DV-asymmetry in maternal Bcd and downstream AP morphogens. The inclusion of a DV-asymmetric Bcd signal led to qualitatively different patterning with P^{1D-F}_{VE} **(Fig. 4, columns 2,4)**. This suggests that the assumption of DV and AP independence in previous modeling studies is violated. Parameter sensitivity remained high; parameter optimization made small changes to parameter values but led to significantly improved RMSE error **(Fig. 4, columns 4,5)**.

Finally, two cases of DV model mismatch suggest modifications that could be incorporated into future models. First, anterior Gt is more highly expressed on the dorsal side of the embryo in vivo, but posterior Gt displays posterior-splaying. This expression localization is not accounted for by Bcd distribution alone and should be addressed in future models that also include input from the DV patterning system downstream of the active Dorsal protein distribution [65], [66]. Second, many protein species display DV-asymmetry in terms of anterior or posterior splaying. E.g., Cad bands anterior to the AP midline are anterior-splayed **(Fig. 4b, column 1)** while bands posterior to the AP midline are posterior-splayed. This pattern is observed for all modeled proteins **(Fig. 4, column 1)**, though it is lacking in DV-symmetric M^{3D}_{Bcd1D} **(Fig. 4, column 2)**. Addition of DV-symmetric Bcd (M^{3D}_{Bcd1D}) restores anterior-splaying aligned with the DV Bcd gradient **(Fig. 4, column 5)**. This suggests that a missing posterior determinant may be responsible for posterior-splaying. The posterior maternal morphogen Nanos is a candidate that has not been included in previous models. With interpolated Cad and Tll, future models will explore the effects of posterior determinants such as Nanos [67] and, as examined in prior 1D models, Huckebein [37].

METHODS

Model Construction

Building on the successful 1D/3D embryonic modeling approach of Umulis et al., [4], [68], we reimplemented the Jaeger et al. model of gap gene regulation (M^{1D-P}) using the finite element method (FEM). This model represents six gene products as state variables: Cad, Gt, Hb, Kr, Kni, and Tll [13]. A seventh protein, Bcd (Bcd), is maintained at a constant concentration during gap gene patterning and is represented as a spatially heterogeneous stationary input [13], [63]. Each of the state variables is represented by a PDE

$$\frac{\partial c_a}{\partial t} = D_a \nabla^2 c_a + R_a \phi_a(u_a) - \lambda_a c_a, \tag{1}$$

where c_a is the concentration of protein a, the first term on the right hand side represents diffusion, the second term represents gene expression, and the third term represents first order decay [13]. D_a is the diffusion constant of protein a and λ_a is the first order decay constant of protein a. R_a is the maximal rate of gene expression of proteins a and Φ_a is a sigmoid function,

$$\phi_a(u_a) = \frac{1}{2}\left(\frac{u_a}{\sqrt{u_a^2+1}}+1\right), \tag{2}$$

which ranges from zero to one and accepts a regulatory argument u_a:

$$u_a = h_a + \sum_b w_{b,a} c_a. \tag{3}$$

Here, h_a is a minimal regulatory threshold for expression, $w_{b,a}$ is an element in the regulatory matrix W representing the influence of protein b on the expression of protein a (ranging from −0.2 to 0.2), and c_b is the local concentration of protein b. There are six PDEs representing protein proteins a=Cad, Gt, Hb, Kr, Kni, Tll (eqn. 1). In each PDE, the regulatory effects of all seven proteins, b=[41 Kr, Kni, Tll, Bcd], control protein expression (eqns. 2–3). PDEs are numerically solved using the FEM implemented in the software package COMSOL Multiphysics 3.5a [69]. Except for GRN parameters $w_{b,a}$, these parameters are fixed at values in Jaeger et al. [13] and may be found in

Note that previous 1D models were simulated by the spatially-discretized ordinary differential equations using the finite difference method: concentrations were tracked at uniformly-distributed nodes (nuclei) along the AP axis and diffusive fluxes across the Δx inter-node distance were modeled as a first-order differential equations. As such, previously reported diffusion parameters ($\check{D}_a$) were in units of inverse time [1/t]. To convert these parameters to diffusion constants (D_a) with units of squared-length-per-time [L^2/t], we multiplied $\check{D}_a$ by $(\Delta x)^2$. To compute Δx, we took into account the length of the original model's domain (0.57 EL) and the number of nodes where the finite difference model was solved (58 nuclei). From these values, we approximated Δx as 0.57EL/57. The model spans 0.35–0.92 or 0.57 EL and is divided into 57 intervals between 58 nodes. In the case of the 3D geometry, we further accounted for the curvature of the embryo in our approximation of Δx. Scaling the embryo length to unity (1 EL), we observed an arc length of 1.14 along the lateral AP. Upon the assumption that curvature was uniformly-distributed along the AP axis, Δx was computed as (0.57/1.14)EL/57. The approach slightly overestimates D_a in the 3D model relative to 1D because most curvature occurs at the AP extrema and not the trunk, but this does not translate to a large impact on AP patterning versus 1D. Whereas finite difference models explicitly modify$\check{D}_a$ values to account for mitotic nuclear division and the halving of Δx, the continuous FEM representation renders diffusion constants independent of nuclear density. It should be noted that this representation does not account for reduced effective diffusivity due to increased nuclear trapping. While nuclear density has been linked with decreased effective diffusivity in some simulations of Bcd diffusion [70], Grimm and Wieschaus found that transcription factor distributions are largely independent of nuclear density [71]. 3D nuclear density distributions have been published [47] and nuclear density-dependent diffusion is an area for further investigation.

We developed two FEM meshes on which to simulate spatiotemporal gap gene evolution. A 1D linear domain represents the 35–92% AP axis, and replicates the domain used in previous models [13]. By scaling diffusion constants and choosing initial conditions, the 1D domain also represents the 0–100% AP length (M^{1D-F}). A 3D mesh modified from the VE geometry represents a realistic embryonic geometry. Though the embryonic syncytium includes the yolk interior of the embryo, nuclei are located within the periplasmic domain of the exterior surface[10], [49]. Cytoplasmic viscosity increases in the embryonic interior and is presumed to limit effective diffusion of gap gene products to the 2D layer in the periplasmic volume containing the nuclei. While some gap gene products may diffuse into yolk, this process may be considered as part of the decay terms, λ_a. We took this into account when constructing the 3D domain. The reaction-diffusion equations (eqns. 1–3) are implemented as weak form PDEs on a 2D manifold; this manifold is "wrapped" around the 3D embryonic geometry in 3D model implementations(M^{3D}_{Bcd1D},M^{3D}_{Bcd3D},$M^{3D}_{Bcd3D-S}$).

Though the 3D domain is a closed surface without AP flux boundaries, the partial (M^{1D-P}) and full (M^{1D-F}) 1D domains are bounded at both termini by zero-flux conditions. These internal boundaries are unrealistic in the case of the partial AP length domain as there are no such physical barriers in the embryo; they were introduced in previous gap gene models to help account for artifacts in previously inferred GRNs [14], [42]–[44]. In full length 1D models the anterior and posterior ends of the embryo are realistically represented by zero-flux boundaries.

Numerical integration of PDEs requires specification of initial conditions as well as boundary conditions. For purposes of model comparison, we chose initial conditions specified in previous models [13]. On both 1D and 3D domains, the proteins Gt, Kni, Kr, and Tll have initial uniform concentrations of zero. Jaeger et al. provide initial non-uniform 1D distributions for Cad and Hb (**Fig. 3a**) [13]. These distributions span the entire AP length and provide initial conditions for both the partial and full length domains. Jaeger et al. also provide a constant exponential 1D Bcd distribution for the full AP length. These 1D distributions were used as initial conditions in the 1D models (M^{1D-F} and M^{1D-P}). They were projected onto the 3D domain to approximate full 3D initial conditions (M^{3D}_{Bcd1D}, **Fig. 3b–d**). This projection was performed using built-in interpolation tools in the Comsol package. Provided AP-coordinates and corresponding concentration values, Comsol created a linear interpolant of DV-symmetric concentration values along the AP-axis of the 3D geometry.

While the Bcd data provided by Jaeger et al. describes the lateral AP distribution of Bcd, it fails to capture the observed DV asymmetry found in embryonic Bcd. Though sufficient for a 1D model (**Fig. 3a**), the resulting 3D distribution (**Fig. 3b**) qualitatively disagrees with VE data (**Fig. 3e**). We therefore built an interpolating function from the VE Bcd data and used this interpolant when simulating the model (M^{3D}_{Bcd3D}). Again, we used Comsol's interpolation functionality. However, this interpolant required full 3D specification of coordinates. We used the coordinates of nuclei and corresponding Bcd concentration values provided in the VE. Because the software does not support interpolation on a 2D boundary (the periplasmic space) in a 3D geometry, we used nearest-neighbor interpolation (**Fig. 3e**). Because this Bcd distribution is represented in the model as a static interpolant, noise in the data (and hence the interpolant) is not smoothed by diffusion and decay. Initial attempts at directly importing VE Bcd data resulted in persistent asymmetries and mottled distributions inconsistent with data (**Fig. 4, column 3**). In an ideal situation, inter-embryo variability would be averaged out of VE data. However, the data set was generated with few replicates (13 embryos for Bcd [49]) and spatial noise remained. To remove this noise from the interpolant, we first fit a steady-state source-diffusion-decay (SDD) model of Bcd production [18] to VE Bcd data on the 3D domain (**Fig. 4a, column 1**). Once we had obtained agreement between this regularized Bcd distribution and the data, we used the solution of the SDD model to create a new interpolant. This smoothed interpolant shown in **Figure 3f** and $M^{3D}_{Bcd3D-S}$'s output (**Fig. 4b–g, columns 4–5**) compares favorably with the results M^{3D}_{Bcd3D} (**Fig. 4b–g, columns 3**).

Spatiotemporal regulation of gap gene expression spans the mitotic nuclear division between nuclear cycle 13 and 14a. For purposes of comparison, we chose to simulate the same time-course as previous models. We begin by simulating the conclusion of cycle 13 for sixteen minutes, mitosis for five minutes, and continue to simulate cycle 14a for the remaining forty-nine minutes [13]. The reaction-diffusion equations (eqn. 1–3) describe the model during interphase. During mitosis, gene expression (the second term in eqn. 1) is set to zero. Molecules may diffuse and decay, but they are not transcribed or translated while the chromatin is compacted for mitotic division.

This set of initial and boundary conditions, coupled with the reaction-diffusion equations and a geometric domain, constitutes a numerically soluble model. To calculate model error, we used a straightforward root mean squared error cost function:

$$J_{RMSE}=\sum_{t=1}^{6}\sum_{a=1}^{6}\sum_{i=1}^{n}\sqrt{\frac{\left(c_{a,\text{mod}}(\theta,i,t)-c_{a,\exp}(i,t)\right)^{2}}{6n}}. \quad (5)$$

Here, θ is the GRN parameter set, n is the number of data points in the 35%–92% EL region of the embryo, a is the index of protein species, i is the index of n nuclear coordinates, and t is the time index. This function sums the root squared error between model output from a given GRN, $c_{a,mod}(\theta,i,t)$, and experimental data, $c_{x,exp}(i,t)$, over data points i, model proteins a, and time t.

M^{1D-P} was originally fit to immunofluorescence data in Jaeger et al. [13]. As a result, both the model's concentration units and GRN parameters are scaled to reflect observed relative intensity ranges of those data. To facilitate fitting between models utilizing Jaeger et al. 's parameters and VE data, we pre-scaled the VE data to agree with the initial conditions reported by Jaeger et al. This was performed by optimizing scaling factors A_a and offsets b_a such that the difference between Jaeger et al.'s initial conditions and the VE data was minimized,

$$\min_{A,b}\sqrt{\left(\left(A_a c_{VE,a}-b_a\right)-c_{Jaeger,a}\right)^2}. \quad (6)$$

The resulting scaling was applied to the VE data, allowing for direct comparison of model outputs. VE protein data is unavailable for Cad and Tll. For the former, we substituted expression data used by Jaeger et al. to fit the original model [13]. For the latter, we substituted Tll mRNA data from the VE and scaled it according to **eqn. 6**.

Optimization

Using the cost function (eqn. 5), we optimized the full 1D and 3D models against scaled VE data using the Optimization Toolbox in MATLAB R2009a [72]. We began with local searches of the GRN weight matrix $\boldsymbol{W}$ (containing 42 parameters) using the constrained nonlinear minimization function fmincon. We initialized these searches at the best-fit inferred GRN parameter set of the original modeling study and bounded all parameters within the interval [−0.2, 0.2] [13]. Parameter and cost function tolerances for stopping criteria were set to zero and the search was allowed to progress for 4200 model evaluations (100 evaluations per parameter), resulting in arrival at local minima. In the case of the DV-asymmetric Bcd model (M^{3D}_{Bcd3D}), we subsequently included this locally optimal GRN in the initial population of a global search using genetic algorithms (GAs).

We used the GA as implemented in MATLAB. The population of size twenty genomes (parameter sets) was initialized with nineteen randomized parameter sets and the locally-optimized parameter set found for M^{3D}_{Bcd3D}. Stopping criteria were specified as a maximum of 100 generations or failure to improve cost function values above a tolerance of 10^{-6}. The latter criterion increments a "stall" counter for each generation that fails to improve the score, ending the GA when the counter reaches fifty [72]. This algorithm was used to search the parameter space while fitting the 3D model incorporating DV-asymmetric Bcd ($M^{3D}_{Bcd3D-S}$).

AUTHOR CONTRIBUTIONS

Conceived and designed the experiments: DU JH AR MG CF. Performed the experiments: DU JH. Analyzed the data: DU JH AR MG. Contributed reagents/materials/analysis tools: CF. Wrote the paper: DU JH AR MG CF.

REFERENCES

1. St Johnston D, Nusslein-Volhard C (1992) The origin of pattern and polarity in the Drosophila embryo. Cell 201–219.
2. Jaeger J (2011) The gap gene network. Cellular and Molecular Life Sciences 243–274.
3. Kopp A, Blackman R, Duncan I (1999) Wingless, decapentaplegic and EGF receptor signaling pathways interact to specify dorso-ventral pattern in the adult abdomen of Drosophila. Development 126: 3495–3507.
4. Umulis D, Serpe M, O'Connor M, Othmer H (2006) Robust, bistable patterning of the dorsal surface of the Drosophila embryo. Proceedings of the National Academy of Sciences of the United States of America 103: 11613–11618.
5. Anderson K (1987) Dorsal Ventral Embryonic Pattern Genes of Drosophila. Trends in Genetics 3: 91–97.
6. Reinitz J, Sharp D (1995) Mechanism of Eve stripe formation. Mechanisms of Development 133–158.
7. Surkova S, Kosman D, Kozlov K, Manu , Myasnikova E, et al. (2008) Characterization of the Drosophila segment determination morphome. Developmental Biology 844–862.
8. Akam M (1987) The Molecular-Basis for Metameric Pattern in the Drosophila Embryo. Development 101: 1–22.
9. Ingham P, Arias A (1992) Boundaries and fields in early embryos. Cell 68: 221–235.
10. Counce S (1961) Analysis of Insect Embryogenesis. Annual Review of Entomology 6: 295–312.
11. Jaeger J, Sharp D, Reinitz J (2007) Known maternal gradients are not sufficient for the establishment of gap domains in Drosophila melanogaster. Mechanisms of Development 108–128.
12. Irish V, Lehmann R, Akam M (1989) The Drosophila posterior-group gene Nanos functions by repressing Hunchback activity. Nature 338: 646–648.
13. Jaeger J, Surkova S, Blagov M, Janssens H, Kosman D, et al. (2004) Dynamic control of positional information in the early Drosophila embryo. Nature 368–371.
14. Jaeger J, Blagov M, Kosman D, Kozlov K, Manu , et al. (2004) Dynamical analysis of regulatory interactions in the gap gene system of Drosophila melanogaster. Genetics 1721–1737.

15. Small S, Kraut R, Hoey T, Warrior R, Levine M (1991) Transcriptional regulation of a pair-rule stripe in Drosophila. Genes & Development 827–839.

16. Driever W, Ma J, Nusslein-Volhard C, Ptashne M (1989) Rescue of Bicoid mutant Drosophila embryos by Bicoid fusion proteins containing heterologous activating sequences. Nature 149–154.

17. Driever W, Nusslein-Volhard C (1989) The Bicoid protein is a positive regulator of Hunchback transcription in the early Drosophila embryo. Nature 138–143.

18. Driever W, Nusslein-Volhard C (1988) The Bicoid protein determines position in the Drosophila embryo in a concentration-dependent manner. Cell 95–104.

19. Lehmann R, Nusslein-Volhard C (1987) Hunchback, a gene required for segmentation of the anterior and posterior region of the Drosophila embryo. Developmental Biology 119: 402–417.

20. Jurgens G, Wieschaus E, Nusslein-Volhard C, Kluding H (1984) Mutations affecting the pattern of the larval cuticle in Drosophila melanogaster 2 Zygotic loci on the 3rd chromosome. Wilhelm Rouxs Archives of Developmental Biology 193: 283–295.

21. Nusslein-Volhard C, Wieschaus E, Kluding H (1984) Mutations affecting the pattern of the larval cuticle in Drosophila melanogaster 1 Zygotic loci on the 2nd chromosome. Wilhelm Rouxs Archives of Developmental Biology 193: 267–282.

22. Wieschaus E, Nusslein-Volhard C, Jurgens G (1984) Mutations affecting the pattern of the larval cuticle in Drosophila melanogaster 3 Zygotic loci on the X-chromosome and the 4th chromosome. Wilhelm Rouxs Archives of Developmental Biology 193: 296–307.

23. Wieschaus E, Nusslein-Volhard C, Kluding H (1984) Kruppel, a gene whose activity is required early in the zygotic genome for normal embryonic segementation. Developmental Biology 104: 172–186.

24. Nusslein-Volhard C, Wieschaus E (1980) Mutations affecting segment number and polarity in Drosophila. Nature 287: 795–801.

25. Jackle H, Tautz D, Schuh R, Seifert E, Lehmann R (1986) Cross-regulatory interactions among the gap genes of Drosophila. Nature 324: 668–670.

26. Nusslein-Volhard C, Frohnhofer H, Lehmann R (1987) Determination of the anteroposterior polarity in Drosophila. Science 238: 1675–1681.

27. Harding K, Levine M (1988) Gap genes define the limits of Antennapedia and Bithorax gene-expression during early development in Drosophila. Embo Journal 7: 205–214.

28. Tautz D (1988) Regulation of the Drosophila segmentation gene Hunchback by 2 maternal morphogenetic centers. Nature 332: 281–284.

29. Hulskamp M, Pfeifle C, Tautz D (1990) A morphogenetic gradient of Hunchback protein organizes the expression of the gap genes Kruppel and Knirps in the early Drosophila embryo. Nature 346: 577–580.

30. Bronner G, Jackle H (1991) Control and function of terminal gap gene activity in the posterior pole region of the Drosophila embryo. Mechanisms of Development 205–211.

31. .Eldon E, Pirrotta V (1991) Interactions of the Drosophila gap gene Giant with maternal and zygotic pattern-forming genes. Development 111: 367.

32. .Kraut R, Levine M (1991) Mutually repressive interactions between the gap genes Giant and Kruppel define the middle body regions of the Drosophila embryo. Development 111: 611–621.

33. Capovilla M, Eldon E, Pirrotta V (1992) The giant gene of Drosophila encodes a B-zip DNA-binding protein that regulates the expression of other segmentation gap genes. Development 114: 99–112.

34. Struhl G, Johnston P, Lawrence P (1992) Control of Drosophila body pattern by the Hunchback morphogen gradient. Cell 69: 237–249.

35. Hulskamp M, Lukowitz W, Beermann A, Glaser G, Tautz D (1994) Differential regulation of target genes by different alleles of the segmentation gene Hunchback in Drosophila. Genetics 138: 125–134.

36. Simpsonbrose M, Treisman J, Desplan C (1994) Synergy between the Hunchback and Bicoid morphogens is required for anterior patterning in Drosophila. Cell 78: 855–865.

37. Ashyraliyev M, Siggens K, Janssens H, Blom J, Akam M, et al. (2009) Gene Circuit Analysis of the Terminal Gap Gene huckebein. Plos Computational Biology 5(10): e1000548.

38. Manu , Surkova S, Spirov A, Gursky V, Janssens H, et al. (2009) Canalization of Gene Expression and Domain Shifts in the Drosophila Blastoderm by Dynamical Attractors. Plos Computational Biology 7(3): e1000049.

39. Manu , Surkova S, Spirov A, Gursky V, Janssens H, et al. (2009) Canalization of Gene Expression in the Drosophila Blastoderm by Gap Gene Cross Regulation. Plos Biology 591–603.

40. Alves F, Dilao R (2006) Modeling segmental patterning in Drosophila: Maternal and gap genes. Journal of Theoretical Biology 241: 342–359.

41. Sanchez L, Thieffry D (2001) A logical analysis of the Drosophila gap-gene system. Journal of Theoretical Biology 212(1): 115–141.

42. Perkins T, Jaeger J, Reinitz J, Glass L (2006) Reverse engineering the gap gene network of Drosophila melanogaster. Plos Computational Biology 417–428.

43. Ashyraliyev M, Jaeger J, Blom J (2008) Parameter estimation and determinability analysis applied to Drosophila gap gene circuits. BMC Systems Biology 2: 83.

44. Fomekong-Nanfack Y, Postma M, Kaandorp J (2009) Inferring Drosophila gap gene regulatory network: a parameter sensitivity and perturbation analysis. BMC Systems Biology 3: 94.

45. Fomekong-Nanfack Y, Kaandorp J, Blom J (2007) Efficient parameter estimation for spatio-temporal

models of pattern formation: case study of Drosophila melanogaster. Bioinformatics 23: 3356–3363.

46. Jaeger J, Reinitz J (2006) On the dynamic nature of positional information. Bioessays 1102–1111.
47. Keranen S, Fowlkes C, Luengo Hendriks C, Sudar D, Knowles D, et al. (2006) Three-dimensional morphology and gene expression in the Drosophila blastoderm at cellular resolution II: dynamics. Genome Biology 7: R124.
48. Luengo Hendriks C, Keranen S, Fowlkes C, Simirenko L, Weber G, et al. (2006) Three-dimensional morphology and gene expression in the Drosophila blastoderm at cellular resolution I: data acquisition pipeline. Genome Biology 7: R123.
49. Fowlkes C, Luengo Hendriks C, Keranen S, Weber G, Rubel O, et al. (2008) A quantitative spatiotemporal atlas of gene expression in the Drosophila blastoderm. Cell 364–374.
50. Vakulenko S, Manu , Reinitz J, Radulescu O (2009) Size Regulation in the Segmentation of Drosophila: Interacting Interfaces between Localized Domains of Gene Expression Ensure Robust Spatial Patterning. Physical Review Letters 103: 168102–5.
51. Gursky V, Jaeger J, Kozlov K, Reinitz J, Samsonov A (2004) Pattern formation and nuclear divisions are uncoupled in Drosophila segmentation: comparison of spatially discrete and continuous models. Physica D-Nonlinear Phenomena 286–302.
52. Nusslein-Volhard C (1991) Determination of the embryonic axes of Drosophila. Development 113: 1–10.
53. Carroll S, Winslow G, Twombly V, Scott M (1987) Genes that control dorsoventral polarity affect gene-expression along the anteroposterior axis of the Drosophila embryo. Development 99: 327–332.
54. Okabe-Oho Y, Murakami H, Oho S, Sasai M (2009) Stable, Precise, and Reproducible Patterning of Bicoid and Hunchback Molecules in the Early Drosophila Embryo. Plos Computational Biology 5(8): e1000486.
55. Deng JY, Wang W, Lu LJ, Ma J (2010) A Two-Dimensional Simulation Model of the Bicoid Gradient in Drosophila. Plos One 5:
56. He F, Saunders T, Wen Y, Cheung D, Jiao R, et al. (2010) Shaping a Morphogen Gradient for Positional Precision. Biophysical Journal 697–707.
57. Jaeger J (2010) A matter of timing and precision. Molecular Systems Biology 6: 427.
58. Jaeger J (2009) Modelling the Drosophila embryo. Molecular Biosystems 5: 1549–1568.
59. Bieler J, Pozzorini C, Naef F (2011) Whole-embryo modeling of early segmentation in Drosophila identifies robust and fragile expression domains. Biophysical Journal 101: 287–296.
60. Munteanu A, Sole R (2008) Neutrality and Robustness in Evo-Devo: Emergence of Lateral Inhibition. Plos Computational Biology 4(11): e1000226.
61. Wagner A (2008) Robustness and evolvability: a paradox resolved. Proceedings of the Royal Society B-Biological Sciences 91–100.
62. Wagner A (2005) Distributed robustness versus redundancy as causes of mutational robustness. Bioessays 176–188.
63. Gregor T, Wieschaus E, McGregor A, Bialek W, Tank D (2007) Stability and nuclear dynamics of the bicoid morphogen gradient. Cell 141–152.
64. Gregor T, Bialek W, van Steveninck R, Tank D, Wieschaus E (2005) Diffusion and scaling during early embryonic pattern formation. Proceedings of the National Academy of Sciences of the United States of America 18403–18407.
65. Kanodia J, Rikhy R, Kim Y, Lund V, DeLotto R, et al. (2009) Dynamics of the Dorsal morphogen gradient. Proceedings of the National Academy of Sciences of the United States of America 21707–21712.
66. Liberman L, Reeves G, Stathopoulos A (2009) Quantitative imaging of the Dorsal nuclear gradient reveals limitations to threshold-dependent patterning in Drosophila. Proceedings of the National Academy of Sciences of the United States of America 22317–22322.
67. Wang C, Lehmann R (1991) Nanos is the localized posterior determinant in Drosophila. Cell 637–647.
68. Umulis D, Shimmi O, O'Connor M, Othmer H (2010) Organism-Scale Modeling of Early Drosophila Patterning via Bone Morphogenetic Proteins. Developmental Cell 18: 260–274.
69. COMSOL Multiphysics® 3.5a. COMSOL AB, Palo Alto, CA, USA.
70. Sample C, Shvartsman S (2010) Multiscale modeling of diffusion in the early Drosophila embryo. Proceedings of the National Academy of Sciences of the United States of America 107: 10092–10096.
71. Grimm O, Wieschaus E (2010) The Bicoid gradient is shaped independently of nuclei. Development 137: 2857–2862.
72. MATLAB® R2009a. The Mathworks, Natick, MA, USA.

CHAPTER 21

EVIDENCE-BASED STRUCTURAL MODEL OF THE STAPHYLOCOCCAL REPRESSOR PROTEIN: SEPARATION OF FUNCTIONS INTO DIFFERENT DOMAINS

Kinga Nyíri[1, 2], Bianka Kőhegyi[1, 2], András Micsonai[3] , József Kardos[3] , Beata G. Vertessy[1,2]

[1] Department of Applied Biotechnology and Food Sciences, Budapest University of Technology and Economics, Budapest, 1111, Hungary,

[2] Institute of Enzymology, Research Centre for Natural Sciences, Hungarian Academy of Sciences, Budapest, 1117, Hungary,

[3] Department of Biochemistry and MTA-ELTE NAP B Neuroimmunology Research Group, Institute of Biology, Eötvös Loránd University, Budapest, 1117, Hungary

ABSTRACT

Horizontal transfer of mobile genetic elements within Staphylococci is of high biomedical significance as such elements are frequently responsible for virulence and toxic effects. Staphylococcus-encoded repressor proteins regulate the replication of these mobile genetic elements that are located within the so-called pathogenicity islands. Here, we report structural and functional characterization of one such repressor protein, namely the Stl protein encoded by the pathogenicity island SaPIbov1. We create a 3D structural model and based on this prediction, we investigate the different functionalities of truncated and point mutant constructs. Results suggest that a helix-turn-helix motif governs the interaction of the Stl protein with its cognate DNA site: point mutations within this motif drastically decrease DNA-binding ability, whereas the interaction with the Stl-binding partner protein dUTPase is unperturbed by these point mutations. The 3D model also suggested the potential independent folding of a carboxy-terminal domain. This suggestion was fully verified by independent experiments revealing that the carboxy-terminal domain does not bind to DNA but is still capable of binding to and inhibiting dUTPase. A general model is proposed, which suggests that among the several structurally different repressor superfamilies Stl-like Staphylococcal repressor proteins belong to the helix-turn-helix transcription factor group and the HTH motif is suggested to reside within N-terminal segment.

INTRODUCTION

Phage mediated mobilization of pathogenicity islands, i.e. genetic elements encoding virulence factors and toxins in *Staphylococcus aureus* (SaPI) has been an intensively studied field in recent years [1]. It has been shown that excision and replication of SaPIs is induced by formation of a repressor:derepressor complex constituting the Staphylococcal master repressor protein and another phage-related protein [2,3]. In the specific case of the SaPIbov1 pathogenicity island repressor Stl (abbreviated as Stl in the present study), the derepressor is a phage-related dUTPase enzyme [3]. dUTPases are important guardians of genome integrity [4,5]. Their physiological role is to deplete the cellular dUTP pool to prevent incorporation of uracil into DNA [6,7]. The enzyme family of dUTPases constitutes two large subfamilies with different protein structure but very similar catalytic function [8,9].

Citation: Nyíri K, Kőhegyi B, Micsonai A, Kardos J, Vertessy BG (2015) Evidence-Based Structural Model of the Staphylococcal Repressor Protein: Separation of Functions into Different Domains. PLoS ONE 10(9): e0139086. doi:10.1371/journal.pone.0139086

In-depth studies on the Stl:dUTPase interaction revealed an additional function of the Stl repressor, namely it has been proven to be an effective inhibitor of the dUTPase enzyme from the Φ11 Staphylococcal phage [10]. In addition, we have recently shown that Stl might be a cross-species general dUTPase inhibitor, which may open new horizons in studying dUTPase cellular function [11]. Moreover, since dUTPase has been proposed as a significant novel target in antimycobacterial drug design [12–14], Stl may also be a possible candidate for designing protein-based dUTPase-inhibitors to figureht *Mycobacterium tuberculosis*.

For the dUTPase enzyme family, detailed studies have already addressed its structural and functional characteristics [8,15–21]. In contrast to the essential role of Stl-like repressors in the patho-mechanism and in the horizontal gene transfer of several toxins, no structural information has yet been reported for any of the SaPI repressor proteins or their complexes.

The SaPIs are generally considered as phage originated genetic elements, and this is supported by their mobilization by phage proteins. Similarly to master repressors of temperate phages Stl is an autoinducer [2], and blocks the expression of genes *int*, *xis* and *str*, which are responsible for SaPI excision and replication [22,23]. It has been shown that one of the DNA-binding sites of Stl resides within the *stl-str* intergenic region, which coincides with the repression of genes downstream to that region [22]. Other binding sites or sequence specificity of Stl have not been identified yet, however the question is intensively studied by our laboratory.

Based on functional similarities, the gene regulation mechanism of Stl may be adequately modeled by the mechanism of the main lifecycle regulator CI repressor of temperate phages. Within this model, repressor proteins are responsible for binding to a specific DNA segment thereby preventing excision and replication of the relevant genetic segments. The CI repressor proteins have separate domains for DNA binding and protein binding [24,25]. The protein binding domain is usually responsible for oligomerization of the repressor and for the interaction with the derepressor protein [24,26–28]. The oligomerization makes the regulation more sensitive to the alteration of protein concentration since it depends on the oligomer monomer equilibrium [28]. Higher order oligomers are suitable for more complex regulation patterns [29,30]. The mechanism of derepression can be reversible or irreversible. Prophage induction in lambdoid phages is achieved through a RecA binding induced specific autoproteolytic inactivation of the maintenance repressor [31]. Several P2 related repressors are inactivated through a noncovalent complex formation with the derepressor protein [32–35]. The DNA binding function of repressors can be performed through several different structural motifs, such as helix-turn-helix (HTH) as in CI repressor protein of lambda phage, winged- helix turn helix (wHTH) as in the MuR repressor protein of Mu phage, and the antiparallel β-strands of the ribbon-helix-helix (RHH) fold as in Arc repressor of P22 phage [36,37].

Based on these models, our focus was to resolve how potential domains within the Stl repressor may be defined. Towards this end, first we created a structural model of the full length Stl protein and investigated its folding by synchrotron radiation circular dichroism (SRCD) measurements. Based on the 3D model verified by CD results, we produced truncated and point mutants and studied their function in DNA and dUTPase binding. We show that the produced carboxy-terminal segment is an independently folded domain, which retains its binding affinity to dUTPase, but shows reduced inhibitory effect. The amino-terminal putatively DNA binding segment was also studied by point mutations. Our experimental results convincingly support the predicted position of helix-turn-helix motif.

MATERIALS AND METHODS

Homology modeling and *in silico* predictions

The 3D homology model of Stl was constructed using the Phyre2 Server in intensive mode. Seven templates were selected by the program (PDB IDs: 1E3O, 4YV9, 2GRM, 4RYK, 2QFC, 2AXZ, 2EBY) to model Stl protein based on heuristics to maximize confidence, percentage identity and alignment coverage [38]. Five out of the seven templates covered >90% of the Stl sequence, while the other two templates provided only partial coverage of the sequence, but with higher local similarity. In the final model, 97% of residues were modeled at >90% confidence (see Table A in S1 File for additional information on the template proteins). Homology prediction was made using HHpred [39], subsequent 3D structure predictions with Modeller was performed with automatic template selection and also with the Phyre2 hits manually selected as templates from the HHpred list [40]. Related structures were searched in the Molecular Modeling Database (MMDB) also referred to as the Entrez Structure database [41]. Functional domain search was performed by Pfam and NCBI Conserved Domain Database [42,43]. The possible position of the helix-turn-helix (HTH) DNA binding motif in the sequence was predicted by NPS@ server [44]. Disorder prediction during construct design was performed by using GeneSilico MetaDisorder service [45]. To compare the homology model to the structural information obtained experimentally by SRCD spectroscopy, the secondary structure composition of the model, including the helix content, was assigned using the DSSP algorithm [46] and the BeStSel and CONTIN secondary structure definitions [47,48].

Cloning and expression of proteins

The cDNA of Stl-CTD was made by PCR amplification from the pGEX-4T-1 vector containing Stl protein (GenBank ID AAG29617.1) described in our previous work [10]. For the amplification of Stl-CTD the Stl-CTD-F (5'-TATTGAATTCAGCCCGACCCTGAACG-3') and the Stl-CTD-R (5'-GGTCCTCGAGTT AGTTGGTATCTTTTTCCAGAATAATT TTTTTCTGATG-3') primers were used. The resulting insert was cloned with EcoRI and XhoI restriction sites of the pGEX-4T-1 vector in frame with the amino-terminal GST tag and the thrombin cleavage site. A stop codon was mutated to the full length sequence to provide a construct for the N-terminal segment of Stl (residues 1–84) by QuikChange site-directed mutagenesis (Stratagene) using mutagenic primers Stl-NT-F 5'-GCGATGAATTTAAAGAAAAAGGCTATTAGCTAACTG AGCCCGACCCT GAACG -3') and Stl-NT-R 5'-CGTTCAGGGTCGGGCTCAGTTAGCTAATAGCCTTTTT CTTTAAATTCATCGC-3'. Stl- Q40A,N41A (Stl-AA) mutant was created from the original vector by QuikChange site-directed mutagenesis (Stratagene) using mutagenic primers Stl-AA-F 5'-CGTTTTCATGGTTGCTAATGGTCGCT GCGCTAAAGCCGGTGCG-3') and Stl-AA-R 5'-CGCA CCGGCTTTAGCGCAGCGACCATTAGCAACC AT GAAA ACG-3'. DNA sequencing for verification of the resulting constructs were performed by Eurofins MWG Operon. Vectors were transformed to into *Escherichia coli* strain BL21 Rosetta (DE3) and propagated in 500 ml LB till exponential growth, then the culture was induced with 0.5 mM iso-propyl-β-D-thiogalactoside. After induction, the cell cultures were grown at 303 K for further 4 h. Finally the cells were harvested by centrifugation and stored at 193 K. Protein over-expression from the created constructs, except the one which encodes the N-terminal segment, was successful.

Purification of proteins

For purification of GST-tagged proteins (Stl, Stl-CTD, Stl-AA), cell pellets were solubilized using Potter-Elvehjem homogenizer in 20 ml buffer A (15 ml Hepes (pH 7.5), 200 mM NaCl) supplemented with 2 mM dithiothreitol (DTT), 1% Triton X-100, ca. 2 μg/ml RNase and DNase and one tablet of Complete ULTRA Tablets, Mini, EDTA-free protease inhibitor. Cell suspensions were sonicated (4 x 60 s), and centrifuged (16000g for 30 min). Supernatant loaded on a pre-equilibrated benchtop glutathione-agarose affinity-chromatography column (GE Healthcare). The column was washed with ten volumes of buffer A (200 mM NaCl). After that 80 Cleavage Units thrombin (GE Healthcare) was added to perform on-column cleavage for the removal of GST tag. After overnight cleavage purified were obtained in the flow-through.

Purification of Φ11dUTPase was performed as described previously [49]. Briefly, supernatant resulting from centrifugation of cell lysate was purified on Q-Sepharose (GE Healthcare) anion-exchange column, followed by gel filtration on a Superdex 75 column (GE Healthcare) using an AKTA Explorer purifier. Protein preparations were used freshly or were flash-frozen in liquid nitrogen, and stored at 193 K. All protein preparations were >95% pure as judged by SDS–PAGE.

Protein quantification

Protein concentration was measured using NanoDrop 2000 UV-Vis spectrophotometer using the following $A^{0.1\%}_{280}$ values 1.087, 1.261, 1.090 and 0.786 ml * mg^{-1}* cm^{-1} for Stl, Stl-CTD, Stl-AA and Φ11dUTPase respectively, calculated based on amino acid composition (http://web.expasy.org/protparam/).

Native polyacrylamide gel electrophoresis

Native gel electrophoresis was performed in 8% polyacrylamide gel. After preparation the gel was subjected to pre-electrophoresis with constant voltages of 100 V. Then 25 μl of the premixed samples was applied on the gel and electrophoresis was performed for 1.5 hours on 150 V in pH 8.7 Tris-HCl buffer. The apparatus was cooled on ice during electrophoresis in order to avoid denaturation caused by the evolving heat. Coomassie-Brilliant Blue dye washed to stain the gel.

Electrophoretic mobility shift assay (EMSA)

EMSA experiments were performed using a 57mer dsDNA oligonucleotide (5'-GCTCATATTATTCCTCTCCTACCATTT TATCTCTAATTGAGATATTTATATTCAGAT-3') based on our previous results. Complementary oligonucleotides were custom synthesized by Eurofins MWG Operon and hybridized by controlled gradual cooling after 5 minutes incubation on 95°C. The investigated proteins were mixed with 100 ng DNA and in 20 μl total volume, concentration of NaCl was set to 100 mM in all the samples. After incubation for 15 min at 4°C, samples were loaded onto 8% polyacrylamide gel. Electrophoresis was performed in Tris- Borate- EDTA (TBE) buffer for 70 min at room temperature, following 1 h pre-electrophoresis of the gel. Bands were detected after staining with GelRed (Biotium), using a Uvi-Tec geldocumentation system (Cleaver Scientific Ltd., Rugby, UK).

Steady-state kinetics experiments

Proton release during the transformation of dUTP into dUMP and PPi was followed using a Jasco V550 spectrophotometer at 559 nm and 293 K [50]. Reaction mixtures contained 20 nM Φ11 dUTPase enzyme in 1 mM HEPES–HCl (pH 7.5) containing 5 mM $MgCl_2$, 150 mM KCl and 40 mM phenol red pH indicator. After preincubation of the two proteins for 5 minutes the reaction was started with the addition of 20 mM dUTP. The initial velocity was determined from the slope of the first 10% of the progress curve. Quadratic binding equation was fitted to the data.

Synchrotron radiation circular dichroism (SRCD) measurements and CD spectrum analysis

SRCD spectrum of Stl was recorded at the DISCO beamline of SOLEIL French Synchrotron Facility (proposal No. 20140646). The Stl concentration was 2.1 mg/ml in a buffer of 50 mM Hepes, 200 mM NaCl, pH 7.5. A CaF_2 cell with a path length of 6.13 μm was used. 38 scans were accumulated in the 180–270 nm wavelength range at 1 nm steps with a lock-in time constant of 300 msec and integration time of 1200 msec. In this wavelength range and path length, the photomultiplier voltage did not exceed the 700 V limit. After baseline subtraction, the spectrum was corrected with the CSA calibration [51].

To estimate the secondary structure content, the CD spectrum was analyzed by the BeStSel [47] and CONTIN methods [48]. These algorithms distinguish two types of spectrally different helical components, helix1 and helix2. Helix1 is the regular, middle part of the helix where all the backbone-backbone hydrogen bonds are formed and helix2, called "distorted helix", consists of the two-two residues at the ends of the helix with unsatisfied H-bonding. The helix2 content together with the helix2/helix1 ratio provide a chance to predict the number and average length of helices in the protein [52]. We have to note that BeStSel defines helix as α-helix while the CONTIN definition includes α-helix and 3_{10}-helix. Usually the 3_{10}-helix content is low or absent and we expect similar results for the two algorithms.

RESULTS AND DISCUSSION

Stl protein is mainly α-helical

In lack of atomic resolution structural information, the structure of Stl was analyzed by *in silico*methods and synchrotron radiation circular dichroism (SRCD) spectroscopy. The predicted homology model as provided by the Phyre2 Server [38] applying intensive mode is shown on figure 1A, while the templates used for modeling are listed in Table A in S1 File. The Stl sequence was entirely covered by alignment with the template sequences, with the exception of a very short part of the N-terminus (residues 1–7), which were modeled by *ab initio* methods. Apart from these first few residues the confidence of the alignment was more than 95%, so we assume that the overall fold and the core of the protein is modeled reliably, although the orientation of the surface loops are less well-defined. Based on this model, the repressor protein has a mostly α-helical (74%) secondary structure thus it is likely to belong to the class of "all-α" proteins (SCOP ID 46456) that includes various protein superfamilies. Related structures found by the Molecular Modeling Database (MMDB) with E-value > 10^{-6} (PDB ID: 2B5A, 3NTG, 2P5T) are also proteins with mostly α-helical fold [41]. Since the predicted fold type is shared by numerous proteins performing vast array of functions and the sequence identity between the templates and Stl are relatively low, we did not analyze the template proteins in detail. To compare this model with others generated by alternative structural annotation servers Stl sequence was submitted also to the HHpred predictor [39]. All templates used by Phyre2 (cf. Table A in S1 File) with the only exception of the human transcription factor Oct-1 were also within the list of closest homologs provided by HHpred. Following the HHpred search 3D structural models were generated by Modeller in two ways, i) with automatic template selection, which optimizes diversities of query and template HMMs, reranks templates and automatically selects best set, and ii) with the Phyre2 hits manually selected from the HHpred list as templates [40]. The first approach provided a well defined structural model only for the N-terminal residues (1–75), while the other resulted a structural model for almost all residues, however the C-terminal 33 residues are unstructured even in that model. Both the Phyre2 and the Modeller 3D structures agreed in that the protein is mostly α-helical and contains an N-terminal HTH motif (S1 figure). However the orientation of the helices except the HTH was quite different in the two types of models.

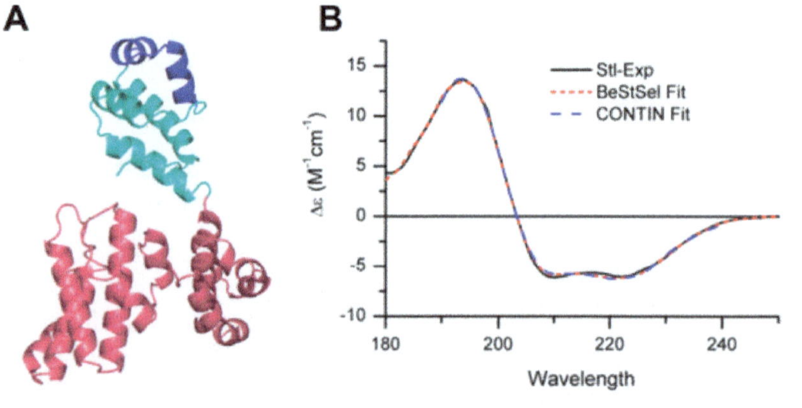

figureure 1: (A) Homology model of the Stl protein. Ribbon representation of the homology model of the *Staphylococcus aureus*pathogenicity island repressor Stl produced by Phyre2 Server [38]. Based on the homology model the protein is highly α helical (74%), and seems to be divided into two segments: the amino terminal segment colored cyan and the carboxy-terminal segment colored hotpink. According to Pfam and NCBI CDD the protein is predicted to contain a helix-turn-helix DNA binding motif. The position of the HTH predicted by NPS@ server is colored to dark blue [44]. **(B) SRCD spectrum of Stl and fitted curves by BeStSel and CONTIN algorithms.** Stl was measured at 2.1 mg/ml concentration in 50 mM Hepes, 200 mM NaCl, pH 7.5. http://dx.doi.org/10.1371/journal.pone.0139086.g001

Secondary structural elements of the full length Stl protein were also experimentally characterized by CD spectroscopy. CD spectroscopy is a frequently used method to assess protein conformation: tertiary structural elements can be well reflected in the near-UV wavelength range (250–380 nm) [53], while secondary structural elements show well-known spectral characteristic in the far-UV wavelength range (180–240 nm) [48,54]. Due to the high far-UV absorption of the 200 mM NaCl in the buffer, these measurements were performed using synchrotron radiation CD (SRCD) providing extended wavelength range and increased signal/noise ratio compared to conventional CD measurements. The observed spectral shape was characteristic of α-helical proteins (figure 1B). The secondary structure content was quantitatively estimated from the CD spectrum by the BeStSel [47] and CONTIN [48] algorithms showing 67.6 and 63.3% α-helix content respectively (Table 1). For both algorithms, the predicted β-sheet content was low, being close to zero within the accuracy of the methods. Only the Phyre2 model was suitable for a comparison to the results of the SRCD measurements since it provided a 3D model for the whole protein (Table 1). The Stl Phyre-predicted 3D model shows similar secondary structural composition to that of the CD analysis within the experimental error supporting the reliability of the model. The two CD analysis methods also provided the possibility of estimation of the length and number of helical segments in the protein, which are in good agreement with the model, as well (Table 1). The accordance of the data obtained from the CD spectra and from the Phyre2 model, led us to consider this 3D model as a useful starting point for construct design and analysis tool for the Stl-like proteins.

Table 1: Secondary structure estimation from the synchrotron radiation CD spectrum of Stl and comparison to the homology model. http://dx.doi.org/10.1371/journal.pone.0139086.t001

	BeStSel		CONTIN	
	CD analysis[a]	Model[b]	CD analysis[a]	Model[b]
Helix1	44.6	40.0	44.5	42.5
Helix2	23.0	21.4	18.8	21.4
β-sheet	3.6	0.0	6.3	0.0
Turn	8.4	7.5	9.0	10.4
Others	20.5	31.1	21.4	25.7
Number of helices	16	15	13	15
Average helix length	11.8	11.5	13.5	11.9

[a]The secondary structure composition from the CD spectrum was estimated by the BeStSel and CONTIN algorithms. The two algorithms use different secondary structure components, however, the overall helix, β-sheet and turn+others contents are comparable.

[b]The secondary structure contents were also calculated for the Phyre2 homology model using the DSSP algorithm [46] and the BeStSel and CONTIN definitions [47,48].

doi:10.1371/journal.pone.0139086.t001

Stl may possess two segments with distinct functions

The 3D homology model revealed that the Stl protein– as numerous other repressors –seemingly consists of two segments, which may fold independently (figure 1A). Pfam and NCBI CDD protein domain annotation engines [42,43] predicted that Stl contains a helix-turn-helix DNA binding motif (HTH) at the amino-terminal part of the protein (between residues 15–68; with expectance value of $7.97 \cdot 10^{-11}$). The HTH was predicted to reside between residues 27–48 with 100% probability by NPS@ server [44]. In addition, we used the MetaDisorder server [45] to estimate flexibility characteristics of the Stl protein (S2 figure). Although these flexibility predictors provided rather different pattern, most agreed in indicating that a somewhat more ordered segment residues between residues 95–150, C-terminal to the HTH motif.

Based on these *in silico* predictions we hypothesized that the Stl, similarly to other well-known repressors, possesses two segments, with putative distinct functions. We hypothesize that the amino-terminal segment encodes DNA binding function, while the carboxy-terminal segment may be responsible for other protein-protein interactions. In the context of the Stl protein, complex formation with another protein partner, namely phage dUTPase was already shown to exist [10]. Hence, we speculated that this C-terminal segment may be involved in binding to phage dUTPase. We set out to investigate these suggestions and asked whether the N-terminal and C-terminal segments may fold independently and may still provide either DNA-binding or dUTPase-interacting function. To define the boundaries of the designed constructs, we relied on the i) the results from the flexibility annotation server (S2 figure) ii) the Phyre2 structural prediction iii) length of the N-terminal protein constructs in experimentally determined three-dimensional structures of DNA-bound bacteriophage repressors [55–60]. With these considerations, a truncated construct encoding the C-terminal segment (residues 84–263) was produced by PCR based cloning and a stop codon was mutated to the full length sequence to obtain a construct for the N-terminal segment of Stl (residues 1–84). The expression of N-terminal segment proved to be unsuccessful even if attaching this segment to a GST-tag, indicating that the N-terminus of Stl, truncated at residue 85, may not fold independently. It was therefore not straightforward to analyze the N-terminal segment on its own. Hence we later took the approach to mutate the key residues in the HTH motif within the full-length context. The expression of the C-terminal segment was successful, indicating that this segment might be considered as an independently folded C-terminal domain (Stl-CTD).

The Stl-CTD domain is not capable of DNA-binding but binds and inhibits dUTPase

To check the suggested domain functions, the DNA binding ability of Stl-CTD was investigated with electrophoretic mobility shift assay. Increasing concentrations (from 2 μM up to 30 μM) of Stl-CTD was added to 100 ng DNA and mixtures were run on native PAGE gel (figure 2A). No DNA shift was observed in any of the Stl-CTD containing samples, while the positive control containing 2 μM Stl clearly showed the expected shift due to complex formation between Stl and DNA (cf figure 3 in the present work and also figure 2D in [10]).

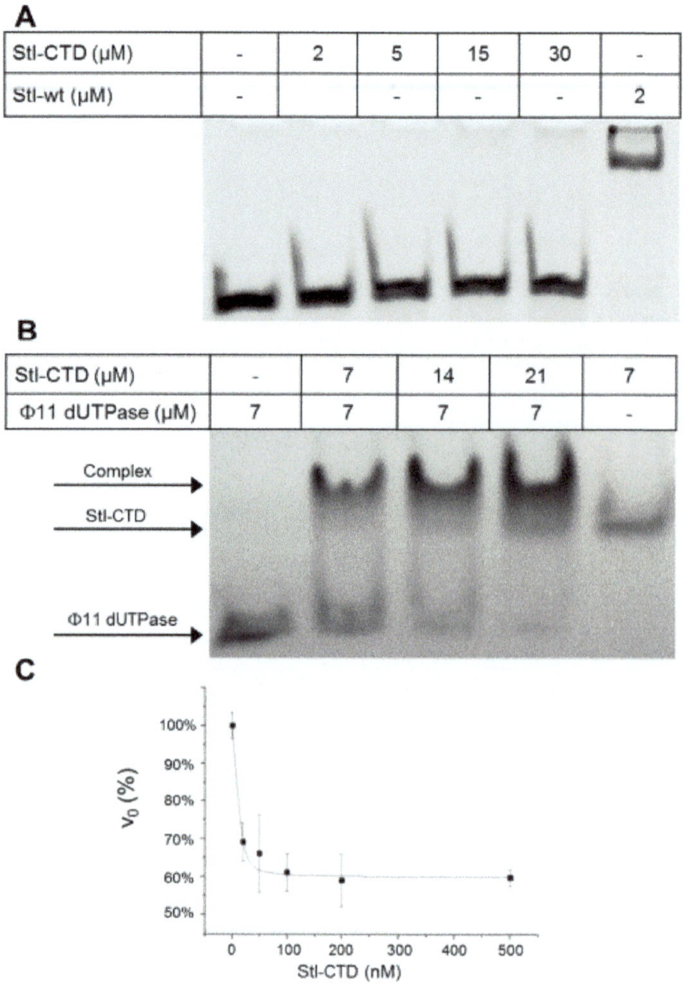

figureure 2: DNA and protein binding ability of Stl-CTD. (A) Electrophoretic mobility shift assay was performed to investigate the DNA binding ability of Stl-CTD. Species and concentrations given in monomers are indicated on the figureure. The band of the dsDNA is only shifted upwards if wild type Stl is added but there is no shift even at high concentrations Stl-CTD. **(B) Native gel electrophoresis**experiment was performed to investigate the dUTPase binding ability of Stl-CTD. Species and concentrations given in monomers are indicated. The mixture of Stl-CTD and Φ11 dUTPase shows up in distinct position comparing to the individual proteins, which clearly indicate the complex formation. **(C) Activity of the Φ11 dUTPase** was measured in the presence and absence of Stl-CTD. Each measurement was repeated three times. The quadratic binding equation was fit to the data resulted in the apparent $K_i = 1.5 \pm 0.5$ nM. The total change in amplitude of the activity was 40%. http://dx.doi.org/10.1371/journal.pone.0139086.g002

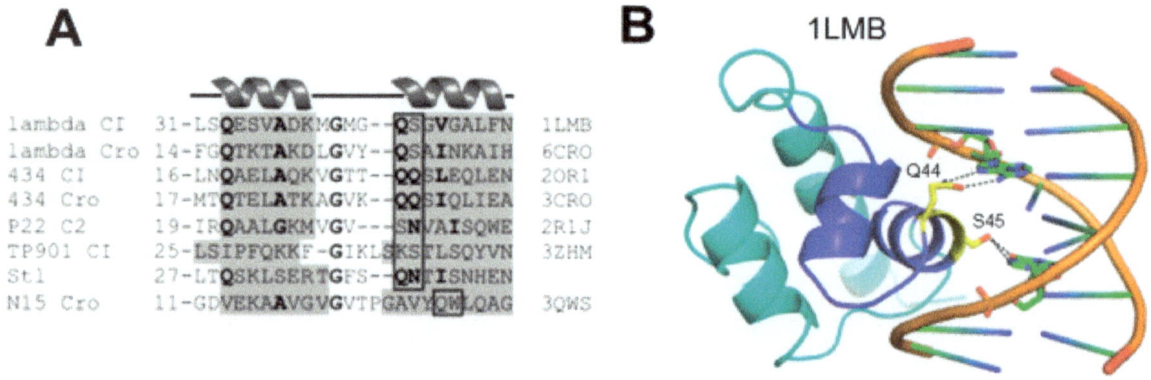

figureure 3: DNA binding domain of bacteriophage repressors. (A) Sequence alignment of the HTH motifs of bacteriophage repressors and Stl.The number before each segment is the amino acid sequence number of the first residue in the sequence. Helices are with gray background, similar residues are in bold, and box highlights residues interacting with DNA nucleobases. PDB ID of the proteins is indicated on the right side of the sequences. (B) Experimentally determined structure of the DNA-bound CI bacteriophage repressor DNA cartoons orange, protein cartoon: dark blue for HTHs, otherwise cyan. DNA bases and DNA interacting amino acid residues are stick representation with atomic coloring (protein carbon yellow, DNA carbon green, oxygen red, nitrogen blue, and phosphorus orange.) The PDB ID of the structure is indicated. Stereo representation of all experimentally determined protein-DNA complex structures represented in the sequence alignment in this figureure are available in S3–S9 figures. http://dx.doi.org/10.1371/journal.pone.0139086.g003

The interaction of Stl-CTD with Φ11 dUTPase was tested by two independent methods. Native polyacrylamide gel electrophoresis experiment was performed with the mixture of the two proteins (figure 2B). Similarly to our previous results with wild type full length Stl [10], the band corresponding to the complex of Stl-CTD: Φ11 dUTPase shows up at a distinct position as compared to the positions of the two individual proteins. The interaction of the two proteins was also investigated by measuring the effect of Stl-CTD on dUTPase activity (figure 2C). Stl-CTD inhibited the dUTPase activity with the apparent inhibitory constant of $K_i = 1.5 \pm 0.5$ nM. However, in contrast to wild type full length Stl, which caused a practically complete loss of dUTPase activity if added in high excess [10], the maximum inhibitory effect exerted by Stl-CTD on dUTPase was about 40% even at saturating inhibitor concentration. The difference in maximal dUTPase inhibition as compared to the wild type Stl suggests that the amino-terminal segment of Stl may also contribute to the interaction between the two proteins. In summary, these data show that the C-terminal segment of Stl lacks the potential for DNA-binding while partially constitutes ability for dUTPase binding and inhibition. These observations suggest that while the C-terminal segment is capable folding on its own, it does not possess full functional capability for either of the two function of the full-length Stl protein.

Specific point mutations within the putative helix-turn-helix motif of the Stl N-terminal segment result in decreased DNA binding ability

To verify the prediction for a helix-turn-helix (HTH) motif, point mutations were performed within the Stl protein to abolish DNA binding. Design of these mutations was based on experimentally determined three-dimensional structures of DNA-bound bacteriophage repressors. As of present, 3D structures for seven such proteins are available in the PDB, among these we have focused on the best resolution structures, for which the PDB IDs are as follow: 1LMB; 6CRO, 2OR1, 3CRO, 2R1J, 3ZHM, 3QWS [55–60]. Besides the structural similarity of these phage repressor HTHs, five of those showed high level of sequence similarity to each other (figure 3A), forming the basis of HTH prediction [44]. These structures together with numerous additional studies have established that the residues responsible for specific DNA binding are situated on the second helix of the HTH [61–63]. As shown on figure 3A, with the exception of N15 phage Cro repressor, two residues could be identified as being conserved in different repressor-related HTH motifs, and moreover, in the published 3D structures, the role of these residues were also well defined. Namely, the first two residues of the second helix of the HTH motif provide H-bonding interaction with DNA nucleobases (figure 3, S3–S9 figures). In the Stl protein, two similar polar residues show up at the same site within the predicted HTH motif (figure 3A). In conclusion, based on the sequence alignment of the predicted Stl HTH with different HTHs and the superimposition of the Stl Phyre2 structural model on the crystal structures of these repressors, mutations of the polar residues Q40-N41 in the second helix were performed. To preserve the helical secondary structure while erasing the potential H-bonding ability, these two residues were exchanged into alanines. A similar double alanine mutational analysis was successfully performed with the TP901 repressor and yielded important insights [60].

According to the expectation, the double mutant (Q40-N41 – A40-A41) Stl construct (termed as Stl-AA) proved to be highly defective in DNA binding based on EMSA experiments (figure 4, S10 figure). To verify that the double mutations within the HTH motif did not perturb dUTPase binding ability, we checked whether the double mutant protein may still form a complex with dUTPase. We showed that the complex between Stl-AA and dUTPase is readily observable on native polyacrylamide gel electrophoresis (S11 figure). Also, dUTPase inhibition by the Stl-AA mutant has practically the same characteristics as compared to the inhibitory effect of wild type Stl (S12 figure): the inhibitory constant has been determined to be 1.4 ± 0.9 nM, and the maximal inhibition was above 90%. Comparing to the wild type Stl the inhibition effect of Stl-AA on Φ11 dUTPase is not perturbed [10]. Based on these findings, the design of an *in vivo* reporter system in in progress in our laboratory to test Stl-DNA and Stl-dUTPase interactions.

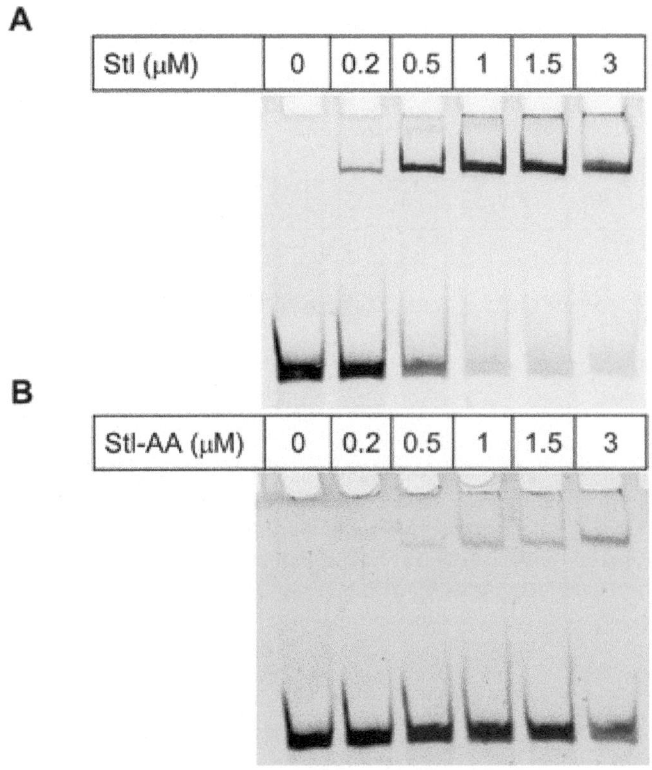

figureure 4: Electrophoretic mobility shift assay (EMSA) for testing the DNA binding ability of Stl constructs. (A) Result of EMSA with the wild type Stl protein. Note that the DNA band is shifted upwards at 1 μM Stl concentration. **(B)** EMSA gel result of Stl-AA mutant protein. Note that even at relatively high (ie. 3 μM) concentration of the Stl-AA mutant construct, most DNA still appears at the lower position on the gel, indicating lack of binding to the mutant protein. http://dx.doi.org/10.1371/journal.pone.0139086.g004

General model for Stl-like repressors in *Staphylococci*

The group of Stl-like repressors has a major regulatory effect on the replication and subsequent horizontal transfer of mobile genetic elements in *Staphyloccoccal* strains. However, their structural and functional traits have not yet been addressed in details. In the present study, we have used the Phyre2 3D structural modeller software and the resulting model structure was in excellent agreement with experimentally determined structural elements via synchrotron radiation CD spectroscopy. Domain prediction based on this model did, in fact, make it possible to design an independently folding truncated construct (Stl-CTD). This domain lost DNA-binding capability but still preserved functionality with respect to binding and inhibition of dUTPase. Another *in silico* prediction method was also used with success in the present study: the HTH motif of Stl could be localized with confidence and this result could be fully ascertained by wetlab experiments (point mutations, EMSA and other assays).

Based on the success of these *in silico* predictions in the context of our Stl-focused study, we made an additional more generalized approach to decide if similar structural/functional elements may be identified in other representatives of the family of Stl-like repressors, as well. Towards this end, we first ran the HTH predictor on the different repressor protein sequences within the diverse *S. aureus* pathogenicity islands. figure 5 shows an alignment for the HTH motif identified in the diverse repressors: out of the 12 repressor sequences, HTH could be identified in 8 proteins with a probability above 50% (Table B in S1 File). In all of these cases, the HTH motif is located at the N-terminal part of the proteins.

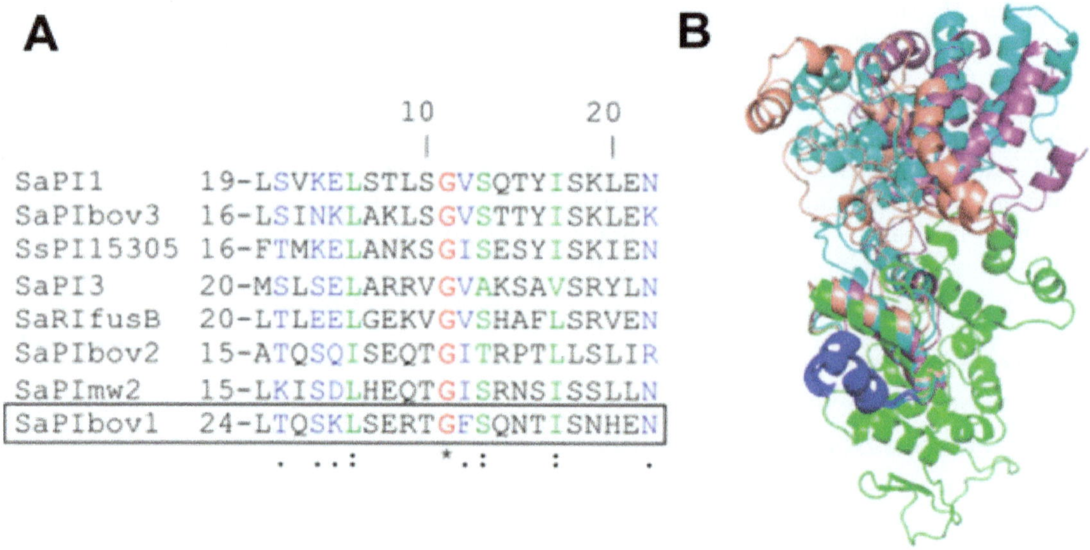

figureure 5: (A) Sequence alignment for the HTH segments of Stl-like repressors. Identical residues are in red (*), strongly similar (:) residues are in green, weakly similar residues are in blue (.). The protein investigated in the current study, termed as Stl throughput the text, is boxed (it is equivalent to the repressor of the SaPIbov1 pathogenicity island). **(B) Superimpositioned structural models of representative Stl-like repressors from different SaPIs.** Proteins are in cartoon representation. SaPIbov1 Stl is cyan, Stl-like repressors of SaPIbov3, SaPI1, and SsPI15305 pathogenicity islands are violetpurple, salmon, and green, respectively. Predicted HTH motifs of all proteins are colored dark blue. Stereo view of these structures is shown inS14 figure. http://dx.doi.org/10.1371/journal.pone.0139086.g005

It is worthwhile to point out that there is considerable sequence conservation within these HTH segments, while the other parts of these repressor proteins show high diversity (cf. figure 5 andS13 figure). Protein Blast analysis of the Stl sequences did not find similarity among these proteins [64], however all of these were annotated as functioning as master repressors [65]. We have also used the Phyre2 modeler to build 3D models of these Stl-like repressors, to decide if these may possess similar 3D structures despite the high sequence diversity. Although these models are all characterized by a high α-helical content, i.e. similar folding pattern, the overall structures are still greatly diverse except for the well-conserved HTH motif (see structural models in figure 5). It is rather probable that the HTH motifs are responsible for sequence-specific DNA binding in all of these repressors. For the case of the Sapibov1 Stl repressor, we have obtained detailed information of the mechanism of de-repression regulated by the dUTPase interacting partner [10]. Based on these data and the similar overall folding pattern, we propose that the carboxy-terminal domains of the other Stl-like repressors may constitute binding surface for interaction with other, yet unidentified proteins.

SUPPORTING INFORMATION

S1 File. Document including Table A and Table B.

Table A Sequence coverage of Stl by templates and confidence values of the match to that template overall in the Phyre2 model [1] a number of computational tools for structure prediction have been developed. It is critical that the biological community is aware of such tools and is able to interpret their results in an informed way. This protocol provides a guide to interpreting the output of structure prediction servers in general and one such tool in particular, the protein homology/analogy recognition engine (Phyre.

Template PDB code	Protein Name and Function	Aligned residues*	Alignment Coverage* (%)	Confidence** (%)	Identity (%)
1E3O	Human transcription factor Oct-1. activator of promoters of genes for some small nuclear RNAs (snRNA), histone H2B and immunoglobulins	2-48	17[#]	96.5	33[#]
4YV9	Rgg protein, transcriptional regulator of *Streptococcus dysgalactiae*	14-263	93	99.7	12
2GRM	Prgx, molecular switch controlling expression of conjugation and virulence genes encoded by plasmid pCF10 of *Enterococcus faecalis*	14-261	92	99.8	12[i]
4RYK	Lmo0325 protein, a putative transcriptional regulator from *Listeria monocytogenes* EGD-e	13-260	92	99.8	12
2QFC	PlcR, major virulence regulator of the *Bacillus cereus* group	12-261	93	99.7	13
2AXZ	Prgx, molecular switch controlling expression of conjugation and virulence genes encoded by plasmid pCF10 of *Enterococcus faecalis*	14-261	92	99.7	13[i]
2EBY	ybaQ, putative HTH-type transcriptional regulator from *E. coli*	11-96	31	99.6	15

* Part of Stl sequence which is covered by alignment with the template sequence.

** Confidence represents the estimated precision, it is the probability that the match between the query sequence and a template is because of true homology [1]a number of computational tools for structure prediction have been developed. It is critical that the biological community is aware of such tools and is able to interpret their results in an informed way. This protocol provides a guide to interpreting the output of structure prediction servers in general and one such tool in particular, the protein homology/analogy recognition engine (Phyre.

[#] This template was included to the final model by the Phyre2 because of higher local similarity.

[i] The difference in the sequence identity of Stl with the same type of protein occurs because the PDB entry 2GRM contains the structure of a longer protein construct and more residues are modeled in that structure, than those in 2AXZ, therefore the Phyre2 alignment differs somewhat in the two cases.

S1 figure. Comparison of the 3D models generated by Phyre2 and Modeller.

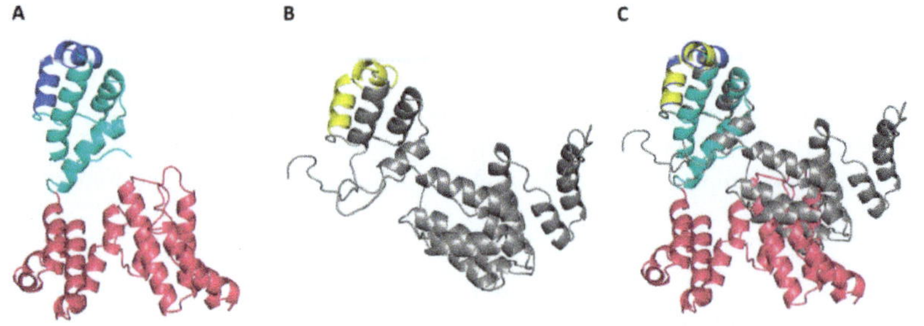

(A) Ribbon representation of the homology model of the *Staphylococcus aureus* pathogenicity island repressor Stl produced by Phyre2 Server [38]. Based on the homology model the protein is highly α helical (74%), and seems to be divided into two segments: the amino terminal segment colored cyan and the carboxy-terminal segment colored hotpink. According to Pfam and NCBI CDD the protein is predicted to contain a helix-turn-helix DNA binding motif. The position of the HTH predicted by NPS@ server is colored to dark blue [44]. **(B)** Ribbon representation of the homology model of Stl obtained by Modeller [40], predicted HTH colored yellow. **(C)** Superimposition of the two models. Both models agreed in that the protein is mostly α-helical and contains an N-terminal HTH motif.

S2 figure. Disorder prediction and construct design.

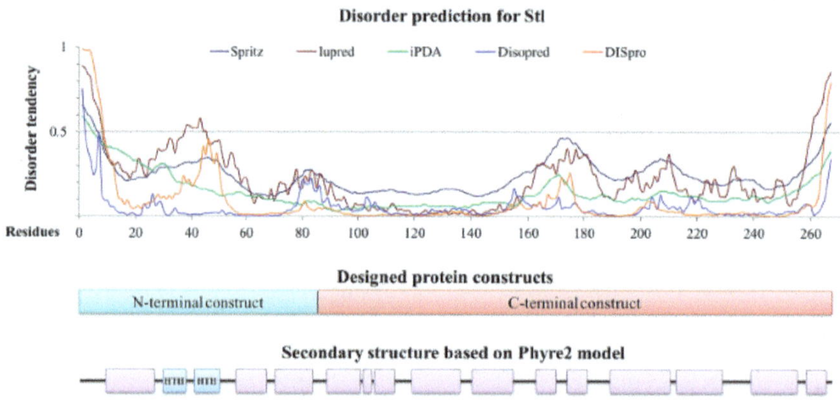

Representative curves from the results of flexibility prediction by MetaDisorder server are shown [45]. The mid-panel indicates our designed constructs represented as horizontal bars. Bottom-panel shows the secondary structure elements along the sequence, helices from the helix-turn-helix motif are in blue other helices are in purple.

S3 figure. Experimentally determined structure of the DNA-bound lambda bacteriophage CI repressor (PDB ID 1LMB).

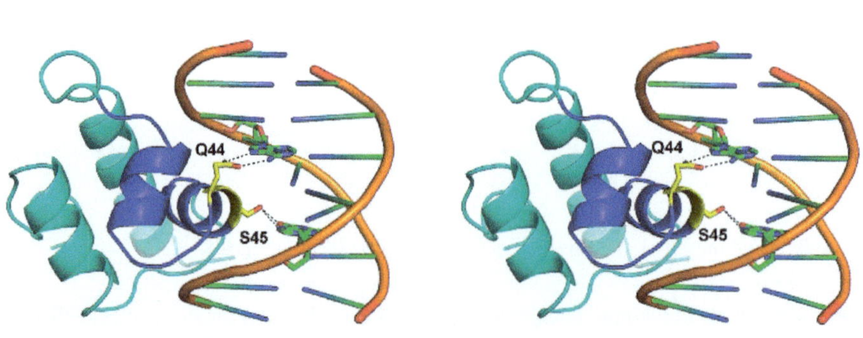

DNA cartoons orange, protein cartoons: dark blue for HTHs, otherwise cyan. DNA bases and DNA interacting amino acid residues are stick representation with atomic coloring (protein carbon yellow, DNA carbon green, oxygen red, nitrogen blue, phosphorus orange.)

S4 figure. Experimentally determined structure of the DNA-bound lambda bacteriophage Cro repressor (PDB ID 6CRO).

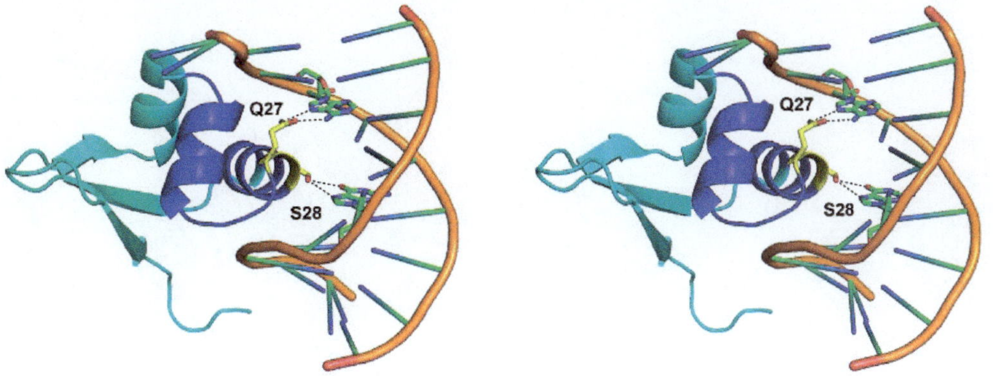

Coloring as in S3 figure.

S5 figure. Experimentally determined structure of the DNA-bound 434 bacteriophage CI repressor (PDB ID 2OR1).

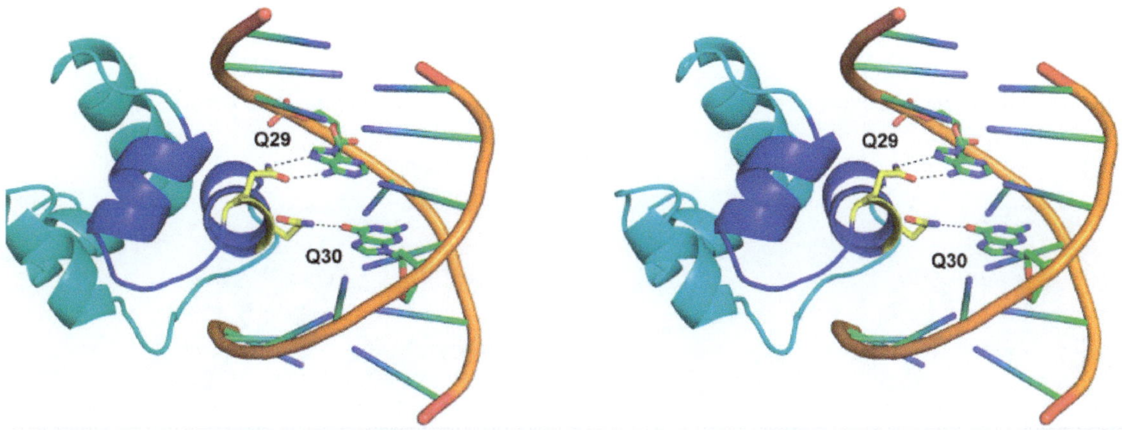

Coloring as in S3 figure.

S6 figure. Experimentally determined structure of the DNA-bound 434 bacteriophage Cro repressor (PDB ID 3CRO).

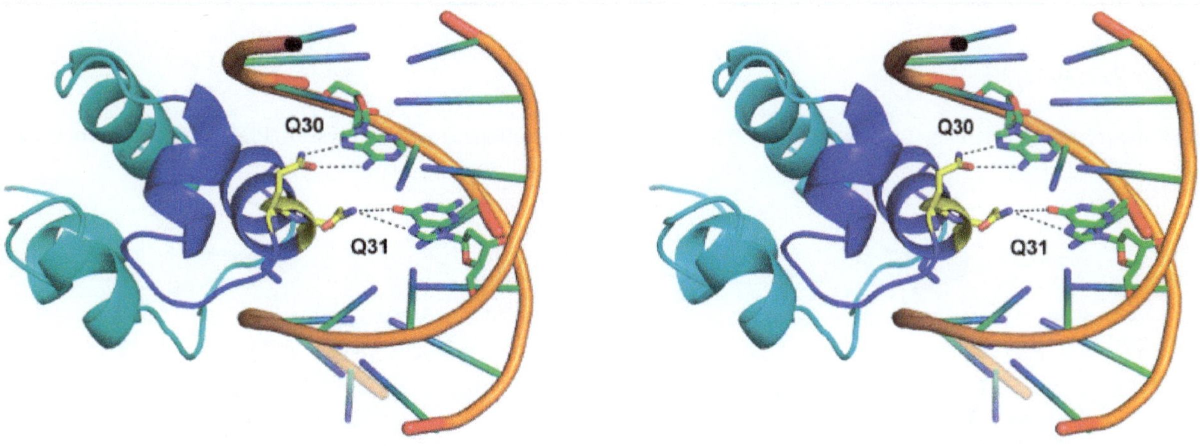

Coloring as in S3 figure.

S7 figure. Experimentally determined structure of the DNA-bound P22 bacteriophage C2 repressor (PDB ID 2R1J).

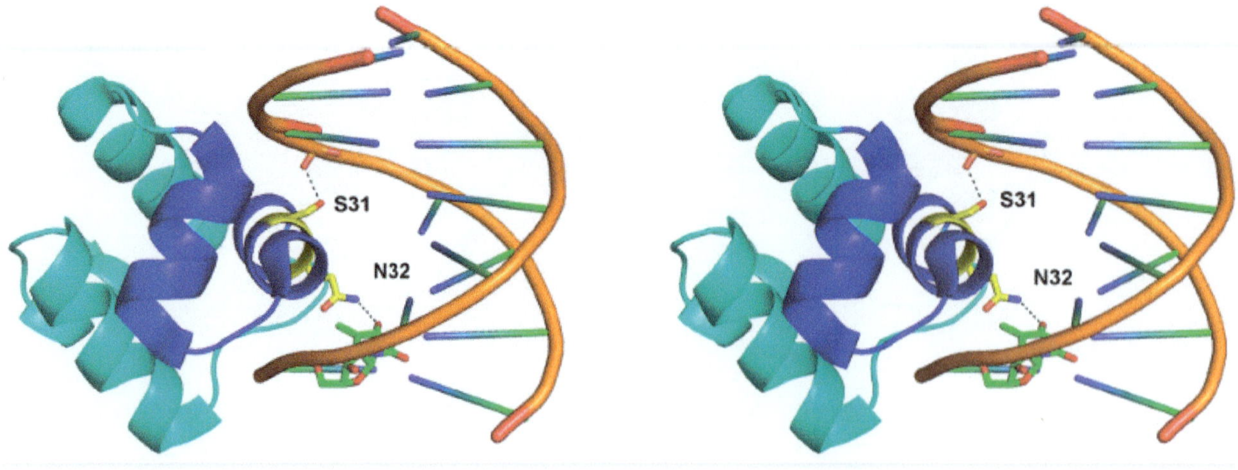

Coloring as in S3 figure.

S8 figure. Experimentally determined structure of the DNA-bound TP901 bacteriophage CI repressor (PDB ID 3ZHM).

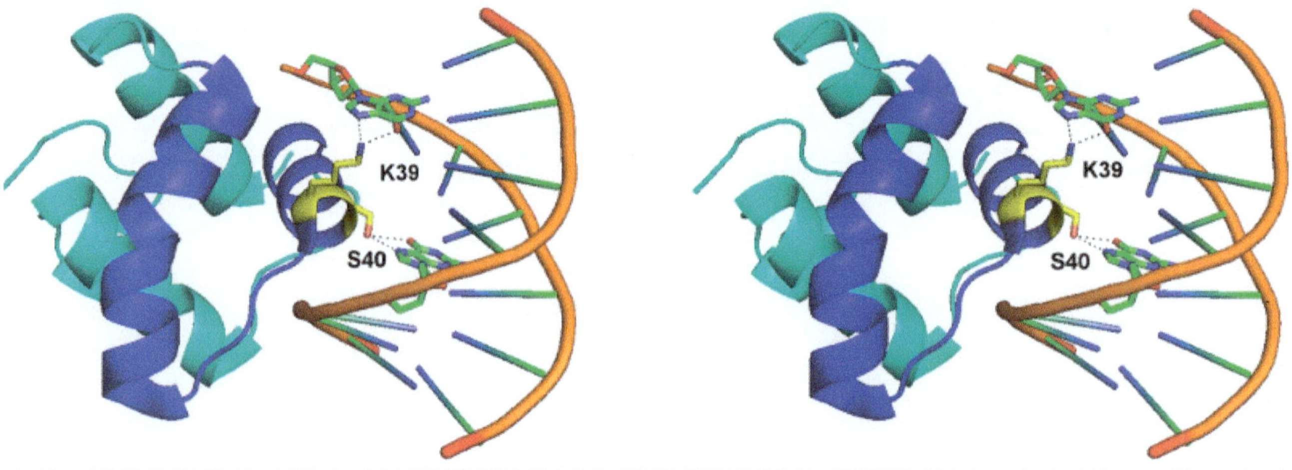

Coloring as in S3 figure.

S9 figure. Experimentally determined structure of the DNA-bound N15 bacteriophage Cro repressor (PDB ID 3QWS) Coloring as in S3 figure.

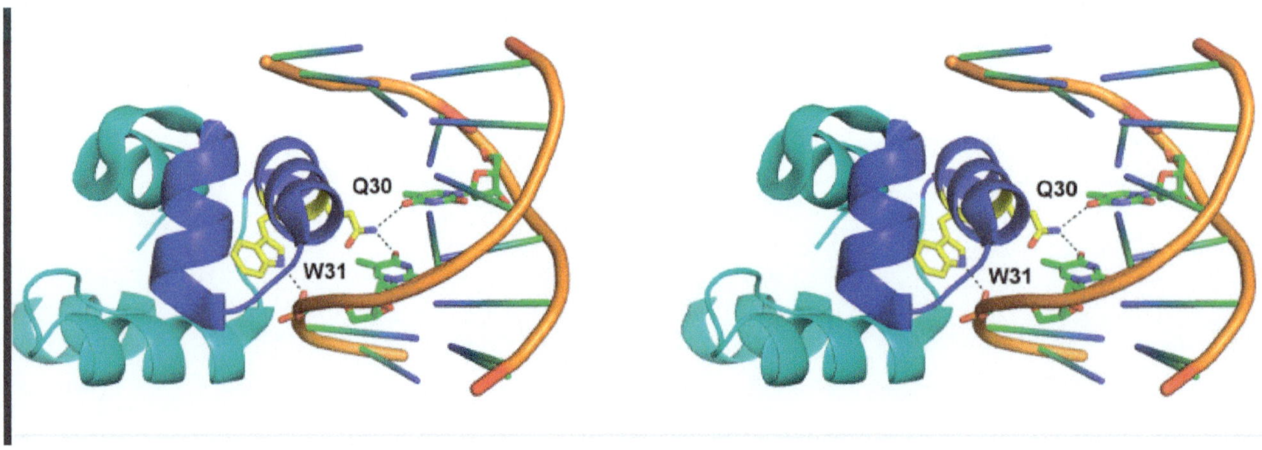

S10 figure. Testing the DNA binding ability of Stl-AA.

Stl-AA (μM)	0	1	3	4.5	6	7.5	9

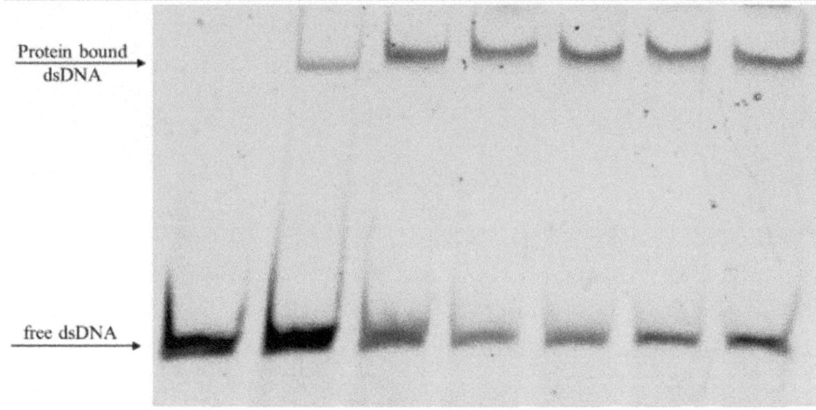

Electrophoretic mobility shift assay was performed to investigate the DNA binding ability of Stl-AA. Species and concentrations given in monomers are indicated on the figureure. The band of the dsDNA is only partially sifted upwards even if high concentrations of Stl-AA. Wild type Stl shows shift of the same amount of DNA at concentration of 1μM (cf. figure 3).

S11 figure. Testing the interaction of Stl-AA with Φ11 dUTPase on native-PAGE.

	Lane 1	Lane 2	Lane 3	Lane 4	Lane 5
Stl-AA (μM)	12	12	12	12	0
Φ11 dUTPase (μM)	0	12	8	4	12

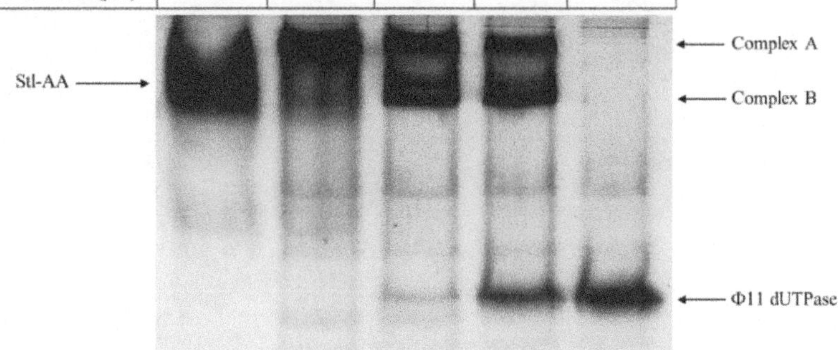

Native gel electrophoresis experiment was performed to investigate the Φ11 dUTPase binding ability of Stl-AA. Species and concentrations given in monomers are indicated. Comparing to the wild type Stl the complex formation of Stl-AA with Φ11 dUTPase is not perturbed (cf. figure 1B. in [10]).

S12 figure. Testing the inhibition of Stl-AA with Φ11 dUTPase.

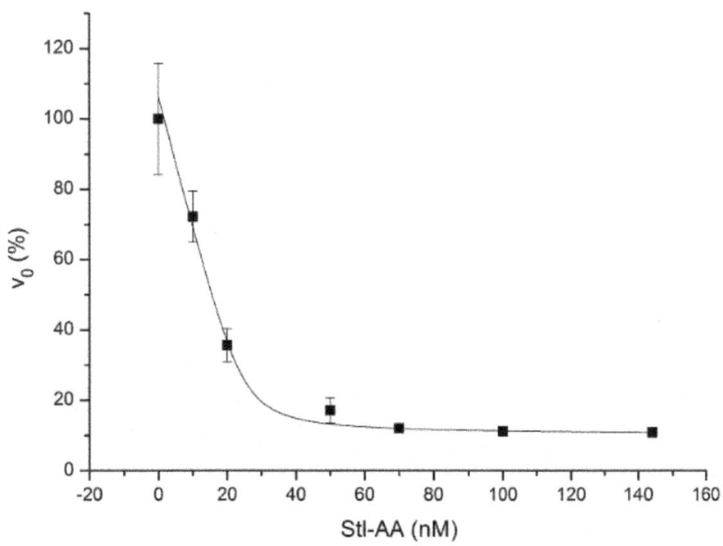

Enzyme activity Φ11dUTPase was measured in mixtures containing different concentrations of Stl-AA. The inhibitory constant has been determined to be 1.4 ± 0.9 nM, and the maximal inhibition was above 90%. Comparing to the wild type Stl the inhibition effect of Stl-AA on Φ11 dUTPase is not perturbed (cf. figure 2A. in [10]).

S13 figure. Alignment of different SaPI Stl proteins.

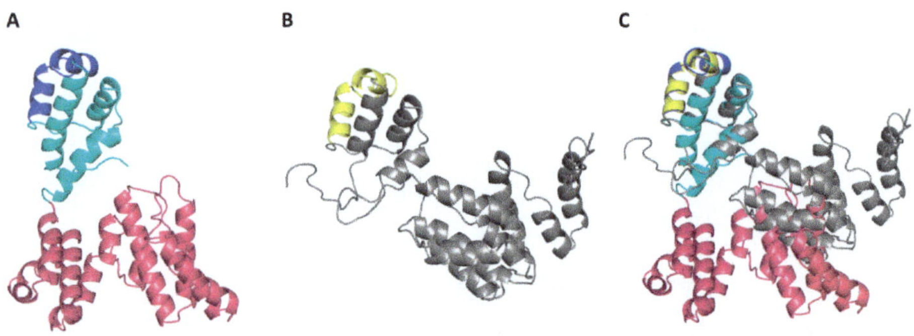

Identical residues are red (*). Strongly similar (:) residues are green, weakly similar residues are blue (.). Residues in the predicted HTH are underlined.

S14 figure. Stereo view of superimpositioned structural models of representative Stl-like repressors from different SaPIs.

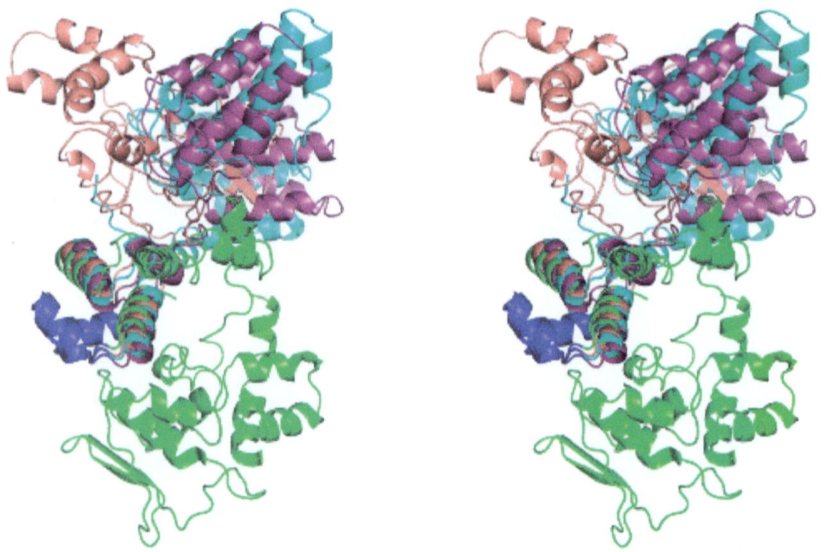

Protein	Color	Number of residues	Identity to Stl SaPIbov1 (%)	Local identity to Stl SaPIbov1 at the HTH (%)
Stl-SaPIbov1	cyan	267	NA	NA
Stl-SaPIbov3	violetpurple	240	14	36
Stl-SaPI1	salmon	247	16	45
Stl-SsPI15305	green	274	16	36

Proteins in cartoon representation. SaPIbov1 Stl is cyan, Stl-like repressors of SaPIbov3, SaPI1, and SsPI15305 pathogenicity islands are violetpurple, salmon, and green, respectively. Predicted HTH motifs of all proteins colored dark blue.

ACKNOWLEDGMENTS

We thank Frank Wien and Matthieu Réfrégiers at DISCO beamline for their kind help in SRCD data collection.

REFERENCES

1. Lindsay JA, Ruzin A, Ross HF, Kurepina N, Novick RP. The gene for toxic shock toxin is carried by a family of mobile pathogenicity islands in Staphylococcus aureus. Mol Microbiol. 1998;29: 527–543. doi: 10.1046/j.1365-2958.1998.00947.x. pmid:9720870
2. Ubeda C, Maiques E, Barry P, Matthews A, Tormo MA, Lasa I, et al. SaPI mutations affecting replication and transfer and enabling autonomous replication in the absence of helper phage. Mol Microbiol. 2008;67: 493–503. doi: 10.1111/j.1365-2958.2007.06027.x. pmid:18086210

3. Tormo-Más MA, Mir I, Shrestha A, Tallent SM, Campoy S, Lasa I, et al. Moonlighting bacteriophage proteins derepress staphylococcal pathogenicity islands. Nature. 2010;465: 779–82. doi: 10.1038/nature09065. pmid:20473284
4. Vértessy BG, Tóth J. Keeping Uracil Out of DNA: Physiological Role, Structure and Catalytic Mechanism of dUTPases. Acc Chem Res. 2009;42: 97–106. doi: 10.1021/ar800114w.Keeping. pmid:18837522
5. Persson R, Cedergren-Zeppezauer ES, Wilson KS. Homotrimeric dUTPases; structural solutions for specific recognition and hydrolysis of dUTP. Curr Protein Pept Sci. 2001;2: 287–300. doi: 10.2174/1389203013381035. pmid:12369926
6. Takács E, Grolmusz VK, Vértessy BG. A tradeoff between protein stability and conformational mobility in homotrimeric dUTPases. FEBS Lett. 2004;566: 48–54. doi: 10.1016/j.febslet.2004.04.039. pmid:15147867
7. Muha V, Horváth A, Békési A, Pukáncsik M, Hodoscsek B, Merényi G, et al. Uracil-Containing DNA in Drosophila: Stability, Stage-Specific Accumulation, and Developmental Involvement. Hawley RS, editor. PLoS Genet. 2012;8: e1002738. doi: 10.1371/journal.pgen.1002738. pmid:22685418
8. Vertessy BG, Zalud P, Nyman PO, Zeppezauer M. Identification of tyrosine as a functional residue in the active site of Escherichia coli dUTPase. Biochim Biophys Acta. 1994;1205: 146–150. doi: 10.1016/0167-4838(94)90103-1. pmid:8142479
9. Hemsworth GR, González-Pacanowska D, Wilson KS. On the catalytic mechanism of dimeric dUTPases. Biochem J. 2013;456: 81–8. doi: 10.1042/BJ20130796. pmid:24001052
10. Szabó JE, Németh V, Papp-Kádár V, Nyíri K, Leveles I, Bendes AÁ, et al. Highly potent dUTPase inhibition by a bacterial repressor protein reveals a novel mechanism for gene expression control. Nucleic Acids Res. 2014;42: 11912–20. doi: 10.1093/nar/gku882. pmid:25274731
11. Hirmondó R, Szabó JE, Nyíri K, Tarjányi S, Dobrotka P, Tóth J, et al. Cross-species inhibition of dUTPase via the Staphylococcal Stl protein perturbs dNTP pool and colony formation in Mycobacterium. DNA Repair (Amst). Elsevier B.V.; 2015;30: 21–27. doi: 10.1016/j.dnarep.2015.03.005.
12. Varga B, Barabás O, Takács E, Nagy N, Nagy P, Vértessy BG. Active site of mycobacterial dUTPase: structural characteristics and a built-in sensor. Biochem Biophys Res Commun. 2008;373: 8–13. doi: 10.1016/j.bbrc.2008.05.130. pmid:18519027
13. Pecsi I, Hirmondo R, Brown AC, Lopata A, Parish T, Vertessy BG, et al. The dUTPase enzyme is essential in Mycobacterium smegmatis. PLoS One. 2012;7: e37461. doi: 10.1371/journal.pone.0037461. pmid:22655049
14. Horváti K, Bacsa B, Szabó N, Fodor K, Balka G, Rusvai M, et al. Antimycobacterial activity of peptide conjugate of pyridopyrimidine derivative against Mycobacterium tuberculosis in a series of in vitro and in vivo models. Tuberculosis. 2015;95: 207–211. doi: 10.1016/j.tube.2015.02.026.
15. Cedergren-Zeppezauer ES, Larsson G, Nyman PO, Dauter Z, Wilson KS. Crystal structure of a dUTPase. Nature. Nature Publishing Group; 1992;355: 740–3. doi: 10.1038/355740a0. pmid:1311056
16. Larsson G, Nyman PO, Kvassman JO. Kinetic characterization of dUTPase from Escherichia coli. J Biol Chem. 1996;271: 24010–24016. doi: 10.1074/jbc.271.39.24010. pmid:8798636
17. Barabás O, Pongrácz V, Kovári J, Wilmanns M, Vértessy BG. Structural insights into the catalytic mechanism of phosphate ester hydrolysis by dUTPase. J Biol Chem. 2004;279: 42907–15. doi: 10.1074/jbc.M406135200. pmid:15208312
18. Takács E, Barabás O, Petoukhov M V, Svergun DI, Vértessy BG. Molecular shape and prominent role of beta-strand swapping in organization of dUTPase oligomers. FEBS Lett. 2009;583: 865–71. doi: 10.1016/j.febslet.2009.02.011. pmid:19302784
19. Pecsi I, Leveles I, Harmat V, Vertessy BG, Toth J. Aromatic stacking between nucleobase and enzyme promotes phosphate ester hydrolysis in dUTPase. 2010;38: 7179–7186. doi: 10.1093/nar/gkq584. pmid:20601405
20. Leveles I, Németh V, Szabó JE, Harmat V, Nyíri K, Bendes ÁÁ, et al. Structure and enzymatic mechanism of a moonlighting dUTPase. Acta Crystallogr Sect D Biol Crystallogr. International Union of Crystallography; 2013;69: 2298–2308. doi: 10.1107/S0907444913021136.
21. Nagy GN, Leveles I, Vértessy BG. Preventive DNA repair by sanitizing the cellular (deoxy)nucleoside triphosphate pool. FEBS J. 2014;281: 4207–23. doi: 10.1111/febs.12941. pmid:25052017
22. Tormo-Más MÁ, Mir-Sanchis I, Shrestha A, Tallent SM, Campoy S, Lasa Í, et al. Moonlighting bacteriophage proteins derepress staphylococcal pathogenicity islands. Nature. 2010;465: 779–82. doi: 10.1038/nature09065. pmid:20473284
23. Mir-Sanchis I, Martínez-Rubio R, Martí M, Chen J, Lasa Í, Novick RP, et al. Control of Staphylococcus aureus pathogenicity island excision. Mol Microbiol. 2012;85: 833–45. doi: 10.1111/j.1365-2958.2012.08145.x. pmid:22742067
24. Stayrook S, Jaru-Ampornpan P, Ni J, Hochschild A, Lewis M. Crystal structure of the lambda repressor and a model for pairwise cooperative operator binding. Nature. 2008;452: 1022–5. doi: 10.1038/nature06831. pmid:18432246
25. Hochschild A, Lewis M. The bacteriophage lambda CI protein finds an asymmetric solution. Current Opinion in Structural Biology. 2009. pp. 79–86. doi: 10.1016/j.sbi.2008.12.008. pmid:19181516
26. Pinkett HW, Shearwin KE, Stayrook S, Dodd IB, Burr T, Hochschild A, et al. The structural basis of cooperative regulation at an alternate genetic switch. Mol Cell. 2006;21: 605–15. doi: 10.1016/j.molcel.2006.01.019. pmid:16507359
27. Carlson PA, Koudelka GB. Expression, purification, and functional characterization of the carboxyl-terminal

domain fragment of bacteriophage 434 repressor. J Bacteriol. 1994;176: 6907–6914. pmid:7961451

28. Donner AL, Carlson PA, Koudelka GB. Dimerization specificity of P22 and 434 repressors is determined by multiple polypeptide segments. J Bacteriol. 1997;179: 1253–1261. pmid:9023209

29. Révet B, Von Wilcken-Bergmann B, Bessert H, Barker A, Müller-Hill B. Four dimers of λ repressor bound to two suitably spaced pairs of λ operators form octamers and DNA loops over large distances. Curr Biol. 1999;9: 151–154. doi: 10.1016/S0960-9822(99)80069-4. pmid:10021390

30. Bell CE, Lewis M. Crystal Structure of the cl Repressor C-terminal Domain Octamer. J Mol Biol. 2001;314: 1127–1136. doi: 10.1006/jmbi.2001.5196. pmid:11743728

31. Roberts JW, Roberts CW. Proteolytic cleavage of bacteriophage lambda repressor in induction. Proc Natl Acad Sci U S A. 1975;72: 147–151. doi: 10.1073/pnas.72.1.147. pmid:1090931

32. Shearwin KE, Brumby AM, Egan JB. The tum protein of coliphage 186 is an antirepressor. J Biol Chem. 1998;273: 5708–5715. doi: 10.1074/jbc.273.10.5708. pmid:9488703

33. Heinzel T, Velleman M, Schuster H. C1 Repressor of phage P1 is inactivated by noncovalent binding of P1 coi protein. J Biol Chem. 1992;267: 4183–4188. pmid:1740459

34. Heinzel T, Velleman M, Schuster H. The c1 repressor inactivator protein coi of bacteriophage P1: Cloning and expression of coi and its interference with c1 repressor function. J Biol Chem. 1990;265: 17928–17934. pmid:2211669

35. Riedel HD, Heinrich J, Heisig A, Choli T, Schuster H. The antirepressor of phage P1. Isolation and interaction with the C1 repressor of P1 and P7. FEBS Lett. 1993;334: 165–169. doi: 10.1016/0014-5793(93)81705-5. pmid:8224242

36. Rohs R, Jin X, West SM, Joshi R, Honig B, Mann RS. Origins of specificity in protein-DNA recognition. Annu Rev Biochem. 2010;79: 233–269. doi: 10.1146/annurev-biochem-060408-091030. pmid:20334529

37. Wojciak JM, Iwahara J, Clubb RT. The Mu repressor-DNA complex contains an immobilized "wing" within the minor groove. Nat Struct Biol. 2001;8: 84–90. doi: 10.1038/83103. pmid:11135677

38. Kelley L a, Sternberg MJE. Protein structure prediction on the Web: a case study using the Phyre server. Nat Protoc. 2009;4: 363–71. doi: 10.1038/nprot.2009.2. pmid:19247286

39. Söding J. Protein homology detection by HMM-HMM comparison. Bioinformatics. 2005;21: 951–60. doi: 10.1093/bioinformatics/bti125. pmid:15531603

40. Eswar N, Webb B, Marti-Renom MA, Madhusudhan MS, Eramian D, Shen M-Y, et al. Comparative protein structure modeling using Modeller. Current Protocols in Bioinformatics. John Wiley & Sons, Inc.; 2006. p. Unit 5.6. doi: 10.1002/0471250953.bi0506s15. pmid:18428767

41. Wang Y, Addess KJ, Chen J, Geer LY, He J, He S, et al. MMDB: annotating protein sequences with Entrez's 3D-structure database. Nucleic Acids Res. 2007;35: D298–300. doi: 10.1093/nar/gkl952. pmid:17135201

42. Finn RD, Bateman A, Clements J, Coggill P, Eberhardt RY, Eddy SR, et al. Pfam: the protein families database. Nucleic Acids Res. 2014;42: D222–30. doi: 10.1093/nar/gkt1223. pmid:24288371

43. Marchler-Bauer A, Derbyshire MK, Gonzales NR, Lu S, Chitsaz F, Geer LY, et al. CDD: NCBI's conserved domain database. Nucleic Acids Res. 2014;43: 222–226. doi: 10.1093/nar/gku1221.

44. Dodd IB, Egan JB. Improved detection of helix-turn-helix DNA-binding motifs in protein sequences. Nucleic Acids Res. 1990;18: 5019–5026. doi: 10.1093/nar/18.17.5019. pmid:2402433

45. Kozlowski LP, Bujnicki JM. MetaDisorder: a meta-server for the prediction of intrinsic disorder in proteins. BMC Bioinformatics. BMC Bioinformatics; 2012;13: 111. doi: 10.1186/1471-2105-13-111. pmid:22624656

46. Kabsch W, Sander C. Dictionary of protein secondary structure: pattern recognition of hydrogen-bonded and geometrical features. Biopolymers. 1983;22: 2577–2637. doi: 10.1002/bip.360221211. pmid:6667333

47. Micsonai A, Wien F, Kernya L, Lee Y-H, Goto Y, Réfrégiers M, et al. Accurate secondary structure prediction and fold recognition for circular dichroism spectroscopy. Proc Natl Acad Sci U S A. 2015;112: E3095–E3103. doi: 10.1073/pnas.1500851112. pmid:26038575

48. Provencher SW, Glöckner J. Estimation of globular protein secondary structure from circular dichroism. Biochemistry. 1981;20: 33–37. doi: 10.1021/bi00504a006. pmid:7470476

49. Leveles I, Róna G, Zagyva I, Bendes Á, Harmat V, Vértessy BG. Crystallization and preliminary crystallographic analysis of dUTPase from the φ11 helper phage of Staphylococcus aureus. Acta Crystallogr Sect F. 2011;67: 1411–3. doi: 10.1107/S1744309111034580.

50. Vertessy BG. Flexible glycine rich motif of Escherichia coli deoxyuridine triphosphate nucleotidohydrolase is important for functional but not for structural integrity of the enzyme. Proteins Struct Funct Genet. 1997;28: 568–579. doi: 10.1002/(SICI)1097-0134(199708)28:4<568::AID-PROT10>3.0.CO;2-E. pmid:9261872

51. Chen GC, Yang JT. Two-Point Calibration of Circular Dichrometer with d-10-Camphorsulfonic Acid. Anal Lett. Taylor & Francis Group; 1977;10: 1195–1207. doi: 10.1080/00032717708067855

52. Sreerama N, Venyaminov SY, Woody RW. Estimation of the number of alpha-helical and beta-strand segments in proteins using circular dichroism spectroscopy. Protein Sci. 1999;8: 370–80. doi: 10.1110/ps.8.2.370. pmid:10048330

53. Orosz F, Kovács J, Löw P, Vértessy BG, Urbányi Z, Acs T, et al. Interaction of a new bis-indol derivative, KAR-2 with tubulin and its antimitotic activity. Br J Pharmacol.

1997;121: 947–954. doi: 10.1038/sj.bjp.0701189. pmid:9222552

54. Whitmore L, Wallace BA. Protein secondary structure analyses from circular dichroism spectroscopy: methods and reference databases. Biopolymers. 2008;89: 392–400. doi: 10.1002/bip.20853. pmid:17896349

55. Beamer LJ, Pabo CO. Refined 1.8 A crystal structure of the lambda repressor-operator complex. J Mol Biol. 1992;227: 177–196. doi: 10.1016/0022-2836(92)90690-L. pmid:1387915

56. Albright RA, Matthews BW. Crystal structure of lambda-Cro bound to a consensus operator at 3.0 A resolution. J Mol Biol. 1998;280: 137–151. doi: 10.1006/jmbi.1998.1848. pmid:9653037

57. Aggarwal AK, Rodgers DW, Drottar M, Ptashne M, Harrison SC. Recognition of a DNA operator by the repressor of phage 434: a view at high resolution. Science. 1988;242: 899–907. doi: 10.1126/science.3187531. pmid:3187531

58. Shimon LJ, Harrison SC. The phage 434 OR2/R1-69 complex at 2.5 A resolution. J Mol Biol. 1993;232: 826–838. doi: 10.1016/0022-2836(91)90568-Q. pmid:8355273

59. Watkins D, Hsiao C, Woods KK, Koudelka GB, Williams LD. P22 c2 repressor-operator complex: Mechanisms of direct and indirect readout. Biochemistry. 2008;47: 2325–2338. doi: 10.1021/bi701826f. pmid:18237194

60. Pedersen M, Lo Leggio L, Grossmann JG, Larsen S, Hammer K. Identification of quaternary structure and functional domains of the CI repressor from bacteriophage TP901-1. J Mol Biol. Elsevier Ltd; 2008;376: 983–96. doi: 10.1016/j.jmb.2007.12.022. pmid:18191944

61. Isackson PJ, Bertrand KP. Dominant negative mutations in the Tn10 tet repressor: evidence for use of the conserved helix-turn-helix motif in DNA binding. Proc Natl Acad Sci U S A. 1985;82: 6226–6230. doi: 10.1073/pnas.82.18.6226. pmid:2994067

62. Brennan RG, Matthews BW. The helix-turn-helix DNA binding motif. J Biol Chem. 1989;264: 1903–1906. pmid:2644244 doi: 10.1016/0092-8674(93)90456-z

63. Aravind L, Anantharaman V, Balaji S, Babu MM, Iyer LM. The many faces of the helix-turn-helix domain: transcription regulation and beyond. FEMS Microbiol Rev. 2005;29: 231–62. doi: 10.1016/j.femsre.2004.12.008. pmid:15808743

64. Johnson M, Zaretskaya I, Raytselis Y, Merezhuk Y, McGinnis S, Madden TL. NCBI BLAST: a better web interface. Nucleic Acids Res. 2008;36: W5–9. doi: 10.1093/nar/gkn201. pmid:18440982

65. Novick RP, Christie GE, Penadés JR. The phage-related chromosomal islands of Gram-positive bacteria. Nat Rev Microbiol. 2010;8: 541–51. doi: 10.1038/nrmicro2393. pmid:20634809

INDEX

W